KV-268-489

BIOTECHNOLOGY IN THE FEED INDUSTRY

This year Alltech lost two of its stalwart supporters: Mr. Frank Palmer and Mr. Brian Hutton, both from Canada.

Frank was our oldest serving colleague in the sales area and at 88 was still going strong.

Brian Hutton, often called the coaching salesman, was a bundle of energy who supported everybody he came into contact with. Cut down prematurely at the age of 47, Brian will be long remembered.

To Brian and Frank we dedicate this year's proceedings. We thank them for the memories and for being there when we needed them.

Front Cover
Fermentation is the backbone of biotechnology. Alltech's recently expanded fermentation facilities provide state-of-the-art growth, concentration and drying processes.

Biotechnology in the Feed Industry

Proceedings of
Alltech's Eleventh Annual Symposium

Edited by TP Lyons and KA Jacques

Nottingham University Press
Manor Farm, Main Street, Thrumpton,
Nottingham NG11 0AX, United Kingdom

NOTTINGHAM

First published 1995

ISBN 1897676565

Typeset by The Midlands Book Typesetting Company, Loughborough, Leicestershire, England
Printed and bound by Redwood Books, Trowbridge, Wiltshire, England

TABLE OF CONTENTS

SECTION 2:
THE NEW FRONTIER: IMMUNE MODULATORS IN NUTRITION

SECTION 3:
A SHOWCASE ON THE DIVERSITY OF AGRICULTURE AROUND THE WORLD

SECTION 5:
ROUNDTABLE INTERACTIVE PROGRAM

BIOTECHNOLOGY IN THE FEED INDUSTRY
A look forward and backward

T.P. LYONS

President, Alltech Inc., Nicholasville, Kentucky, USA

Introduction

Alltech, founded 15 years ago, has prospered and grown on the basis of one principle: to scientifically investigate problems facing animal production world-wide and produce marketable products based upon the results. This approach requires a near-market philosophy of ensuring that research is market driven. Furthermore, problems ascertained in the field need first to be fully investigated by the technical support staff who, in many cases, have begun to formulate a solution before involving the research staff. A close and continued involvement of field and laboratory personnel is required so that the project stays on course and on time.

Fifteen years into the Alltech program, the company now employs more than 250 people and was recently ranked 89th in the top 100 fastest growing global companies in the United States (Table 1). Ironically, Alltech was the only company in that 100 involved in biotechnology and animal production; the others being, not surprisingly, in computer applications and telecommunications. The heartbeat of Alltech's progress has always been research laboratories, now located in two Bioscience Centers with a third opening in Beijing, China in 1995. These centers, working with many universities and companies through strategic alliances, offer students an opportunity to experience research relevant to the industry in a creative atmosphere. The students in turn provide the company an almost unlimited supply of talented, young, inquisitive scientists. With the opening of the Beijing Bioscience Center, full-time research students will number over 45.

Table 1. Alltech profile.

Founded	1980
Staff	250
Sales	60% export
Research staff	45+
Bioscience Centers	3

Figure 1. North American Biosciences Center.

Figure 2. European Biosciences Center.

Figure 3. Bejing Biosciences Center.

In order to realize the success of this market driven approach, one has only to consider some of the previous results and the on-going projects. Selected projects are presented in this chapter in summary form and others are detailed in some of the following chapters. All are addressed in the Problem/Solution fashion in which they were presented. Solutions to some field problems may emerge rapidly while others involve many years of research. As with any scientific project, all are considered to be on-going and approaches used are constantly upgraded as new findings emerge. If, however, science is to be alive and yet responsive to the marketplace, one must be prepared to go with best judgment answers and allow the marketplace to judge the effectiveness of the solution. The gap between laboratory and marketplace is constantly in danger of widening, and the key is to keep it narrow.

Current projects at Alltech target a wide range of industry problems. Though they might be categorized as environmental, health or animal performance topics, from our perspective they are very much interrelated (Table 2). For example, improving ways to use enzymes in animal feeds means better feed utilization; but it also reduces the animal waste burden on the environment. Similarly, improved utilization of zinc, copper, selenium and chromium and phosphorus reduces mineral output in manure and has a large impact on animal health and nutrient utilization.

It is equally important to realize that these projects ultimately speak to important consumer issues. Livestock feed quality and animal health translate to the consumer in terms of human food quality and human health whether the issue is aflatoxin in feed, salmonella, antibiotic usage, lean meat or by-product and manure utilization. Whether approaching these problems in the laboratory or on the farm, it is critically important that we never lose site of our ultimate market, the consumer.

Table 2. Alltech projects targeting agricultural problems.

The environment and agriculture	Animal health	Feed quality and utilization
Offal utilization	Rumen acidosis	Mycotoxins in feed
Phosphate pollution	Chromium and metabolism	Diet specific enzymes
Selenium utilization	Immunity and nutrition	Protecting amino acids
Manure odor, decomposition	Microbial probiosis: *Saccharomyces cerevisiae* var. *boulardii*	Mineral proteinates

The environment and agriculture

ODOR ARISING FROM LIVESTOCK MANURE

Animal production world-wide has been rightly or wrongly classified as a major polluter of both the atmosphere and the waterways. Nutritionists constantly hear about new legislation written to define how manure must be disposed in order to prevent phosphate or nitrogen discharge into waterways or legislation to reduce nitrogen emissions. Directly or indirectly such regulation limits the number of animals which can be raised on a given land area.

In the United States the recent rapid consolidation of pig units has precipitated adverse press even in publications typically friendly to agriculture. An example is the recent series of articles in the *Des Moines Register* entitled 'The Hog Wars'. Even in the heart of pig-producing Iowa, restrictions on production are in place or being discussed. Three focal points of pollution have to be quickly addressed: odor, manure nitrogen (N) and manure phosphate (P). Odor has been described by the North Carolina State University Task Force on animal pollution as 'the number one problem'. Offensive odor from livestock facilities consists of many noxious gases including hydrogen sulfide and ammonia. The latter has been the focus of our own research and the use of a standardized plant-derived glycocomponent extract has great potential. The key to its success was development of a quality control process with which the natural extract could be standardized. The glycocomponent, extracted from the *Yucca schidigera* plant, has been shown to bind ammonia (Headon *et al.*, 1991; Killeen *et al.*, 1994). This in turn stimulates microbial populations in manure to more efficiently use manure N and thereby reduce its emission as ammonia by some 50% from pig and poultry manure. The additional benefits of improved animal health and performance are quite marked and include improvements in weight gain and reductions in mortality and morbidity due to ammonia-related problems. In broilers, significantly fewer ascites losses are reported while improvements in efficiency have been recorded in growing/finishing pigs (Tables 3 and 4).

Importantly, the odor arising from slurry spread on pastures and cropland is also reduced as a result the yucca extract. Extension specialists at Cornell tested effects of Yucca extract on manure odor in an experiment where cows were fed 3g De-Odorase per head per day. The lagoon was also treated with De-Odorase. The result was a 40–50% reduction in ammonia released into the air (Figure 4). Furthermore, neighbors no longer complained when manure was pumped from the lagoon and spread on nearby fields. A further benefit was the reduction in solids in the treated lagoon which reduced the time and energy costs associated with pumping and spreading (irrigation) of the manure.

Table 3. Effect of De-Odorase on performance of broilers at 21 and 42 days*.

	At 21 days		At 42 days	
	Control	De-Odorase	Control	De-Odorase
Weight, g	532[b]	553[a]	1640	1614
Feed intake, g	755[b]	767[a]	2873	2864
FCR	1.53[b]	1.49[a]	1.80	1.82
Mortality, %				
General	5.54[b]	3.88[a]	14.88[b]	12.13[a]
Ascites	0.55[a]	0.64[a]	6.69[b]	5.00[a]

[ab]Means differ, $P < 0.05$
*Arce and Avila, 1994

Table 4. Effect of De-Odorase (*Yucca schidigera* extract) on performance of pigs and ammonia content of the finishing barn.

	Control	De-Odorase
Initial NH_3	30	30
Final NH_3	30	20
Weight in, kg	32	30.23
Weight out	91.67	91.88
Number of days	73	73
Daily gain, g	820	850

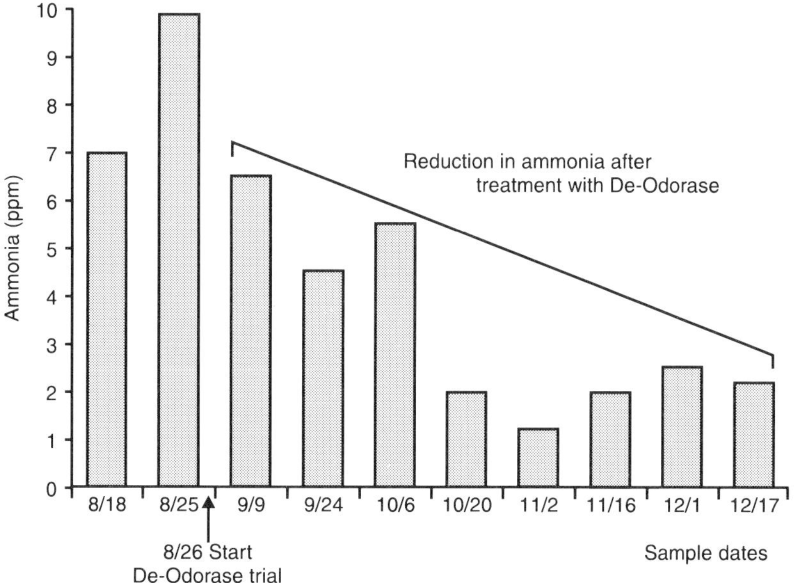

Figure 4. Effect of De-Odorase on ammonia arising from a dairy lagoon (Weaver, 1993).

UTILIZATION OF PHYTATE PHOSPHORUS

Phosphate pollution can also be reduced by the use of a novel enzyme complex consisting of phytase and cellulase. With these enzyme activities combined, dicalcium phosphate in the diet can be reduced by 0.1% leading to a substantial reduction in phosphate wastage. Cantor *et al.* (1994) calculated the increase in available phosphorus due to using yeast acid phosphatase, *Aspergillus niger* acid phosphatase and *A. niger* phytase. Based on a linear regression equation for tibia ash (tibia ash had the most linear response to graded additions of Ca and P) vs percentage available phosphorus in the diet, the equivalent available phosphorus was calculated for the various enzyme supplements. Supplementing the basal diet with equal amounts of the phytase sources increased the available phosphorus in the diet by approximately 0.10%, which corresponds to roughly 40% of the phytate phosphorus. Thus, it appears that enzyme supplementation can be an effective method of replacing some of the supplemental phosphorus in poultry diets. The removal of the antinutritive effect of phytin also leads to an improvement in utilization of amino acids and minerals.

OFFAL UTILIZATION: AN OPPORTUNITY FOR ENZYMES

The harsh reality of animal production is that a significant quantity of waste is produced. For broilers, about 35% of a 2 kg bird is by-product which must be processed (Figure 5). There are about 185 g in feather

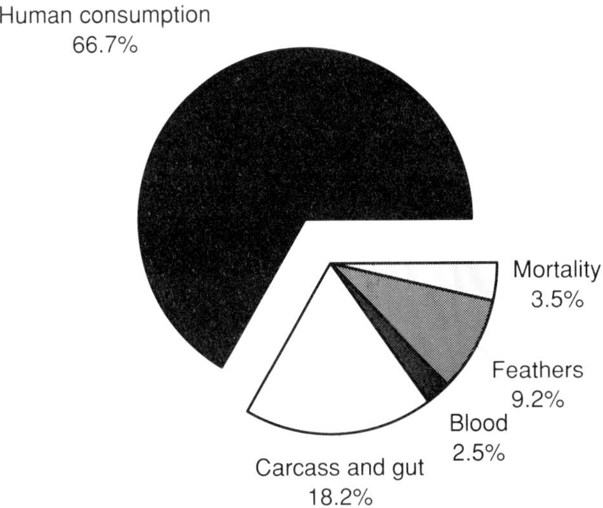

Figure 5. The proportions of poultry used for human consumption and for animal by-product processing (Woodgate, 1993).

dry weight for disposal. For large integrated operations the quantities of waste generated can be enormous. When mortality figures are added in, sometimes running as high as 6%, the total comes to some 12.5 million tons annually.

Enzyme processes and fermentation techniques have now been developed to the point that proteins, with very controlled amino acid profiles, can be produced from offal. When the protein source is fed back to animals, the net result is an improvement in performance and a reduction in costs. Feathers are of particular interest. Traditionally they have been processed using an energy inefficient system that not only destroys amino acid digestibility but causes pollution. The new enzyme system saves energy and improves amino acid digestibility to produce a consistent feed ingredient that can be used by non-ruminants and ruminants alike. Woodgate (1995) summarized the benefits of the enzymatic process of digesting offal as being one which gave higher nutrient density and proven cost benefits. One such product was claimed to improve broilers feed efficiency from 1.95 to 1.84 (Table 5).

Table 5. Effect of Protagen, an enzyme/offal-derived feed protein source, on final weight and feed efficiency of broilers.

	Control	Protagen
Final weight, kg	2.025	2.035
Feed efficiency, kg/kg	1.950	1.840*

*Savings per million birds $20 000

IMPROVING BIOACTIVITY AND BIOAVAILABILITY OF SELENIUM: ENVIRONMENTAL AND ANIMAL HEALTH CONCERNS

Metabolism of inorganic and organic Se

Differences in how organic selenoproteins and sodium selenite are metabolized have implications for both environmental impact and animal health. Against the background that many animals across the world suffer from selenium (Se) deficiency at one or more points in the growth and production curve is the growing specter of legislation driven by both pollution and safety concerns against the use of sodium selenite in animal diets. Forced to re-examine the role of sodium selenite, researchers have demonstrated that it is not necessarily the nutritionally ideal source of Se, and indeed in the case of a ruminant most is actually reduced to unabsorbable form in the rumen and subsequently excreted (Figure 6). While absorption by the non-ruminant is higher, none the less alternative sources need to be evaluated. Fermentation technology can provide us with good organic sources of Se. By using selected microorganisms, selenoproteins such as selenomethionine can be produced. Selenium provided in these forms is not only more biologically available (and therefore less is excreted) but is more biologically active.

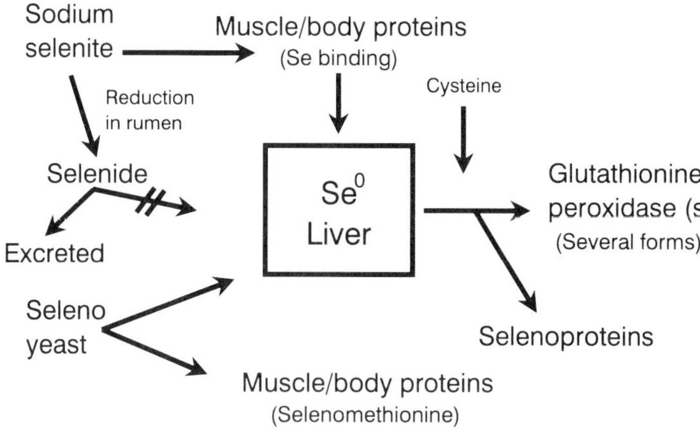

Figure 6. Selenium metabolism in ruminants.

Unlike inorganic forms of Se which are quickly absorbed then lost to a great extent via urine, selenomethionine is apparently digested and absorbed as an amino acid. The selenomethionine in Sel-Plex 50 (the predominant form of selenium in selenium yeast) is rapidly taken up by tissues where it is available for cellular use (Figure 7). Less selenium is excreted by a factor of almost three in the case of pigs (Figure 8, Mahan, 1994). For ruminants the organic form is also more active in terms of improving fertility and in reducing somatic cell counts in dairy cows.

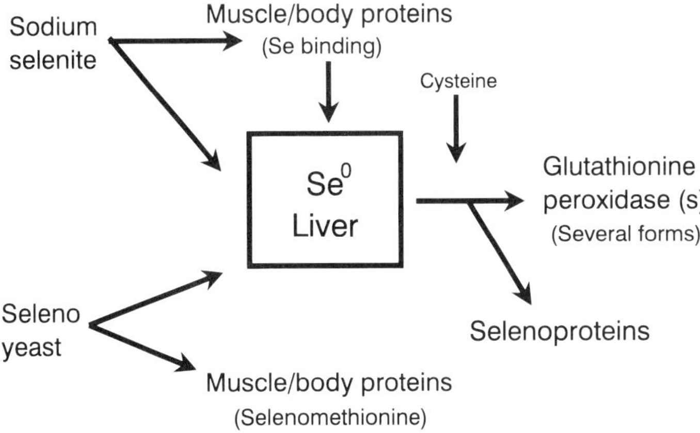

Figure 7. Inorganic and organic Se metabolism in pigs.

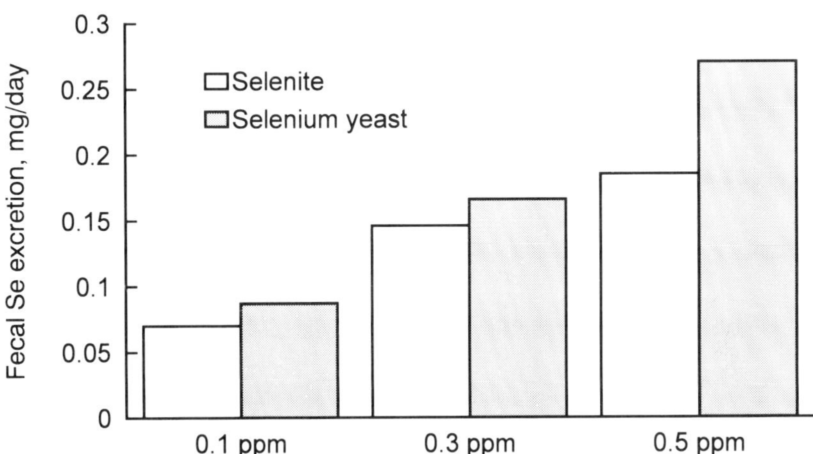

Figure 8. Effect of Se source and dietary Se on fecal Se excretion of Se by growing pigs.

Selenium and thyroid hormone function

It is now known that selenium has many non-GSH functions. Amongst these, perhaps the most important is the role of Se in thyroid hormone activity and in fertility. Selenium is present in the form of selenocysteine at the active site of the enzyme type I deiodinase (IDI) which is involved in the conversion (deiodination) of thyroxin (T4) to 3,3',5–triiodothyronine (T3) in the kidney and liver (Arthur, 1993; Figure 9). In Se-deficient rats and cattle, plasma T4 is increased and T3 decreased. The liver and kidney can use peripheral T3, but brain, pituitary and brown adipose tissue need locally-produced T3. This is generally achieved by type II IDI which, although not a selenoprotein, is reduced in activity in

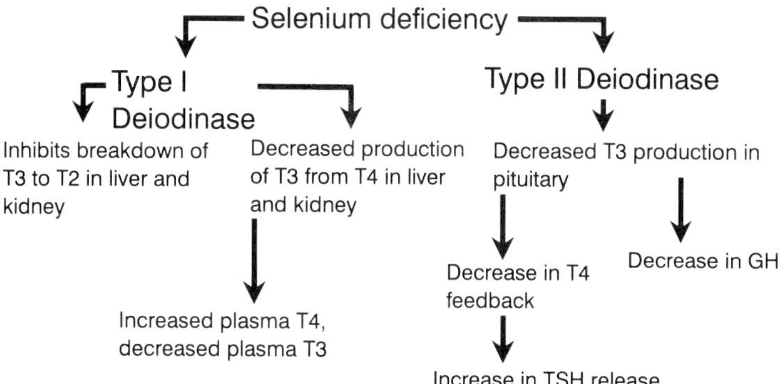

Figure 9. Effects of selenium deficiency on thyroid hormone metabolism (adapted from Arthur, 1993).

pituitary (brain) and brown adipose tissue during Se deficiency because signals for activity depend on circulating T4. A characteristic effect of Se deficiency on thyroid hormone metabolism is continued or increased thyroid-stimulating hormone (TSH) concentration despite elevated T4 which normally suppresses TSH. Low levels of IDI caused by selenium deficiencies lead to reduced pituitary growth hormone secretion and may explain the inhibitory effects of Se deficiency on growth. Since brown adipose tissue relies on locally-produced T3 by type II IDI for normal thermogenic activity, Se deficiency may impair the ability of some animals, notably neonatal ruminants, to withstand cold stress.

Thyroidal concentrations of iodine, T4 and T3 are decreased in Se-deficient rats as a result of increased plasma TSH. These effects can be reversed with small doses of Se which are insufficient to restore normal GSH-Px activity but can normalize plasma TSH. This underscores the need for an adequate Se supply for thyroid hormone metabolism and the need to divert Se toward thyroid gland function when Se supply is limited (Arthur, 1993).

Selenium and immune response

The role of Se in the GSH-Px enzyme underscores its importance in immune response. Reduced immune response is a particular concern in regions where plant deficiencies make obtaining adequate Se status difficult even when adding maximum allowable levels of inorganic Se. Researchers across the world have examined a large number and variety of aspects of immunocompetence in relation to Se nutrition, however studies in Finland with fish point out both immune response benefits and environmental advantages of using a more bioavailable Se source. In various studies at the University of Kuopio over several years challenge trials have demonstrated increased response to Vibrio vaccine (Figure 10) and earlier vaccine response (Kurkela, personal communication).

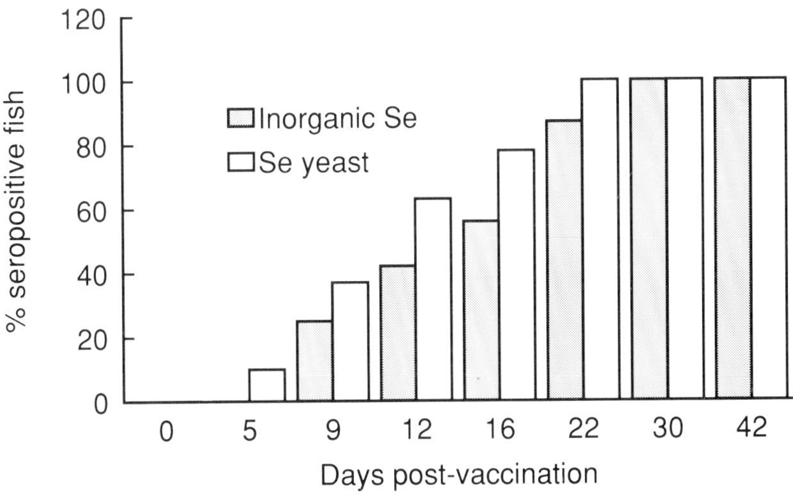

Figure 10. Comparative effects of inorganic and organic Se sources on development of specific immune response in rainbow trout following challange vaccination.

Practical application of organic selenium

In practical terms, both for humans and animals, the possibility of using a natural biologically available form of selenium has already been reflected in higher fertility, more effective immune response, less drip loss in meat and reduced mortality. Since the selenium is 'buried' in the yeast cell, which can also be prilled to avoid dust, the new form is also safer to handle in the feedmill. Indeed, in certain parts of the world it is promoted as 'the safer form of selenium' for these very reasons.

Mahan (personal communication) recently recommended that for all diets involving replacement or reproducing non-ruminants that all inorganic selenium added to the diet be replaced by organic selenium. In the case of grower/finsher pigs where maintainance of health, not progeny, is the main concern, a 50/50 mixture of sodium selenite and organic selenium is sufficient. For ruminants, especially those in Se-deficient regions, only organic selenium should be used.

Animal health research topics

COUNTERING RUMEN ACIDOSIS WITH NATURAL BUFFERING SYSTEMS

Stability of rumen fermentation is still the single most important key to successful forage and concentrate utilization in the high-producing dairy or beef animal. Despite the widespread use of chemical buffers, instability is often the norm. The late 1980s saw the emergence of yeast cultures, driven mainly by the success of one strain (1026), as a means to ensure rumen stability. More recently, yeast culture production has shifted towards targeting specific rumen bacterial populations whose activities limit digestion rate or rumen stability on specific diets. Alltech, together with the University of Kentucky, have developed yeast strains for specific diet categories, Yea-Sacc^{1026} for mixed grain and forage-based diets and Yea-Sacc^{8417} for corn silage and concentrate-based diets.

Yea-Sacc^{1026} modifies the rumen environment by stimulating the production of fiber digesting bacteria like *Ruminococcus albus, Fibrobacter succinogens and Butyrivibrio fibrisolvens.* In contrast, Yea-Sacc^{8417} stimulates lactic acid-utilizing species such as *Selenomonas ruminantium* to a greater degree (Table 6). In both cases the result is a stabilization of rumen pH and consequently of the conditions needed to optimize fiber digestion and reduce ruminal lactic acid. The increased fiber-digesting bacterial numbers stimulate fiber digestion rate, improve rate of passage and ultimately dry matter intake. Not only is intake increased, but forage utilization and feed efficiency are improved. Amongst the effects of Yea-Sacc^{1026} are improvement in milk and component yield and flow of microbial protein (Figures 11 and 12).

Table 6. Effects of two yeast culture strains on rumen microbial populations.

Rumen bacteria	Yea-Sacc^{1026}	Yea-Sacc^{8417}
Concentration of anaerobes	Increased 58%	Increased 30%
Cellulolytic bacteria	Increased 100%	Increased 51%
Lactate-utilizing bacteria	Increased 42%	Increased 77%
Ammonia concentration	Decreased 12%	Decreased 8%

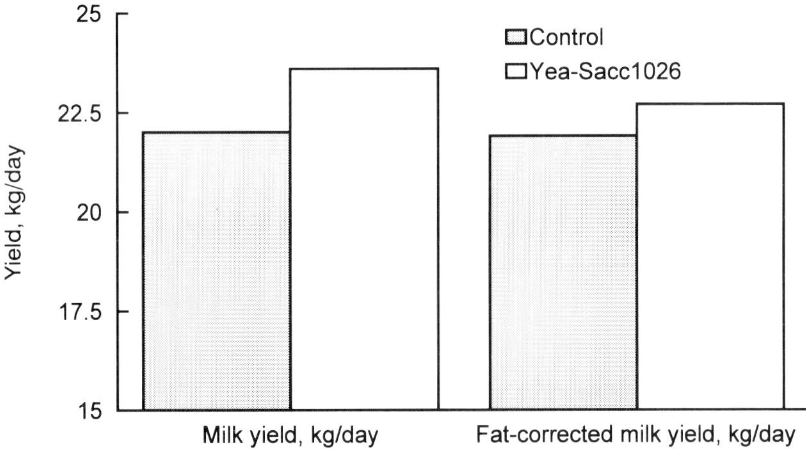

Figure 11. Effect of Yea-Sacc^{1026} on yield of milk and fat-corrected milk of Holstein cows (Harris and Smith, 1993).

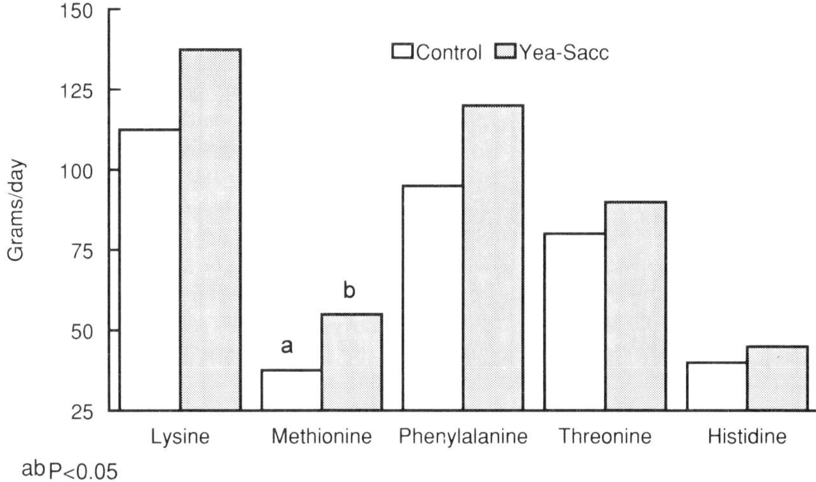

Figure 12. Effect of viable yeast culture on duodenal flow of amino acids considered limiting for dairy cattle (Erasmus *et al.*, 1992).

Yea-Sacc[8417] has been specially selected to stimulate bacterial populations in the rumen responsible for lowering ruminal lactate levels and stabilizing rumen pH in cattle fed high lactic acid-producing diets. A recent study demonstrated the ability of this yeast culture to stabilize pH in ruminal fluid (Figure 13, Klopfenstein, personal communication). The increase in buffering capacity of the rumen has been associated with increased milk production (Girard *et al.*, 1993, Figure 14).

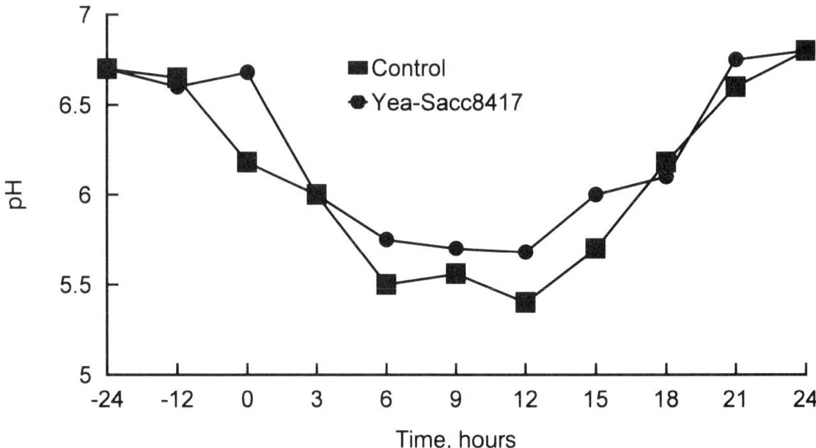

Figure 13. Effect of Yea-Sacc[8417] on pH of ruminal fluid (Klopfenstein, personal communication).

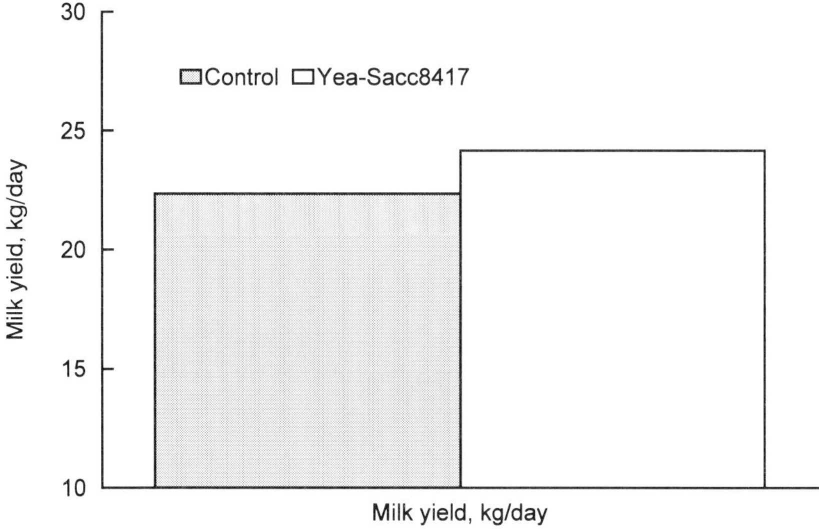

Figure 14. Effect of Yea-Sacc[8417] on milk production in a commercial dairy herd (Girard *et al.*, 1993).

CHROMIUM, AN ESSENTIAL NUTRIENT FOR OPTIMUM NUTRIENT METABOLISM: LEAN GROWTH, STRESS AND REPRODUCTIVE EFFICIENCY

World-wide there is a trend towards reducing carcass fat content in animal products. Virtually all sectors of the food animal industry have been closely following research with trivalent chromium (Cr^{+3}) due to enhanced lean gain noted in some experiments (e.g. Cr yeast and Cr picolinate effects on carcass composition, Wenk, *et al.* 1995, Table 7). Trivalent chromium (Cr^{+3}) was first determined to be the active constituent of the glucose tolerance factor (GTF) isolated by Schwartz and Mertz in 1959. This substance could restore glucose tolerance in rats

Table 7. Effects of the Cr supplements on carcass characteristics

	Control	Cr Yeast	Cr picolinate
Quality score*	2.1	2.3	2.4
Relative, %	100	110	114
Backfat thickness, cm			
Croupe (thinnest point)	1.6	1.7	1.6
Back	1.8	2.0	1.9
Longissimus dorsi			
Area, cm^2	49.5	53.8	48.6

*Quality score: 2, normal; 3, high meat content.

fed mineral deficient diets. This work and subsequent studies in both animals and humans suggested that Cr is a cofactor in GTF needed to potentiate insulin in moving glucose from circulation into peripheral tissues (Anderson and Mertz, (1977).

The exact structure of the glucose tolerance factor is as yet unknown, but it is thought to be a nicotinic acid-Cr^{+3}–nicotinic acid axis with ligands of glutamic acid, glycine and cysteine (Mertz *et al.*, 1974; Figure 15). Likewise, the precise mechanism by which GTF potentiates insulin is not known, but research by Mooradian and Morley in 1987 suggested that GTF enhanced the binding of insulin to its specific receptors on target cells (Figure 16). Heretofore, the GTF factor was too expensive for practical use in livestock feeds, but developments have changed this

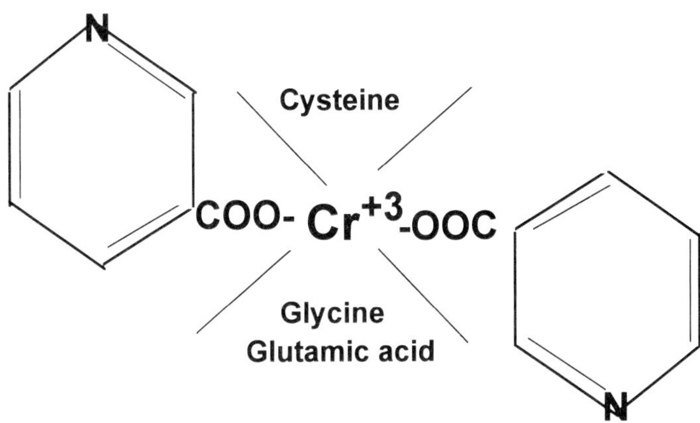

Figure 15. Proposed structure of the glucose tolerance factor.

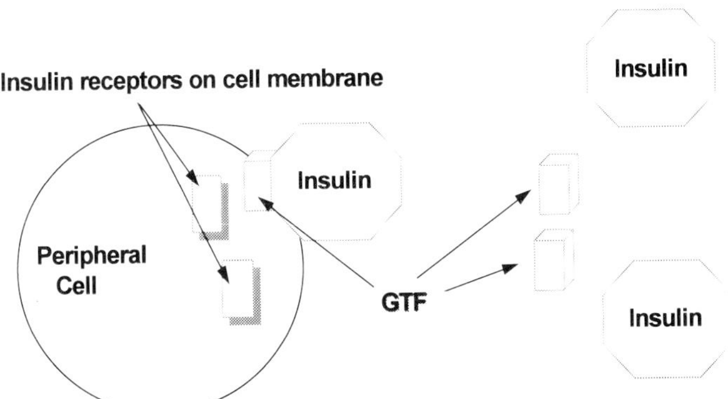

Figure 16. Method by which GTF Cr is thought to interact with insulin in moving glucose into peripheral cells.

so that standardized products with 1000 ppm organic chromium are available.

Recent work with Cr has focused on effects of organic Cr on reproductive efficiency. Lindemann *et al.* (1995) reported that first parity sows given 200 ppb organic Cr through the growth and reproductive cycles had larger litter sizes both with respect to numbers born and numbers weaned (Table 8). They suggested that the effect of Cr on fecundity might be mediated through the effect of Cr in increasing tissue sensivity to insulin. The metabolic state of the sow is sensed by the hypothalmic-pituitary-ovarian axis for stimulation/inhibition of various reproductive functions. The pattern of leutenizing hormone (LH) release from the pituitary in response to GnRH from the hypothalmus has a key role in establishing the estrous cycle after weaning. Exogenous insulin has been shown to increase the frequency of leutenizing hormone release and to increase follicular development and ovulation rate (Cox, 1995). Cox found that treating first parity sows with insulin effectively countered the 'second litter slump' and increased litter size by an extra pig. As the negative energy balance of the lactating sow is associated with decreased insulin, possibly the signal transmitted to the hypothalmus/pituitary/ovarian axis by increased tissue sensitivity to insulin explains in part the association of Cr supplementation with increased fecundity (Lindemann *et al.*, 1995).

Other researchers have found relationships among Cr^{+3} and hormonal regulatory mechanisms. Page *et al.* (1993) authors found that although results were inconsistent, growth hormone levels of pigs supplemented with Cr tended to be increased. Heifers fed 5–6 mg Cr in organic form had increased IGF levels (Mowat, 1994), along with milk production responses of up to 13% greater than controls. Additionally, the receptors for insulin and IGF-1 are similar and can interact with both molecules at high levels. IGF-1 has insulin-like effects on glucose, amino acid and lipid metabolism. Insulin, IGF-1 and growth hormone are related through complex interactions and regulatory mechanisms (White *et al.*, 1994).

Pagan and Jackson (1995) found that strenuously exercised horses had improved glucose clearance from blood with lower insulin and cortisol compared with horses fed the same amount of grain without added Cr.

Table 8. Effect of dietary Cr on litter size of first parity sows*.

	Trt 1	Trt 2	Trt 3		*P* value	
Growth phase Cr, ppb	0	200	500/1000			
Reproductive phase Cr, ppb	0	200	0	SD	Overall	Trt 1 vs Trt 2
Litter size						
Total born	9.58	11.82	10.50	3.01	0.09	0.03
Born live	8.93	11.25	10.07	2.93	0.07	0.02
Day 21	8.15	10.30	9.34	2.77	0.08	0.03
Weaning	8.10	10.25	9.25	2.69	0.07	0.02

*Adapted from Lindemann *et al.*, 1995.

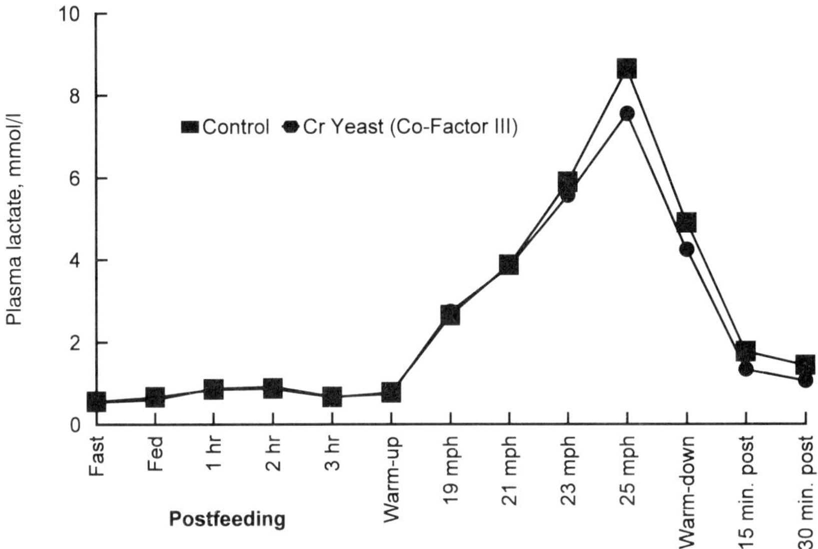

Figure 17. Effect of Cr^{+3} supplementation on plasma lactate concentration of exercising horses (Pagan and Jackson, 1995).

Additionally, plasma lactate levels in supplemented horses were lower both at the faster exercise speeds and post-exercise (Figure 17). These data suggested that adding organic Cr to diets fed strenuously exercised horses could enhance endurance capacity.

IMMUNITY: OPTIMIZING IMMUNE RESPONSE WITH MANNANOLIGOSACCHARIDES

The immune system is the body's extremely complex defense system. All organisms depend on the ability of the immune system to distinguish between self and foreign cells and respond quickly and effectively to pathogen challenge. In recent years there has been increasing interest in better understanding how feed quality and various feed ingredients affect immunocompetence of livestock. This interest reflects both the declining availability of 'new' antibiotics and the recognition that animal health and stress are major problems in modern systems of intensively reared livestock and poultry. We have been examining mannanoligosaccharide derived from cell wall material of selected yeast strains as a dietary supplement added for its ability to stimulate immune response. Mannanoligosaccharide can play a role in defense against pathogens at both the gut level and as a non-specific immunostimulant. Many of the common enteric pathogens including *Escherichia coli* and many species of salmonella attach to the intestinal epithelium via lectins that recognize mannose sugars. Adding mannose to the diet reduces the likelihood that pathogens will colonize the gut and allows the pathogen to pass through the gut. Additionally, bacterial cell wall mannan activates the complement system via the alternative pathway for cascade activiation (Figure 18). It is products of the complement that increase the effectiveness of phagocytic cells such as macrophages to both speed clearance of antigens and promote the inflammatory response. This mode of action explains the increased effectiveness of phagocytic cells observed in response to Bio-Mos.

In Poland it has been recognized now by the authorities as a integral part of an anti-salmonella program. By using relatively high levels (3 kg/tonne) salmonella can effectively be purged out.

Mannanoligosaccharides have proven themselves in commercial production both in terms of the impact on disease resistance and effects on performance parameters of a variety of species. Reid (1994) found that addition of Bio-Mos to weanling rabbit diets helped alleviate losses due to the combined stresses of enteritis and coccidia (Figure 19). Turkey producers report enhanced growth with feed savings of some 1.3 kg of feed per bird (Figure 20, Olsen, personal communication) while pigs scour less and gain weight more quickly. It will be interesting to see where this exciting technology goes.

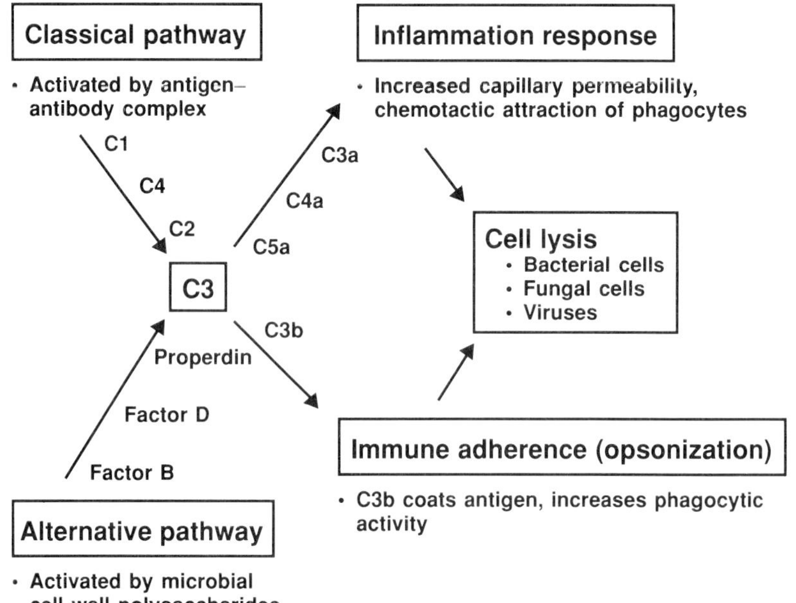

Figure 18. The complement system.

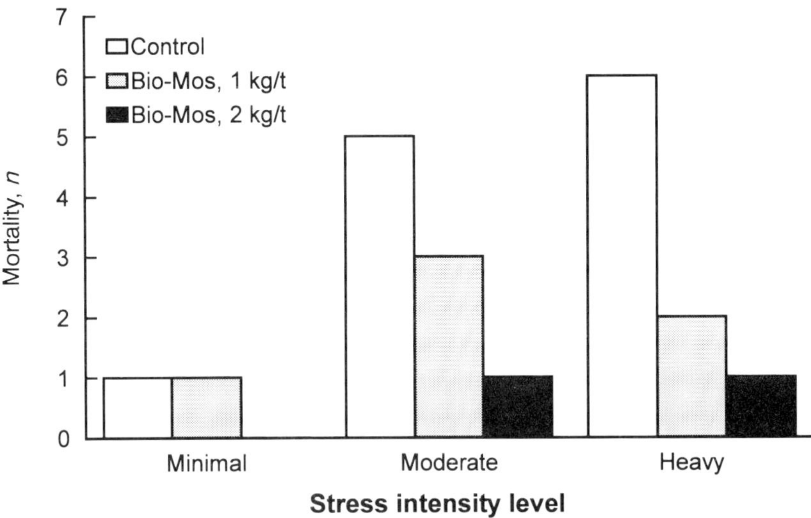

Figure 19. Effect of Bio-Mos and stress level on mortality in weanling rabbits (Stress level was defined by coccidia presence and added corn in the diet. Minimally stressed rabbits had no coccidia and no added corn; moderately stressed rabbits had coccidia via natural exposure and 2 tbls. added corn per day; Heavily stressed rabbits were naturally exposed and challenged with coccidia and given 3 tbls. added corn per day).

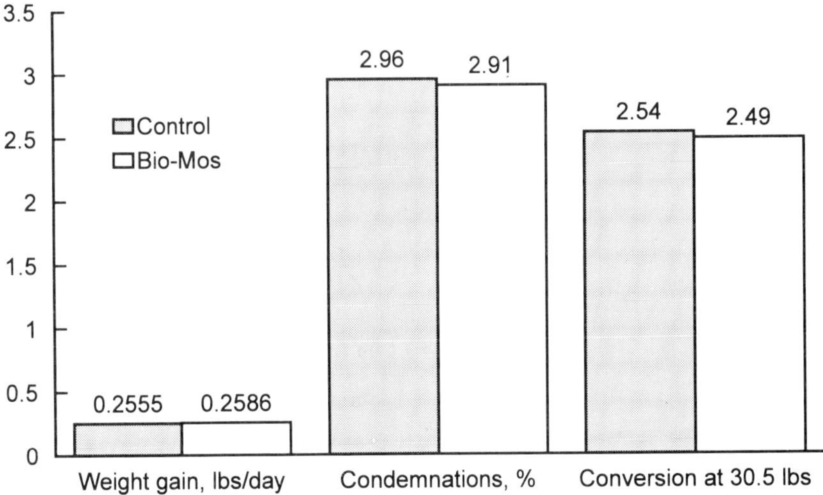

Figure 20. Effect of Bio-Mos on weight gain, condemnation and feed conversion of commercial turkeys.

MICROBIAL PROBIOSIS: *SACCHAROMYCES CEREVISIAE* VAR. *BOULARDII* IN MONOGASTRIC SPECIES

Strains of yeast, due to differing metabolic or cell wall characteristics, vary in potential as useful gut microbial modifiers. While *S. cerevisiae* 1026 and 8417 have been shown to stimulate certain critical bacterial populations in the ruminant, *S. cerevisiae* var. *boulardii* (SCB) may prove more useful in monogastric animals and poultry. Administration of this SCB to human patients undergoing oral antibiotic therapy prevented proliferation of pathogenic species (Surawicz *et al.*, 1989). Such disruptions of the normal intestinal flora are also common in livestock species following therapeutic treatment; and may serve to explain some of the improved performance responses seen when yeast culture products have been added to the diet. Bradley *et al.* (1994) hypothesized that alterations in gut microflora might be accompanied by changes in gut epithelial microstructure (Figure 21a and b). These investigators examined effects of SCB added at 0.01, 0.02 and 0.06% of a corn/soy diet fed to poults. Body weight at 21 days was signifiantly increased by all three treatments. Additionally, histological examination revealed that the number of goblet cells per millimeter of villus height and crypt depth were reduced in poults fed the 0.02% SCB diet. Mucin produced by goblet cells serves as both a habitat and a nutrient source for gut microbes. There were higher numbers of goblet cells in the intestinal villi of germ-free rats (Larson, 1989). The crypts of Lieberkühn produce epithelial cells by mitosis. Decreased crypt depth may reflect decreased epithelial cell rate possibly resulting from fewer toxin-producing bacteria or the ability of SCB to suppress these toxic metabolites. Energy conserved by decreasing gut cell turnover may be utilized for lean tissue growth (Bradley *et al.*, 1994).

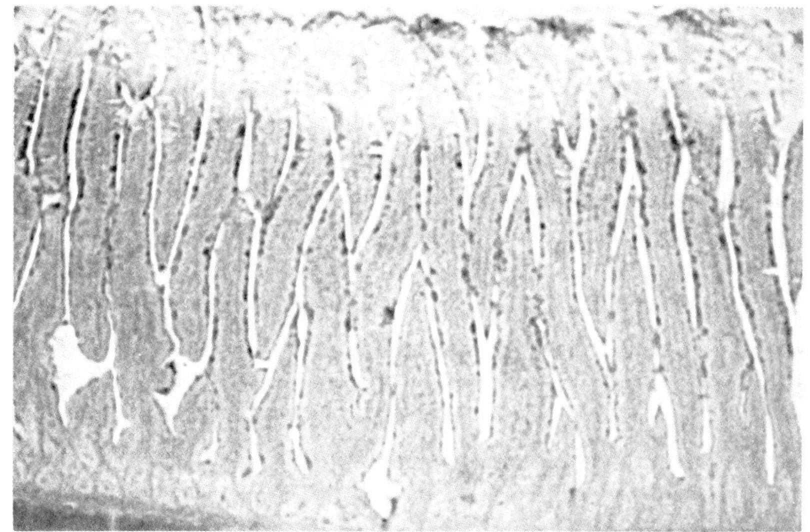

Figure 21a. Gut epithelial microstructure.

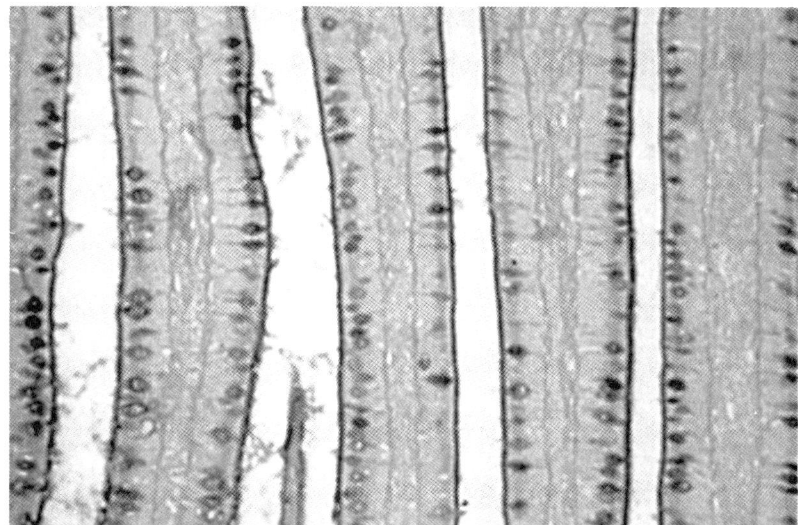

Figure 21b. Goblet cells lining intestinal villi of the turkey poult.

Feed quality and nutrient utilization topics

TRACE MINERAL PROTEINATES

Over the past five years, a quiet revolution has occurred in the area of mineral supplementation. Nutritionists have moved away from inorganic sources toward organically-complexed Bioplexes or trace mineral proteinates. While a number of factors contribute to this change including environmental protection and improved productivity, it is also significant that in recent years considerable efficacy data have appeared in scientific journals. At this point both farmers and nutritionists know a good deal more about trace mineral supplement options and trace mineral metabolism problems than in the not-too-distant past.

While it is true that there is more general knowledge in the industry about the roles of trace minerals in metabolism and immunity than previously, the topics of evaluation of organic mineral complexes, rumen stability and absorption of chelates are often discussed with more enthusiasm than solid information. Data have begun to accumulate, however, to answer questions about comparative absorption of inorganic and organic forms. Du and co-workers at the University of Kentucky showed previously that absorption of copper (Cu) from Bioplex Cu was much higher than that from inorganic forms for rats (Du *et al.*, 1993; 1994). Additionally, their work demonstrated that Cu supplied in proteinate form resulted in higher organ Zn concentrations. This suggested that inorganic and organic Cu were absorbed by different mechanisms. In a recent study by the same researchers with lactating cows given inorganic or Bioplex Cu with or without excess iron (Fe),

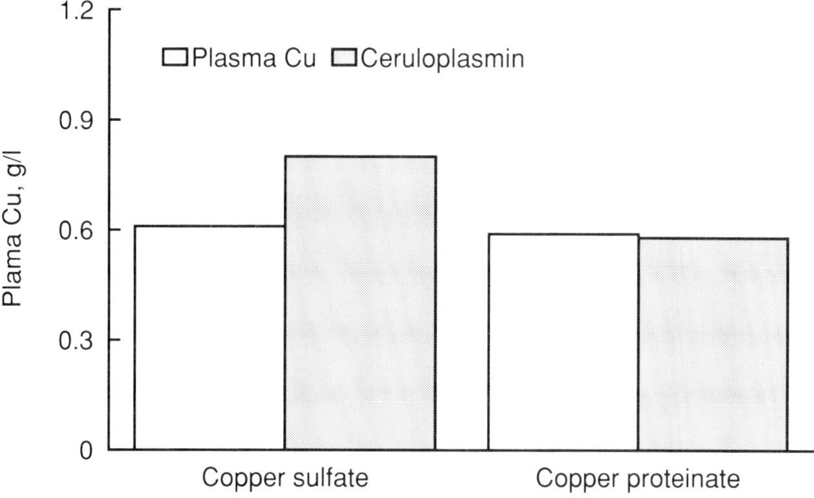

Figure 22. Plasma Cu concentration and ceruloplasmin on day 90 in cows fed Cu proteinate and $CuSO_4$.

hepatic and plasma Cu were unaffected by either Cu source or Fe supplementation; however Cu proteinate increased hepatic iron content at day 90 compared with $CuSO_4$. This suggested that Cu proteinate did not interfere with absorption of Fe when compared to $CuSO_4$. Additionally, cows fed Bioplex Cu had lower plasma ceruloplasmin activity (Figure 22) at day 90 than cows fed $CuSO_4$ while plasma Cu was unaffected. This indicated that the ratio of Cu in ceruloplasmin form to other forms of plasma Cu was lower for cows fed Bioplex Cu than for those fed inorganic Cu. The increased hepatic Fe and lower plasma Cu ratio suggested that Cu proteinate was absorbed in the organic form and transported in blood without binding to ceruloplasmin.

There are also data to better explain the improvements in reproductive efficiency noted in response to Bioplexes fed high-producing dairy cows and embryo transfer donor/recipients. O'Donoghue and Boland (1995,

Table 9. Effect of organic mineral supplementation on reproduction in dairy cows.

	Control	Bioplex
Days to 1st dominant follicle (DF)	9.3	7.8
% which ovulated 1st DF	60	60
% which ovulated 2nd DF	10	30
% which ovulated 3rd DF	20	10
% which ovulated 4th DF	10	–
Days post-partum to ovulation	25.3±3.1	20.4±1.4
Days to first service	75.4±6.1	68.8±3.8
Conception rate to first service, %	60	65.2

this volume) evaluated fertility in dairy cows using ultrasound scanning 25–30 days after first service. Addition of a mixture of Bioplex Cu and Zn reduced the days to appearance of the first follicle (Table 9). Similar numbers ovulated at the first follicle, however 30% and 10% of those given the supplement ovulated at the 2nd and 3rd dominant follicles, respectively. In comparison only 10% of the control group ovulated at the 2nd follicle with 20% at the 3rd and the other 10% at the 4th follicle. There were five fewer days post-partum to ovulation, six fewer days to first service and an improvement in conception rate to first service.

Practical applications for Bioplexes in non-ruminants have also developed in recent years, particularly as a result of the environmental and absorption interference concerns presented by using high levels of copper and/or zinc in pig diets. Cole (personal communication) has suggested combinations for inorganic and organic trace mineral forms in practical diets fed growing pigs (Table 10).

Table 10. Suggested total, inorganic and organic trace mineral concentrations in diets fed young pigs through 50 kgs (110 lbs).

	Total	Inorganic mg/kg total diet	Organic
Zinc	120	80	40
Manganese	50	25	25
Copper	20	14	6
Cobalt	0.1	0.1	–
Iron	100	70	30
Selenium	0.2	–	0.2
Chromium, ppb	200	–	200

MYCOTOXINS IN FEED: A HIDDEN KILLER

Recent surveys have demonstrated that as much as 25% of the world's cereal grains may be contaminated with mycotoxins. While the molds that produce these mycotoxins can be controlled by the use of high levels of acids and more extensive drying, none the less preformed mycotoxin concentrations in feed often remain. The predominant mycotoxins of concern are aflatoxin, zearalenone, vomitoxin and fumonisin. By examining the structure of these toxins, new biological adsorbents are

Table 11. Binding agents found useful with various toxins.

Mycotoxin	Binding agent effective
Aflatoxin	Silicate, yeast mannan
Zearalenone	Yeast mannan, cholestyramine
Ochratoxin	Cholestyramine, charcoal
Vomitoxin	
Fumonisin	

being developed which remove the mycotoxin but do not affect valuable nutrients such as vitamins and minerals. A number of potential binding agents have been examined (Table 11). Modified mannan sugars prove promising as toxin adsorbents, and initial studies indicate success with aflatoxin and zearalenone. Trenholm and co-workers (1994) found that mannanoligosaccharides bound aflatoxin and zearalenone (Figure 23).

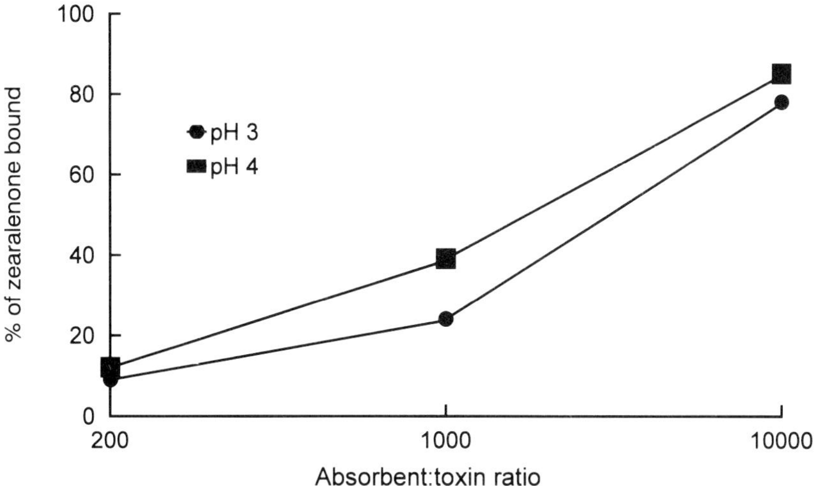

Figure 23. Percentages of zearalenone bound by Graingard (mannan sugars) at pH 3 and pH 4.

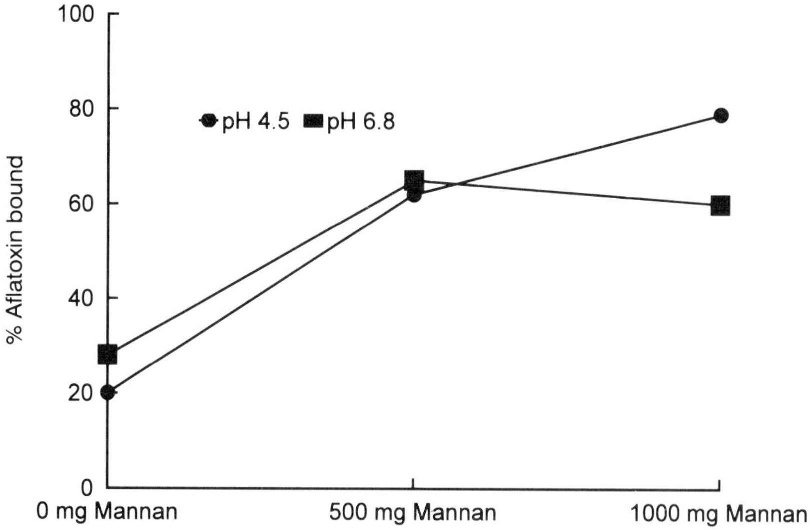

Figure 24a. Effect of pH on percentages of aflatoxin B_1 bound by 500 or 1000 mg mannanoligosaccharide at 250 ppb aflatoxin.

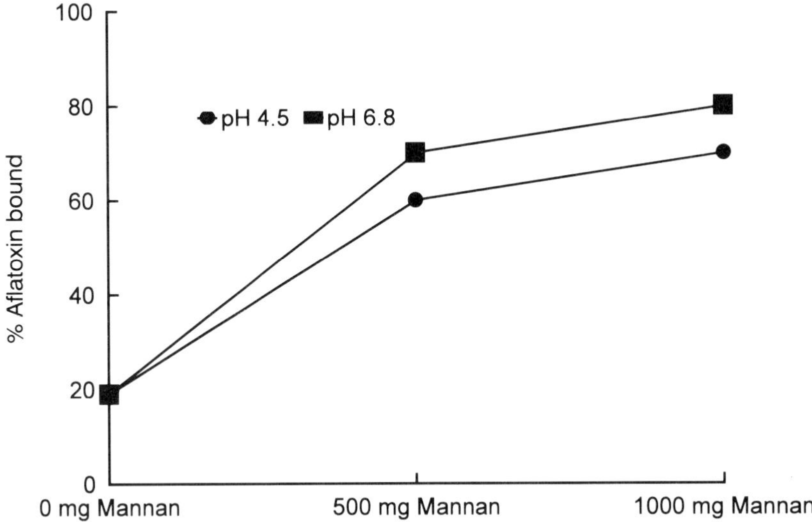

Figure 24b. Effect of pH on percentages of aflatoxin B_1 bound by 500 or 1000 mg mannanoligosaccharide at 500 ppb aflatoxin.

Devegowda (personal communication) confirmed that aflatoxin B_1 was bound by mannan and found that while at pH 4.5 similar percentages were bound whether initial toxin concentration was 250 or 500 ppb, at pH 6.8 there was a gradual increase in the amount of toxin bound (Figures 24a and b). Studies continue with vomitoxin. By using this range of binding agents, perhaps in combination, prospects for a valid antitoxin pack could become reality.

DIET SPECIFIC ENZYMES: IMPROVING UTILIZATION OF PROTEIN FEED INGREDIENTS

The feed industry has come to accept the value of carbohydrase enzymes in improving energy utilization of cereal grains after several years of both research and field experience. In much of the world there is routine use of betaglucanase and pentosanase where barley and wheat (and related cereals) are fed to poultry. The net result has been the ability of the feed company to switch among raw materials without losing efficiency. Attention is turning now toward improving utilization of both protein and energy in commonly-used protein ingredients. It is particularly important, however, to again remember the need to carefully match enzyme with substrate when customizing an enzyme complement to a particular ingredient. Pugh and Charlton (personal communication) found an increase in total metabolizable energy (corrected for nitrogen) (TME (N)) when a specially formulated enzyme supplement (Allzyme VegPro I) was added to a broiler diet containing oilseed rape. Changes in the formula to better fit the structure of soy protein resulted in an

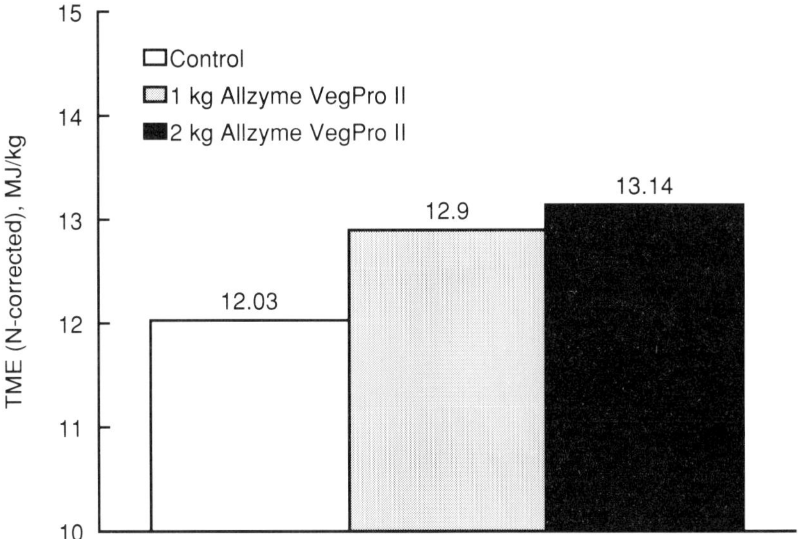

Figure 25. Effect of Allzyme VegPro II level on nitrogen-corrected total metabolizable energy level of soybean meal fed broilers.

increase of 7.2% in TME (N) of a diet based on soybean meal when added at 1 kg/tonne and 9.2% when added at 2 kg/tonne (Figure 25).

PROTECTING AMINO ACIDS THROUGH THE RUMEN

Rumen fermentation, while necessary for the digestion of fiber, can be wasteful in terms of utilization of essential amino acids such as methionine and lysine. The key is to supply these amino acids to the intestine, but since they must run the gauntlet of degradation in the rumen, very often the supply of these essential amino acids is not well balanced. The search is on to discover how to provide an amino acid which will not be degraded at a pH of 6–7, but will become available at a pH of 4. Two approaches have been developed.

1. *The use of the yeast culture Yea-Sacc[1026]*. It has been observed that this yeast culture can selectively stimulate certain populations of bacteria in the rumen. Work by Erasmus *et al.* (1992) has demonstrated that the resulting biomass presented to the small intestine is richer in lysine and methionine (see Figure 12). In effect, the rumen microbes are producing their own bypass amino acids.
2. An alternative approach is to use the yeast's own ability to survive through the rumen as a vehicle by which metabolites can be protected. Lysine and methionine-enriched strains of yeast have now been produced and are being commercially tested. One strain has methionine and lysine present in a ratio of 1:3 with total levels of the amino acids in excess of 10%. This product will probably be commercialized next year.

Conclusions: the future

In many countries including the USA and most of Europe, new products such as these described must be registered. As many of these products are non-traditional types of feed ingredients, the procedures for evaluation and registration are new, as well. Analytical techniques have been or are being developed to assay the active ingredients both in the product and in the subsequent feed. Already radial diffusion techniques for enzymes have been perfected and methods for detecting the active component of the *Yucca schidigera* plant have been established. The enzyme assay procedure has also been used to confirm that enzymes are indeed able to survive feed pelleting.

Alltech has set itself the goal of expanding its business to 750 people and opening Bioscience centers on all continents by the year 2000. These centers will support local production and draw from local university expertise. The potential is enormous, the tasks daunting and exciting. However, if history is any indication, then the strategy of using young scientific minds in an open challenging way will allow us to fulfill the company's goals and solve many major animal production problems. If to be blessed is to live in exciting times, then at least scientifically we are truly blessed.

References

Arce, J. and E. Avilla. 1994. Effect of De-Odorase (*Yucca schidigera* extract) on performance and mortality due to ascites in heavy broilers. Poultry Sci. 73(Suppl. 1):122.

Anderson, R.A. and W. Mertz. 1977. Glucose tolerance factor: an essential dietary agent. Trends Biochem. Sci. 2:277.

Arthur, J.R. 1993. The biochemical functions of selenium: relationships to thyroid metabolism and antioxidant systems. In: The Rowett Research Institute Annual Report. Rowett Research Institute, Aberdeen, Scotland, UK.

Bradley, G.L., T.F. Savage and K.I. Timm. 1994. The effects of supplementing diets with *Saccharomyces cerevisiae* var. *boulardii* on male poult performance and illeal morphology. J. Poultry Sci. 73: 1766–1770.

Cantor, A.H., E. Moore, A.J. Pescatore, M.L. Straw and M.J. Ford. 1994. Improvement of phosphorus utilization in broiler chickens with acid phosphatases from genetically modified microorganisms. Poultry Sci. 73 (Suppl. 1): 78.

Du, Z., R.W. Hemken and T.W. Clark. 1993. Effects of copper chelates on growth and copper status of rats. J. Dairy Sci. (Suppl.) 76:306.

Du, Z., R.W. Hemken and S. Trammell. 1994. Comparison of bioavailabilities of copper in copper proteinate, copper lysine and cupric sulfate and their interaction with iron. J. Anim. Sci. (Suppl.) 72:273.

Erasmus, L.J., P.M. Botha and A. Kistner. 1992. Effects of yeast culture supplementation on production, rumen fermentation and duodenal nitrogen flow in dairy cows. J. Dairy Sci. 75:3056–3065.

Headon, D.R., K.A. Buggle, A.B. Nelson and G.F. Killeen. 1991. Glycofractions of the yucca plant and their role in ammonia control. In: Biotechnology in the Feed Industry-7. T.P. Lyons (Ed.). Alltech Technical Publications, Nicholasvile, Kentucky, pp 95–108.

Killeen, G.F, K.A. Buggle, M.J. Hynes, G.A. Walsh, R.F. Power and D.R. Headon. 1994. Influence of *Yucca schidigera* preparations on the activity of urease from *Bacillus pasteurii*. J. Sci. Food Agric. 65:433–440.

Larson, G. 1989. The normal microflora and glycosphingolipids. In: The Regulatory and Protective Role of the Normal Microflora. Wenner-Gren International Symposium series, Volume 52. R. Grubb, T. Midtvedt and E. Norin (eds). Stockton Press, New York, NY.

Lindemann, M.D., C.M. Wood, A.F. Harper, E.T. Kornegay and R.A. Anderson. 1995. Dietary chromium picolinate additions improve gain:feed and carcass characteristics in growing-finishing pigs and increase litter size in reproducing sows. J. Anim. Sci. 73:457–465.

Mahan, D.C. 1994. Organic selenium sources for swine – how do they compare to inorganic selenium sources? In: Biotechnology in the Feed Industry: Proceedings of the 10th Annual Symposium. T.P. Lyons and K.A. Jacques (Eds). Nottingham University Press, Loughborough, Leics. UK.

Mertz, W., E.W. Toepfer, E.E. Roginski and M.M. Polansky. 1974. Present knowledge of the role of chromium. Fed. Proc. 33:2275.

Mooradian, A.D. and J.E. Morley. 1987. Micronutrient status in diabetes mellitus. Amer. J. Clin. Nutr. 45:877.

Pagan, J. and S.J. Jackson. 1995. The effect of Cr supplementation on metabolic response to exercise in Thoroughbred horses. Proc. Equine Nutr. and Physiol. Symp.

Page, T.G., L.L. Southern, T.L. Ward and D.I. Thompson. 1993. Effect of chromium picolinate on growth and serum and carcass traits of growing-finishing pigs. J. Anim. Sci. 71:656–662.

Schwartz, K. and W. Mertz. 1959. Chromium (III) and the glucose tolerance factor. Arch. Biochem. Biophys. 85:292.

Smith W.A., B. Harris, Jr, H.H. Van Horn and C.J. Wilcox. 1993. Effect of forage type on production of dairy cows supplemented with whole cottonseed, tallow and yeast. J. Dairy Sci. 76:205.

Surawicz, C.M., G.W. Elmer, P. Speelman, L.V. McFarland, J. Chinn and G. van Belle. 1989. Prevention of an antibiotic-associated diarrhea by *Saccharomycese boulardii*: a prospective stu y. Gastroenterology 96:981.

Weaver, D. 1993. Manure Odor Control Project. In: Farm News of Erie and Wyoming Counties, Cornell Coop. Ext. Serv., April.

Wenk, C., S. Gebert and H.P. Pfirter. 1995. Chromium supplements in the feed for growing pigs: influence on growth and meat (quality). Arch. Anim. Nutr.

White, M., J. Pettigrew. J. Zollitsch-Stelzl and B. Crooker. 1994. Chromium in Swine Diets. 54th Minnesota Nutrition Conference and National Renderers Technical Symposium. Minnesota Extension Service; Bloomington, MN.

Woodgate, S.L. 1993. The case for recycling: possibilities for profitable nutritional upgrading. In: Biotechnology in the Feed Industry. Proceedings of the 9th Annual Symposium. T.P. Lyons (Ed.). Alltech Technical Publications, Nicholasville, KY.

Woodgate, S.L. 1995. Focus On Animal By-products: A Necessary Waste or Valuable Opportunity. Beacon Research Limited, Greenleigh, Kelmarsh Road, Clipston, Market Harborough, Leics LE16 9RX, United Kingdom.

MYCOTOXINS: A WORLDWIDE PROBLEM AFFECTING 25% OF OUR CEREAL GRAINS: LIGHT AT THE END OF THE TUNNEL

MOLDS, MYCOTOXINS, AND THE PROBLEMS THEY CAUSE

ROGER D. WYATT

Department of Poultry Science, University of Georgia, Athens, Georgia, USA

Introduction

Microbial deterioration of grain, feedstuffs and animal feed is of utmost concern to all segments of the animal industries. Transmission of bacteria such as *Salmonella*, *Listeria*, and pathogenic *Escherichia coli* to livestock through consumption of contaminated feedstuffs emphasizes the importance of controlling and/or eliminating these microorganisms in feed and feed ingredients. Molds are another group of microorganisms of extreme concern to the animal industries. Molds can cause various infections in livestock (i.e., aspergillosis and candidiasis) as well as intoxications due to the formation of mycotoxins in feedstuffs that have undergone 'molding'. Of all these microorganisms, molds, and the roles they play in the contamination of feeds, are perhaps the most poorly understood.

Mold growth in stored feeds and grains

Molds reproduce by forming spores on a specialized structure known as an aerial mycelium. This structure serves to raise the spores above the material upon which the mold is growing. The spores are then carried by air currents to other environments where spore germination followed by rapid mycelial growth and sporulation take place. On a single aerial mycelium, literally thousands of spores can be produced. Due to the widespread distribution of these spores, molds are regarded as ubiquitous, that is, they are present in virtually all niches in our environment.

Mold spores are considered 'resistant' because they can remain viable under extremely dry conditions. On the other hand, mold spores are relatively susceptible to heat. For example, mold spores in a sample of feed will be drastically reduced due to exposure to typical pelleting conditions used in the manufacture of animal feed. Although the spores will not be eliminated completely from the feed, this reduction points to the relative susceptibility of mold spore to moist heat. Even though mold spores can be destroyed during pelleting, recontamination of feed with mold spores can occur in feed manufacturing equipment, storage bins, bulk feed trucks and on-the-farm feed bins. When this recontamination occurs, subsequent germination and 'molding' can occur at a rapid rate.

Viable mold spores can germinate, grow, and reproduce provided that a suitable substrate, oxygen, warmth, and moisture are available. Most animal feeds, stored under routine conditions, provide all of these requirements. Starches and fats in feed ingredients provide the molds with an abundant supply of energy. Protein from both animal and plant sources ensures that a full complement of amino acids is present. Endogenous and supplemental vitamins meet yet other nutritional needs of molds. During the manufacture, transport and storage of poultry feed, a warm environment and a readily available supply of oxygen are almost always present. Most often, the limiting requirement for mold growth is moisture. Theoretically, if the moisture level of feed is maintained at 11.5% or less, mold growth will not occur. However, for a variety of reasons, this 'low' moisture level cannot always be maintained. A pellet cooler that is not functioning properly can permit high moisture levels in pelleted feed. The introduction of 'warm' or 'hot' feed into a storage container (feed bin or truck) can lead to high moisture levels in portions of the feed. Storage containers in a poor state of maintenance can leak and allow water seepage into the bin. Finally, the phenomenon known as 'moisture migration' can cause the moisture content within certain areas of a feed bin to reach extremely high levels. Usually moisture collects along the periphery of the bin, resulting in the commonly observed buildup of caked moldy material on the bin walls. This material must be removed periodically and then destroyed. Excessive demands upon labor, costly down-time, and a general attitude of '. . . out of sight, out of mind . . ." often discourages this important part of an effective feed management program.

Mycotoxins

The process of molding of feedstuffs was considered for decades to be nothing more than an unsightly nuisance. Further, it was assumed that if animals were fed the moldy material, no harm would come to them. We now know this is not the case. Mold growth on feedstuffs is known to 1) alter the nutrient profile of the feedstuff, and 2) result in the formation of toxic substances known as mycotoxins in the feedstuff. Obviously, either one or both of these situations will have an adverse impact on the performance of animals consuming such feed.

With specific regard to mycotoxins, several key points should be kept in mind. Other than the fact that mycotoxins are, by definition, highly toxic substances produced by molds, few similarities exist among the hundreds of mycotoxins known to occur in nature. For example, in some cases, poultry are quite susceptible to certain mycotoxins, whereas cattle and swine are resistant. With specific regard to other mycotoxins, poultry are rather resistant, whereas, these same mycotoxins can prove lethal to cattle and horses.

The formation of mycotoxins in nature is considered a global problem, however, in certain geographical areas some mycotoxins are produced more readily than others. Such an example can be seen with aflatoxin and some Fusarium-produced mycotoxins. Aflatoxin is known to be produced predominantly in the southeastern and southwestern regions of the United States, whereas zearalenone and many of the trichothecene mycotoxins predominate in somewhat cooler climates such as the upper midwestern regions of the United States and the grain producing areas of Canada.

The biological manifestation in livestock experiencing mycotoxicosis is dependent upon the specific mycotoxin involved in the toxicosis. For example, aflatoxin is a known hepatotoxin whereas ochratoxin is a known nephrotoxin. Both oosporein and ochratoxin have a devastating effect upon kidney tissue of affected animals, however only oosporein is known to cause visceral and articular gout. Although ochratoxin affects primarily the kidney, glycogen accumulation in the liver is a specific effect of this mycotoxin. Most trichothecene mycotoxins can cause necrosis and inflammation of the oral cavity of animals consuming trichothecene-contaminated feedstuffs. This effect is rather specific for the trichothecenes and can occur with little or no clinical change in either the liver or kidney in affected animals. These and many other idiosyncracies inherent to mycotoxins often lead to extreme difficulty in diagnosing such problems and can lead to a very confusing and perplexing disease situation for the animal producer.

Assessing mold damage in feeds

An obvious means of minimizing these adverse effects is for the grain producer, feed manufacturer and animal producer to employ management techniques that will retard or stop mold growth in grain and feed. Although it sounds rather simple and straightforward, this task is not accomplished easily. From a researcher's point of view, if mold growth is to be stopped or at least minimized in feedstuffs, a technique must be available that will permit the accurate and rapid monitoring of mold growth in feedstuffs. This should permit the development of technology to enable one to study the phenomenon of 'molding' and permit the assessment of techniques to minimize or stop the 'molding' process. Again, this is more difficult than it sounds. Unlike most bacteria and yeasts, molds grow as a continuous, intertwined body called the mycelial mass. Following this vegetative (i.e., non-reproductive) growth, mold spores are formed in a secondary phase referred to as sporulation. In essence, the development of the mycelial mass, not sporulation, represents mold growth. This is why mold spore plate counts performed on feedstuffs may not reflect accurately the amount of mold growth that has occurred in the feedstuff. On the other hand, the traditional mold spore plate count typically serves as an indication of the degree of sporulation that has occurred with regard to the molds found in the feedstuff and not a direct measurement of the degree of 'molding' of the feedstuff. Since this mycelial mass is not characterized by individual cells which can be counted, other means must be used to assess the progression of mold growth in a complex substrate such as complete animal feeds. Only when this assessment can be made accurately can the utility of techniques aimed at minimizing or stopping mold growth in animal feeds be determined.

One fundamental concept in microbiology states that '. . . the growth of a microorganism can be indirectly monitored by either the disappearance of a substrate used by the microbe or the appearance of an end-product resulting from growth of the microbe . . .'. Historically, respirometry has been used to assess the growth of microorganisms by measurement of carbon dioxide (an end-product) production under a variety of conditions. The recent advent of a respirometer capable of simultaneous measurement of both carbon dioxide production and oxygen consumption has permitted a more detailed assessment of mold growth, specifically in complex matrices such as animal feeds.

For example, the data shown in Figure 1 compare the growth of *Aspergillus parasiticus* (as measured directly as dry mycelial weight/flask) in a liquid microbiological medium with the production of carbon dioxide and the consumption of oxygen. Note that there is a very high correlation between growth of this mold and either carbon dioxide production or oxygen consumption. This simple experiment offers verification of respirometry as a valid means to measure mold growth.

The recent development of a respirometer with the capability of

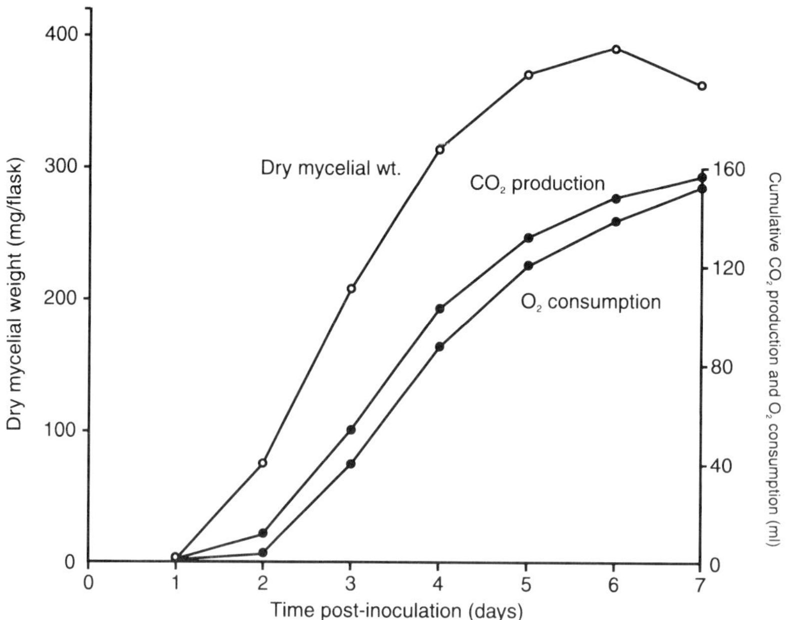

Figure 1. Comparison of dry mycelial weight and cumulative carbon dioxide production and oxygen consumption of *Aspergillus parasiticus* in liquid medium.

measuring continuously and simultaneously both carbon dioxide (end-product formation) and oxygen (substrate) consumption in an aerobic culture of animal feed has permitted the extension of our capability and understanding of mold deterioration of animal feedstuffs. Depending on the experimental design, the feed used with this respirometer may contain either a single mold, a variety of molds, the normal microflora or a mixture of molds, yeasts and bacteria. The basic principle of respirometry is that the rate of carbon dioxide generation and the rate of oxygen consumption are directly related to the rate of mold (or microbial) growth in the substrate.

The respirometer used in our laboratory is computer controlled and has a unique feature that permits 'refreshment' of the atmosphere in the cultures. Refreshment ensures that oxygen does not become limiting to the microbes and that carbon dioxide does not reach toxic or growth-suppressing concentrations. Furthermore, humidifiers connected to each of the incubation vessels used with this respirometer help to maintain a constant moisture level in the substrate throughout a typical 7–14-day incubation period. Additionally, data are collected as the rate of specific gas consumption or generation and as the cumulative gas consumption or generation with respect to time. The goals of research using respirometry include: 1) development of basic information about various methods, including respirometry, for accurate and rapid assessment of mold growth in poultry feeds; 2) determination of the influence of moisture and dietary composition of poultry feed on mold growth, and 3) application of these results to a comprehensive assessment of the efficacy of various mold retardants in poultry feeds.

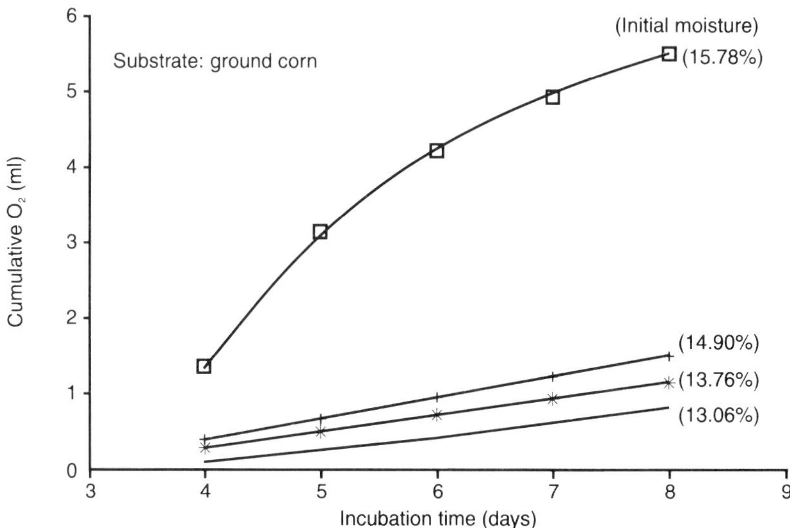

Figure 2. Cumulative oxygen consumption by *Aspergillus parasiticus* in ground corn at various moisture levels.

Results from numerous studies using respirometry indicate that detectable mold growth can occur in corn at moisture levels between 13 and 14%. Additionally, very rapid mold growth can occur in corn at moisture levels in excess of 15% (Figure 2). Mold growth at these very low moisture levels has been suspected, but until recently had not been documented conclusively. It is believed that the ability of molds to grow at these moisture levels may offer a partial explanation for poor performance of poultry and livestock that consume feed manufactured from grain with a moisture level higher than average, but acceptable according to current grain standards. Furthermore, these data point to the need for controlling moisture levels in either feed ingredients or complete feeds during storage of literally any length of time.

Another interesting observation is that oxygen consumption appears to be a more sensitive indicator of mold growth in complete animal feeds compared with only carbon dioxide formation. Currently, numerous techniques employed by the poultry and allied industries rely on only carbon dioxide formation as the indicator of mold or total microbial growth. Based on the results of our research, oxygen consumption may indeed be preferable to carbon dioxide generation as the primary trait to assess mold growth in feedstuffs.

The animal industries are keenly aware of the impact of the utilization of moldy feedstuffs. Consequently, efforts are made to quantify the 'moldiness' of feedstuffs and the mycotoxin concentration in feedstuffs. Unfortunately, neither characteristic is easy to quantify. For example, the list of known mycotoxins continues to increase in length. Furthermore, many of these mycotoxins are known to have a detrimental impact on animal health, however, in many cases, there are no suitable analytical methods available for routine application to feedstuffs. Given the fact that mycotoxin formation in feedstuffs can not always be detected or measured, the animal industries have turned to alternative means to assess 'moldiness'. The most commonly used analysis is the mold spore plate count. The mold spore plate count can be less time-consuming than specific mycotoxin analysis, is available in numerous laboratories, and is almost always less expensive than mycotoxin analyses. However, interpretation of mold spore counts is very difficult.

Perhaps the most relevant question is 'What does the mold spore count of a particular feedstuff really mean?' Usually it is assumed that the higher the mold spore count the more mold has grown in the feedstuff. However, this interpretation of this test can be erroneous. For example, in a recent experiment a sample of commercially manufactured poultry feed was obtained from a local feed mill. The feed sample was examined with the aid of a scanning electron microscope. This examination indicated that the feed had no obvious mold growth. The moisture content of this feed was increased to approximately 17% (a moisture level typical of localized areas within a feed bin) and respirometric measurements were made over a 13-day incubation period. The data in Figure 3 demonstrate that while mold growth (as indicated by both carbon dioxide formation and oxygen consumption) increased dramatically over the incubation period, the mold spore count was virtually unchanged over the 13-day period from the initial count.

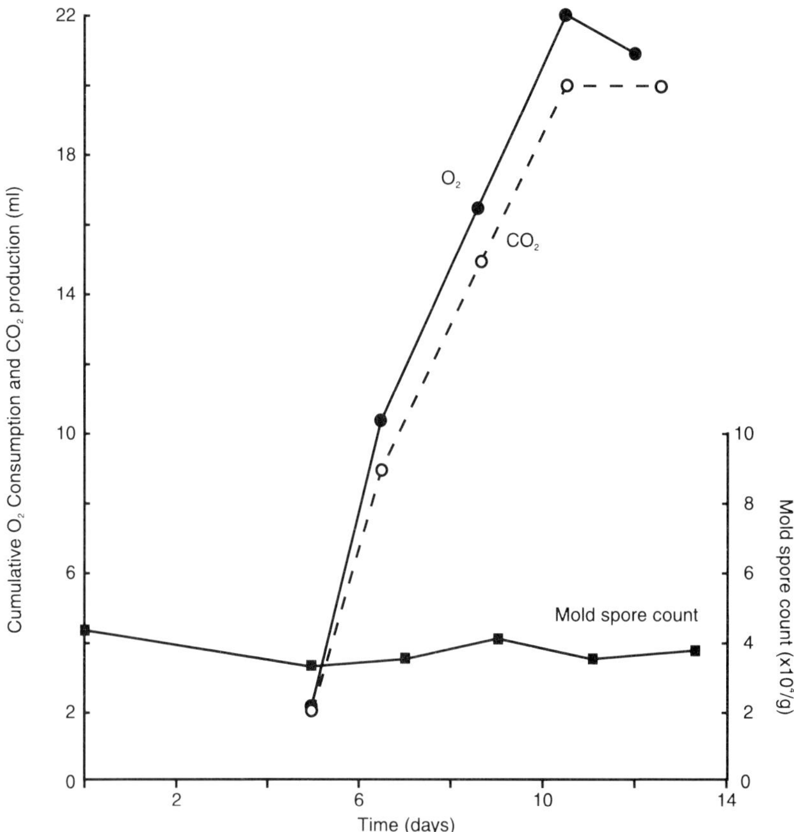

Figure 3. Comparison of mold growth to mold spore counts in moist poultry feed containing a natural microflora.

These data clearly cause one to question the use of the mold spore plate count as even an indicator of the degree of moldiness of a feedstuff.

It appears that respirometry, particularly respirometry that can maintain aerobic conditions during an incubation period, is an innovation that may prove extremely valuable in aiding in the development of information to better understand the role of microbial degradation of feedstuffs.

MYCOTOXINS: THEIR ORIGIN, IMPACT AND IMPORTANCE: INSIGHTS INTO COMMON METHODS OF CONTROL AND ELIMINATION

L.L. CHARMLEY[1], H.L. TRENHOLM[2] and D.B. PRELUSKY[2]

[1]*Amherst, Nova Scotia, Canada*
[2]*Centre for Food and Animal Research, Agriculture and Agrifood Canada, Ottawa, Ontario, Canada*

Origin of mycotoxins

Mycotoxins are secondary metabolites produced by fungi that grow on a wide range of agricultural commodities including cereals, nuts, soybeans and other crops. Since mycotoxin-producing fungi can contaminate commodities destined for use in animal feed and human food products, both populations can be affected by them. Moreover, when species tolerant to specific mycotoxins are given contaminated feed, the potential carryover of mycotoxins into animal products such as meat, milk and eggs becomes a consideration from a food safety viewpoint.

Aflatoxins are of concern in warm (tropical and sub-tropical) regions where aflatoxin-producing fungi, primarily *Aspergillus flavus* and *A. parasiticus*, can grow on cereals, oilseeds, tree nuts, peanuts and dried figs. Because aflatoxin-contaminated food poses a serious health hazard, the United States Food and Drug Administration (US FDA) enforces an action level of 20 μg/kg in foods for human consumption. However, consumer demand and the increasing world export market for commodities susceptible to aflatoxin contamination are pushing towards zero tolerance. In colder, more temperate regions, such as Canada and northern USA, aflatoxins are not found on indigenous crops. Economically in these regions, the most important mycotoxins are deoxynivalenol, zearalenone, ochratoxin A, diacetoxyscirpenol, T2–toxin, and HT-2 toxin. The recently discovered fumonisins, in particular fumonisin B_1 are of concern in specific geographical regions. These mycotoxins have been found in a wide range of commodities from small-grained cereals and corn to banana fruits.

The economic impacts of fungal and mycotoxin contamination are felt by crop and animal producers, food and feed processors and the national economy as a whole (Charmley *et al.*, 1994). In terms of reduced financial returns and increased costs for crop and animal producers, and from the extent of infestation and contamination it has been estimated that multimillion dollar losses occur each year in Canada alone due to such contamination (Trenholm *et al.*, 1985b).

Impact and importance: toxicology

Mycotoxins cause a wide variety of adverse clinical signs depending on the nature and concentration of mycotoxin present, duration of exposure, and the animal species, its age and nutritional and health status at the time of exposure to contaminated feed (Prelusky *et al.*, 1994). Overt toxicosis, morbidity and death occur infrequently with most economic losses being due to subtle non-specific effects associated with reduced animal performance and increased disease incidence (Thompson, 1991).

TOXIC EFFECTS OF MYCOTOXINS ON ANIMALS

Aflatoxins

Aflatoxins have been the most studied group of mycotoxins. Numerous reviews have been published on the effect of aflatoxins on animals (Bryden, 1982; Norred, 1986). Main effects relate to liver damage, impaired productivity (including reduced growth rate and milk production), adverse effects on egg shell and carcass quality, immunosuppression, and carcinogenicity. The transfer of aflatoxin M_1 (a metabolite of aflatoxin B_1) to milk has serious health implications. Therefore, the US FDA action levels require milk aflatoxin M_1 concentrations to be less than 0.5 μg/kg. However, in western European countries action levels require milk aflatoxin M_1 concentrations to be 0.05 μg/kg or less (van Egmond, 1993).

Aflatoxins have had serious economic effects for poultry producers (Boutrif, *in press*), pig producers (Nichols, 1983), and in a few cases cattle producers (McKenzie *et al.*, 1981). Poultry are extremely sensitive to the adverse effects of aflatoxins with levels as low as 0.2 mg/kg causing decreased feed consumption and body weight gain (Johri and Sadagopan, 1989). Although cattle are somewhat more tolerant than poultry to the presence of aflatoxin in their diet, there have been a few reports of significant losses due to aflatoxicosis on individual cattle farms in Australia (Blaney and Williams, 1991).

Deoxynivalenol

Deoxynivalenol, a trichothecene mycotoxin produced by *Fusarium* species of fungi, in pigs causes incremental reductions in feed intake as levels increase and feed refusal and vomiting at high concentrations (Pier, 1981; Young *et al.*, 1983; Lun *et al.*, 1985; Trenholm *et al.*, 1988), leading to reduced body weight gain or body weight loss. Deoxynivalenol may cause immunosuppression and affect reproduction (Reotutar, 1989). Poultry (Moran *et al.*, 1987), cattle (Trenholm *et al.*, 1985a; Cote *et al.*, 1986; Charmley *et al.*, 1993) and sheep (Harvey *et al.*, 1986; Prelusky *et al.*, 1987) are relatively more tolerant than pigs. Moreover, there appears to be little carryover of deoxynivalenol into eggs, tissues, or milk.

T-2 toxin, HT-2 toxin, and diacetoxyscirpenol
These trichothecenes are more toxic, but fortunately are less widespread and more rarely encountered than deoxynivalenol. At low to moderate dietary concentrations (1 to 8 mg/kg) T-2 toxin causes a moderate, but significant reduction in feed consumption and body weight gain in pigs. At higher concentrations (10 to 12 mg/kg) significant adverse effects on body weight gain, blood chemistry and fertility may be observed (Harvey *et al.*, 1990) with experimentally induced T-2 toxicosis resulting in infertility, small litter size, small piglets, and abortion (Weaver *et al.*, 1978b). In addition, oral lesions have been described in pigs exposed to feed containing elevated levels of T-2 toxin and/or diacetoxyscirpenol (Weaver *et al.*, 1978a, 1981). In poultry T-2 toxin causes reduced feed consumption, egg production, shell thickness, and hatchability (James, 1987). However, combinations of toxins, as found naturally in moldy feed, cause greater adverse signs than feed containing a single toxin (Mannion and Blaney, 1988; Kubena *et al.*, 1989). Cattle are relatively more sensitive to T-2 toxin than they are to deoxynivalenol with cases of lethal T-2 toxicosis being reported (Hsu *et al.*, 1972). Severe health problems also have been found in some field situations in which, unfortunately, the dietary T-2 toxin levels were not defined (Petrie *et al.*, 1977).

Zearalenone
Pigs (and in particular, young females) are extremely sensitive to the presence of zearalenone in the diet (James and Smith, 1982; Kuiper-Goodman *et al.*, 1987; Friend and Trenholm, 1988) with reproductive dysfunctions reported in gilts fed diets containing as little as 1 to 3 mg zearalenone/kg and vulvovaginitis (enlargement or swelling of the vulva) at even lower dietary levels (Friend *et al.*, 1990). Serious breeding problems have been observed in gilts fed diets containing 5 to 6 mg zearalenone/kg. Clinical signs include: vulvovaginitis, swelling of the mammary glands, atropy of the ovaries, pseudopregnancy, constant estrous, reduced litter sizes, pregnancy loss, increased fetal mortality, and agalactia (no milk production). In extreme cases vaginal and rectal prolapse may be seen (Sydenham *et al.*, 1988). Zearalenone ingestion may adversely affect immature males (Ruhr, 1979) and boars (Young and King, 1986).

Cattle are less sensitive than pigs to dietary zearalenone, with adverse effects being observed only when dietary concentrations (natural contamination) are 10 to 15 mg/kg or more (Mirocha *et al.*, 1968; Kallela and Ettala, 1984). In cattle and sheep clinical manifestations of zearalenone ingestion include: restlessness, diarrhea, udder enlargement, decreased milk yield, vaginitis, mucoid vaginal discharge, continuous estrous, infertility, and abortion. Precocious mammary development and sterility may occur in prepubertal heifers (Coppock *et al.*, 1990). However, it seems likely that the most drastic effects on cattle reproduction and fertility observed under natural conditions are due to zearalenone in combination with other mycotoxins (Schuh and Baumgartner, 1988; Coppock *et al.*, 1990).

Poultry are extremely resistant to the effects of zearalenone with signs being observed only at very high levels of contamination (800 to several thousand mg pure zearalenone/kg) (Chi *et al.*, 1980a, b; Allen *et al.*, 1981).

Fumonisins

Fumonisins are a recently discovered group of mycotoxins primarily produced by *Fusarium moniliforme*. Although the magnitude of the fumonisin-induced problem is unknown, *F. moniliforme* is one of the most prevalent fungi found to contaminate certain cereal grains and corn in many parts of the world (Marasas *et al.*, 1984). Consequently, the fumonisins are increasingly viewed as a major concern from human and animal health, and food and feed safety perspectives (Kellerman *et al.*, 1990; Wilson *et al.*, 1990). Fumonisins, in particular fumonisin B_1, have been implicated in a variety of clinical syndromes in several species including equine leukoencephalomalacia (ELEM), porcine pulmonary edema (PPE) (Colvin and Harrison, 1992; Haschek, 1992) and a condition referred to as 'spiking mortality' or 'toxic feed syndrome' in poultry. ELEM, a potentially fatal disease of horses, is characterized by extensive damage to the brain and in some cases damage to the liver and kidneys (Haliburton and Buck, 1986; Marasas *et al.*, 1988b; Kellerman *et al.*, 1990; Wilson *et al.*, 1990, 1991). PPE, at low levels of fumonisin contamination, is characterized by ill defined clinical signs sometimes involving the liver, but at high levels of contamination, rapid death due to massive pulmonary edema or hydrothorax is observed (Colvin and Harrison, 1992; Haschek *et al.*, 1992). Spiking mortality has been linked to feed contaminated with fumonisin B_1 at 10 to 25 mg/kg diet. Clinical signs include extended legs and neck, ataxia, paralysis, wobbly gait, dyspnea, gasping, and poor growth. High concentrations of fumonisins have been associated with higher than normal incidence of human esophageal cancer in China and South Africa (Yang, 1980; Marasas *et al.*, 1988a). However, although evidence implicates the fumonisins as carcinogenic agents in rats (Gelderblom *et al.*, 1988, 1991; Voss *et al.*, 1990) their role as causative agents in human esophageal cancer remains to be proven conclusively.

Ochratoxin A

Ochratoxin A is a nephrotoxic mycotoxin produced primarily by the *Aspergillus* and *Penicillium* groups of fungi that have been found as contaminants on grain in North America and parts of Europe. Adverse effects have been observed in pigs and poultry given naturally occurring levels of ochratoxin A (2 mg/kg or less). In pigs signs include adverse effects on kidney tissue and kidney function (Elling *et al.*, 1985; Carlton and Tuite, 1986). In some cases the clinical signs can be so subtle (Mortensen *et al.*, 1983, Lippold *et al.*, 1987) that occurrences of chronic ochratoxin A toxicosis may be evident only upon post-mortem inspection of the kidneys. Visual signs of damage to the kidneys include swelling and discolouration with histological abnormalities (Golinski *et al.*, 1984; Tapia and Seawright, 1985; Cook *et al.*, 1986). Such damage may occur at dietary ochratoxin A concentrations as low as 0.2 mg/kg (Krough *et al.*, 1974). At higher concentrations (2 mg/kg or more) decreased performance and weight gain may be seen (Huff *et al.*, 1980; Tapia and Seawright, 1985; Harvey *et al.*, 1989), accompanied by damage to the liver and urinary bladder.

In poultry, low concentrations (2 mg/kg diet) of ochratoxin A can have adverse effects on growth rate, feed conversion, and egg production (Dwivedi and Burns, 1984; Gibson *et al.*, 1989; Rotter *et al.*, 1990). At higher levels (4 mg/kg or more) increases in mortality (Jayakumar *et al.*, 1988; Sreemannarayana *et al.*, 1989; Gibson *et al.*, 1989,1990), skeletal abnormalities (Duff *et al.*, 1987), avian nephropathy (Hamilton *et al.*, 1982); and/or damage to other organs and tissues (Jayakumar *et al.*, 1988; Gibson *et al.*, 1989; Sreemannarayana *et al.*, 1989) may be seen. Ochratoxin A also elicits immune suppression. Consequently, ochratoxin A has been associated with significant production losses in the poultry industry. Moreover, turkeys appear to be more sensitive than chickens to this toxin (Burditt *et al.*, 1984).

Since ruminants are relatively more tolerant of the effects of ochratoxin A, documented cases of ochratoxicosis in cattle and sheep are rare. It appears that ochratoxin A is efficiently hydrolysed enzymatically to non-toxic metabolites by rumen microflora. Therefore, ruminants are tolerant to dietary ochratoxin A at levels typically found in contaminated feeds (Patterson *et al.*, 1981). However, preruminant calves are much more sensitive to this toxin. Ochratoxicosis can be induced experimentally in ruminants at very high ochratoxin A concentrations (approximately 800 mg/kg). Signs include, reduced feed consumption and body weight gain, diarrhea, decreased milk production, kidney and liver damage, dehydration associated with impaired renal function, and death at very high doses (Pier *et al.*, 1976; Lloyd, 1980; Chu, 1984). When such tolerant species are fed ochratoxin A contaminated diets the potential for the carryover of toxin into milk and meat becomes a major concern for human health and food safety.

Insights into common ways to control and eliminate mycotoxins: decontamination

Prevention of mycotoxin contamination of agricultural commodities is of utmost importance. However, under certain conditions of temperature and humidity contamination is unavoidable and during certain seasons can be quite extensive. Strategies to reduce the impact of mycotoxins include plant breeding for mold resistance (Snijders, 1994), efficient harvesting and storage practices to minimize contamination, and the development of potentially commercially applicable techniques for de-contaminating such commodities. Many decontamination methods have been tried (Charmley and Prelusky, 1994) and can be broadly categorized as physical, chemical or biological (Table 1). Some of these methods have proved to be very successful, others less so, and even some of the most successful ones would be difficult to implement on a commercial basis.

Table 1. Physical, chemical and biological methods used in reducing toxin levels in feeds.

Physical	Chemical	Biological
Cleaning/washing	Calcium hydroxide monomethylamine	Diluting contaminated grain
Dehulling	Sodium bisulfite	Improving the nutritional content of the diet
Polishing	Moist and dry ozone	Addition of mold inhibitors
Separation of contaminated from non-contaminated kernels	Chlorine gas	Addition of flavoring agents
Heat treatments	Hydrogen peroxide	Addition of potential mycotoxin binding agents
	Ascorbic acid	
	Ammonium hydroxide	
	Hydrochloric acid	
	Sulfur dioxide gas	
	Formaldehyde (vapour form)	
	Ammonium hydroxide	
	Ammonia	

PHYSICAL METHODS

Physical methods include cleaning and washing, dehulling, polishing, separation of contaminated from non-contaminated kernels and heat treatments. The success of these procedures depends on the initial degree of contamination and the distribution of mycotoxins throughout the grain. Cleaning and washing methods include screening, cleaning with removal of dockage, dust, screenings, small and broken kernels, reddog and fines, scouring, dehulling, polishing, and washing with water or sodium carbonate solutions and water and have been found to reduce deoxynivalenol, zearalenone, and/or nivalenol concentrations in a variety of grains by 7 to 100%. However, washing does not physically remove the outer contaminated layer of the kernel and may lead to the requirement of an additional, costly drying step to the processing of grain. Moreover, in some cases although cleaning may significantly reduce the deoxynivalenol concentrations, this may not necessarily result in a reduction in toxicity when this grain is fed to pigs (Patterson and Young, 1992b, 1993). The distribution of deoxynivalenol in the milling fractions of wheat depends on the degree of fungal penetration of the endosperm. For example, in some cases milling causes significant (15 to 100%) reductions in deoxynivalenol, zearalenone, and nivalenol concentrations in the flour fractions of wheat (Young *et al.*, 1984; Seitz *et al.*, 1985; Tanaka *et al.*, 1986), but in others results in a relatively even distribution of deoxynivalenol among the milling fractions (Hart and Braselton, 1983; Scott *et al.*, 1983; Seitz *et al.*, 1986) with a tendency towards slightly higher concentrations in the shorts, reddog, dockage and bran fractions.

In some cases, mold damaged, mycotoxin contaminated grains exhibit differing physical properties from normal ones and can be separated by density segregation in liquids or fractionation by specific gravity table. In wheat, corn, or sorghum removal of kernels buoyant in water and 30% sucrose solution (Huff and Hagler, 1985) or water and saturated sodium chloride solutions (Babadoost *et al.*, 1987) or fractionation on a specific gravity table (Tkachuk *et al.*, 1991) led to reductions in deoxynivalenol, or deoxynivalenol and zearalenone concentrations of 40 to 100%. In the peanut industry physical segregation of moldy aflatoxin contaminated kernels is widely used, but primarily limited to hand and electronic color sorting (Dickens and Whittaker, 1975). The efficiency of electronic color sorting is variable with, on average, 70% of the aflatoxin contaminated kernels being removed. Hand sorting, although more efficient, is more time consuming, monotonous and commercially impractical, and may involve the loss of normal peanuts, or because some highly contaminated kernels may appear normal, may result in a contaminated product. Density segregation of aflatoxin contaminated peanuts may be an alternative decontamination method (Henderson *et al.*, 1989), although in this case losses may be somewhat high. An inverse (but not strictly linear) relationship was found between individual peanut density and aflatoxin concentration (Gnanesekharan and Chinnan, 1992) with high aflatoxin concentrations being significantly correlated with low kernel densities. However, since non-contaminated peanuts may have low densities, there is a potential for loss of some normal peanuts. Water flotation for separating contaminated peanuts from sound ones has been patented in the USA, but has not gained wide acceptance because of the requirement of an additional, costly, drying step following treatment. Density segregation also has been tried for aflatoxin-contaminated corn and cottonseed (Huff, 1980; Cole, 1989). Density segregation in hydrogen peroxide solutions can effectively separate aflatoxin-contaminated from normal peanuts; but unless the initial degree of contamination is less than 200 μg/kg and reaction conditions are strictly optimized (with respect to reaction time and peroxide concentration) this treatment is not very effective (Clavero *et al.*, 1993).

Mycotoxins may be destroyed by heat treatments. Baking cookies, yeast free doughnuts (Young *et al.*, 1984), and bread (Seitz *et al.*, 1986) made from deoxynivalenol contaminated flour (0.2 to 0.9 mg/kg) resulted in a 20 to 40% reduction in toxin concentration. In contrast, baking bread (Scott *et al.*, 1983,1984; Seitz *et al.*, 1986; Boyacioglu *et al.*, 1993), cookies and doughnuts (Scott *et al.*, 1984), sponge cakes (Tanaka *et al.*, 1986), or Egyptian bread (El Banna *et al.*, 1983) made from more highly contaminated flour (1 to 7 mg/kg) caused little reduction in deoxynivalenol concentration.

Autoclaving (Young *et al.*, 1987), microwave oven heating (Young *et al.*, 1986; Stahr *et al.*, 1987) and convection oven heating (Young *et al.*, 1986) reduced deoxynivalenol concentrations in corn; while roasting reduced concentrations in wheat (Stahr *et al.*, 1987). Microwave oven heating and roasting reduced T-2 toxin concentrations in corn (Stahr *et al.*, 1987). However, flame roasting, while reducing mold count by 48 to 98%, did not reduce significantly the deoxynivalenol and zearalenone concentrations in corn (Hamilton and Thompson, 1992).

Ultraviolet radiation was found to reduce aflatoxin concentrations in dried figs (Altug *et al.*, 1990) and to degrade aflatoxin M_1 in milk (Yousef and Marth, 1986), the latter being enhanced by the addition of hydrogen peroxide.

CHEMICAL METHODS

Although numerous chemicals have been tested for reducing mycotoxin concentrations in a variety of grains and grain products, only some have proved successful and very few are used commercially. Chemicals found to be effective against mycotoxins include calcium hydroxide monomethylamine in T-2 toxin-, diacetoxyscirpenol-, and zearalenone-contaminated corn meal (Bauer *et al.*, 1987); sodium bisulfite in deoxynivalenol- (Swanson *et al.*, 1984; Young *et al.*, 1987; Boyacioglu *et al.*, 1993), and aflatoxin- (Moerk *et al.*, 1980) contaminated corn, and aflatoxin-spiked figs (Altug *et al.*, 1990); moist and dry ozone, chlorine gas and ammonia in deoxynivalenol-contaminated corn (Young *et al.*, 1986); chlorine gas in aflatoxin B_1 spiked copra meal (Samarajeewa *et al.*, 1991); hydrogen peroxide, ascorbic acid, ammonium hydroxide, hydrochloric acid, and sulfur dioxide gas (moistened) in deoxynivalenol-contaminated wheat (Young *et al.*, 1987); formaldehyde (vapor form) and ammonium hydroxide in zearalenone-spiked corn grits and naturally contaminated corn meal (Bennett *et al.*, 1980); and ammonia in ochratoxin-contaminated grain (Chelkowski *et al.*, 1981, 1982; Madsen *et al.*, 1983) and aflatoxin-contaminated groundnut cakes (Frayssinet and Lafarge-Frayssinet, 1990) and corn (Moerk *et al.*, 1980; Hammond, 1991). Ammonia can decrease aflatoxin concentrations by over 99% in corn, peanut meal cakes, whole cottonseed, and cottonseed products and, if the reaction is given sufficient time to proceed, the detoxification process is irreversible. Treatment of aflatoxin-contaminated cottonseed meal with ammonia has been approved for use on a commercial basis in Arizona, California, and Texas (Park *et al.*, 1988; Price *et al.*, 1993) and in other countries such as France (Park *et al.*, 1988), and for aflatoxin-contaminated corn in Texas (Park, 1993; Price *et al.*, 1993). Moreover, because it is effective, inexpensive, and can be done on the farm at a relatively low cost, this treatment may be eventually approved by the FDA to allow the interstate shipment of ammonia-treated aflatoxin-contaminated commodities. No toxic lesions related to this ammoniation process have been observed in feeding trials; and in Arizona the use of ammonia to treat cottonseed for lactating cows has had significant benefits in keeping the milk supply free of aflatoxin M_1.

In some instances physical and chemical treatments have been used in combination to result in a more effective decontamination. Chemical procedures enhanced by heat include formaldehyde treatment of zearalenone in corn or corn grits (Bennett *et al.*, 1980); calcium hydroxide monomethyamine treatment of T-2 toxin, diacetoxy scirpenol and zearalenone in corn meal (Bauer *et al.*, 1987); sodium bisulfite treatment of deoxynivalenol in corn (Young *et al.*, 1987), and aflatoxin in figs (Altug *et al.*, 1990). Heat and treatment with lime water was found to reduce significantly deoxynivalenol, zearalenone and 15–acetyl deoxynivalenol concentrations in corn (Abbas *et al.*, 1988), heat and sodium hydroxide treatment was found to reduce ochratoxin concentrations in grains (Madsen *et al.*, 1983), and ammonia combined with heat and pressure treatment was found to reduce fumonisin concentrations in corn (Park *et al.*, 1992).

BIOLOGICAL METHODS

An alternative approach to the decontamination of mycotoxin-contaminated grain is to minimize the effect of the toxins on the animal by modifying the diet. These methods include diluting contaminated grain with normal grain, improving the nutritional content of the diet, addition of mold inhibitors or flavoring agents, addition of potential mycotoxin binding agents to the diet to reduce absorption in the gastrointestinal tract, and other experimental techniques.

Diluting contaminated grain is one of the most effective, widely used methods for counteracting the effects of a mycotoxin-contaminated diet on animals, particularly for improving feed intake and weight gain in pigs fed *Fusarium* mycotoxin contaminated diets (Patterson and Young, 1991). However, the success of this method depends on the initial degree of contamination, the dilution achievable, and the ready availability of a source of suitable non-contaminated grain.

The addition of mold inhibitors (McNear *et al.*, 1981; Foster *et al.*, 1987) and flavoring agents (McNear *et al.*, 1981) did not counteract the adverse effects of a moldy corn or a deoxynivalenol-contaminated corn-based diet in pigs. However, when ammonium propionate was used as a mold inhibitor growth rates improved in pigs fed non-contaminated and deoxynivalenol-contaminated diets (Foster *et al.*, 1987).

Increasing the energy, crude protein, and mineral and vitamin content of a deoxynivalenol-contaminated diet by 20% improved weight gain in pigs, but only if feed intake was reduced by 20% or less (Chavez and Rheaume, 1986).

A number of mycotoxin binding agents have shown limited success depending on the mycotoxin involved (Table 2). These include alfalfa, synthetic anion exchange zeolite, bentonite, spent canola oil bleaching

Table 2. Mycotoxin-binding agents with reported efficacy in various diets.

Binding agent	Toxin bound (species)
Alfalfa	Zearalenone (rats, pigs), T-2 toxin (rats)
Canola oil-bleaching clays	T-2 toxin (rats)
Sodium bentonites	Aflatoxin (pigs)
Synthetic zeolite	Zearalenone (rats)
Activated charcoal	Ochratoxin and T-2 (rats), aflatoxin (poultry)
Cholestyramine	Ochratoxin (rats), zearalenone (mice)
HSCAS	Aflatoxin (poultry, lambs, dairy cows)
Yeast	Aflatoxin (poultry)

clays, activated charcoal, cholestyramine, a hydrated sodium calcium aluminosilicate (HSCAS), and yeast cell wall products.

Alfalfa overcame the adverse effects of a diet containing high concentrations of zearalenone (Smith 1980a; James and Smith, 1982; Stangroom and Smith, 1984), and T-2 toxin (Carson and Smith, 1983b) in rats, but had no effect on uterine enlargement in gilts fed a diet containing 50 mg zearalenone/kg (Smith, 1980a). Alfalfa was partially effective in gilts fed a diet containing only 10 mg zearalenone/kg. Similarly, synthetic anion exchange zeolite alleviated the effects of zearalenone in rats (Smith, 1980b). Bentonite and spent canola oil bleaching clays overcame the adverse effects of a diet containing a relatively high concentration of T-2 toxin in rats (Carson and Smith, 1983a; Smith, 1984). Sodium bentonites (Volclay powder and FD-181) improved body weight gain, feed consumption, and clinical indications in pigs fed diets containing aflatoxin (Lindermann *et al.*, 1993). Activated charcoal was found to reduce ochratoxin A absorption (Rotter *et al.*, 1989), to be efficacious in preventing T-2 toxicosis in rats (Bratich *et al.*, 1990), and in some cases to increase weight gain and feed intake in poultry fed aflatoxin-contaminated diets (Dalvi and Ademoyero, 1984; Dalvi and McGowan, 1984). However, activated charcoal (at 0.5% of the diet) did not protect chickens from the adverse effects of diets containing 5 or 7.5 mg aflatoxin B_1/kg (Kubena *et al.*, 1990).

Cholestyramine was found to reduce ochratoxin A concentrations in the blood, and its cumulative excretion in urine and feces of rats fed an ochratoxin A-contaminated diet (Madyastha *et al.*, 1992). Cholestyramine (at 0.2% of the diet) reduced ochratoxin A bioavailability by 56%. This agent reduced the estrogenic effects of zearalenone in mice (Underhill *et al.*, 1995).

HSCAS at 0.5% of the diet diminished significantly many of the adverse effects of aflatoxin B_1 or aflatoxin in chickens (Phillips *et al.*, 1988; Kubena *et al.*, 1990, 1993), turkeys (Kubena *et al.*, 1991), pigs (Haydon *et al.*, 1990) and growing lambs (Harvey *et al.*, 1991a). This level of HSCAS also reduced aflatoxin M_1 concentrations in milk of cows (Harvey *et al.*, 1991b). HSCAS (1%) appeared to improve body weight gains in chicks fed moldy corn-based diets containing low concentrations (trace to 0.35 mg/kg) of T-2 toxin, deoxynivalenol and zearalenone. However, HSCAS did not improve feed intake and feed efficiency significantly in pigs fed diets containing deoxynivalenol and zearalenone (Orr, 1987) or deoxynivalenol (Patterson and Young, 1992a, 1993; Trenholm *et al.*, unpublished); and was found to be ineffective against diacetoxyscirpenol (Kubena *et al.*, 1993), T-2 toxin (Kubena *et al.*, 1990), and fumonisin B_1 (Brown *et al.*, 1992) in chick diets. Explanations of the mechanism of action of some of these compounds include ionic attraction between binding agent and mycotoxin, alterations in the enterohepatic circulation of bile acids and entrapment of the mycotoxin within the matrix of the binding agent. *Saccharomyces cervisiae* (a yeast) at 0.1% of the diet was found to be effective in alleviating some of the adverse effects of a diet containing 5 mg/kg aflatoxin in chickens (Stanley *et al.*, 1993). It was suggested that the possible mechanism of action of this yeast was to supplement the diet with enzymes that increase feed utilization

(Day *et al.*, 1987) or that this agent was chelating aflatoxin eliminated from the gastrointestinal tract (Cooney, 1980). The capabilities of a yeast culture (Yea Sacc) and yeast cell wall material (Bio Mos) to bind zearalenone, deoxynivalenol and fumonisin *in vitro* were recently tested (Trenholm *et al.*, unpublished). Although incapable of binding deoxynivalenol and fumonisin, both products bound zearalenone in a non-linear manner (Figure 1). Cholestyramine was found to be four to six times more effective at binding this mycotoxin than Bio Mos or Yea Sacc, respectively; but the high cost of this product would make its commercial use prohibitively expensive. However with research and commercial interest increasing in this area, it is likely that cost-effective compounds capable of binding a variety of mycotoxins will soon be on the market.

Treatment of moldy corn (containing approximately 5 mg deoxynivalenol/kg) with microbial inoculum from the digestive tract of poultry reduced the deoxynivalenol concentration by approximately 55% and partially alleviated its toxic effects on feed intake and body weight gain in young pigs (Ping *et al.*, 1992; He *et al.*, 1993).

Administration of monoclonal antibodies specific for T-2 toxin was found to neutralize the *in vitro* inhibitory effects of the toxin on protein synthesis in human B-lymphoblastoid cultures and protect rats from lethal T-2 toxicosis (Feuerstein *et al.*, 1988). The monoclonal antibodies

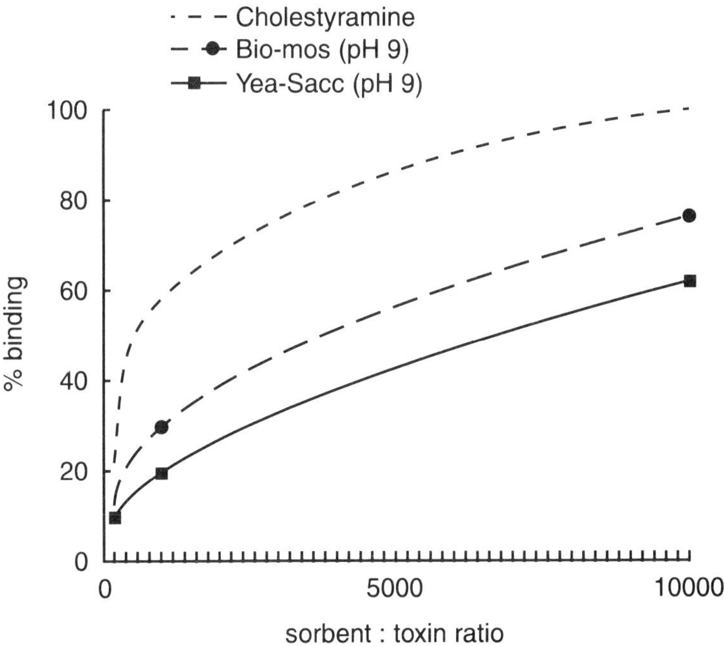

Figure 1. Binding of zearalenone to cholestyramine and two yeast products *in vitro* (Trenholm *et al.*, unpublished).

caused a net efflux of toxin from poisoned human B-lymphoblastoid cells *in vitro* and restored protein synthesis to normal. Moreover, administration of monoclonal antibodies 30 minutes before an infusion of a lethal dose of T-2 toxin to rats caused sequestration of the toxin in the plasma (Hunter *et al.*, 1990), and when administered 35 minutes after facilitated migration of the toxin back into the plasma from the tissues thereby reducing its toxic effects.

Conclusions

Mycototoxin contamination of crops continues to be a major economic problem for the agricultural and food industries of the world. The fungi and the mycotoxins produced by them not only may cause reductions in crop production and quality, but may also result in a variety of syndromes in the animals exposed to these toxins. Although prevention is the primary goal, at present under certain environmental temperature and humidity conditions contamination is unavoidable. Therefore, there has been an increased impetus to discover effective decontamination procedures that may be implemented on a commercial basis. However, because contaminated grain may contain a complex mixture of mycotoxins each with differing chemical characteristics (including heat stability, solubility, and adsorbent affinity) it is difficult to find a single method that is equally effective against all mycotoxins. Some treatments, such as screening and air separation, dilution of contaminated grain, roasting, certain chemical treatments, and the use of binding agents have had commercial applications already. Our knowledge of the chemistry and toxicology of mycotoxins, and advances in our understanding of which absorbents bind certain mycotoxins is still far from complete. However, the widespread nature of mycotoxin contamination of crops worldwide is providing opportunities for commercial development of binding agents in feed to reduce production losses and illness in farm animals. Until more reliable, cost-effective, and commercially applicable methods are universally available, problems associated with mycotoxins will continue to plague the agricultural and food industries of the world.

References

Abbas, H.K., C.J. Mirocha, R. Rosiles and M. Carvajal. 1988. Decomposition of zearalenone and deoxynivalenol in the process of making tortillas from corn. Cer. Chem. 65:15.

Allen, N.K., C.J. Mirocha, G. Weaver, S. Aakhus-Allen and F. Bates. 1981. Effects of dietary zearalenone on finishing broiler chickens and young turkey poults. Poult. Sci. 60:124.

Altug, T., A.E. Yousef and E.H. Marth. 1990. Degradation of aflatoxin B_1 in dried figs by sodium bisulfite with or without heat, ultraviolet energy or hydrogen peroxide. J. Food Protect. 53:581.

Babadoost, M., W.M. Hagler and D.T. Bowman. 1987. Field contamination of sorghum with zearalenone and deoxynivalenol in North

Carolina: density segregation to remove mycotoxins. In: Biodeterioration Research 1. G.C. Llwewllyn, C.E. O'Rear (Eds) Plenum Press, New York and London, 99.

Bauer, J., M. Gareis, W. Detzier, B. Gedek, K. Heinritzi and G. Kabilka. 1987. Detoxification of mycotoxins in animal feeds. Tierarztliche Umschau 42:70.

Bennett, G.A., O.L. Shotwell and C.W. Hesseltine. 1980. Destruction of zearalenone in contaminated corn. J. Amer. Oil Chem. Soc. 57:245.

Blaney, B.J. and K.C. Williams. 1991. Effective use in livestock feeds of mouldy and weather-damaged grain containing mycotoxins – case histories and economic assessments pertaining to pig and poultry industries of Queensland. Australian J. Agric. Res. 42:993.

Boyacioglu, D., N.S. Hettiarachchy and B.L. Dappolonia. 1993. Additives affect deoxynivalenol (vomitoxin) flour during breadbaking. J. Food Sci. 58:416.

Bratich, P.M., W.B. Buck and W.M. Haschek. 1990. Prevention of T-2 toxin-induced morphologic effects in the rat by highly activated charcoal. Archives of Toxicol. 64:251.

Brown, T.P., G.E. Rottinghaus and M.E. Williams. 1992. Fumonisin mycotoxicosis in broilers: performance and pathology. Avian Dis. 36:450.

Bryden, W.L., 1982. Aflatoxin and animal produce: an Australian perspective. Food Tech. In Australia 34:216.

Burditt, S.J., W.M.Jr. Hagler and P.B. Hamilton. 1984. Feed refusal during ochratoxicosis in turkeys. Poult. Sci. 63:2172.

Carlton, W.L. and J. Tuite. 1986. Comparative pathological changes in ochratoxicosis and citrinin toxicosis. In: J.L. Richard, J.R. Thurston (Eds), Diagnosis of mycotoxicoses. Martinus Nijhoff Publishers.

Carson, M.S. and T.K. Smith. 1983a. Role of bentonite in prevention of T-2 toxicosis in rats. J. Anim. Sci. 57:1498.

Carson, M.S. and T.K. Smith. 1983b. Effect of feeding alfalfa and refined plant fibers on the toxicity and metabolism of T-2 toxin in rats. J. Nutr. 113:304.

Charmley, E., H.L. Trenholm, B.K. Thompson, D. Vudathala, J.W.G. Nicholson, D.B. Prelusky and L.L. Charmley. 1993. Influence of level of deoxynivalenol in the diet of dairy cows on feed intake, milk production, and its composition. J. Dairy Sci. 76:3580.

Charmley, L.L. and D.B. Prelusky. 1994. Decontamination of *Fusarium* mycotoxins. In: Mycotoxins in Grain, Compounds Other than Aflatoxin. J.D. Miller, H.L. Trenholm (Eds) Eagan Press, Ma, USA, 421.

Charmley, L.L., A. Rosenberg and H.L. Trenholm. 1994. Factors responsible for economic losses due to *Fusarium* mycotoxin contamination of grains, foods, and feedstuffs. In: Mycotoxins in Grain, Compounds Other than Aflatoxin. J.D. Miller, H.L. Trenholm (Eds) Eagan Press, Ma, USA, 471.

Chavez, R.R. and J.A. Rheaume. 1986. The significance of the reduced feed consumption observed in growing pigs fed vomitoxin-containing diets. Can. J. Anim. Sci. 66:277.

Chelkowski, J., P. Golinski, B. Godlewska, W. Radomyska, K. Szebiotko and M. Wiewiorowska. 1981. Mycotoxins in cereal grain. Part IV Inactivation of ochratoxin A and other mycotoxins during ammoniation. Nahrung 25:631.

Chelkowski, J., K. Szebiotko, M. Buchowski, B. Godlewska, W. Radomyska and M. Wiewiorowska. 1982. Mycotoxins in cereal grain. Part V Changes of cereal grain biological value after ammoniation and mycotoxins (ochratoxins) inactivation. Nahrung 26:1.

Chi, M.S., C.J. Mirocha, H.J. Kurtz, G.A. Weaver, F. Bates, T. Robison and W. Shimoda. 1980a. Effect of dietary zearalenone on growing broiler chicks. Poult. Sci. 59:531.

Chi, M.S., C.J. Mirocha, G.A. Weaver and H.J. Kurtz. 1980b. Effect of zearalenone on female White Leghorn chickens. Appl. Environ. Mircrobiol. 39:1026.

Chu, F.S. 1984. Studies on ochratoxins. CRC Crit. Rev. Toxicol. 2:499.

Clavero, M.R.S., Y.C. Hung, L.R. Beuchat and T. Nakayama. 1993. Separation of aflatoxin-contaminated kernels form sound kernels by hydrogen-peroxide treatment. J. Food Protect. 56:130.

Cole, R.J. 1989. Technology of aflatoxin decontamination. In: Mycotoxins and Phycotoxins '88. S. Natori, K. Hashimoto, Y. Ueno (Eds) Elsevier Scientific Publishing, Amsterdam, 177.

Cook, W.O., G.D. Osweiler, T.D. Anderson and J.L. Richard. 1986. Ochratoxicosis hydrothorax in swine. Mycopathologia 117:79.

Cooney, D.O. 1980. Activated charcoal: antidotal and other medical uses. Marcel Dekker, Inc., New York, NY.

Cote, L.M., A.M. Dahlem, T. Yoshizawa, S.P. Swanson and W.B. Buck. 1986. Excretion of deoxynivalenol and its metabolite in milk, urine and feces of lactating dairy cows. J. Dairy Sci. 69:2416.

Colvin, B.M. and L.R. Harrison. 1992. Fumonisin-induced pulmonary edema/hydrothorax in swine. Mycopathologia 117:79.

Coppock, R.W., M.S. Mostrom, C.G. Sparling, B. Jacobsen and J.C. Ross. 1990. Apparent zearalenone intoxication in a dairy herd from feeding spoiled acid-treated corn. Vet. Human Toxicol. 32:246.

Dalvi, R.R. and A.A. Ademoyero. 1984. Toxic effects of aflatoxin B_1 in chickens given feed contaminated with *Aspergillus flavus* and reduction of the toxicity by activated charcoal and some chemical agents. Avian Dis. 28:61.

Dalvi, R.R. and C. McGowan. 1984. Experimental induction of chronic aflatoxicosis in chickens by purified aflatoxin B_1 and its reversal by activated charcoal, phenobarbital, and reduced glutathione. Poult. Sci. 63:485.

Day, E., J.B.C. Dilworth and S.Omar. 1987. Effect of varying levels of phosphorus and live yeast culture in caged laying diets. Poult. Sci. 66:1402.

Dickens, J.W. and T.B. Whitaker. 1975. Efficacy of electronic color sorting and hand picking to remove aflatoxin contaminated kernels from commercial lots of shelled peanuts. Peanut Sci. 2:45.

Duff, S.R.I., R.B. Burns and P. Dwivedi. 1987. Skeletal changes in broiler chicks and turkey poults fed diets containing ochratoxin A. Res. Vet. Sci. 43:301.

Dwivedi, P. and R.B. Burns. 1984. Pathology of ochratoxicosis A in young broiler chicks. Res. Vet. Sci. 36:92.

El-Banna, A.A., P.-Y. Lau and P.M. Scott. 1983. Fate of mycotoxin during processing of foodstuffs. II. Deoxynivalenol (vomitoxin) during making of Egyptian bread. J. Food Prot. 46:484.

Elling, F., J.P. Nielsen, E.B. Lillehoj, M.S. Thomassen and F.C. Stormer. 1985. Ochratoxin A-induced porcine nephropathy: Enzyme and ultrastructure changes after short term exposure. Toxicon 23:247.

Feuerstein, G., J.A. Powell, A.T. Knower and K.W. Hunter. 1988. Monoclonal antibodies to T-2 toxin. *In vitro* neutralization of protein synthesis inhibition and protection of rats against lethal toxemia. J. Clin. Invest. 76:2134.

Foster, B.C., H.L. Trenholm, D.W. Friend, B.K. Thompson and K.E. Hartin. 1987. The effect of a propionate feed preservative in deoxynivalenol (vomitoxin) containing corn diets fed to swine. Can. J. Anim. Sci. 67:1159.

Frayssinet, C. and C. Lafarge-Frayssinet, 1990. Effect of ammoniation on the carcinogenicity of aflatoxin-contaminated groundnut oil cakes – long term feeding study in the rat. Food Addit. Contam. 7:63.

Friend, D.W. and H.L. Trenholm. 1988. Mycotoxins in pig nutrition. Pig News and Information. C.A.B. International 9:395.

Friend, D.W., H.L. Trenholm, B.K. Thompson, K.E. Hartin, P.S. Fiser, E.K. Asem and B.K. Tsang. 1990. The reproductive efficiency of gilts fed very low levels of zearalenone. Can. J. Anim. Sci. 70:635.

Gelderblom, W.C.A., K. Jaskiewicz, W.F.O. Marasas, P.G. Thiel, R.M. Horak, R. Vleggaar and N.P.J. Kriek. 1988. Fumonisins: novel mycotoxins with cancer-promoting activity produced by *Fusarium moniliforme*. Appl. Environ. Microbiol. 54:1806.

Gelderblom, W.C.A., N.P.J. Krick, W.F.O. Marasas and P.G. Thiel. 1991. Toxicity and carcinogenicity of the *Fusarium moniliforme* metabolite, fumonisin B_1 , in rats. Carcinogenesis 12:1247.

Gibson, R.M., C.A. Bailey, L.F. Kubena, W.E. Huff and R.B. Harvey. 1989. Ochratoxin A and dietary protein. 1. Effects on body weight, feed conversion, relative organ weight, and mortality in three-week-old broilers. Poult. Sci. 68:1658.

Gibson, R.M., C.A. Bailey, L.F. Kubena, W.E. Huff and R.B. Harvey, 1990. Impact of L-phenylalanine supplementation on the performance of three-week-old broilers fed diets containing ochratoxin A. 1. Effects on body weight, feed conversion, relative organ weight, and mortality. Poult. Sci. 69:414.

Gnanasekharan, V. and M.S. Chinnan. 1992. Density distributions of aflatoxin-contaminated peanuts in naturally and laboratory infected seeds. Amer. Soc. Ag. Eng. 35:631.

Golinski, P., K. Hult, J. Grabarkiewicz-Szczesna, J. Chelkowski, P. Kneblewski and K. Szebiotko. 1984. Mycotoxic porcine nephropathy and spontaneous occurrence of ochratoxin A residues in kidneys and blood of Polish swine. Appl. Environ. Microbiol. 47:1210.

Haliburton, J.C. and W.B. Buck. 1986. Equine leucoencephalomalacia: An historical review. In: Diagnosis of Mycotoxicoses: Current Topics in Veterinarian Medicine and Animal Science. J.L. Richard, J.R. Thurston (Eds.) Martinus Nijhoff Publishers, Dordrecht, 75.

Hamilton, P.B., W.E. Huff, J.R. Harris and R.D. Wyatt. 1982. Natural occurrences of ochratoxicosis in poultry. Poult. Sci. 61:1832.

Hamilton, R.G. and B.K. Thompson. 1992. Chemical and nutrient content of corn (*Zea mays*) before and after being flame roasted. J. Sci. Food Agric. 58:425.

Hammond, W.C. 1991. Techniques used to ammoniate Aflatoxin-contaminated corn in the field. In: Aflatoxin in Corn, New Perspectives. O.L. Shotwell, C.R. Hurburg (Eds) Iowa Agriculture and Home Economics Experiment Station, North Central Research Publication 329:377.

Hart, L.P. and W.E. Braselton. 1983. J. Agric. Food Chem. 31:657.

Harvey, R.B., L.F. Kubena, D.E. Corrier, D.A. Witzel, T.D. Phillips and N.D. Heidelbaugh. 1986. Effects of deoxynivalenol in a wheat ration fed to growing lambs. Amer. J. Vet. Res. 47:1630.

Harvey, R.B., W.E. Huff, L.F. Kubena and T.D. Phillips. 1989. Evaluation of diets contaminated with aflatoxin and ochratoxin fed to growing pigs. Amer. J. Vet. Res. 50:1400.

Harvey, R.B., L.F. Kubena, W.E. Huff, D.E. Corrier, G.E. Rottinghaus and T.D. Phillips. 1990. Effects of treatment of growing swine with aflatoxin and T-2 toxin. Amer. J. Vet. Res. 51:1688.

Harvey, R.B., L.F. Kubena, T.D. Phillips, D.E. Corrier, M.H. Elissalde and W.E. Huff. 1991a. Diminution of aflatoxin toxicity to growing lambs by dietary supplementation with hydrated sodium calcium aluminosilicate. Amer. J. Vet. Res. 52:152.

Harvey, R.B., T.D. Phillips, J.A. Ellis, L.F. Kubena, W.E. Huff and H.D. Petersen. 1991b. Effects on aflatoxin M_1 residues in milk by addition of hydrated sodium calcium aluminosilicate to aflatoxin contaminated diets of dairy cows. Amer. J. Vet. Res. 52:1556.

Haschek, W.M., G. Motelin, D.K. Ness, K.S. Harlin, W.F. Hall, R.F. Vesonder, R.E. Peterson and V.R. Beasley. 1992. Characterisation of fumonisin toxicity in orally and intravenously dosed swine. Mycopathologia 117:83.

Haydon, K.D., R.W. Beaver, D.M. Wilson, B.M. Colvin and L.T. Sangster. 1990. Special Publication – Georgia College of Agriculture Experiment Stations 67:42.

He, P., L.G. Young and C. Forsberg. 1993. Microbially detoxified vomitoxin-contaminated corn for young pigs. J. Anim. Sci. 71:963.

Henderson, J.C., S.H. Kreutzer, A.A. Schmidt, C.A. Smith and W.R. Hagen. 1989. Flotation separation of aflatoxin contaminated grain or nuts. US Patent No. 4,795,651.

Hsu, I.-C., E.B. Smalley, F.M. Strong and W.E. Ribelin. 1972. Identification of T-2 toxin in moldy corn associated with a lethal toxicosis in dairy cattle. Appl. Microbiol. 24:685.

Huff, W.E. 1980. A physical method for the segregation of aflatoxin contaminated corn. Cereal Chem. 57:236.

Huff, W.E., J.A. Doerr, P.B. Hamilton, D.D. Hamann, R.F. Peterson and A. Ciegler. 1980. Evaluation of bone strength during aflatoxicosis and ochratoxicosis. Appl. Environ. Microbiol. 40:102.

Huff, W.E. and W.M. Hagler. 1985. Density segregation of corn and wheat naturally contaminated with aflatoxin, deoxynivalenol and zearalenone. J. Food Prot. 48:416.

Hunter, K.W., A.A. Brimfield, A.T. Knower, J.A. Powell and G. Feuerstein. 1990. Reversal of intracellular toxicity of the trichothecene mycotoxin T-2 with monoclonal antibody. J. Pharmacol. Exp. Ther. 255:1183.

James, L.J. 1987. Mycotoxins and poultry performance. Chic Chat 3:1.

James, L.J. and T.K. Smith. 1982. Effect of dietary alfalfa on zearalenone toxicity and metabolism in rats and swine. J. Anim. Sci. 55:110.

Jayakumar, P.M., K.V. Valsala and A. Rajan. 1988. Testicular pathology in experimental ochratoxicosis in ducks. Indian J. Vet. Pathol. 12:37.

Johri, T.S. and V.R. Sadagopan. 1989. Aflatoxin occurrence in feedstuffs and its effect on poultry production. J. Tox. Rev. 8:281.

Kallela, K. and E. Ettala. 1984. The oestrogenic *Fusarium* toxin (zearalenone) in hay as a cause of early abortions in the cow. Nord. Veterinaermed. 36:305.

Kellerman, T.S., W.F.O. Marasas, P.G. Thiel, W.C.A. Gelderblom, M. Cawood and J.A.W. Coetzer. 1990. Leukoencephalomalacia in two horses induced by oral dosing with fumonisin B_1. Ondersspoort J. Vet. Res. 57:269.

Krogh, P., N.H. Axelson, F. Elling, N. Gyrd-Hansen, B. Hald, J. Hyldgaard-Jansen, A.E. Larsen, A. Madsen, H.P. Mortensen, T. Moller, O.K. Petersen, U. Ravnskov, M. Rostgaard and O. Aalund. 1974. Experimental porcine nephropathy: changes of renal function and structure induced by ochratoxin A-contaminated feed. Acta. Pathol. Microbiol. Scand. A246:21.

Kubena, L.F., R.B. Harvey, W.E. Huff, D.E. Corrier, T.D. Phillips and G.E. Rottinghaus. 1990. Efficacy of a hydrated sodium calcium aluminosilicate to reduce the toxicity of aflatoxin and T-2 toxin. Poult. Sci. 69:1078.

Kubena, L.F., R.B. Harvey, W.E. Huff, M.H. Ellisalde, A.G. Yersin, T.D. Phillips and G.E. Rottinghaus. 1993. Efficacy of a hydrated sodium calcium aluminosilicate to reduce the toxicity of aflatoxin and diacetoxyscirpenol. Poult. Sci. 72:51.

Kubena, L.F., W.E. Huff, R.B. Harvey, T.D. Phillips and G.E. Rottinghaus. 1989. Individual and combined toxicity of deoxynivalenol and T-2 toxin in broiler chicks. Poult. Sci. 68:622.

Kubena, L.F., W.E. Huff, R.B. Harvey, A.G. Yersin, M.H. Ellisalde, D.A. Witzel, L.E. Giroir, T.D. Phillips and H.D. Petersen. 1991. Effects of a hydrated sodium calcium aluminosilicate on growing turkey poults during aflatoxicosis. Poult. Sci. 70:1823.

Kuiper-Goodman, T., P.M. Scott and H. Watanube. 1987. Risk assessment of the mycotoxin zearalenone. Regul. Toxicol. Pharmacol. 7:253.

Lindermann, M.D., D.J. Blodgett, E.T. Kornegay and G.G. Schurig. 1993. Potential ameliorators of aflatoxicosis in weanling/growing swine. J. Anim. Sci. 71:171.

Lippold, C.C., S.C. Stothers, A.A. Frohlich and R.R. Marquardt. 1987.

Effect of ochratoxin-A (OA) contaminated barley on young growing pigs (Abstr.) Can. J. Anim. Sci. 67:1205.

Lloyd, W.E. 1980. Citrinin and ochratoxin toxicosis in cattle in the United States. Proc. Int. Symp. Vet. Lab. Diagn. 2nd, vol. 3:436.

Lun, A.K., L.G. Young and J.H. Lumsden. 1985. The effects of vomitoxin and feed intake on the performance and blood characteristics of young pigs. J. Anim. Sci. 61:1178.

Madhyastha, M.S., A.A. Frohlich and R.R. Marquardt. 1992. Effects of dietary cholestyramine on the elimination pattern of ochratoxin A in rats. Food Chem. Toxicol. 30:709.

Madsen, A., B. Hald and H.P. Mortensen. 1983. Feeding experiments with ochratoxin A contaminated barley for bacon pigs. 3. Detoxification by ammoniation, heating + NaOH or autoclaving. Acta Agric. Scand. 33:171.

Mannion, P.F. and B.J. Blaney. 1988. Responses of meat chickens offered 4–deoxynivalenol- and zearalenone-containing wheat, naturally infected with *Fusarium graminearum*. Aust J. Agric. Res. 39:533.

Marasas, W.F.O., K. Jaskiewicz, F.S. Venter and D.J. Van Schalkwyk. 1988a. *Fusarium moniliforme* contamination of maize in oesophageal cancer areas in Transkei. S. Afr. Med. J. 74:110.

Marasas, W.F.O., T.S. Kellerman, W.C.A. Gelderblom, J.A.W. Coetzer, P.G. Thiel and J.J. Van der Lugt. 1988b. Leukoencephalomalacia in a horse induced by fumonisin B_1 isolated from *Fusarium moniliforme*. Onderspport J. Vet. Res. 55:197.

Marasas, W.F.O., P.E. Nelson and T.A. Toussoun (Eds). 1984. Section Liseola. In: Toxigenic *Fusarium* Species: Identity and Mycotoxicology. The Pennsylvania State University Press, University Park.

McKenzie, R.A., B.J. Blaney, M.D. Connole and L.A. Fitzpatrick. 1981. Acute aflatoxicosis in calves fed peanut hay. Aust. Vet. J. 57:284.

McNcar, R.M., R.F. Wilson and G.A. Stitzlein. 1981. Short term swine appetite responses to dietary additives in moldy corn. Ohio Swine Research and Industry Report Animal Science Series 81–2:79.

Mirocha, C.J., J. Harrison, A.A. Nichols and M. McClintock. 1968. Detection of fungal estrogen (F-2) in hay associated with infertility in dairy cattle. Appl. Microbiol. 16:797.

Moerck, K.E., P. McElfresh, A. Wohlman and B.W. Hilton. 1980. Aflatoxin destruction in corn using sodium bisulfite, sodium hydroxide, and aqueous ammonia. J. Food Prot. 43:571.

Moran, E.T., P.R. Ferket and A.K. Lun. 1987. Impact of high dietary vomitoxin on yolk yield and embryonic mortality. Poult. Sci. 66:977.

Mortensen, H.P., B. Hald, A.E. Larsen and A. Madsen. 1983. Ochratoxin A contaminated barley for sows and piglets. Pig performance and residues in milk and pigs. Acta Agric. Scand. 33:349.

Nichols, T.E. 1983. Economic effects of aflatoxin in corn. In: Aflatoxin and *Aspergillus flavus* in Corn. U. Diener, R. Asquith, J. Dickens (Eds.) Southern Coop Service Bulletin 279, Auburn University, Auburn, Al, USA, 67.

Norred, W.P. 1986. Occurrence and clinical manifestations of aflatoxicosis. In: Diagnosis of Mycotoxicoses. J.L. Richard, J.R. Thurston (Eds) Martinus Nijhoff Publishers, 11.

Orr, D.E. 1987. Field studies on swine with a selected aluminosilicate. In: Proceedings, Recent Developments in the Study of Mycotoxins. Sponsored by Kaiser Aluminum and Chemical Corporation, Rosemont, IL. Dec 17, G1.

Park, D.L. 1993. Perspectives on mycotoxin decontamination procedures. Food Addit. Contam. 10:49.

Park, D.L., L.S. Lee, R.L. Price and A.E. Pohland. 1988. Review of the decontamination of aflatoxins by ammoniation: current status and regulations. J. Assoc. Off. Anal. Chem. 71:685.

Park, D.L., S.M. Rua, C.J. Mirocha, E-S.A.M. Abd-Alla and C.Y. Weng. 1992. Mutagenic potentials of fumonisin contaminated corn following ammonia decontamination procedure. Mycopathologia 117:105.

Patterson, D.S.P., B.J. Shreeve, B.A. Roberta, S. Berrett, P.J. Brush and E.M. Glancy. 1981. Effect on calves of barley naturally contaminated with ochratoxin A and groundnut meal contaminated with low concentrations of aflatoxin B_1. Res. Vet. Sci. 31:213.

Patterson, R. and L.G. Young. 1991. Diluting mouldy corn still best bet. The Market Place. Fall 1991 issue 36.

Patterson, R. and L.G. Young. 1992a. Screening moldy corn to alleviate the effects of vomitoxin in moldy corn diets. Ontario Swine Research Review. O.A.C. Publication no. 0292:20.

Patterson, R. and L.G. Young. 1992b. Using Novasil^R to alleviate the effects of vomitoxin in moldy corn diets. Ontario Swine Research Review. O.A.C. Publication No. 0292:18.

Patterson, R. and L.G. Young, 1993. Efficacy of hydrated sodium-calcium aluminosilicate, screening and dilution in reducing the effects of mold contaminated corn in pigs. Can. J. Anim. Sci. 73:615.

Petrie, L., J. Robb and A.F. Stewart. 1977. The identification of T-2 toxin and its association with haemorrhagic syndrome in cattle. Vet. Rec. 101:326.

Phillips, T.D., L.F. Kubena, R.B. Harvey, D.R. Taylor and N.D. Heidelbaugh. 1988. Hydrated sodium calcium aluminosilicate: a high affinity sorbent for aflatoxin. Poult. Sci. 67:243.

Pier, A.C. 1981. Mycotoxins and animal health. Adv. Vet. Sci. Comp. Med. 25:185.

Pier, A.C., S.J. Cysewski, J.L. Richards, A.L. Baetz and L. Mitchell. 1976. Experimental mycotoxicoses in calves with aflatoxin, ochratoxin, rubratoxin, and T-2 toxin. Proc. U.S. Anim. Health Assoc. 80:130.

Ping, H.E., L.G.Young and C. Forsberg. 1992. Microbial detoxification of vomitoxin contaminated corn for pigs. Ontario Swine Research Review. O.A.C. Publication no. 0292:21.

Pohland, A.E. and G.E. Wood. 1987. Occurrence of mycotoxins in food. In: Mycotoxins in Foods. P. Krough (Ed.) Academic Press, New York, 35.

Prelusky, D.B., B.A. Rotter and R.G. Rotter. 1994. Toxicology of mycotoxins. In: Mycotoxins in Grain, Compounds Other than Aflatoxin. J.D. Miller, H.L. Trenholm. (Eds) Eagan Press, Ma, USA, 359.

Prelusky, D.B., D.M. Veira, H.L. Trenholm and B.C. Foster. 1987. Metabolic fate and elimination in milk, urine and bile of deoxynivalenol following administration to lactating sheep. J. Environ. Sci. Health B22:125.

Price, W.D., R.A. Lovell and D.G. McChesney. 1993. Naturally occurring toxins in feedstuffs – Center for Veterinary Medicine Perspective. J. Anim. Sci. 71:2556.

Reotutar, R. 1989. Swine reproductive failure syndrome mystifies scientists. J. Amer. Vet. Med. Assoc. 195:425.

Rotter, R.G., A.A. Frohlich and R.R. Marquardt. 1989. Influence of dietary charcoal on ochratoxin A toxicity in leghorn chicks. Can. J. Vet. Res. 53:449.

Rotter, R.G., R.R. Marquardt and A.A. Frohlich. 1990. Ensiling as a means of reducing ochratoxin A concentrations in contaminated barley. J. Sci. Food Agric. 50:155.

Ruhr, L.P. 1979. The effect of the mycotoxin zearalenone on fertility in the boar. PhD thesis, University of Missouri, Columbia.

Samarajeewa, U., A.C. Sen, S.Y. Fernando, E.M. Ahmed, C.I. Wei. 1991. Inactivation of aflatoxin B_1 in corn meal, copra meal and peanuts by chlorine gas treatment. Food and Chem. Toxicol. 29:41.

Schuh, M. and W. Baumgartner. 1988. Microbiological and mycotoxicological contaminated feedstuffs as disease causing agents in cattle. Wien. Tierartzl. Monatsschr. 75:329.

Scott, P.M., S.R. Kanhere, J.E. Dexter, P.W. Brennan and H.L. Trenholm. 1984. Distribution of the trichothecene mycotoxin deoxynivalenol (vomitoxin) during the milling of naturally contaminated Hard Red Spring wheat and its fate in baked products. Food Addit. Contam. 1:313.

Scott, P.M., S.R. Kanhere, P.-Y. Lau, J.E. Dexter and R. Greenhalgh. 1983. Effects of experimental flour milling and breadbaking on retention of deoxynivalenol (vomitoxin) in hard red spring wheat. Cer. Chem. 60:421.

Seitz, L.M., W.D. Eustace, H.E. Mohr, M.D. Shogren and W.T. Yamazaki. 1986. Cleaning, milling, and baking tests with Hard Red Winter wheat containing deoxynivalenol. Cer. Chem. 63:146.

Seitz, L.M., W.T. Yamazaki, R.L. Clements, H.E. Mohr and L. Andrews. 1985. Distribution of deoxynivalenol in soft wheat mill streams. Cer. Chem. 62:467.

Smith, T.K. 1980a. Influence of dietary fiber, protein and zeolite on zearalenone toxicisis in rats and swine. J. Animal Sci. 50:278.

Smith, T.K. 1980b. Effect of dietary protein, alfalfa and zeolite on excretory patterns of 5′,5′,7′,7′-[3H] zearalenone in rats. Can. J. Physiol. Pharmacol. 58:1251.

Smith, T.K. 1984. Spent canola oil bleaching clays: Potential for treatment of T-2 toxicosis in rats and short-term inclusion in diets for immature swine. Can. J. Anim. Sci. 64:725.

Snijders, C.H.A. 1994. Breeding for resistance to Fusarium in wheat and maize. In: Mycotoxins in Grain-Compounds Other than Aflatoxin. J.D. Miller, H.L. Trenholm (Eds) Eagan Press, Ma, USA, 37.

Sreemannarayana, O., R.R. Marquardt, A.A. Frohlich, D. Abramson and G.D. Phillips. 1989. Organ weights, liver constituents, and serum components in growing chicks fed ochratoxin A. Arch. Environ. Contam. Toxicol. 18:404.

Sreenivasamurthy, V., H.A.B. Parpia, S. Srikanta and A.S. Murti. 1967. Detoxification of aflatoxin in peanut meal by hydrogen peroxide. J. Assoc. Off. Anal. Chem. 50:350.

Stahr, H.M., G.D. Osweiler, P. Martin, M. Domoto and B. Debey. 1987. Thermal detoxification of trichothecene contaminated commodities. In: Biodeterioration Research 1. G.C. Llewellyn, C.E. O'Rear (Ed.) Plenum Press NY and London, 231.

Stangroom, K.E. and T.K. Smith. 1984. Effect of whole and fractionated alfalfa meal on zearalenone toxicosis and metabolism in rats and swine. Can. J. Physiol. Pharmacol. 62:1219.

Stanley, V.G., R.O.S. Woldesenbet and D.H. Hutchinson. 1993. The use of *Saccharomyces cerevisiae* to suppress the effects of aflatoxicosis in broiler chicks. Poult. Sci. 72:1867.

Steiner, W.E., R.H. Rieker and R. Battaglia. 1988. Aflatoxin contamination in dried figs: distribution and association with fluorescence. J. Agric. Food Chem. 36:88.

Swanson, S.P., W.M. Hagler and H.D. Rood. 1984. Destruction of deoxynivalenol (vomitoxin) with sodium bisulfite. Abstr. Ann. Meet. Amer. Soc. Microbiol. 84:192.

Sydenham, E.W., P.G. Thiel and W.F.O. Marasas. 1988. Occurrence and chemical determination of zearalenone and alternariol monomethyl ether in sorghum-based mixed feeds associated with an outbreak of suspected hyperestrogenism in swine. J. Agric. Food Chem. 36:621.

Tanaka, T., A. Hasegawa, Y.M. Yamamoto and Y. Ueno. 1986. Residues of Fusarium mycotoxins nivalenol, deoxynivalenol and zearalenone, in wheat and processed food after milling and baking. J. Food Hyg. Soc. Jpn. 27:653.

Tapia, M.O. and A.A. Seawright. 1985. Experimental combined aflatoxin B_1 and ochratoxin A intoxication in pigs. Aust. Vet. J. 62:33.

Tkachuk, R., J.E. Dexter, K.H. Tipples and T.W. Nowicki. 1991. Removal by specific gravity table of tombstone kernels and associated trichothecenes from wheat infected with *Fusarium* head blight. Cereal Chem. 68:428.

Thompson, L.J. 1991. Fusarium mycotoxins and animal effects. In: Plant Pathology Extender. Fusarium Molds and Mycotoxins Associated with Corn. G.C. Bergstrom, L.J. Thompson (Eds) Plant Pathology Extension Report 91–1, Cornell Cooperative Extension, Ithaca, NY, 2.

Trenholm, H.L., D.B. Prelusky, J.C. Young and J.D. Miller. 1988. Reducing mycotoxins in animal feeds. Agriculture Canada Publication 1827E.

Trenholm, H.L., B.K. Thompson, K.E. Hartin, R. Greenhalgh and A.J. McAllister. 1985a. Ingestion of vomitoxin (deoxynivalenol)-contaminated wheat by nonlactating dairy cows. J. Dairy Sci. 68:1000.

Trenholm, H.L., B.K. Thompson, J.F. Standish and W.L. Seamen. 1985b. Mycotoxins in feeds and feedstuffs. In: Mycotoxins: A Canadian Perspective, P.M. Scott, H.L. Trenholm, M.D. Sutton (Eds). National Research Council of Canada, Ottawa, On. 22848:43.

Underhill, K.L., B.A. Rotter, B.K. Thompson, D.B. Prelusky and H.L. Trenholm. 1995. Effectiveness of cholestyramine in the detoxofication of zearalone as determined by mice. Bull. Environ. Contam. Toxicol. 54:128.

van Egmond, H.P. 1993. Rationale for regulatory programs for mycotoxins in human foods and animal feeds. Food Addit. Contam. 10:29.

Voss, K.A., R.D. Plattner, C.W. Bacon and W.P. Norred. 1990. Comparative studies of hepatotoxicity and fumonisin B_1 and B_2 content of water and chloroform/methanol extracts of *Fusarium moniliforme* strain MRC 826 culture material. Mycopathologia 112:81.

Weaver, G.A., H.J. Kurtz, F.Y. Bates, M.S. Chi, C.J. Mirocha, J.C. Behrens and T.S. Robison. 1978a. Acute and chronic toxicity of T-2 mycotoxin in swine. Vet. Rec. 103:531.

Weaver, G.A., H.J. Kurtz, C.J. Mirocha, F.Y. Bates, J.C. Behrens and T.S. Robison. 1978b. Effect of T-2 toxin on porcine reproduction. Can. Vet. J. 19:310.

Weaver, G.A., H.J. Kurtz, F.Y. Bates, C.J. Mirocha, J.C. Behrens and W.M. Hagler. 1981. Diacetoxyscirpenol toxicity in pigs. Res. Vet. Sci. 31:131.

Wilson, T.M., P.F. Ross, D.L. Owens, L.G. Rice, R.D. Plattner, C. Reggiardo, T.H. Noon and J.W. Pickrell. 1990. Fumonisin B_1 levels associated with an epizootic of equine leukoencephalomalacia. J. Vet. Diagn. Invest. 2:213.

Wilson, T.M., P.F. Ross and P.E. Nelson. 1991. Fumonisin mycotoxins and equine leukoencephalomalacia. J. Amer. Vet. Med. Assoc. 198: 1104.

Wyatt, R.D. 1987. The relationship of Fusarium and other mold produced toxins and a selected aluminosilicate. In: Proc. Recent Developments in the Study of Mycotoxins. Sponsored by Kaiser Aluminum and Chemical Corporation, Rosemont, IL. Dec 17, D1.

Yang, C.S. 1980. Research on esophageal cancer in China: A review. Cancer Res. 40:2633.

Young, J.C. 1985. Decontamination in Mycotoxins: A Canadian perspective. National Research Council Scientific Criteria Document 11848; National Research Council: Ottawa, 119.

Young, J.C. 1986. Formation of sodium bisulfite addition products with trichothecenes and alkaline hydrolysis of deoxynivalenol and its sulfonate. J. Agric. Food Chem. 34:919.

Young, J.C., R.G. Fulcher, J.H. Hayhoe, P.M. Scott and J.E. Dexter. 1984. Effect of milling and baking on deoxynivalenol (vomitoxin) content of eastern Canadian wheats. J. Agric. Food Chem. 32:659.

Young, L.G. and G.J. King. 1986. Low concentrations of zearalenone in diets of boars for a prolonged period of time. J. Anim. Sci. 63:1197.

Young, L.G., L. McGuire, V.E. Vallee, J.H. Lumsden and A. Lun. 1983. Vomitoxin in corn fed to young pigs. J. Anim. Sci. 57:655.

Young, J.C., L.M. Subryan, D. Potts, M.E. McLaren and F.H. Gobran. 1986. Reduction in levels of deoxynivalenol in contaminated wheat by chemical and physical treatment. J. Agric. Food Chem. 34:461.

Young, J.C., H.L. Trenholm, D.W. Friend and D.B. Prelusky. 1987. Detoxification of deoxynivalenol with sodium bisulfite and evaluation of the effects when pure mycotoxin or contaminated corn was treated and given to pigs. J. Agric. Food Chem. 35:259.
Yousef, A.E. and E.H. Marth. 1986. Use of ultraviolet energy to degrade aflatoxin M_1 in raw or heated milk with and without added peroxide. J. Dairy Sci. 69:2243.

THE EUROPEAN PERSPECTIVE ON MYCOTOXINS

JOSEF LEIBETSEDER

Institute of Nutrition, University of Veterinary Medicine, Vienna, Austria

Introduction

In the history of mycotoxins Europe plays a remarkable role. In the Middle Ages hundreds of thousands of people died due to ergot intoxication, forty thousand for example in Limoges (France) in 934. A chronicle reports 'Crying, wailing and bending people collapsed on streets, some got up during dinner and mangled through the room like a wheel, some fell down slavering and with epileptic spasms, others vomited and showed signs of sudden insanity, most of them screamed "Fire – I am burning".'

The reason for the devastating epidemic was the predominant position of rye in human nutrition in Northern and Western Europe before the potato was introduced and before wheat became the most common cereal. In years with unfavorable climatic conditions, one third to half of the rye harvest consisted of kernels contaminated with *Claviceps purpurea*. The last intoxication of large extent happened at the French town Pont St. Esprit in 1951 when 200 people fell ill and four of them died.

From 1942 to 1947 about 10% of the population of Orenburg County in the USSR suffered from alimentary toxic aleukia, a disease of unknown etiology at that time. About 60% of the people affected died. Years later the epidemic was identified as fusariotoxicosis.

The outbreak of 'Turkey X Disease' in Great Britain in 1960 changed the general attitude toward the fungal contamination of food and feed and prompted an awareness of the scope of the problem. The discovery of aflatoxin in 1961 led to a vast amount of research in chemistry, biochemistry, mycology, toxicology, nutrition and also food and feed technology. The term 'mycotoxicosis' was introduced in 1962 by Forgacs and Carll defined as 'poisoning of the host following entrance into the body of toxic substances of fungal origin'.

Present situation in Europe

Because of the differences in climatic conditions in Northern, Middle and Southern Europe, different types of fungi and different mycotoxins became important in respective European regions during the last two decades. In the maize-growing areas of Southern and Middle Europe mainly fusariotoxins cause illness and poor performance of livestock and poultry whereas ochratoxin A(OTA) is of greater importance in the northern part of Europe.

FUSARIOTOXINS

An unknown disease of fattening pigs was first observed in the fall of 1977 in Europe in the southeast part of Austria, a maize-growing area with intensive pig production based on corn cob mix. On many farms in this area feed intake and weight gain of pigs were reduced; and on some farms feed refusal and vomiting were the main symptoms. All kinds of treatments were ineffective; only changing the feed improved the situation. Farmers were very anxious about this new porcine disease and the concomitant tremendous economic loss. In many cases signs of hyperestrogenism were observed in sows as well as in young fattening pigs. Searching the literature for diseases exhibiting symptoms of vomiting and feed refusal led us to the fusariotoxin vomitoxin (deoxynivalenol, DON) while the hyperestrogenism symptom led to information about zearalenone. The chemical analysis of feed samples for the suspected mycotoxins confirmed the presumptive diagnosis.

In the following years we analyzed several thousand feed samples for DON and zearalenone, mainly cereals and mixed feed. The results are shown in Table 1. Other trichothecenes (T2– and HT2–toxin) could be detected only rarely. As a result of our experience grain was investigated for fungal infection and analyzed for mycotoxins in some other European countries including Denmark, France, Germany, Hungary and Sweden. Some of the results are shown in Tables 2–4. More recent investigations in Germany (Bauer, 1988) demonstrated that

Table 1. Deoxynivalenol (DON, vomitoxin) and zearalenone in Austrian cereals and mixed feed.*

Toxin	DON	Zearalenone
Number of samples	1913	2311
positive, *n*	1053	663
%	55.0	28.7
Concentration (mg/kg)		
< 0.1,*n*	144	461
%	13.6	69.5
0.1–1.0, *n*	683	173
%	64.9	26.1
> 1.0, *n*	226	29
%	21.5	4.4

*Leibetseder, 1994; analyzed at the Institute of Nutrition, University of Veterinary Medicine, Vienna, in 1979–1985, detection limits: DON: 30 µg/kg, zearalenone: 5 µg/kg

Table 2. Fusarium infection of oats in Bavaria, Germany, in 1985–1988 and relative occurrence of Fusarium species most frequently detected (Anonymous, 1990).

	Year			
	1985	1986	1987	1988
Infected samples, %	6.3	10.0	20.4	18.3
Fusarium graminearum, %	19	16	65	24
Fusarium avenaceum, %	28	52	22	24
Fusarium poae, %	53	32	13	52

Table 3. Average content (mg/kg) of the most important fusariotoxins in samples of infected cereals harvested in Germany in 1982.*

Toxin	Wheat	Barley	Oats	Maize
Zearalenone	0.015–2.000	0	0.010–0.075	0.020–0.025
HT2–toxin	0.100–0.220	0.075–0.450	0.100–0.660	0
T2–toxin	0.200–0.450	0.220–0.450	0.100–0.660	0.070–0.220
Deoxynivalenol	0.010–1.300	0.010–10.120	0.150–0.530	0.030–0.120

*Anonymous, 1990.

Table 4. Occurrence of deoxynivalenol in Swedish cereals collected randomly in 1982 (A), in the field in 1984 (B) and delivered for hygienic analysis (C).*

Sample		Analyzed	Positive		μg/kg	
		n	*n*	%	Range	Mean
A	Barley	32	4	13	60–150	90
	Oats	32	11	34	40–260	140
B	Wheat	14	8	57	110–1180	400
	Barley	14	2	14	80–160	120
	Oats	13	0			
	Mixed grains	1	0			
C	Wheat	29	23	79	60–360	190
	Barley	6	1		50	
	Oats	6	3		420–520	470
	Mixed grains	11	4	36	90–200	145

Detection limit: 20 μg/kg
*Pettersson *et al.*, 1986

8.4% of 1000 feed samples were positive for trichothecene. Fifteen grain samples out of 966 contained zearalenone (mean = 0.1 mg/kg feed; contaminated samples were maize: 27%, barley: 15%; Gedek, 1988).

Besides the trichothecenes, other fusariotoxins were detected in cereals. Nivalenol has been analyzed in Swedish cereals 1987–1990 and was found in 35% of oat samples (144–4700 μg/kg, in barley at 13% (59–700 μg/kg) and in wheat at 4% (60–85 μg/kg) (Pettersson *et al.*, 1992). Ványi *et al.* (1992) presented recent results for Austria and Hungary (Table 5). Lew (1994) surveyed the *Fusarium* species and associated mycotoxins

Table 5. Occurence of mycotoxins in feedstuffs harvested in Austria and Hungary in 1991.*

Country	Austria (n = 212)			Hungary (n = 706)		
	n	%	µg/kg	n	%	µg/kg
Positive	185	87		651	92	
Zearalenone	85	40	50–610	31	4	50–290
T2-toxin	n.a.			113	16	50–520
HT2-toxin	41	19	50–250	39	6	50–300
Nivalenol	5	2	50–120	75	11	50–500
Deoxynivalenol	112	53	50–1250	354	50	50–850
Fusarenon X	2	1	50–120	12	2	50–300
Diacetoyscirpenol	6	3	50–240	12	2	50–400
Ochratoxin A	11	5	50–950	15	2	10–480

n.a. = not analyzed
* Ványi *et al.*, 1992

Table 6. Dominant (A) and less frequent (B) Fusarium species and their most important toxins in Austrian cereals.*

Maize	Wheat	Oats	Toxins
(A) *F. sacchari* var. *subglutinans*			Moniliformin
F. graminearum	*F. graminearum*	*F. graminearum*	Zearalenone Deoxynivalenol 15–Acetyldeoxy-nivalenol 3–Acetyldeoxy-nivalenol
	F. avenaceum	*F. avenaceum*	Moniliformin
	F. poae	*F. poae*	Nivalenol
	F. culmorum		Zearalenone Deoxynivalenol 3–acetyldeoxy-nivalenol
(B) *F. avenaceum*			Moniliformin
F. poae			Nivalenol
F. equiseti	*F. equiseti*	*F. equiseti*	Zearalenone T2–toxin HT2–toxin Fusarochromanone
F. proliferatum			Moniliformin Fumonisine
F. tricinctum	*F. tricinctum*	*F. tricinctum*	Moniliformin
F. culmorum		*F. culmorum*	Zearalenone Deoxynivalenol 3–Acetyldeoxy-nivalenol
F. cerealis (*F. crookwellense*)	*F. cerealis*	*F. cerealis*	Zerealenone Nivalenol
F. verticillioides (*F. moniliforme*)	*F. verticillioides*	*F. verticillioides*	Fumonisine
F. oxysporum	*F. oxysporum*	*F. oxysporum*	Moniliformin
	F. sacchari var.*subglut.*	*F. sacchari* var.*subglut.*	Moniliformin
F. sporotrichioides	*F.sporotrichioides*	*F. sporotrichioides*	T2–toxin HT2–toxin

*Lew, 1994.

occurring in Austria (Table 6). These results seem to be representative for the situation in Middle Europe. DON, zearalenone and OTA were reviewed recently (Böhm, 1992).

OCHRATOXIN A

Ochratoxin A(OTA) obviously occurs more frequently and in higher concentrations in the northern part of Europe (Denmark, Poland), but also in former Yugoslavia. In Denmark, OTA was found in 19 out of 33 samples of cereals; mainly in barley taken from three districts with a high incidence of porcine nephropathy (Krogh *et al.*, 1973). The levels of OTA ranged up to 27.5 mg/kg; and over half of the contaminated samples contained more than 0.2 mg/kg. In Germany OTA was found in 13.3% of wheat, barley and oat samples ($n = 225$, mean: 0.14 mg/kg, maximum: 1 mg/kg) in the period of 1982 to 1987 (Anonymous, 1990). In 1990 frequency and concentration were lower (Table 7). The occurrence of OTA in Austrian cereals and in mixed feed is shown in Table 8.

Table 7. Ochratoxin A in feedstuffs analyzed in Germany in 1990.*

Feedstuff	*n*	Positive		μg/kg
		n	%	
Cereals after harvest	292	1	<1	0.4
Cereals from farms				
without clinical signs	1209	29	2	0.1–2370
with clinical signs	253	11	4	0.1–441
Concentrates from farms				
without clinical signs	154	24	16	0.1–55
with clinical signs	66	14	21	0.2–64
Milling byproducts	30	5	17	0.2–181
Hay/grass silage	55	1	2	0.75
Soybean meal	13	0	0	–
Field beans/peas	27	0	0	–
Others (tapioca, flakes)	8	0	0	–

Detection limit: 0.1 μg/kg
*Thalmann, 1994

Table 8. Ochratoxin A in feedstuffs analyzed in Austria 1979 – 1987.*

Feedstuff	*n*	positive	<0.1 mg/kg	0.1–1.0 mg/kg
Barley	44	7	7	0
Oats	48	22	20	2
Maize	27	3	2	1
CCM	13	4	4	0
Rye	41	18	18	0
Wheat	41	3	3	0
Concentrates	41	12	9	3

Detection limit: 5 μg/kg
*Leibetseder, 1994

Because toxin formation takes place more intensively after grinding, concentrates contain OTA more frequently and at higher levels than grains. Occurrence of OTA in grains explains why OTA residues are found in animal tissue and blood more often than expected.

The only country in Europe with legal regulations for maximum levels of OTA in pig meat is Denmark where 10–25 μg/kg in liver or kidney results in confiscation of these organs and greater than 25 μg/kg causes confiscation of the carcass. In a representative study in Austria (*n* = 191) the average OTA concentrations in liver and kidney were 2.01 and 2.45 μg/kg, respectively. In 60.7% of the pig carcasses sampled OTA was detected in liver or kidney (detection limit: 0.01 μg/kg). In 4.4% of contaminated carcasses the content was higher than 25 μg/kg liver or kidney (Köfer *et al.*, 1990).

Research studies found OTA in 69 out of 104 kidneys confiscated in Austrian slaughterhouses because of pathological changes with an average concentration of 0.75 (0.11–3.04) μg/kg. The results of a survey in Nordrhein-Westfalen, a province in northern Germany, in 1990 (1074 serum samples and 223 macroscopically changed kidneys of pigs) are shown in Table 9.

Table 9. Frequency (%) of ochratoxin A in serum and kidneys of pigs slaughtered in Nordrhein-Westfalen, Germany, in 1990.*

	Serum		Kidneys
μg/kg	February–April *n*=1074	May–August *n*=627	February–August *n*=223
n.d.	62.2	31.1	71.7
0.1–0.2			16.6
0.2–0.9	23.9	39.5	8.1
1.0–4.9	9.1	24.2	3.1
5.0–9.9	3.6	3.6	
⩾ 5.0			0.5
⩾ 10.0	1.4	1.6	

n.d. = not detectable (serum: < 0.2 μg/kg, kidney: < 0.1 μg/kg)
*Schotte, 1992

AFLATOXINS

Aflatoxins are not considered to be a problem in Europe, except in imported feedstuffs grown in southern climates. In 1980 we analyzed a few samples of groundnut and soybean meal and detected a very high concentration in one of the groundnut meal samples (Table 10). The immediate consequence was that the inclusion of groundnut meal in dairy diets was forbidden by law. Subsequently limits (10, 20 or 50 μg/kg) were set for different feedstuffs in Austria and later on in the various countries of the European Union.

Table 10. Total aflatoxins in feedstuffs analyzed in Austria in 1980.*

Feedstuff	*n*	Positive	μg/kg		
Origin			Mean	Min.	Max.
Groundnut meal					
Origin unknown	20	20	3023	10	10680
Argentina	1	1	120		
Brazil	12	10	1408	19	5570
India	5	0			
Soybean meal					
Brazil	12	3	93	25	190
USA	4	0			

Detection limit: 5 μg/kg
*Leibetseder, 1994

ERGOT ALAKALOIDS

Since the middle 1980s ergot infection of rye, wheat and barley as well as of different varieties of grass (especially wild grasses at the non-cultivated edges of fields) has increased. Unfortunately many of the infected kernels are not larger than non-infected ones so it is difficult to sieve them. Consequently the frequency of ergotism in animals is increasing.

Reducing risks of mycotoxicosis

The most desirable and effective measure to minimize the risk of mycotoxicosis is prevention of fungal infection and mycotoxin contamination of feedstuffs. Additionally required are quality control, good milling practices, and the use of chemical preservatives by millers as excellently described by Trenholm *et al.* (1988).

DETOXIFICATION

Müller (1982, 1989) reviewed the available methods of detoxification. Because mycotoxins are mostly not uniformly distributed in kernels, mechanical separation by milling techniques leads to fractions with different mycotoxin concentrations. Some mycotoxins can be extracted by different solvents; most of the toxins (except ergot alkaloids) are heat resistant; some are decomposed to an extent by intensive sunlight or UV irradiation. Experiments on various methods of chemical inactivation are not very promising.

ADSORPTION

Another possibility for reducing harmful effects of mycotoxins is the inhibition of absorption by binding agents. Clay minerals such as bentonite and montmorillonite, also zeolites, activated charcoal and polymers like polyvinylpolypyrrolidone have been used with varying efficacy in practice. *In vitro* studies were performed at the author's institute with different adsorbing agents (Ehebruster, 1979; Becvar, 1984; Böhm, 1990; Leibetseder and Böhm, 1990). Recently a combination of highly effective adsorbents and enzymes degrading particular functional groups of mycotoxins such as the 12,13–epoxy group in trichothecenes or the lactone ring in zearalenone was introduced (Pasteiner, 1994). Adsorption in general is based on the hydration hull formed by the adsorbent. Particles must be polar to be attracted towards this hull and consequently to be bound to the adsorbent.

DEGRADATION OF MYCOTOXINS

Degradation of some of the mycotoxins takes place by ruminal fermentation (e.g. hydrolysis of OTA to ochratoxin a). Various microorganisms are obviously able to decompose mycotoxins. This could also be demonstrated for *Saccharomyces cerevisiae* (in Yea-Sacc) which is able to degrade aflatoxin *in vitro* and to improve performance in broilers and ducklings fed diets containing aflatoxin (Devegowda *et al.*, 1994). Because of the general ability of yeast to adsorb particles like bacteria and certain toxic compounds and to stimulate the immune response, the protective effects of *Saccharomyces cerevisiae* against mycotoxins should be studied in the future.

Conclusion

Mycotoxins cause a tremendous economic loss in animal production and to some extent pose a health risk for consumers in Europe although types and concentrations of mycotoxins in feedstuffs might be different from those on other continents. In spite of intensive instruction of farmers in ways to reduce fungal infection of crops and of feed companies in ways to minimize mycotoxin formation by optimizing storage conditions, preserved crops contain mycotoxins in variable amounts depending on the climatic conditions at harvest. For this reason all measures to reduce intake and/or absorption of mycotoxins are very welcome and can be recommended even if the efficacy may vary under different conditions. Besides all other available methods, adsorbent agents are therefore also used in Europe as on the other continents. Adsorbing ability of mannan and the degrading efficacy of *Saccharomyces cerevisiae* need to be demonstrated more clearly in additional experiments, but nevertheless it would be wise to use it even now for this purpose.

References

Anonymous. 1990. Natürliche Gifte in Getreide. Integrierter Pflanzenbau, Heft 6.

Bauer, J. 1988. Krankheit und Leistungsdepression in der Schweinehaltung durch Mykotoxine. Tierärztliche Praxis Suppl. 3: 40–47.

Becvar, B. 1984. Untersuchungen über die Bindung von Mykotoxinen an Adsorbentien. Thesis, Univ. of Vet. Med., Vienna.

Böhm, J. 1990. Untersuchungen über die Mykotoxinbindung an Adsorbentien. Proc. 44. Tagung Ges. f. Ernährungsphysiol., Göttingen, p. 35.

Böhm, J. 1992. Über die Bedeutung der Mykotoxine Desoxynivalenol, Zearalenon und Ochratoxin A für landwirtschaftliche Nutztiere. Arch. Anim. Nutr. 42:95–111.

Devegowda, G., B.I.R. Aravind, K. Rajendra, M.G. Morton, A. Baburathna and C. Sudarshan. 1994. A biological approach to counteract aflatoxicosis in broiler chickens and ducklings by the use of *Saccharomyces cerevisiae* cultures added to feed. Proc. Alltech's 10th Ann. Symp. T.P. Lyons and K.A. Jacques (Eds). Nottingham University Press, pp. 235–245.

Ehebruster, J. 1979. Versuche zur Detoxifikation von mykotoxinkontaminierten Futtermitteln durch Toxinadsorption. Thesis, Univ. of Vet. Med., Vienna.

Forgacs, J. and W.T. Carll. 1962. Mycotoxicosis. Adv.Vet.Sci. 7:274–382.

Gedek, B. 1988. Pilzgifte lauern im Futtergetreide. Tierzüchter 40:26–28.

Köfer, J., M. Schuh, and K. Fuchs. 1990. Ochratoxin-A- Kontamination bei steirischen Schlachtschweinen. Tierärztl. Umschau 46:657–660.

Krogh, P., B. Hald, and E.J. Pedersen. 1973. Occurrence of ochratoxin A and citrinin in cereals associated with mycotoxic porcine nephropathy. Acta Pathol. Microbiol. Scand. Sect. B 81:689–695.

Leibetseder, J. 1994. Mykotoxine und Mykotoxikosen in Österreich. Proc. 16. Mycotoxin-Workshop und Symposium: Mykotoxine in der Nahrungskette. LAF-Informationen, Sonderheft 1:26–49.

Leibetseder, J. and J. Böhm. 1990. Verminderung von Mykotoxinen im Nutztier durch Fütterungsmaβnahmen. Proc. Int. IFF-Symp.: Unerwünschte Stoffe in Futtermitteln, Mykotoxine in Getreide und Futtermitteln, Maβnahmen zur Beseitigung. Braunschweig, pp. 133–160.

Lew, H. 1994. Zur Taxonomie, Häufigkeitsverteilung und Toxigenität der Getreidefusarien. Proc. 16. Mycotoxin-Workshop und Symposium: Mykotoxine in der Nahrungskette. LAF- Informationen, Sonderheft 1:77–80.

Müller, H.-M. 1982. Entgiftung von Mykotoxinen. I. Physikalische Verfahren. Übers. Tierernährg. 10:95–122.

Müller, H.-M. 1989. Maβnahmen zur Minderung von Mykotoxinbildung und –anreicherung in Futtermitteln. Dtsch. Tierärztl. Wschr. 96:363–368.

Pasteiner, S. 1994. Mycotoxins in Animal Husbandry. Biomin, St. Pölten, Austria.

Pettersson, H. K-H. Kiessling and K. Sandholm. 1986. Occurence of the trichothecene mycotoxin deoxynivalend (vomitoxin) in Swedish-Grown Cereals. Swedish J. Agric. Res. 16:179–182.

Pettersson, H., R. Hedman, B. Engström, K. Elvinger and O. Fossum. 1992. Nivalenol in Swedish Cereals – Occurrence, Production and Toxicity towards Chickens. Proc. VIII Int. IUPAC Symp. on Mycotoxins and Phycotoxins, Mexico City, p. 47.
Schotte, M. 1992. Untersuchungen zum Vorkommen von Ochratoxin A im Serum und in Nieren von Schlachtschweinen aus Nordrhein-Westfalen. Proc. 14. Mykotoxin-Workshop, Gießen, pp. 34–43.
Thalmann, A. 1994. Vorkommen von Ochratoxin A in Getreide und daraus gewonnenen Lebens- und Futtermitteln. Proc. 16. Mycotoxin-Workshop und Symposium: Mykotoxine in der Nahrungskette. LAF-Informationen, Sonderheft 1:57–59.
Trenholm, H.L., D.B. Prelusky, J.C. Young, and J.D. Miller. 1988. Reducing mycotoxins in animal feeds. Agriculture Canada Publication 1827E, Communications Branch, Agriculture Canada, Ottawa.
Ványi, A., J. Böhm, and A. Bata. 1992. Occurrence of mycotoxins on feedstuffs in Middle-Europe (Hungary and Austria). Proc. VIII Int. IUPAC Symp. on Mycotoxins and Phycotoxins, Mexico City, p. 145.

THE NEW FRONTIER: IMMUNE MODULATORS IN NUTRITION

THE IMMUNE SYSTEM: NATURE'S DEFENSE MECHANISM – MANIPULATING IT THROUGH NUTRITION

KYLE NEWMAN

North American Biosciences Center, Nicholasville, KY, USA

Immunology is not a new science. In 430 BC the historian Thucydides noted that only individuals that had recovered from the plague could aid the sick because they would not contract the disease a second time. In 1718 the English physician Edward Jenner became intrigued with the fact that milkmaids who contracted cowpox were subsequently immune to the more debilitating and disfiguring smallpox. He reasoned that it might be possible to protect people from smallpox by inoculating them with the fluid from a cowpox pustule. Jenner then set forth to test his hypothesis by inoculating an 8-year-old boy with cowpox fluid and later intentionally infecting the child with smallpox. The child did not develop smallpox and knowledge was gained regarding vaccination (and possibly the first malpractice suit was filed).

The first insights into the mechanisms of immunity were provided by Emil von Behring and Shibasaburo Kitasato in 1890. These individuals demonstrated that serum from animals previously immunized to diphtheria could transfer this immunity to unimmunized animals. An antitoxin in the serum was hypothesized and further investigations by other researchers in the 1930s showed that a single substance called an antibody was responsible for neutralization and precipitation of toxins and agglutination (clumping) and lysis of bacteria. Since these antibodies were found in body fluids (known at the time as humors), the name humoral immunity was used to describe this activity.

At the same time antibody activity was characterized, Elie Metchnikoff observed that certain white blood cells were able to engulf microorganisms. The term phagocytes was used to describe these cells because of their ability to ingest foreign material. Controversy ensued as to whether antibodies (humoral immunity) or phagocytic cells (cellular immunity) were responsible for immunity. It was not until the 1950s that the lymphocyte was identified as the cell responsible for both cellular and humoral immunity.

Immunity has both acquired (specific) and innate (non-specific) components. Innate immunity refers to naturally occurring defense mechanisms that protect an animal from disease. As the name implies, this resistance does not exhibit specificity. There are a number of innate defense mechanisms which provide protection against a variety of stresses. Defense barriers that comprise innate immunity include anatomic, physiological, phagocytic and inflammatory factors. Specific immunity reflects the presence of a functional immune system capable of recognizing and selectively eliminating foreign material. Unlike innate immunity, specificity, diversity, memory and self-recognition are displayed. Acquired and innate immunity interact – macrophages of the innate system are involved in activation of the acquired system. The interaction between the two systems works to effectively eliminate foreign invaders.

Defense mechanisms

SKIN, MUCOUS MEMBRANES: ANATOMIC DEFENSES

The first line of defense in the animal is intact skin which provides a physical barrier to prevent the entry of most microorganisms. In order to maintain tissue integrity, zinc is required. It is generally thought that 2% of the total body zinc is present in the skin and hair in humans and cattle and improvements in hair and skin have been demonstrated with organic zinc supplementation (Spain, 1993). The skin also has a low pH which inhibits many microorganisms. Most bacteria and many viral and fungal agents are susceptible to low concentrations of organic acids such as lactic and fatty acids (the components of sebum) produced by sebaceous and sweat glands. Sebum maintains the pH of the skin between 3 and 5. The respiratory, urinary, and gastrointestinal tracts are covered by mucous membranes and a number of non-specific defense mechanisms serve to prevent pathogen entry into the body. Saliva, tears and mucous secretions contain antibacterial and antiviral substances and serve to neutralize or wash away potential infecting substances. Mucus is a viscous fluid secreted by epithelial cells of mucous membranes which entraps foreign microorganisms. The mucous membranes in the respiratory and gastrointestinal tracts are covered with cilia which move foreign material out of these tracts. Some organisms have developed methods to avoid these defense mechanisms. For example, the influenza virus is able to attach to cells in the mucous membrane thereby preventing the ciliated epithelial cells from sweeping it out. Another example involves the hair-like lectins or fimbriae on certain bacteria which attach to certain sugars, glycoproteins or glycolipids on some epithelial cells. The specificity of these lectins for certain sugars explains why some tissues are susceptible to bacterial invasion and others are not (Newman, 1994).

PHYSIOLOGICAL DEFENSE MECHANISMS

Chickens are naturally immune to anthrax because their body temperature inhibits the growth of the causative organism (*Bacillus anthracis*). Temperature, pH, oxygen levels and soluble proteins such as lysozyme, lactoferrin, lactoperoxidase, interferons and complement are all physiological factors involved in immunity. Gastric acidity prevents many potential pathogens from causing infection because very few organisms can survive the low pH of the stomach. This is one of the reasons why newborns are especially susceptible to enteric infections. Stomach acidity takes time to develop and during this time pathogens can flourish. In addition, the presence of bile salts in the intestinal tract has been shown to be inhibitory to many microorganisms. Bile salts are detergents which can disrupt certain bacterial membranes. The resident microflora in the gastrointestinal tract also play a protective role in pathogen challenge. In fact, nearly half the volume of the colon contents can be due to bacteria. The intestinal microflora helps to protect the mucosa by occupying binding sites and competing with pathogens for nutrients. The principle of competitive exclusion is based on these factors. Figure 1 shows some of the anatomic and physiological defenses in the gastrointestinal tract.

A number of immunoglobulins have been identified and characterized in terms of function and are described elsewhere (Hyde and Patnode, 1978). IgA represents the predominant immunoglobulin in external secretions such as breast milk, saliva, tears and mucus in the respiratory, gastrointestinal and genito-urinary tracts. IgA-secreting plasma cells are concentrated along mucus membrane surfaces. In the jejunum in humans there are more than 2.5×10^{10} IgA-secreting plasma cells. Each day humans secret 5–15 g of secretory IgA into mucous secretions (Kuby, 1994). Secretory IgA has been shown to provide an important line of defense against bacteria such as *Salmonella, Vibrio cholerae, Neisseria*

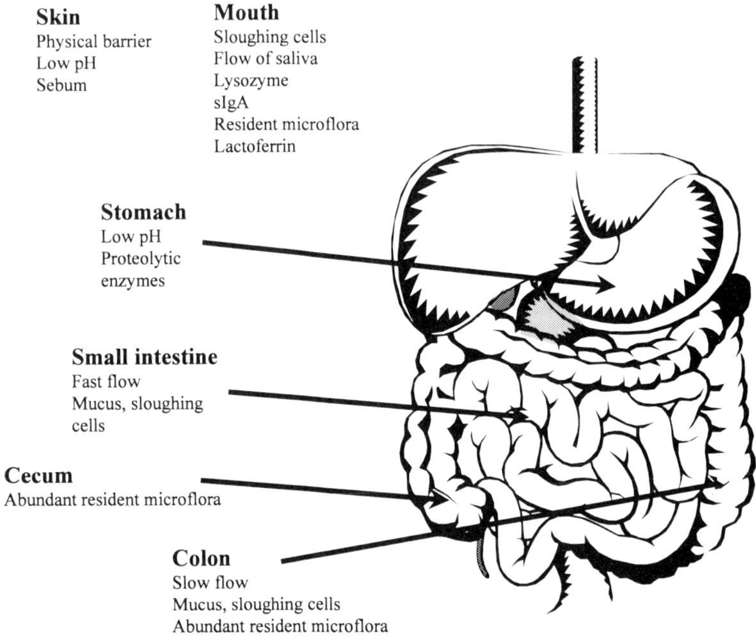

Figure 1. Defenses of the gastrointestinal tract.

gonorrhoea and viruses such as polio and influenza. The secret to vaccination through oral means is to ensure that the antigen remains intact long enough to be taken up by M cells in mucosal epithelia. The use of certain immune modulators administered via animal feed also has to satisfy this requirement. In general, these compounds are resistant to mammalian enzymatic hydrolysis and therefore are able to provide their response in the lower gastrointestinal tract. The exact mode(s) of action of immune modulators such as mannanoligosaccharide are still under investigation but we will explain what is known about immune response when mannanoligosaccharide is included in the diet later in this chapter.

THE COMPLEMENT SYSTEM

Complement involves a group of serum proteins that circulate throughout the body in an inactive (proenzyme) state. Complement can be activated by a variety of specific and non-specific immunological mechanisms. Once activated, complement components are involved in a controlled enzymatic cascade that results in membrane damage and cell lysis (Figure 2). From Figure 2 it is clear that certain cell wall components can activate the alternate pathway of complement. The complex polysaccharides found in microbial cell walls may serve to cause the non-specific immune stimulation that is noticed in many species with Bio-Mos (mannanoligo-saccharide) inclusion in the feed.

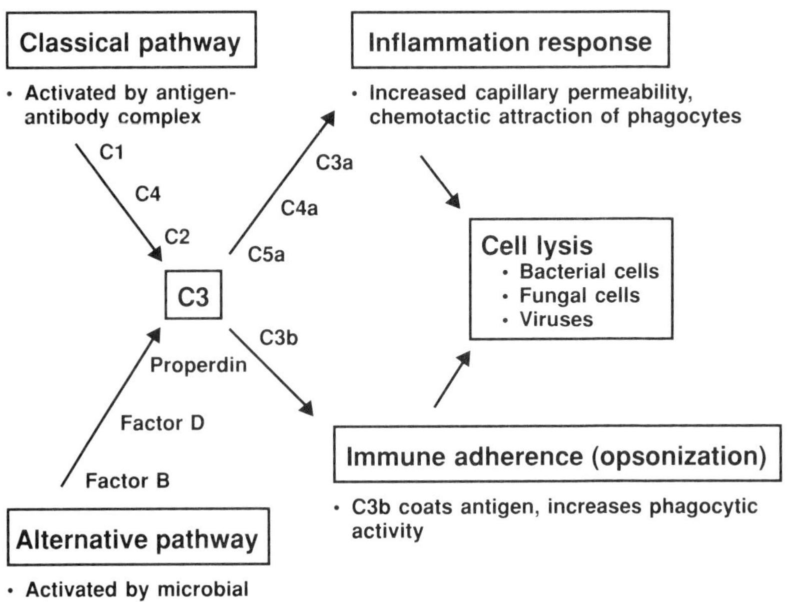

Figure 2. The complement system.

PHAGOCYTIC DEFENSE MECHANISMS

Phagocytosis involves the ingestion of particles that are foreign to the body. Several nutritional factors are involved in enhancing immune

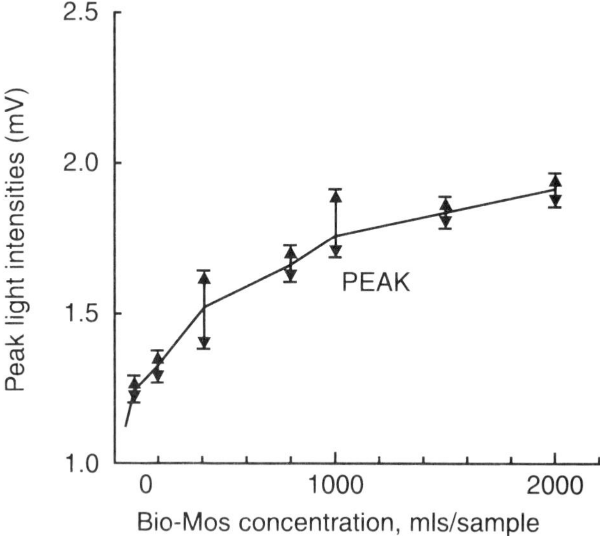

Figure 3. Luminol-enhanced chemiluminescense of rat phagocytes activated with increasing concentrations of mannanoligosaccharide (Bio-Mos).

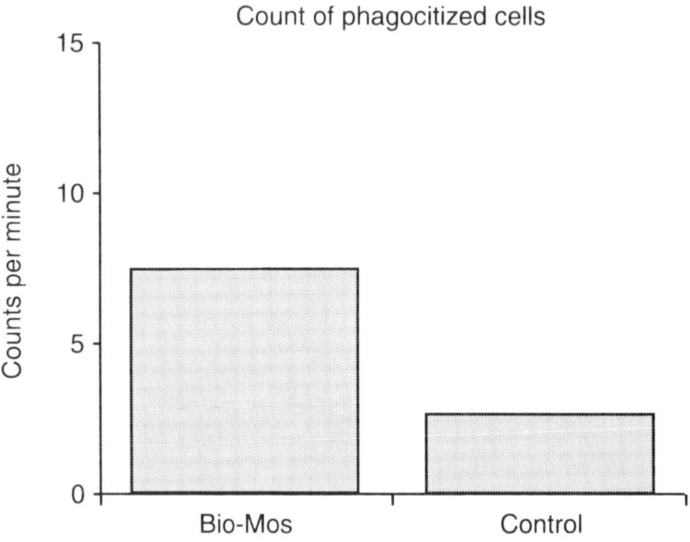

Figure 4. Macrophage phagocytosis potential (measured by chemiluminescense) of spleen derived monocytes from mice given yeast cell wall material orally (Zennoh, personal communication).

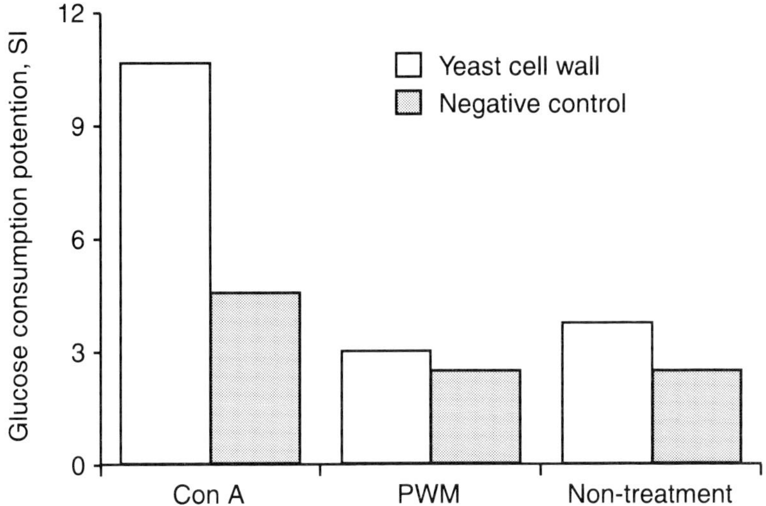

Figure 5. Effect of yeast cell wall material on stimulation of monocytes exposed to either a T-cell mitogen (Con A, Concanavalin A), a B-cell mitogen (PWM, pokeweed mitogen) or control (Zennoh, personal communication).

function and phagocytosis. Very recently, the ability of certain complex sugars such as mannanoligosaccharides have been shown to stimulate phagocytosis and macrophage activity in rats (Figure 3). These results have been substantiated by other investigators (Figure 4) who also determined that stimulation of concanavalin A (Con A, a T-cell mitogen) was significant in animals supplemented with Bio-Mos. No effect of pokeweed mitogen (a B-cell mitogen) was noted with Bio-Mos supplementation (Figure 5, Zennoh, personel communication).

Feed ingredients as antigens

Since most antigens are proteinaceous compounds, it is not surprising that immune response to certain protein containing feed ingredients have been noted. Lectins (carbohydrate-binding proteins) are also present as components of many plants. Most of these lectins are resistant to proteolysis during passage through the gastrointestinal tract. Binding of these lectins to carbohydrates on epithelial cells, just as is seen with bacterial lectins, is common. After binding, the plant lectins are endocytosed and can subsequently cause an immune response (Pusztai *et al.*, 1991). Phytohemaglutinin (PHA) is a powerful oral immunogen and produces a high titer of monospecific anti-PHA antibody in many animal species including ruminants (Pusztai, 1989).

The role that certain feed ingredients, which are essentially non-digestible, have on immune function is another area which is generating excitement. Vaccination through oral administration has been practiced for a number of years. A classical example of this vaccination method is the polio vaccine. Polio is caused by a virus which may enter the body through either the gastrointestinal or respiratory tracts. These virus cells have a high affinity for nerve cells and subsequently kill motor nerve cells and cause paralysis. Popular vaccination in the US against polio is accomplished by drinking an orange-flavored drink or eating a sugar cube containing living non-pathogenic strains of the virus. Once ingested, the antigens bind to M cells along the mucous membranes of the digestive tract, the viral antigens are transported across the M cells into a basolateral pocket containing groups of B cells, T cells and macrophages. This serves to activate the B cells which, in turn, differentiate into plasma cells secreting dimeric IgA antibody. IgA is then transported across the epithelia and released as secretory IgA into the lumen where it provides protection against a later challenge by a virulent polio virus. The secretory component of IgA prevents enzymatic cleavage allowing the secretory IgA to exist for a longer period of time in the protease-rich mucosal environment (Kuby, 1994).

THE ROLE OF TRACE MINERALS IN IMMUNITY

There is considerable evidence that zinc deficiency can severely depress immunological function (Table 1; Fernandes *et al.*, 1979; Gross *et al.*, 1979; Chandra and Au, 1980). At wound sites, accelerated uptake and increased retention of zinc are observed. This is probably due to zinc's role in cell replication and DNA synthesis although the role of zinc in immune function is also a strong consideration. The skin of zinc deficient animals is often scaly and brittle and cracks in the skin can allow the entry of pathogens by compromising a basic anatomic defense mechanism. In addition, it has been demonstrated that prolonged zinc deficiency results in an impairment of B-cell response (Zwickl and Fraker, 1980). In moderately deficient rats, Gross *et al.* (1979) observed reduced blastogenic responses to T-cell mitogens suggesting that T-cell function is more sensitive to zinc deficiency compared with B-cell function (Table 1). More recent data have shown that the form of mineral supplementation also influences immunity. Organic zinc and manganese supplementation in turkeys were shown to enhance both cellular and humoral immunity (Ferket, and Qureshi, 1992). Increases in IgG and percentage of phagocytic macrophages were noted with organic mineral supplementation compared with sulfates (Figure 6).

Dietary iron also plays a role in infection. Severe iron deficiency has

Table 1. Effect of dietary zinc on immune function in rats.

	PHA Response	Con A Response	PWM Response
Zinc deficient	270.9[a]	221.4[a]	30.8
Zinc supplemented	140[b]	90.3[b]	15.5[a]

[a,b]Values differ $P<0.05$.
Adapted from Gross *et al.*, 1979.

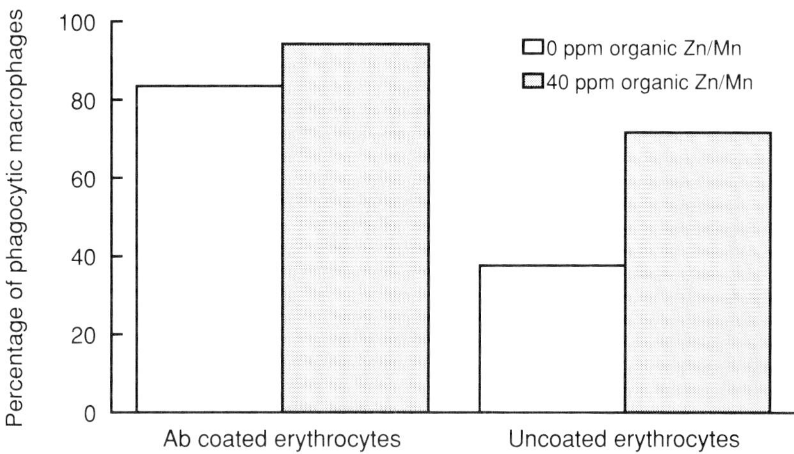

Figure 6. Effect of zinc and manganese source on the percentage of phagocytic macrophages (adapted from Ferket and Qureshi, 1992)

been associated with an increased incidence of infections. One reason for this is the enzyme myeloperoxidase which is found in neutrophils. This enzyme catalyzes the conversion of hydrogen peroxide to hypochlorous acid which is a strong oxidizing agent. In addition, lymphocyte replication is dependent on the iron enzyme ribonucleotide reductase (Brody, 1994).

In cases where extremely high levels of iron are obtained, the host is more susceptible to bacterial infections. This increase in susceptibility occurs due to two factors: high concentrations of iron make it easier for pathogens to satisfy their iron requirements, and chemotactic and phagocytic activity of neutrophils and monocytes is reduced in high iron situations.

Lactoferrin is a protein that has a high affinity for binding iron. At normal iron concentrations, bacteria are unable to compete with this protein for free iron so bacterial growth is limited. Lactoferrin is found in tears, semen, bile, mucosal secretions and milk of all mammals and it is an integral part of the pathogen defense of the neonate. Lactoperoxidase is an enzyme that produces superoxide radicals. Superoxide is a highly reactive form of oxygen that is toxic to many cells. Superoxide dismutase (SOD) is a copper containing enzyme which catalyzes the reaction:

$$2O_2^- + 2H^+ \text{---->} H_2O_2 + O_2$$

This reaction is necessary for protection from membrane oxidation by free radicals produced by neutrophils and macrophages during an immune response. SOD concentrations have been shown to be adversely affected by copper deficiency. The hydrogen peroxide formed is converted to water and oxygen by the enzyme catalase which is an iron-containing enzyme.

Selenium status has also been implicated as a factor in immunity. In ruminants, Boyne and Arthur (1981) demonstrated neutrophils from selenium deficient calves had a decreased ability to kill ingested *Candida albicans*. The authors linked this defect with decreased glutathione peroxidase (GSH-Px) activity in the neutrophil. The active site of glutathione peroxidase contains selenocysteine therefore it is easy to see why a deficiency of selenium would impair GSHPx.

Summary

The basic health of the animal depends on the quality and quantity of the feed that the animal consumes. Environmental stresses will always present challenges to the animal. The ultimate impact of these challenges depends on immunocompetence. The interrelationship between immunity and nutrition is far more complex than can be mentioned in this text. Many of the interactions are just beginning to be understood. By more closely examining the role that certain nutrients and micronutrients play in animal nutrition we will continue to shed light on nutritional immunology. The traditional list of ingredients such as copper, iron, zinc and protein (which provide the basic building blocks for the immune system) has grown to include minerals such as selenium and chromium as valuable 'immune boosters'. Other compounds such as oligosaccharides are also making a strong case for inclusion as powerful immunomodulators.

References

Boyne, R. and J.R. Arthur. 1981. J. Comp. Path. 91:271.

Brody, T. 1994. In: Nutritional Biochemistry. pp. 535.

Chandra, R.K. and B. Au. 1980. Amer. J. Clin. Nutr. 33:736.

Ferket, P.R. and M.A. Qureshi. 1992. Poultry Sci. 71 (Suppl.1):60.

Fernandes, G., M. Nair, K. Onoe, T. Tanaka, R. Floyd and R.A. Good. 1979. Proc. Natl. Acad. Sci. 76:457.

Gross, R.L., N. Osdin, L. Fong and P.M. Newberne. 1979. Am. J. Clin. Nutr. 32:1260.

Hyde, R.M. and R.A. Patnode. 1978. In: Immunology. Reston Publishing Co.

Kuby, J. 1994. Immunology. 2nd edn. Freeman & Co, New York.

Lyons, 1994. Biotechnology in the feed industry: 1994 and beyond. In: Biotechnology in the Feed Industry. T.P. Lyons and K.A. Jacques (Eds). Nottingham University Press, Nottingham, UK.

Newman, K.E. 1994. In: Biotechnology in the Feed Industry. T.P. Lyons and K.A. Jacques (Eds). Nottingham University Press, Nottingham, UK.

Pusztai, A. 1989. In: Toxicants of Plant Origin. P.R. Cheeke (Ed.). Vol. III. CRC Press, Boca Raton, FL. p. 29.

Pusztai, A., S.W.B. Ewen, A.F.F.U. Carvalho, G. Grant, J.C. Stewart and S. Bardocz. 1991. Proceedings of the interdisciplinary conference on: Effects of Food on the Immune and Humoral Systems. 20–24.

Spain, J. 1993. In: Biotechnology in the Feed Industry. T.P. Lyons (Ed.). Alltech Technical Publications, Nicholesville, KY, pp. 53–60.

Zwickl, C.M. and P.J. Fraker. 1980. Immunol. Commun. 9:611.

SYSTEMIC INDUCED RESISTANCE AS A NOVEL STRATEGY FOR PLANT DISEASE CONTROL

JOSEPH KUĆ

Department of Plant Pathology,
University of Kentucky, Lexington, Kentucky, USA

Introduction

To survive the selection pressure by pathogens during coevolution, plants have acquired effective disease resistance mechanisms. Susceptibility to disease is an exception in nature, resistance is the rule and survival is the *sine qua non* (Dawkins, 1976). Both plants and pathogens survive. Reports from many laboratories indicate that plants, whether susceptible or resistant to certain diseases, contain genetic information for effective defense mechanisms (Madamanchi and Kuć, 1991; Kuć and Strobel, 1992; Kuć, 1992; 1993; Staub *et al.*, 1993; Strobel and Kuć, 1994). Resistance can be systemically induced in susceptible plants against many viral, bacterial and fungal diseases by restricted inoculation with pathogens or non-pathogens as well as treatment with chemicals. Research in my laboratory centers on molecular mechanisms for induced systemic resistance (ISR) using tobacco and cucumber as model systems, and on the development of ISR technology for use as part of an integrated disease control strategy, together with resistant cultivars and judicious use of pesticides. Using ISR, susceptible cultivars become resistant to pathogens, and resistant cultivars become more resistant, e.g. induced resistance to TMV on N-gene tobacco cultivars. ISR has been reported in at least twenty crop plant species (Kuć and Strobel, 1992). In tobacco and cucumber, ISR persists for weeks or even months and is very effective in both greenhouse and field tests (Madamanchi and Kuć, 1991; Kuć and Strobel, 1992; Kuć, 1992; 1993; Staub *et al.*, 1993; Strobel and Kuć, 1994).

ISR is similar to human immunization in terms of its activation by induction, and it is often referred to as plant immunization. However, unlike the antibody–antigen aspect of the human immune response, ISR is non-specific with respect to both the inducing agent and challenge pathogen. Many biotic and abiotic agents induce very similar, if not identical, defense-related responses in plants, and ISR caused by one agent is effective against many diseases (Madamanchi and Kuć, 1991; Kuć and Strobel, 1992; Kuć, 1992; 1993; Staub *et al.*, 1993; Strobel and Kuć, 1994). The non-specificity of ISR would be an advantage in the application of the technology to agriculture, because induction with one agent would protect plants against multiple diseases. This non-specificity could have tremendous practical significance. Different diseases, whether caused by bacteria, fungi or viruses, need not be treated differently and individually, and chemical pesticides for the economic control of viral diseases are not available. In combination with other disease control measures, ISR has the potential to reduce the cost of disease control and dependence on pesticides.

The mechanism for ISR is multicomponent. Many biochemical and cellular changes occur with ISR (Madamanchi and Kuć, 1991; Kuć and Strobel, 1992; Kuć, 1992; 1993; Staub *et al.*, 1993; Strobel and Kuć, 1994). Particular attention has recently been directed to a diverse group of low molecular weight proteins, called pathogenesis-related proteins (PR-proteins), the systemic accumulation of which is associated with ISR to some pathogens (Van Loon, 1985; Boller, 1987; Ye *et al.*, 1989; Bol and Linthorst, 1990; Pan *et al.*, 1991; White and Antoniw, 1991). Acidic PR proteins, including acidic β-1,3–glucanases and chitinases, are secreted into intercellular spaces where they would be encountered by, and act against, fungal and bacterial pathogens at an early stage of the infection process (Van Loon, 1985; Boller, 1987; Ye *et al.*, 1989; Bol and Linthorst, 1990; Pan *et al.*, 1991; White and Antoniw, 1991). Basic β-1,3–glucanases and chitinases, which accumulate in the vacuole, may interact with pathogens at a later stage of infection, during host cell deterioration (Boller, 1987; Mauch and Staeheln, 1989). Not all PR-proteins have been demonstrated to have hydrolytic activity. The constitutive high level expression of PR-1a in transgenic tobacco results in tolerance to disease caused by two oomycete pathogens, *Peronospora tabacina* and *Phytophthora parasitica* var. *nicotianae* (Alexander *et al.*, 1993). The mode of action of PR 1a is not reported. It appears unlikely, however, that PR proteins alone account for all the observed effects of ISR against many unrelated fungal, bacterial and viral diseases. Identified PR proteins are not likely to be important in ISR to TMV local lesion formation (Fraser, 1986; Pennazio and Roggero, 1988; Ye *et al.*, 1990; 1992). Cytokinins, antiviral factors, and inhibitors of viral replication have all been implicated in ISR to localized TMV infection (Balazs *et al.*, 1977; Szirak *et al.*, 1980; Speigel *et al.*, 1989; Edelman *et al.*, 1990). A wide variety of other putative defense compounds and processes have also been associated with the expression of ISR including phytoalexins and lignin (Madamanchi and Kuć, 1991; Kuć, 1992). High Pr protein levels are not reliable markers for induced resistance to fungi in cucumber and tobacco.

The multicomponent nature of ISR may explain its activity against diverse pathogens. Because ISR is multicomponent, it may also be durable, in contrast to single gene resistance and single site-specific systemic fungicides. Therefore, ISR offers promise for the control of multiple plant diseases with consequent reduced dependence on pesticides.

The practical application of ISR as a component of IPM strategies is dependent upon having: 1) a convenient, reliable, economical and effective means of delivering ISR; 2) an in-depth assessment of the extent to which ISR can reduce disease development, and hence alter pesticide use patterns or efficacy, under field conditions.

Genetic engineering for resistance

Genetic engineering offers a promising technology for the protection of plants against disease. Constitutive expression of PR protein genes in transgenic plants represents one of the first efforts toward genetic engineering for disease control. PR proteins were selected as promising disease control compounds, because: 1) their accumulation is associated with the development of ISR; 2) some PR proteins have antimicrobial activity; 3) their accumulation in intercellular spaces (acidic PR proteins) make them good candidates for plant defense mechanisms; 4) PR proteins are primary gene products and their overproduction can be relatively easily achieved with plant transformation.

However, in most cases, constitutive expression of individual PR protein genes in transgenic plants has not resulted in a high level of disease resistance. The failure of constitutive expression of a single putative defense-related gene to produce plants highly resistant to disease is consistent with the hypothesis that effective plant defense against diseases depends upon a combination of multiple preformed (physical and chemical barriers) and inducible defense mechanisms (Lamb *et al.*, 1989; Madamanchi and Kuć, 1991). It has been demonstrated *in vitro* that chitinase and β-1,3–glucanase individually may exhibit little antifungal activity, but a combination of the two hydrolases can be highly antifungal (Mauch *et al.*, 1988). Thus, constitutive overproduction of a single component of the defense response may not be effective against diseases, whereas constitutive overproduction of combinations of two and more defense compounds may prove more effective for disease control. However, overproduction of many defense compounds constitutively may result in yield loss and undesirable agronomic consequences due to the diversion of energy and carbon precursors and the perturbation of normal physiological processes of the plant. Particularly, overproduction of resistance factors which also participate in normal metabolism is likely to have adverse physiological effects as shown by tobacco transformation with genes for cell wall-associated peroxidases and a bean phenylalanine ammonia lyase gene (Elkind *et al.*, 1990; Lagrimini *et al.*, 1990).

As an alternative to genetic engineering for direct overexpression of PR protein genes, genetic engineering for disease resistance may be more successfully achieved by focusing efforts on the manipulation of signalling processes involved in activation, or sensitization for activation after infection, of the expression of multiple defense mechanisms already encoded in the plant genome.

Signal transduction processes

The induction of systemic resistance responses, sometimes by only a single local lesion (Kuć and Richmond, 1977), indicates a massive systemic amplification of defense responses to the consequences of localized infection. One explanation of this phenomenon is that an immunity signal(s) released or synthesized at the induction site is systemically translocated to, and then transduced and amplified in the targeted leaves, where it conditions the plant for resistance (Kuć, 1987). The presence of such a systemic signal(s) has been demonstrated by girdling tobacco and cucumber phloem above and below the induction sites, removal of inducer leaves in tobacco and in cucumber, and by reports of the signal's graft transmissibility from root stock to scion (Madamanchi and Kuć, 1991; Kuć and Strobel, 1992; Kuć, 1992; 1993; Staub *et al.*, 1993; Strobel and Kuć, 1994). The signal(s) is likely of plant origin since protection can be induced by diverse agents including not only fungi, bacteria and viruses, but also simple compounds such as oxalate and basic phosphates (Madamanchi and Kuć, 1991; Kuć and Strobel, 1992; Kuć, 1992; 1993; Staub *et al.*, 1993; Strobel and Kuć, 1994). Removal of inducer leaves at various time intervals after inoculation with TMV in tobacco suggests a dosage response for the signal(s). That the signal(s) is not cultivar or species specific is suggested by grafting experiments between cucumber, watermelon and muskmelon plants. The signal(s) is unlikely to be phytotoxic at physiologically active concentrations because the growth of immunized cucumber, watermelon, muskmelon and tobacco is not adversely affected. The signal(s) is also unlikely to be antifungal by itself because host responses are required for protection.

Two endogenous plant compounds with potential signal activity have recently received attention. Jasmonic acid, a product derived from linolenic acid through the lipoxygenase pathway, and salicylic acid, believed to be synthesized via the shikimate pathway and β-oxidation, are suggested to be endogenous signals for ISR.

Increased lipoxygenase activity in inducer leaves is associated with ISR in tobacco and cucumber (Avidiushko *et al.*, 1993). Lipoxygenase activity in inducer leaves is markedly increased after induction and increased lipoxygenase activity precedes the development of ISR (Avidiushko *et al.*, 1993). The increase in lipoxygenase activity is paralleled by a decrease in total polyunsaturated fatty acid content of inducer leaves. Though treatment of plants with jasmonate induces systemic accumulation of proteinase inhibitors, which have a role in plant resistance to insect pests (Gundlach *et al.*, 1992; Farmer and Ryan, 1990; Ryan, 1992; Raskin, 1992), the role of proteinase inhibitors and jasmonate in ISR is not evident.

Salicylic acid (SA) is a natural constituent of plants (Raskin, 1992; Pierpoint, 1994). Systemic increase in SA is associated with ISR in tobacco and cucumber. A dramatic increase of endogenous SA in the inducer leaves coincides with host cell necrosis at the induction site, though a considerably lower systemic increase of SA is detected prior to or coincidentally with the development of ISR. SA is detected in the phloem sap of induced cucumber and tobacco plants, an indication of possible systemic phloem transport for SA. Application of exogenous SA elicits PR protein patterns similar to those induced by pathogens, and the leaves treated with SA develop resistance to some viral, bacterial and fungal pathogens. Thus, SA appeared to be a promising candidate as a signal for ISR.

However, evidence is increasing which indicates that SA may not be a signal, the only signal, or the primary, signal for ISR (Rasmussen *et al.*, 1991; Gaffney *et al.*, 1993; Delaney *et al.*, 1994; Pierpoint, 1994; Vernooij *et al.*, 1994). Exogenous application of SA by foliar spray or by repeated stem injection is a poor inducer of ISR in tobacco and cucumber. Plants unable to accumulate SA (Gaffney *et al.*, 1993) have resistance induced systemically by 2,6,dichloroisonicotinic acid (Delaney *et al.*, 1994). SA also may not have good commercial potential as a chemical agent for plant protection because the safety margin between protection and phytotoxicity by SA is very narrow.

Induced resistance

The use of chemical inducers eliminates potential problems associated with induction by plant pathogens. Chemicals are readily produced, distributed and stored. Application of chemical inducers could be made with the techniques and equipment of mechanized agriculture to minimize labor and expense involved in resistance induction. Chemical inducers of plant resistance are different from systemic fungicides. Firstly, they have no or low *in vitro* toxicity to the pathogens. Secondly, chemically induced resistance is non-specific, being effective against a wide spectrum of fungal, bacterial and viral diseases, whereas systemic fungicides are often specific to targeted pathogens. Therefore, the major advantage of chemical inducers over fungicides is that application of one chemical inducer would protect plants against many diseases. Because chemical inducers work through the activation of plant defense mechanisms, this type of protection is expected to be durable. Since the mechanisms activated for defense are similar to those reported for resistant plants developed by breeders, ISR would be expected to be as safe as resistant plants.

ISR appears to contribute to the efficacy of several pesticides not specifically developed as inducers, namely 2,2–dichloro-3,3–dimethylcyclopropane carboxylic acid (Cartwright and Langcake, 1980), probenazole (Iwata *et al.*, 1980), and tricyclazole (Nikolaev *et al.*, 1990). Further, benzimidazole fungicides exhibit cytokinin-like activities in some bioassays (Person *et al.*, 1957), and the ergosterol biosynthesis-inhibiting fungicide triademifon is reported to increase the cytokinin content of cucumber plants (Fletcher and Arnold, 1986). Elevated content of endogenous cytokinins has been proposed to play a role in ISR (Balazs *et al.*, 1977; Szirak *et al.*, 1980; Sarhan *et al.*, 1991).

The development of chemical inducers of disease resistance is in its infancy and has been undertaken primarily by two research groups: our group at the University of Kentucky, and Ciba-Geigy. Chemical inducers that have been reported include oxalate (Doubrava *et al.*, 1988), di- and tri-basic phosphates (Gottstein and Kuć, 1989), β-ionone (Salt *et al.*, 1986), and 2,6–dichloroisonicotininc acid (INA) and its methylester derivative (Staub *et al.*, 1993). Other inducers of resistance include microbial cell wall extracts including those from yeast (Newton *et al.*, 1993). Yeast extracts reduced mildew and brown rust in barley 93% and 73%, respectively.

Seed treatment with chemicals is an attractive means of achieving ISR without gross damage to plant tissues. In comparison with foliar spraying of inducers or soil treatment under field conditions, lesser amounts of inducer would be required, and inducer uptake should be less variable. Furthermore, seed treatment can be performed under controlled environmental conditions. Resistance induced with seed treatments might assist in protection against seed decay and pre- and post-emergence damping off, as well as diseases which develop later. An amazing long-lived induced resistance of rice to blast disease was reported by Arimoto and co-workers (Arimoto *et al.*, 1991). They treated seed with solutions of the dodecyl ester of DL-alanine hydrochloride. Surprisingly, resistance induced by this means persisted through several generations of rice plants. Transmission of induced resistance through seed was not observed in our studies with tobacco or cucumber. The mechanisms underlying the seed transmissibility of induced resistance observed by Arimoto *et al.* remain to be elucidated.

Integration of induced resistance

For ISR to find widespread commercial application, there must be a high probability that its implementation will be of economic benefit to farmers. In the short term, a likely benefit of ISR would be improvement of plant disease control where induced resistance is simply superimposed upon standard disease management programs. In the longer term, epidemiological studies may indicate that ISR can reduce dependence on conventional pesticides for disease control.

Implementation of ISR without a change in pesticide schedules may provide economic benefits in situations where standard control measures typically provide incomplete protection against economic losses due to plant disease. In such situations, the effects of ISR might be analogous to that of cultivar resistance introduced by traditional plant breeding methods. This added degree of protection may be important under circumstances in which sub-optimal pesticide residues are present on or within leaves (e.g. between pesticide sprays, especially during rainy periods in which pesticides are removed and their re-application delayed). Combining ISR with pesticides may also be helpful where environmental factors promote a high level of disease pressure. Additionally, because ISR is typically active against a broad spectrum of diseases, it may afford economically meaningful levels of protection from the effect of diverse 'minor' pathogens (including those affecting roots). These 'minor' pathogens are uneconomical to control with separate pesticides, but may collectively reduce crop yields. The integration of induced resistance with conventional disease management strategies is likely to be the easiest to evaluate experimentally, and the simplest means by which farmers might begin to utilize ISR.

ISR may be capable of interacting with conventional pesticides in an additive and/or synergistic manner to decrease plant disease. Horizontal (multigenic) resistance, which ISR appears to mimic in some respects, can reduce the dosage of fungicides needed to control phytopathogenic fungi in both greenhouse and field environments. Reduced pathogen vigor due to pesticides would make it less likely that the pathogen would be able to overcome host defense mechanisms, particularly where expression of these defense mechanisms has already been activated in plants with ISR. Reduced growth rate of pathogens due to the effect of pesticides would allow additional time for the mobilization of plant defenses. Because defense responses to pathogens are earlier and greater in induced than non-induced plants, the effect of pesticides would likely be greater in induced plants. ISR could, therefore, reduce the quantity of pesticide needed.

The decreased usage of pesticides, which ISR may permit, could reduce the rate at which plant pathogens develop resistance to fungicides, and thus prolong their useful life. This could benefit not only farmers, but also pesticide producers, who would enjoy extended periods of profitability from these materials. Because of the apparent multi-component nature of ISR mechanisms, it would seem unlikely that plant pathogens would readily evolve to circumvent ISR. The advent of genetic engineering, and use of chemical inducers of ISR, render possible the transformation of ISR from 'an interesting phenomenon' to a practical tool for plant disease control within an IPM context.

References

Alexander, D., R. Goodman, M. Gut-Rella, C. Glascock, K. Weyman, L. Friedrich, D. Maddox, P. Ahl-Goy, T. Luntz, E. Ward and J. Ryals. 1993. Proc. Natl. Acad. Sci. USA 90:7327.

Arimoto, Y., Y. Homma, R. Yoshino and S. Saito. 1991. Ann. Phytopath. Soc. Japan 57:522.

Avidiushko, S.A., X.S. Ye, D. Hildebrand and J. Kuć. 1993. Physiol. Mol. Plant Pathol. 42:83.

Balazs, E., I. Sziraki and Z. Kiraly. 1977. Physiol. Plant Pathol. 11:29.

Bol, J.F. and H.J.M. Linthorst. 1990. Ann. Rev. Phytopathol. 28:113.

Boller, T. 1987. Hydrolytic enzymes in plant disease resistance. In: Plant-Microbe Interactions, Molecular and Genetic Perspectives, Vol. 2. T. Kosuge and E.W. Nester (Eds). Macmillan, N.Y., 385.

Cartwright, D.W. and P. Langcake. 1980. Physiol. Plant Pathol. 17:259.

Dawkins, R. 1976. The Selfish Gene. Oxford Univ. Press, N.Y. 224 pp.

Delaney, T., S. Uknes, B. Vernooÿ, L. Friedrich, K. Weymann, D. Negrotto, T. Gaffney, M. Gut-Rella, H. Kessman, E. Ward and J. Ryals. 1994. Science 266, 1247.

Doubrava, N.S., R.A. Dean and J. Kuć. 1988. Physiol. Mol. Plant Pathol. 33:69.

Edelman, O., N. Ilan, G. Grafi, N. Sher, Y. Stram, D. Novick, N. Tal, I. Sela and M. Rubinstein. 1990. Proc. Natl. Acad. Sci. USA 98:588.

Elkind, Y., R. Edwards, M. Mavandad, S.A. Hedrick, O. Ribak, R.A. Dixon and C.J. Lamb. 1990. Proc. Natl. Acad. Sci. USA 87:9057.
Farmer, E.E. and C.A. Ryan. 1990. Proc. Natl. Acad. Sci. USA 87:7713.
Fletcher, R.A. and V. Arnold. 1986. Physiol. Plant. 656:197.
Fraser, R.S.S. 1986. Physiol. Plant Pathol. 19:69.
Gaffney, T., L. Friedrich, B. Vernooij, D. Negrotto, G. Nye, S. Uknes, E. Ward, H. Kessman and J. Ryals. 1993. Science 261:754.
Gottstein, H.D. and J. Kuć. 1989. Phytopathology 79:176.
Gundlach, H., M.J. Muller, T. Kutchan and M.H. Zenk. 1992. Proc. Natl. Acad. Sci. USA 89:2389.
Iwata, I., Y. Suzuki, T. Watanabe, S. Mase and Y. Sekizawa. 1980. Ann. Phytopath. Soc. Japan 46: 297.
Kuć, J. 1987. Ann. N.Y. Acad. Sci. 494: 221.
Kuć, J. 1992. Antifungal compounds from plants. In: Phytochemical Resources for Medicine and Agriculture. H.N. Nigg and D. Seigler (Eds). Plenum Press, N.Y., 159.
Kuć, J. 1993. Non-pesticide control of plant disease by immunization. In: Proc. of the 10th International Symposium on Systemic Fungicides. H. Lyr and C. Polter (Eds). Enger Ulmer, Stuttgart, 225.
Kuć, J. and S. Richmond. 1977. Phytopathology 67: 533.
Kuć, J. and N.E. Strobel. 1992. Induced resistance using pathogens and nonpathogens. In: Biological Control of Plant Diseases, Progress and Challenges for the Future. E.C. Tjamos, G.C. Papavizas and R.J. Cooke (Eds). Plenum, N.Y., 295.
Lagrimini, L.M., S. Bradord, S. and S. Rothstein. 1990. Plant Cell 2:7.
Lamb, C.J. M.A. Lawton, M. Dron and R.A. Dixon. 1989. Cell 56:215.
Madamanchi, N.R. and J. Kuć, J. 1991. Induced resistance in plants. In: The Fungal Spores and Disease Initiation in Plants and Animals. G.T. Cole and H.C. Hoch (Eds). Plenum, N.Y., 347.
Mauch, F. and L.A. Staehelin. 1989. Plant Cell 1:447.
Mauch, F., B. Mauch-Mani, B. and T. Boller. 1988. Plant Physiol. 88:936.
Newton, A., G. Lyon and T. Reglinski. 1993. Project report 78 of the Scottish Crop Research Institute, Dundee DD2 5DA, Scotland.
Nikolaev, O.N., A.A. Aver'yanov, V.P. Lapikova and V.G. Dzhavakhiya. 1990. Sov. Plant Physiol. 37: 124.
Pan, S.Q., X.S. Ye and J. Kuć. 1991. Physiol. Mol. Plant Pathol. 39:25.
Person, C., D. Samborski and F.R. Forsyth. 1957. Nature (London) 180:1294.
Pierpoint, W.S. 1994. Adv. Botanical Res. 20:163.
Pennazio, S. and P. Roggero. 1988. J. Phytopathol. 121: 255.
Raskin, I. 1992. Ann. Rev. Plant Physiol. Plant Mol. Biol. 43:349.
Rasmussen, J.B., R. Hammerschmidt and M.N. Zook. 1991. Plant Physiol. 97:1342.
Ryan, C.A. 1992. Plant Mol. Biol. 19:123.
Sarhan, A.R.T., Z. Kiraly, I. Sziraki and V. Smedegaard-Petersen. 1991. J. Phytopathol. 131:101.
Salt, S., S. Tuzun and J. Kuć. 1986. Physiol. Mol. Plant Pathol. 28:287.

Speigel, S., A. Gera, R. Salomon, P. Ahl, S. Harlap and G. Loebenstein. G. 1989. Phytopathology 79:258.
Staub, T. P. Ahl Goy and H. Kessman. 1993. Chemically induced disease resistance in plants. In: Proc. of the 10th International Symposium on Systemic Fungicides. H. Lyr and C. Polter (Eds). Eugen Ulmer, Stuttgart, 239.
Strobel, N.E. and J. Kuć. 1994. Development of environmental safe chemicals as inducers of disease resistance in crop plants. In: Proc. of the Fourth National Conference on Pesticides (D.L. Weigman (Ed.). Virginia Water Resources Research Center, Blacksburg, 519.
Szirak, I., E. Balazs and Z. Kiraly. 1980. Physiol. Plant Pathol. 16: 277.
Van Loon, L.D. 1985. Plant Mol. Biol., 4:111.
Vernooij, B., L. Friedrich, A. Morse, A., R. Reist, R. Kolditz-Jawar, E. Ward, S. Uknes, H. Kessman and J. Ryals. 1994. Plant Cell 6: 959.
White, R.F. and J.F. Antoniw. 1991. Crit. Rec. Plant Sci. 9:443.
Ye, X.S., S.Q. Pan and J. Kuć, J. 1989. Physiol. Mol. Plant Pathol. 35: 161.
Ye, X.S., S.Q. Pan and J. Kuć. 1990. Physiol. Mol. Plant Pathol. 36:523.
Ye, X.S., S.Q. Pan and J. Kuć. 1992. Plant Sci. 84: 1.

USE OF IMMUNOSTIMULANTS IN AGRICULTURAL APPLICATIONS

FIONA MACDONALD

Vetrepharm Ltd, Fordingbridge, Hampshire, UK

Work over a number of years has identified a group of both biological and synthetic compounds which influence and enhance the non-specific defence mechanisms in animals. These substances can be grouped under the general heading of immunostimulants, although they may be derived from a widely diverse range of sources. These substances range from nutritional products such as cream (which when injected triggers a non-specific immune response), yeast cell wall derivatives including glucans and mannanoligosaccharides, mushroom derivatives (again polysaccharide in nature) and chicken egg products, through infectious agents to synthetic compounds (Engstad and Robertsen; Sakai; *et al.*, 1993).

The infectious agents form an interesting group, ranging from work with an inactivated avain pox virus used in mammals to 'Freund's complete adjuvant' made from an oily suspension of killed cells of *Mycobacterium tuberculosis*. In fact, a range of bacteria have been investigated for immunostimulant properties – *Streptococcus pyogenes*, various *Mycobacterium* spp. and *Nocardia* spp.

One of the better known synthetic immunostimulants is levamisole, which is a long-established agricultural anthelmintic, but in addition it has well demonstrated and documented non-specific immune stimulatory properties.

The purpose of this chapter is to examine various immunostimulant compounds.

Infectious agents

Work over many years has established that injection into mammals of cell wall preparations of various bacteria results in the following:

1. increase in macrophage production,
2. increase in phagocytic activity and rate,
3. increase in natural killer cell activity,
4. induced production of cytokine proteins such as interleukins and interferons.

Indeed some work from the human field has indicated that subsequent to the injection of such preparations human malignant cancer patients have shown improvement of symptoms with increased survival time. When these preparations are examined it becomes clear that the effects in the non-specific immune response can be directly related to multi-peptide fragments as well as lipopolysaccharides and lipopeptides.

Nutritional agents

The active structures in the range of nutritional agents which can influence the immune response are usually large molecules such as polysaccharides or immunoactive peptides (Sakai *et al.*, 1993).

Extensive investigative work has been carried out particularly in the field of aquaculture to not only demonstrate efficacy of these immunostimulants, but specifically to determine how they work in relation to cellular response. Glucans in particular have attracted a great deal of interest and investigation, and are known to stimulate defence mechanisms in a range of higher organisms. In plants they stimulate production of phytoalemins, which are low molecular weight antimicrobials. In invertebrates glucans activate a defence enzyme called polyphenoloxidase (Engstad and Robertsen; Raa *et al.*, Yoshida *et al.*, 1993).

However, the majority of the work has been carried out on vertebrates, and in particular aquatic vertebrates such as farmed fish and crustacea such as shrimp or prawns. Elements of the non-specific defence mechanisms in fish, as in other vertebrates, include the phagocytic cells, neutrophils and macrophages, complement, lysozyme and interferon (Raa *et al.*).

YEAST CELL WALL PRODUCTS: MANNANOLIGOSACCHARIDES AND GLUCANS

One source of mannanoligosaccharides and glucan is yeast cell wall material. The basic composition of the wall consists of mannan (30%), glucan (30%) and protein (12.5%). While the ratio of one component to another remains relatively constant from strain to strain, the degree of mannan phosphorylation and the interaction among the mannan, glucan and protein components varies. The glucan is a polysaccharide with glucose bonds connected in β1,6 and β1,3 linkages (Peat *et al.*, 1958). Glucan is thought to make up the matrix of the cell wall which is covered by another layer of mannose sugars (Figure 1). These mannose sugars are arranged in a highly branched chain of manno-pyranoside residues. The linkages in the backbone of this chain are α-1,6 with side chains bound by α-1,2 and α-1,3. The cell wall has powerful antigenic stimulating properties, and it is well-established that this property is a characteristic of the mannan chain (Ballou, 1970). Differences in mannan structure exist among strains, and yeast strains can be differentiated based on the antigenicity of extracted mannan. This antigenicity can be

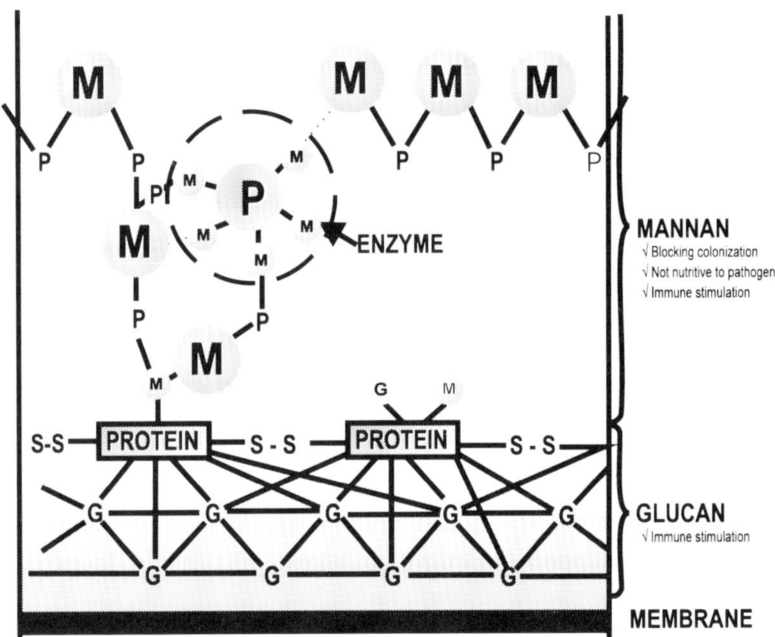

Figure 1. Structure of the yeast cell wall.

increased by the addition of acetyl groups or by increasing the degree of phosphorylation.

Trials have been carried out on fish in the UK by Vetrepharm to establish the immunostimulatory benefits of Vetregard, a mannanoligosaccharide compound derived from yeast cell wall.

Effect of Vetregard (mannanoligosaccharide) on immune function in African sharp-toothed catfish

A study was commissioned at the Institute of Aquaculture University of Stirling, Scotland to investigate the comparative immunostimulatory effects of Macrogard, a β 1,3/β 1,6 glucan product, and Vetregard, a mannanoligosaccharide (MOS) product (Yoshida *et al.*, in press). The objectives were to investigate under controlled laboratory conditions the mechanisms associated with immunostimulation, with a comparison between the two products. The parameters examined were phagocyte activation, activated neutrophil and monocyte production, and bacterial clearance from the tissues of the host, in this case African catfish (*Clarias gariepinus*).

Materials and methods

C. gariepinus (mean body weight 20 g) were maintained in a recirculating water system at 25 ± 1°C. The test substances, glucan and mannanoligosaccharide (MOS), were used at 1 g per kg of feed coated with cod liver oil at a rate of 40 ml per kg. Control diets were treated with cod liver oil but without glucan or MOS. Fish were fed once daily at a rate of 5% body weight. Blood samples were obtained at 0, 12, 30 and 45 days from the caudal artery into a heparinized syringe. Glass adherence and the nitroblue tetrazolium (NBT) reduction assays were used to evaluate phagocytic function of neutrophils and monocytes.

After 14 days feeding with glucan or MOS, catfish were injected intraperitoneally with *Aeromonas hydrophila* (MCMB1134 strain) at a density of 5×10^5 CFU per gram fish body weight. Three fish from each group were sampled at 6, 12 and 24 h after infection. Bacterial counts in the blood and spleen over time were used to evaluate pathogen clearance.

Results

Numbers of glass adherent NBT positive cells in fish treated with MOS peaked at 12 days ($P<0.05$), had slightly decreased by 30 days ($P<0.05$), and dropped to the baseline level at 45 days at which point there were no significant differences between control and treated fish. The peak number of NBT positive cells in glucan-treated fish was at 30 days and then fell to the no significance level compared with control fish by 45 days (Table 1).

Table 1. Numbers of glass-adherent NBT positive cells from *C. gariepinus* given glucan or mannanoligosaccharide supplements.

Days	Glass-adherent NBT positive cells, *n*		
	Control	Glucan	Mannan
0	2.5±1.5	2.5±1.5	2.5±1.5
12	2.0±0.9	5.0±3.3	24.2±6.5*
30	1.7±1.4	11.0±5.0*	16.5±3.5*
45	3.3±2.0	4.7±2.1	5.0±1.0

*Significant differences ($P<0.05$) between control and experimental groups.

Bacterial counts in blood samples from both groups of treated fish were lower than in controls at 12 and 24 h (Table 2). Significantly lower numbers of bacteria were detected in spleen samples from MOS-treated fish compared with controls at 6 and 12 h after infection. Although the average numbers of bacteria in spleen samples from the glucan treated fish were lower than in controls at 12 and 24 h, this was not statistically significant.

These data demonstrate that both products enhanced phagocytic function and also had a marked effect on elimination of bacterial pathogens. However, this study also demonstrated very clearly that Aquamos (Vetregard) produced a more rapid effect on the immune measurements and also a more sustained effect. In addition, bacteria

Table 2. Bacterial counts (mean ± standard error) in the blood and spleen samples from *C. gariepinus* fed glucan or mannan-supplemented diets for 14 days.

Time, hours post infection	Control	Glucan	Mannan
Blood (CFU/ml)			
6	4.1(±3.0)×10³	2.5(±0.8)×10³	1.2(±0.2)×10³
12	1.1(±0.6)×10⁴	2.7(±0.8)×10³*	1.8(±1.4)×10³*
24	3.3(±2.8)×10⁴	2.3(±1.5)×10³*	2.0(±1.6)×10³*
Spleen (CFU/0.1g spleen)			
6	4.8(±2.3)×10⁵	5.9(±4.0)×10⁵	4.9(±3.5)×10⁴*
12	5.2(±4.5)×10⁵	1.1(±0.3)×10⁵	6.3(±2.1)×10⁴*
24	3.2(±2.7)×10⁵	5.0(±3.6)×10⁴	3.5(±2.3)×10⁴

*Significant differences ($P<0.05$) between control and experimental groups.
Bracket value=standard error of means

were cleared from the blood and the spleen demonstrably more rapidly by Aquamos than by the glucan.

At the moment, further work is commissioned on the exact mode of action by Aquamos (Vetregard) in the intestinal cell by electron microscopy studies. In addition, field studies are being conducted to test effects of the mannanoligosaccharide against a range of fish pathogens to determine efficacy under commercial conditions.

MANNANOLIGOSACCHARIDES IN THE UK POULTRY INDUSTRY

There is also considerable interest in products with immunostimulant properties in the UK poultry industry. Whilst the overall efficiency, particularly in the area of broiler production, continues to develop and improve, the area of specific disease control and general morbidity is still cause for concern. Concern has been further heightened by two main factors: 1) loss of nitrofurans, a major therapeutic group, in early 1994, and 2) reduced efficacy in vaccines due to strain variations in organisms.

Nitrofuran antibacterials had been used for many years especially during the first week of life of poultry as broad-spectrum antimicrobials with effects against both Gram-positive and Gram-negative bacteria including *Salmonellae* and other coliforms. They were a very useful management tool, since bacterial resistance to them developed only very slowly. Administration was via the feed, and since they were poorly absorbed from the gastrointestinal tract they were very effective against any enteric infection. However, there have been some misgivings expressed by various organizations including the World Health Organization as to their possible long-term effects in humans. Pending resolution of this situation through evaluation of additional toxicological studies carried out by nitrofuran manufacturers, marketing authorizations have been withheld for the majority of these products. This has left a significant gap in the treatment schedule for this class of livestock, since no other antibacterial has the range or the prolonged life in terms of resistance development of the nitrofuran group.

Although a number of serious poultry diseases are now routinely prevented by vaccination, there are some problems arising in this area. During the few weeks of life, the broiler chick receives a number of different vaccinations. Although in principle the preventative approach is the only way to control a number of poultry diseases, particularly viral diseases, there are some problems associated with this. There are strain variations in some of the infectious agents, for instance with Gumboro, which render the vaccine unable to offer the appropriate protection. In addition, some of the avian viruses have an immunosuppressive effect which reduces all immune response.

The other problem is the sheer number of vaccinations which must be administered over a very short time in the bird's life. These multiple procedures can themselves be stressful to the birds and hence produce a negative effect on health and performance.

Based on these problems within the UK poultry industry, trials are underway to evaluate Bio-Mos, also a mannanoligosaccharide product, in poultry diets.

In a recent field trial Bio-Mos was added to broiler diets (1 kg/tonne) in a house with a history of performance and mortality problems. A similar house served as a control; and performance of the previous flocks in both houses served as a further reference. Both mortality and feed efficiency were improved in response to Bio-Mos (Table 3); and it was noted that birds in the Bio-Mos group gained 150 g during the final 3 days of the 47-day feeding period. These field data suggest the potential of this approach to performance improvement in broiler diets.

Table 3. Effect of Bio-Mos on performance and mortality of broilers at 47 days.

	House A	House B
Previous performance		
Mortality, %	3.34	5.21
Average EPEF	256	256
	Control	Bio-Mos
Test		
Morality, %	3.65	2.99
FCR	1.99	1.94
Bodyweight, kg	2.709	2.655
EPEF	253	266

References

Ballou, C.E. 1970. J. Biol. Chem. 245:1197.

Engstad, R.E. and B. Robertsen. Recognition of yeast cell wall glucan by Atlantic salmon (*Salmon salar* L.) macrophages. Dev. Comp. Immunol. 17: 319.

Jorgensen, J.B., G.J.E. Sharp, C.J. Secombes and B. Robertsen. Effect of yeast cell wall glucan on the bactericidal activity of rainbow trout macrophages. Fish and Shellfish Immunol. 3:267.

Peat, S., W. Whelan and T. Edwards. 1958. J. Chem. Soc. p. 3862.

Phaff, H.J. and C.P. Kurtzman. 1984. In: The Yeasts, a Taxonomic Study. Elsevier Biomedical Press, Amsterdam. pp. 252–262.

Raa, R., G. Rorstad, T. Engstad and B. Robertsen. Glucan (MacroGard) from the yeast increases the resistance of salmon to microbial infections. Diseases in Asian Aquaculture 1:39.

Sakai, M., T. Otubu, S. Atsuta and M. Kobayashi. 1993. Enhancement of resistance to bacterial infection in rainbow trout, *Oncoryhynchus mykiss* (Walbaum), by oral administration of bovine lactoferrin. J. Fish Dis. 16:293–247.

Yoshida, T., M. Sakaib, T. Kitao, S.M. Khlil, S. Arakic, R. Saitohc, T. Inenod and V. Inglise. 1993. Immunodulatory effects of the fermented product of chicken egg, EF203, on rainbow trout, *Oncorhynchus mykiss*. Aquaculture 109:207–214.

Yoshida, T., R. Kruger, and V. Inglis. Augmentation of non-specific protection in African sharptooth catfish, *Clarias gariepinus* (Burchell) by the long-term oral administration of immunostimulants. J. Fish Dis. (in press)

NON-TRADITIONAL USES OF YEAST AND ITS PRODUCTS: THE PAST FIFTEEN YEARS

GRAHAM G. STEWART

International Centre for Brewing and Distilling, Heriot-Watt University, Riccarton, Edinburgh, UK

Introduction

Yeasts are the most significant microorganisms utilized by humans (Rose and Harrison, 1993). Yeast production exceeds millions of tonnes per annum and most of this use falls into three well documented areas namely: brewing (Hammond, 1993), distilling (Watson, 1993) and baking (Rose and Vijayalaskashmi, 1993). In recent years a number of novel applications have emerged which utilize specific yeast strains with unique characteristics (Lyons *et al.*, 1993). These applications are unconventional and involve animal production.

The traditional use of yeast for fodder purposes goes back for at least a century (Harrison, 1993). Food and fodder yeasts are traditionally bulk products in dry form (slurries are sometimes employed) intended for either the nourishment of humans or animals. Their primary nutritional function is as a protein carrier. The qualititative differences between food and fodder yeasts lie in the standards laid down by official bodies and manufacturers with respect to such variables as microbiological and other contaminants, vitamin content, flavour and colour. The limits for food yeasts are more stringent than those for fodder yeast (Reed and Nagadowithana, 1991). Although production processes and strains vary in detail, the composition of food and fodder yeasts, particularly in terms of the content and structure of proteins in the final product, remains reasonably uniform. Yeast employed for nutritional purposes can be produced by one of the following processes: (1) aerobically on carbohydrates; (2) anaerobically on carbohydrates and (3) aerobically on hydrocarbons. In spite of the commercial use of different substrates, production processes and yeast species, there is considerable uniformity in the pattern of related groups of compounds and of individual molecular forms. Typical properties by weight (based on total dry matter) are for the main elements: carbon, 45%; oxygen, 31%; nitrogen, 9%; hydrogen, 6%; and inorganic compounds, 9%.

The bulk manufacture of yeast by aerobic methods is a long-established and well-understood process. It is based on simple concepts of propagation employing a genetically stable microorganism under vegetative growth conditions, an appropriate source of assimilable sugars at an economic cost, and significant cost efficiency advances are difficult to envisage. Intracellular biochemical reactions involved in the metabolism are based on sucrose, glucose and fructose and are well understood. Consequently, in the traditional methods of industrial production on molasses or grain mashes, easily assimilated sources of carbon, hydrogen and oxygen are provided which, with nitrogen supplied in the form of ammonium salts, make 90% of yeast dry matter. Therefore, the basic technology is today substantially the same as it was many years ago. The basic problem is one of economies. Unless suitable new materials can be converted into yeast in the large quantities required, at a cost compatible with other fodder suplies, the manufacture of fodder yeasts has an uncertain future. In addition, growth of yeast in quantity on hydrocarbons has serious disadvantages (Harder and Veenhuis, 1989), not the least being the variable cost of petroleum supplies. The alternative feedstock is cellulose from wood and other vegetable sources but this has significant economic problems (Kosaric *et al.*, 1980). Consequently the long-term development of the feed-yeast industry will require new concepts and ideas if it is to remain viable as an important contributor to the world's nutritional needs.

In recent years, four novel applications of yeast in animal production have emerged which are outside the conventional uses. These are:

- Yeast being used specifically for one of its metabolic products. Namely the use of *Phaffia rhodozyma* for carotenoids;
- The ability of yeast to influence the normal microbial population within the rumen of ruminant animals and the functional caecum of species such as horses, pigs and rabbits;
- The role of some yeasts as a modifier of the gut microflora of livestock and to stimulate the immume system;
- The use of *Saccharomyces cerevisiae*, when added to feed, to counteract aflatoxicosis in broiler chickens and ducklings.

Carotenoid production by *Phaffia rhodozyma*

Phaffia rhodozyma is a carotenoid-producing fermentative yeast. It is a potentially useful dietary supplement in salmonids and poultry feeds because of its high content of the carotenoid pigment astaxanthin (3,3^1-dihydoxy-β,β-carotene-4^1-dione). When hydrolysed cells are added to the diet, astaxanthin is readily absorbed from the gut and effectively enhance the pink to orange colour of the flesh of pen-reared salmonids or the egg-yolk and flesh of poultry. A distinctive red colour is of prime importance to acceptance of trout and salmon by consumers. Crustaceans or crustacean-processing wastes are used as a source of astaxanthins in salmonid diets. However, the astaxanthin content in crustacean shells is only about 10–100 μg/g and the shells are also high in minerals, which complicates feeding formulations when these materials are included. Although synthetic astaxanthin and canthaxanthin are now manufactured for use as additives to fish feed, they have to date not been approved by the US Food and Drug Administration for inclusion in the rations of cultured fish. Currently, farm rearing of salmon exceeds 250 000 tonnes annually and 15 000 kg of oxycarotenoids are required for feeding purposes. *Phaffia rhodozyma* is useful to the aquaculture industry as a natural source of astaxanthin and as a food source of unsaturated lipids, protein, vitamins and other nutrients required for growth. The feasibility of commercial production of *Phaffia rhodozyma* biomass and astaxanthin has been demonstrated using synthetic media, alfalfa juice, molasses and whey permeate in single culture or synthetic media in mixed culture with the Gram positive bacteria *Bacillus circulans*. The growth of *Phaffia rhodozyma* in mixed culture improves its digestibility and consequently the availability of astaxanthin for feeding purposes because the lytic enzymes produced by *Bacillus circulans* hydrolyse the *Phaffia rhodozyma* cell wall. With a bacterial inoculum four times that of yeast, over 80% of the total astaxanthin was extractable in 48 hours.

Although growing the yeast on complex media will increase the astaxanthin content of cells, mutagenesis of *Phaffia rhodozyma* can yield cultures with much higher astaxanthin levels. Mutants consistently producing concentrations of astaxanthin concentrations ten times that of the parent have been produced as a result of *N*-methyl-*N*-nitro-*N*-nitrosoguanine (NTG) treatment. When compared physiologically with the parent this mutant grew more slowly when either ammonia, glutamate or glutamine were the nitrogen source, and also had lower cell yields on several carbon sources. Nevertheless, these mutants remain attractive sources of astaxanthin for the aquaculture industry.

Animal pigmentation experiments with *Phaffia rhodozyma* indicate that high levels of astaxanthin deposition have been achieved in rainbow trout, salmon, lobsters and chicken egg-yolks. However, adequate preparation of the yeast before being incorporated into the feed is of paramount importance. As discussed previously the yeast cell wall is not digested by salmonids which would prevent extraction of the pigment with common solvents. Astaxanthin is best incorporated into the muscle of salmonid or into egg-yolks of poultry after the yeast cells are mechanically disrupted or after digestion of the cell wall with hydrolytic enzymes (for example, co-culture with *Bacillus circulans*). However, intact freeze-died cells or cells treated with cell wall-degrading enzymes do not pigment the flesh of salmonids when incorporated into their diet. Pigmentation is significantly better with mechanically broken or milled yeast than with a carotenoid extract of a culture. This difference is likely due to higher palatability or feed efficiency since fish fed unruptured yeast have a significantly higher weight gain and lower flesh moisture content than fish fed carotenoid extract. Recently (Lyons *et al.*, 1983) excellent pigmentation of trout was achieved when feeding intact spray-dried red yeast cells. This latest development is of major significance with regard to the development of a commercial feed adjunct for aquaculture.

The influence of yeast on the microbial populations of ruminants and non-ruminants

RUMINANTS

The use of yeast culture in the feed industry is now widespread due to the numerous demonstrations of its efficacy. There is an increasing body of evidence that supports the use of yeast culture supplements in dairy cattle, beef cattle, sheep, goats, swine, horses, poultry and dogs. The production responses associated with yeast culture supplementation vary with the animal species, the diet and the type of yeast supplement.

The use of yeast culture probably originates from the observation of some pig farmers in Germany 60 years ago that dry yeast added to cooked waste products being fed to pigs improved performance. The practice of using yeast culture is widespread today. The question of how yeast culture works to improve livestock performance has attracted the attention of a number of researchers in recent years and reasonable understanding of mode of action has emerged from work with ruminants. Two important aspects of rumen fermentation, common to all systems, are the requirements for efficent cellulolysis and for optimum microbial protein synthesis. With breakdown of cellulose, the animal receives energy in the form of volatile fatty acids (VFA), while protein in the form of microbial biomass is passed further down the gastrointestinal tract to supply needed amino acids. The key to efficient breakdown of cellulose by ruminants is pregastric fermentation in the rumen. Fermentation patterns in the rumen change with such factors as hay versus grain and the various types of forage and grain. The starch found in grains is an energy-rich material which promotes rapid fermentation, the production of organic acids and a lower pH in the rumen. In contrast, the cellulose and other structural carbohydrates in forage are fermented more slowly and with less potential for marked changes in the pH value of the rumen environment. Differences in feeding patterns, for instance whether forages or grains are blended to form a total mixed ration or are fed separately, also affect rumen fermentation patterns. Differences in how yeast culture affects ruminal fermentation among various diets and feeding systems have been important in understanding its ability to improve milk or meat production from animal feed sources.

Chemical means of modifying rumen fermentation are used extensively to improve efficiency of rumen fermentation. Addition of ionophores and antibiotics to ruminant feeds has been demonstrated to influence patterns of fermentation, to affect postruminal nutrient absorption and even gut and tissue metabolism. The notion of adding a microorganism to the rumen which might not otherwise be present has only recently been accepted. Certain yeasts and aerobic fungi are normal inhabitants of the rumen, but most species isolated are considered to be transient and non-functional. They usually enter the rumen via the feed. Anaerobic fungi, saprophytic on animal digestion, have only recently been observed in the rumen with the isolation of anaerobic fungi from bovine rumen contents.

The beneficial responses obtained from the addition of fungal microorganisms to the rumen environment have centred on the activity of *Saccharomyces cerevisiae* when used as a dietary supplement to the diets of cattle or sheep. The key to the ruminal effects of live yeast culture appears to be its ability to remain metabolically active in the rumen. However, experiments reporting on reproductively active yeast viable under rumen conditions are contradictory. It has been concluded that *Saccharomyces cerevisiae* was unable to maintain a productive population within the rumen ecosystem, and that centrifuged and autoclaved rumen fluid was not an optimum medium for yeast growth. Furthermore, at rumen temperature (39°C), yeast viability was substantially reduced.

Both rate and extent of feed digestion are important in the ability of ruminants to convert feed to meat or milk. Optimum efficiency of feed conversion requires that the digestive process extracts the maximum amount of nutrients from a given amount of feed. In the rumen of a high-producing dairy cow a compromise must be struck between digestion rate and the digestion extent owing to the slow rate at which fibrous feeds ferment. In order to permit space for additional feed to be consumed, residues from previous meals must be removed from the rumen. Though a longer period of exposure to fibre-depending organisms in the rumen might maximize the extent of fibre degrading, removal of partially digested residues will allow maximum intake. While total digestibility (as a percentage of feed consumed) is slightly depressed, the dairy cow has a greater quanitity of nutrients available due to a higher feed intake. Offer (1990) demonstrated how live yeast culture influenced the rate of digestion and could therefore help increased feed intake in a series of digestion rate measurements using sheep fed diets ranging in concentrate:forage ratio from 6:4 to 1:9. In these experiments a known quantity of feed dry matter was suspended in Terylene bags in the rumens of cannulated sheep. The disappearance of dry matter after 24 hours was measured and disappearance rates were calculated. It is concluded that while there was no consistent effect on the digestibilities of organic matter, gross energy or neutrual detergent fibre, the disappearance of the organic matter of hay from Terylene bags after incubation for 24 hours in the rumen was significantly increased in response to viable yeast culture. In subsequent experiments, the addition of a live yeast culture to diets consistently increased the 24-hour disappearance of hay but had no effect on disappearance at 48 hours. A highly relevant feature of the results was that a 4 to 5 day period was required from the beginning of yeast culture supplementation before maximal improvement in degradability was obtained, and there was a 4 to 5 day carry-over from the withdrawal of yeast culture before the degradation rates returned to pretreatment values.

Virtually all of the work on the effects of yeast cultures in the rumen have involved the measurement of concentrations and proportions of VFAs as indicators of microbial activity. Williams and Newbold (1990) found that live yeast cultures significantly affected the ratio of acetate to propionate in the rumen fluid of steers. The efficiency with which rumen microbes ferment ration nutrients is also related to methane production. It was found that when the diet was supplemented with yeast culture methane production was depressed. Methane production can represent up to a 12% loss in the energy content of the diet. The small changes noted after supplementation with microbial cultures are unlikely to represent a major energy savings, however they confirm changes in fermentation stoichiometry and indicate a shift from acetate production, which yields methane as a by-product, towards propionate production. Inconsistencies in the ability of yeast cultures to alter fermentation patterns in the rumen in a predictable manner may be related to differences in dietary treatments.

A model which could explain the action of yeast culture in the rumen has been developed from observations made with cultures of animal bacteria and in animals fed various yeast-supplemented diets (Figure 1). The model is based on the ability of certain strains of *Saccharomyces* to stimulate the growth and activities of specific groups of gastrointestinal bacteria. While the exact nature of the stimulatory activities are not completely defined, the model explains many of the *in vivo* and *in vitro* activities of yeast culture supplements. The stimulation of lactic acid-utilizing bacteria could account for the observed yeast-induced action and decreases in ruminal lactic acid concentration and moderated ruminal pH value in animals receiving diets containing high concentrations of readily fermented carbohydrates. Similarly, stimulation of certain groups of cellulolytic bacteria could account for enhanced animal digestion, increased numbers of cellulolytic bacteria and increased microbial protein synthesis.

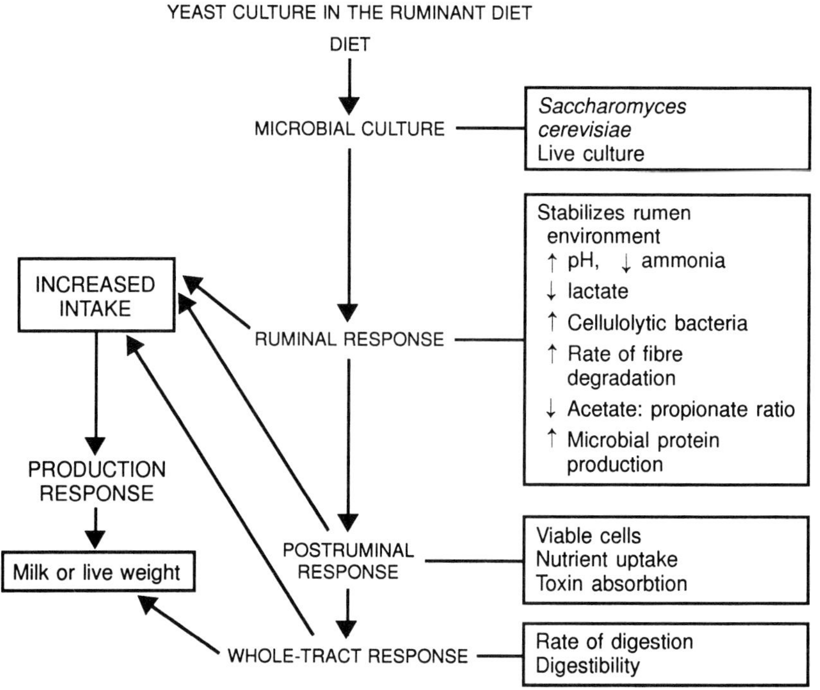

Figure 1. Summary of the action of a yeast culture in the rumen (Lyons *et al.*, 1993).

YEAST CULTURE AND NON-RUMINANTS

Production responses related to improved feed nutrient availability have been reported in a number of other domestic livestock species, including horses, rabbits, pigs and poultry. Although less is known about the mode of action of yeast culture in single-stomached animals, it is possible that a stimulatory effect on hindgut fermentation, similar to the ruminal response, contributes to the digestibility response. Production responses to live yeast culture have been demonstrated for pigs and poultry. It has been shown (Wenk, 1990) that yeast culture supplements in pig feeds have a significant beneficial impact on digestion of neutral detergent fibre and absorption of energy from the supplied nutrients. Pigs appear to benefit through a stimulation of appetite together with an increase in the efficiency of feed conversion. The result is a faster rate of gain at a lower cost per unit.

Models describing the effects of yeast on animal production are currently based on the ability of yeast strains to stimulate the growth and activities of gastrointestinal bacteria, but this stimulatory characteristic may not be common to all strains of yeast. Several studies have demonstrated differences in the ability of various yeast strains to stimulate the activities of specific groups of ruminal bacteria under standard conditions. Recent studies suggest that it may be possible to define specific applications for individual yeast strains. Yeast culture supplements designed specifically to stimulate the growth of lactic acid-utilizing bacteria may be especially useful for beef and dairy cattle fed high-concentrate rations. Alternatively, specific strains which stimulate cellulolytic bacteria may be more beneficial for animals receiving poor-quality forages or hay-based diets.

The role of some yeasts as modifiers of the intestinal microflora of animals and to stimulate the immune system

Control of intestinal pathogens has long been a goal of both livestock producer and microbiologist (Lyons, 1994). Methods used to achieve this goal include addition of antibiotics to feed, acidification of drinking water, competitive exclusion of pathogens using probiotics or chemical probiosis with selected sugars, and antimicrobial levels of minerals such as zinc oxide and copper sulphate.

More recently complex carbohydrates based on fructose, mannose and galactose have been investigated for their impact on gut microbiology with the goal of lessening the effect of pathogen challenge to the young animal or bird. While the response of these complex sugars by the animal is not entirely understood, three possible modes of action have been proposed:

- stimulation of immune response
- blocking colonization of pathogens
- provision of nutrients that cannot be used by pathogens

Carbohydrates occur in nature in a remarkable number of forms. It is well documented how carbohydrates act as metabolic sources of energy or as structural elements involved in maintaining the morphological features of cells and organs. Only in recent years has it emerged that carbohydrates also possess additional biological functions (Parekh, 1993). These functions include: (1) adhesion of viruses to microbes or target cells, and (2) modulating the shape of proteins specific to their functional and immunological activity.

Carbohydrates physically dominate the surfaces of eukaryotic cells and the whole extracellular matrix. Virtually all cell surfaces are glycosylated with attached carbohydrates being the dominant structural components of the resulting glycoproteins. Yeast cell wall oligosaccharides have been found to stimulate immunity and modify gut microflora. Interest in this application centres on the cell wall with glucan, mannan and chitin being the wall's main components. Yeast cell wall material is remarkably stable to acid digestion and fractions are known to survive passage through the stomach to the lower gut. It is this ability to pass through acid digestion undisturbed that may account for this material's biological activity in such a wide range of species. For example, feeding of a mannan-free cell wall fraction (i.e., a glucan-protein complex) in a salmon feed for 6, 8 or 12 weeks resulted in mortalities being reduced by 28%. Both response to the challenge model and reduced mortality suggest enhanced immunity. Studies with trout fry in commercial testing ponds also point towards enhanced immunity in response to glucomannan in the diet.

The uniqueness of mannan as a possible immunomodulator also resides in the fact that each strain of yeast has a characteristic mannan. The immunodominant side chains of mannan consist of four types: mannotetraose, mannotriose, mannobiose and mannose. These differences can now be employed to develop a range of mannanoligosaccharides to suit different feeds and to bring about different responses. A study of the ability of mannanoligosaccharides from a strain of *Saccharomyces cerevisiae* to stimulate immune response examined both the effects of the mannan on phagocytic activity *in vitro* and the response of broiler chickens challenged with *Salmonella*. Significant increases in phagocytic activity were observed when mannanoligosaccharides were incubated in peripheral blood from 3-month-old male Wistar rats. The mannan clearly exerted an immune response.

This immune response could also assist chicks in surviving a *Salmonella* challenge. It was observed that birds fed the mannanoligosaccharide were better able to withstand the challenge. In addition, it has been found that insoluble fractions of yeast act to stimulate phytoalexins (disease resistance mechanisms) against mildew in agricultural crops (Newton *et al.*, 1993). Mannan fractions were sprayed onto barley leaves 24 hours before inoculation with mildew. The control leaves developed the mild infection but the treated leaves did not. Additionally, the rates of papilla formation and phenylalanine ammonia lyase activity were increased in treated leaves. In field trials the mannanoligosaccharides reduced mildrew infection and increased the yield of both winter and spring barley.

The use of *Saccharomyces cerevisiae* to counteract aflatoxicosis in broiler chickens and ducklings

Aflatoxins are toxic metabolites produced by the storage fungi *Aspergillus flavus* and *Aspergillus parasiticus* which are frequently encountered in a variety of foods and feedstuffs. Contamination of corn and other grains with significant levels of aflatoxin continues to be a major problem in many parts of the world and is generally more severe in developing countries than in developed countries due to tropical and subtropical climatic conditions coupled to poor handling during harvest and post-harvest leading to large economic losses.

Numerous strategies for the detoxification of alfatoxins in feeds and feedstuffs have been tried including chemical methods, physical separation, indicator and thermal inactivation. However, practical and cost effective methods of detoxification beyond ammoniation or addition of hydrated sodium calcium aluminosilicate or sodium aluminosilicate clays to the diet are not available (Barmase and Devegowda, 1990). Rapid developments in biotechnology have presented an avenue for detoxification (Devegowda *et al.*, 1994). A trial with inclusion of *Saccharomyces cerevisiae* in broiler diets containing 5 ppm aflatoxin resulted in improvement in body weight and feed efficiency. Recent work has studied the effects of aflatoxins on body weight, mortality and feed efficiency in broiler chickens and ducklings. Addition of 0.1% yeast culture reduced the adverse effects of the toxin (Stanley *et al.*, 1993; Devegowda *et al.*, 1994). It has been suggested that yeast cells can influence the health of animals by adsorbing toxins and pathogenic bacteria onto the cell wall or biodegradation of the toxin.

Conclusions

The traditional uses of yeast in the production of alcoholic beverages, bread, carbon dioxide and single cell protein are well documented (Berry *et al.*, 1987). Over the past decade the uses of yeast have been expanded to include a number of diverse but specialized uses in animal nutrition and fish culture. Use of hydrolysed *Phaffia rhodozyma* added to the diet of salmonids offers a means of providing the carotenoid pigment astaxanthin which can be employed to colour the flesh of farm-raised fish. Certain live yeast culture strains stabilize the rumen environment of livestock. The pH value of rumen contents is not depressed to the same extent in the presence of live yeast culture when feed concentrates are given. Many of the responses to yeast culture in the rumen environment are mediated through effects on the cellulolytic bacteria. The number of such bacteria is increased in the presence of live yeast culture, as is the microbial protein supply to the host animal. This will result in an increased rate of cellulose degradation which will stimulate intake resulting in a greater nutrient supply for production of meat or milk by ruminants. Non-ruminants also benefit from improved availability of nutrient in the feed when live yeast culture is added to the diet.

However, the mechanism is not understood. Fibre digestion and mineral availability are improved in horses, while pigs and poultry benefit from similar digestion improvements in addition to enhanced feed conversion efficiency.

The absorbtive properties of the yeast cell wall are being employed, particularly the mannan component, to modify the gut microflora and modulate the immune response. The cell wall has powerful antigenic stimulating properties, and it is well-established that this property is a characteristic of the mannan chain. The interaction between oligosaccharides and the immune system is only just beginning to be understood. It is clear that mannan has the ability to bind certain bacteria and prevent these from colonizing the gastrointestinal tract. In addition, mannans and glucans elicit an immunomodulating effect. The immunological benefit is not species specific because of the diversity of species that demonstrate a benefit from supplementation with mannanoligosaccarides. The interrelationships betweeen nutrition and immunity are far from understood!

References

Barmase, B.S. and G. Devegowda. 1990. Reversal of aflatoxicosis through dietary adsorbents in broiler chicks. Proc. 13th Annual Poultry Science Conference and Symposium, Bombay, India.

Berry, P.R., I. Russell and G.G. Stewart. 1987. Yeast Biotechnology. Allan and Unwin, London.

Devegowda, G., B.I.R. Aravind, K. Rajendra, M.G. Morton, A. Baburathna and C. Sundarshan. 1994. A biological approach to counteract aflatoxicosis in broiler chickens and ducklings by the use of *Saccharomyces cerevisiae* cultures added to feed. In: Biotcchnology in the Feed Industry. T.P. Lyons and K.A. Jacques (Eds) Nottingham University Press, Nottingham, 235–245.

Hammond, J.R.M. 1993 Brewer's yeasts. In: The Yeasts Volume 5. A.H. Rose and J.S. Harrison (Eds) Academic Presss, London, 7–67.

Harder, W. and M. Veenhuis. 1989. Hydrocarbon utilising yeasts. In: The Yeasts, Volume 3. A.H. Rose and J.S. Harrison (Eds) Academic Press, London, 289–316.

Harrison, J.S. 1993. Food and fodder yeasts. In: The Yeasts, Volume 5. A.H. Rose and J.S. Harrison (Eds) Academic Press, London, 399–434.

Kosaric, N., D.C.M. Ng, I. Russell and G.G. Stewart. 1980. Ethanol for fuel. Advances in Applied Microbiology 26: 148–182.

Lyons, T.P. 1994. Biotechnology in the feed industry: 1994 and beyond. In: Biotechnology in the Feed Industry. T. P. Lyons and K. A. Jacques (Eds) Nottingham University Press, Nottingham, 1–48.

Lyons T.P., K.A. Jacques and K.A. Dawson. 1993. Miscellaneous products from yeast. In: The Yeasts, Volume 5. A.H. Rose and J.S. Harrison (Eds), Academic Press, London, 293–324.

Newton, P.C., G.D. Lyons and T. Reglingski. 1993. Development of a new crop protection system using yeast extracts. Home Grown Cereals Association Project Report No. 1978. Great Britain.

Offer, N.W. 1990. Maximising fibre digestion in the rumen: the role of yeast culture. In: Biotechnology in the Feed Industry. T. P. Lyons (Ed.) Alltech, Nicholasville, Ky, 66–72.

Parekh, R. 1993. Carbohydrate engineering in modern drug discovery. In: The Biotechnology Report. Campden Publishing Ltd., London, 135.

Reed, G. and T.W. Nagodawithana. 1991. Yeast Technology. 2nd edition. Var Nostrand, New York.

Rose, A.H. and J.S. Harrison. 1993. Introduction. In: The Yeasts, Volume 5. A.H. Rose and J.S. Harrison (Eds) Academic Press, London, 1–6.

Rose, A.H. and G. Vijayahkeshima. 1993. Baker's yeast. In: The Yeasts, Volume 5. A.H. Rose and J.S. Harrison (Eds) Academic Press, London, 357–397.

Stanley, U.G., R. Ojo, S. Woldesenbat and D.H. Hutchinson. 1993. The use of *Saccharomyces cerevisiae* to suppress the effects of aflatoxicosis in broiler chicks. Poultry Science 72:1867–1872.

Watson, D.C. 1993. Yeasts in distilled alcoholic-beverage production. In: The Yeasts, Volume 5. A.H. Rose and J.S. Harrison (Eds) Academic Press, London, 215–244.

Wenk, C. 1990. Yeast cultures, lactobacilli and a mixture of enzymes in diets for growing pigs and chickens under Swiss conditions: Influence on utilization of energy. In: Biotechnology in the Feed Industry. T. P. Lyons (Ed.). Alltech, Nicholasville, Ky, 520–528.

Williams, P.E.V. and P.J. Newbold. 1990. The effects of novel micro-organisms on rumen fermentation and rummant productivity. In: Recent Advances in Animal Nutrition. 1990. D. Cole and T.S. Haresigh (Eds). Butterworth, London, 211–216.

IMMUNOSTIMULANTS: MAXIMIZING THE HEALTH AND EFFICIENCY OF ANIMALS THROUGH PLANT-DERIVED BIOMOLECULES

P.J. HYLANDS and A.A. POULEV

Phytopharmaceuticals Inc., Belmont, California, USA

Introduction

Research on immunomodulating drugs is a new field in pharmacology which has been developed only in about the last 20 years. It is providing both theoretical significance and therapeutic value in many diseases, including cancer, viral infections, autoimmune diseases, organ transplantation and ageing, conditions which are caused by deficiency or imbalance of the immune functions of patients. Development of a safe and effective immunomodulator for clinical use has become a major goal of many pharmaceutical investigators. Animal experiments and clinical trials have shown that a number of plants are immunologically active.

Up till now, study on plant medicines that have been used for centuries as general tonics has been particularly successful. Much of this effort has come from work on Chinese medicines, where prophylactic treatment with 'health-promoting' plants and extracts has a long history of use. Table 1 shows a list of plants (principally from the ancient Chinese pharmacopoeia) with immunological activity.

The immune system comprises a complex network of cells including T-cells, B-cells, macrophages and natural killer cells. Many kinds of cytokines, immunoglobulins, interferons and complement molecules play pivotal roles in up and down regulation of immune functions.

Pharmacological and chemical studies to be discussed later have led to the identification of a range of chemical compounds having demonstrable immunological effects. Lead structures discovered so far are both low and high molecular weight compounds.

The most prominent members of the low molecular weight compounds reported by Wagner (1993) include some alkaloids, quinones, isobutylamides, phenol carboxylic acid esters and terpenoids. Close structure-activity relationships have not been observed. However, it seems that many of these compounds are also known as potent cytotoxic agents in the treatment of cancer, bacterial or viral diseases if used at relatively high concentrations. The same compounds, however, may act as immunostimulators when administered in low doses (Wagner *et al.*, 1988). The phenomenon that a drug shows reversal effects depending on the

Table 1. Some plants and fungi with immunological activity (mainly after Xiao *et al.*, 1993).

Acanthopanax senticosus	*Ganoderma lucidum*
Achyranthes bitentata	*Glycyrrhiza uralensis*
*Achyrocline satureioides**	
Aconitum carmichaeli	*Lentinus edodes*
Actinidia chinensis	*Liquisticum wallichii*‡
Allium sativum	*Ligustrum lucidum*
Angelica sinensis	*Lilium brownii* var. *viridulum*
Arnica montana†	*Lycium barbarum*
Artemisia annua	
Asparagus cochinchinensis	*Morinda officinalis*
Astragalus sp.	*Morus alba*
Atractylodes macrocephala	
Auricularia auricula	*Ophiopogon japonicus*
Baptisia tinctoria†	*Paeonia lactiflora*
Bupleurum chinense	*Panax* sp.
	Phellodendron chinense
Camelia sinensis	*Phytolacca acinosa*
Cinnamomum cassia‡	*Polygonatum sibiricum*
Cistanche deserticola	*Polyporus umbellatus*
Codonopsis pilosula	*Poria cocos*
Cordyceps sinensis	*Psoralea corylifolia*
Coriolus versicolor	
Cornus officinalis	*Rehmannia glutinosa*
Curculigo orchioides	*Rheum palmatum*
Cuscuta chinensis	
Cynanchum auriculatum	*Sophora japonica*
	Spatholobus suberectus
Dendrobium nobile	
Dioscorea opposita	*Taraxacum mongolicum*
	Tremella fuciformis
Echinacea sp.	*Tribulus terrestris*
Eclipta prostrata	*Tripterygium wilfordii*
Epimedium sagittatum	*Typha angustifolia*
Eupatorium perfoliatum†	
	Viscum sp.
Fagopyrum botrys	
Ficus pumila	*Zizyphus jujuba*

* – from Puhlmann *et al.* (1992); † – from Wagner and Jurcic (1991a); ‡ – from Zee-Cheng (1992).

dosage is not absolutely new, but has never been practically utilized in therapy. Typical reversal effects for some quinones like plumbagin (Figure 4), and alkaloids such as vincristine and bryostatin (Figure 4) from *Budula neritina* have been found by Wagner (1993). In the light of these new findings an assumption has been proposed by the same author that many cancer drugs of plant origin like 'lapacho' from *Tabebuia avellanedae* probably exert their antitumor activities at least partially by an immune-induced mechanism of action.

High molecular weight substances like polysaccharides and glycoproteins have been reported to enhance the non-specific immune system by activating the phagocytic activity of granulocytes and macrophages, or by inducing cytokine production or influencing complement factors. It is almost impossible to determine clear structure-activity relationships and to indicate which structural features are essential for an optimal immunostimulating activity. Regarding compounds showing activity on phagocytosis, many glucuronic acid-containing arabinogalactans and xylans have been detected (Wagner, 1993). Most of them are highly branched with anionic structural units and molecular weights in the range of 20 000 to 100 000 and higher. They derive from primary cell walls and possess pectic or protopectic properties and some are viscous gums or mucilages. An acidic rhamnoarabinogalactan isolated by Wagner *et al.* (1988) from cell suspension cultures from *Echinacea purpurea* (Figure 3), which has demonstrated a series of remarkable immunostimulating activities (Lüttig *et al.*, 1989; Roesler *et al.*, 1991a; Roesler *et al.*, 1991b) can be regarded as a prototype of these polysaccharides (Wagner, 1993).

Among the most studied medicinal plants with influence on the mammalian immune system are ginseng (*Panax* sp.) and *Echinacea* sp., and these two merit special attention in this chapter.

Pharmacological activities of ginseng

The drug ginseng has been derived from various species of the genus *Panax* (*P. notoginseng*, *P. quinquefolium*, *P. ginseng*, *P. pseudo-ginseng*, *P. zingiberensis*, *P. japonicus*, *P. japonicus* var. *angustifolius*, *P. japonicus* var. *major*, *P. japonicus* var. *bipinnatifidus*, *P. stipuleanatus*, *P. vietnamensis*) grown in different regions in Central, East and South-East Asia and North America (Fujimoto *et al.*, 1991; Liu and Xiao, 1992; Duc *et al.*, 1993). It has been a famous Chinese traditional medicine for centuries. Some of the most explored pharmacological functions of ginseng are depicted in Table 2.

Table 2. Clinical and pharmacological activities of ginseng, after Liu and Xiao (1992).

Anti-stress activity
Anti-circulatory shock effects and modulation of cardiovascular activities
Improvement or facilitation of learning and memory processes
Modulation of neuro-endocrine system activities, hypothalamic-adrenal-gonadal system
Modulation of cellular metabolic processes on carbohydrate, fat and protein metabolism
Promotion of haematopoiesis
Modulation of immune functions and activities
Protection against radiation and liver toxicities

A mixture of saponins isolated from *P. notoginseng* caused a marked increase in serum complement and haemolysin levels, especially in immunodefficient animal models, such as in experimental allergic encephalomyelitis, which is a T-lymphocyte-mediated autoimmune disease in rats and guinea pigs (Xiao-Yu, 1991). Polysaccharide fractions of the same species stimulated T-lymphocyte proliferation both *in vitro* and *in vivo* and antagonized the action of the T-cell suppressor cyclosporin A.

It also promoted antibody formation and interleukin production with at the same time remarkably low toxicity.

Early studies revealed that ginseng possessed biomodulatory effects on the higher centers of the central nervous system, facilitating both physical and mental activities. It has a noteworthy effect on the endocrine system in regulating the blood sugar level as demonstrated in alloxan diabetes. Recent experimental and clinical studies performed by Liu and Xiao (1992) have concluded that it has a wide range of effects, such as an anti-shock effect in circulatory failure, modulatory effects on the immune functions, modulation of neuro-endocrine system activities, improvement of learning and memory processes and others.

An acute hypoxia due to negative air pressure was induced in rodents in order to study the effect of ginseng against hypoxia stress and analyse the mechanism of action (Liu and Xiao, 1992). The results showed that ginsenosides isolated from the root, stems and leaves of *P. ginseng* significantly increased the survival in mice and significantly delayed the time of survival of cortex electroencephalogram in non-anaesthesized rats in acute hypoxia. The anti-hypoxia effect is probably related to the effect of improving or raising cerebral resistance to hypoxia and reducing cerebral consumption of oxygen in acute hypoxia. When mice were exposed to a hypobaric environment for 5 min, mitochondria of their heart and brain cells were seriously damaged and brain neurotransmitters noradrenaline, dopamine, 5-hydroxy-tryptamine (serotonin) and acetylcholine were significantly decreased, which were preserved and diminished by ginseng root saponin applied 10 mg/kg i.p. Ginseng saponins increased the level of serum corticosterone and prolonged hypobaric survival time of mice, but it had no effect on adrenalectomized mice. This indicates that the protective action on hypobaric hypoxic mice may be related to the action of the saponins on pituitary-adrenal gland (Liu and Xiao, 1992). Ginseng also alleviated rectal temperature precipitation fall in mice subjected to hypoxia and did not change the brain acetylcholine concentration.

Another aspect of the anti-stress activity of ginseng is its anti-cold and anti-heat stress properties. When mice were put in –2°C environment for 1 h, the rectal temperature fell significantly, while 50 or 100 mg/kg i.g. of ginseng root saponin caused only a slow fall in rectal temperature in mice at the same conditions. The brain noradrenaline, serotonin and 5-OH-indoleacetic acid concentrations were decreased in mice in a –4°C environment for 1 h, and increased after 100 mg/kg i.g. ginseng root saponin. After treatment with 70 mg/kg i.p. ginseng root saponin, the rectal temperature of rats put at –2°C for 1 h showed no change, the brain acetylcholine and plasma corticosterone increased apparently, but no change in brain g-aminobutyric acid, glutamine and asparagine could be detected (Liu and Xiao, 1992).

After exposure of mice to 45°C for 15 min, the core body temperature and serum corticosterone increased, brain serotonin and noradrenaline decreased and brain dopamine was unchanged. 100 and 200 mg/kg i.g. ginseng root saponins attenuated the increase in body temperature and corticosterone and the decrease in brain serotonin and noradrenaline, but did not alter brain dopamine in stressed mice. Ginseng decreased body temperature in unstressed mice as well, i.p. injection 25 mg/kg of reserpine abolished the saponins' effects on lowering body temperature in both stressed and unstressed mice.

Widespread immunosuppression can be induced by several kind of stresses, including surgical stress (Yang and Yu, 1990). There are experimental results from a study from the same authors suggesting that ginseng can antagonize the suppression of the natural killer cell activity, and can completely antagonize surgically-induced inhibition of phagocytosis by peritoneal macrophages in mice.

Another activity of ginseng concerns the effect on neurotransmitters in pithed animals. It was found by Liu and Xiao (1992) that 30 mg/kg i.v. ginseng root saponin did not affect the pressor response of exogenous noradrenaline in pithed rats. However, ginseng significantly attenuated the pressor action of noradrenaline released by electric stimulation on spinal T7–13. Since yohimbine is a selective prejunctional a2-receptor blocker, it inhibited the praesynaptic negative feedback on noradrenaline release, 50 mg/kg i.v. yohimbine augmented the pressor response of spinal electric stimulation, and ginseng root saponins blunted significantly the augmentation. It is proposed that ginseng root saponin serves as a praesynaptic a2–receptor agonist, and its hypotensive effect is attributed to the reduction of transmitter release of sympathetic nerves. Its hypotensive effect is explained as a result of less selective action on postsynaptic a1- and possibly a2-receptors localized in vascular muscles. Other *in vivo* experiments in a study by the same authors with ginseng root saponin suggest its selective modulating activity on the circadian variations of brain serotonin, 5-OH-indoleacetic acid, noradrenaline and dopamine as a function of the duration of the light-dark cycle. It has been also found that ginsenosides influence the circadian variation on plasma corticosterone and liver glycogen in rats, which led to the conclusion that the circadian rhythm may be an important factor in pharmacological response to ginseng saponins.

Earlier studies (1920–1950) have demonstrated that ginseng possesses a wide range of cardiovascular pharmacological activities including effects on heart, heart rate, blood pressure and vasculature, which has also been confirmed by a series of recent studies (Liu and Xiao, 1992). Together with positive effects on cardiac performance and haemodynamics demonstrated by *in vivo* experimental models with dogs, ginsenosides possess also protective action against myocardial infarction. It has also been demonstrated by *in vitro* experiments with isolated arteries that ginsenosides show a prejunctional excitatory effect on sodium release and postjunctional inhibitory effect on histamine and noradrenaline response which involves interference with calcium influx process. Based on results from different experiments, it is generally agreed that the protective effect of ginsenosides on myocardial ischemia is related to the slowing of heart rate, reducing oxygen consumption and decreasing peripheral blood vascular resistance, improving of the myocardial metabolism by correcting sugar and fat metabolic processes, promoting prostaglandin release and inhibiting thromboxane A_2 production and reducing release and inhibiting calcium influx (Liu and Xiao, 1992). In conjunction with the intensive investigations in the past 20 years in the field of the prevention of myocardial infarction, stroke, postoperative deep thrombosis, vascular graft occlusion, peripheral arterial disease etc., the interest in drugs with inhibitory effects on platelet aggregation has enormously increased. Different *in vitro* and *in vivo* tests reported by Liu and Xiao (1992) have shown that ginsenosides inhibited rabbit platelet aggregation induced by adenosine diphosphate, arachidonic acid, collagen and thrombin. Despite the intensive investigations, the mechanism of action of ginsenosides is still not completely understood. Nevertheless, based on numerous observations, two mechanisms of action are most probable, including either the rise of platelet cyclic adenosine monophosphatase, or the inhibition of prostacyclin production.

Several diseases, such as sickle cell anaemia, diabetes mellitus and stroke are associated with impaired red blood cell deformability. Improvement of the deformability of the red blood cells is considered as the mechanism of action of drugs used in the therapy of occlusive vascular diseases. A purified extract fraction of *P. pseudo-ginseng* containing mainly a triacylglycerol was shown by Hong *et al.* (1993) to improve deformability of red blood cells.

One of the most remarkable pharmacological activities of ginseng is the modulatory effect of standardized mixtures of ginseng saponins on the mammalian immune system. It has been demonstrated by Liu and Xiao (1992) that these compounds markedly stimulated phagocytosis of the reticuloendothelial system, the production of antibodies, a substantial increase of the spleen weight of experimental animals and the clearance rate of carbon particles.

In numerous experiments the potentiating effect of ginseng on the phagocytic activity of the plaque forming cells has been demonstrated (Liu and Xiao, 1992; Yang and Yu, 1990). After either enteral or parenteral administration of ginseng and after harvesting of the plaque forming cells a significant increase of the phagocytosis, together with an increase of the lysozyme content was detected by Yang and Yu (1990). In other experiments with both normal and immunodepressed mice, ginseng was shown to significantly enhance mitogenesis of T- and B-lymphocytes primed by different mitogens. The mechanism of action has been related to the altered ratio of cGMP to cAMP in immunocompetent cells. It has been demonstrated by *in vitro* experiments that ginseng promotes the natural killer cells interferon-interleukin-2 regulatory system (Yang and Yu, 1990). After addition of ginseng into the medium of a culture of murine spleen cells, the activity of the natural killer cells was substantially enhanced, compared with control samples. Ginseng also enhances the production of interferon and interleukin-2. In an experiment with spleen cell cultures it was shown that at a dose of 10 mg/ml ginseng induces the release of both interferon and interleukin-2 in the presence of a mitogen with a 2– and 3–fold increase respectively compared with the controls. Another *in vivo* experiment demonstrated that on the eighth day after treatment with ginseng the natural killer cells activity and the production of interferon and interleukin-2 in tumor-transplanted mice were at higher levels than in controls. In a similar experiment with tumor-transplanted mice with different administration pattern of ginseng

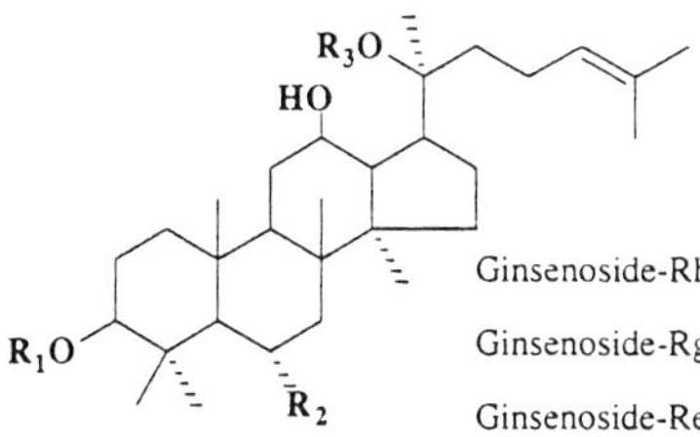

	R_1	R_2	R_3
Ginsenoside-Rh:	H	-O-glu	H
Ginsenoside-Rg:	H	-O-glu	-glu
Ginsenoside-Re:	H	-O-glu-rham	-glu
Ginsenoside-Rd:	-glu^2-glu	H	-glu
Ginsenoside-Rb_2:	-glu^2-glu	H	-glu^6-ara(p)
Ginsenoside-Rb_1:	-glu^2-glu	H	-glu^6-glu
Notoginsenoside-R1:	H	-O-glu^2-xyl;	-glu
Notoginsenoside-Fa:	-glu^2-glu^2-xyl	H	-glu^6-glu
Pseudo-ginsenoside-RS_1:	H	-O-glu^2-rham (^{6}acetate)	-glu

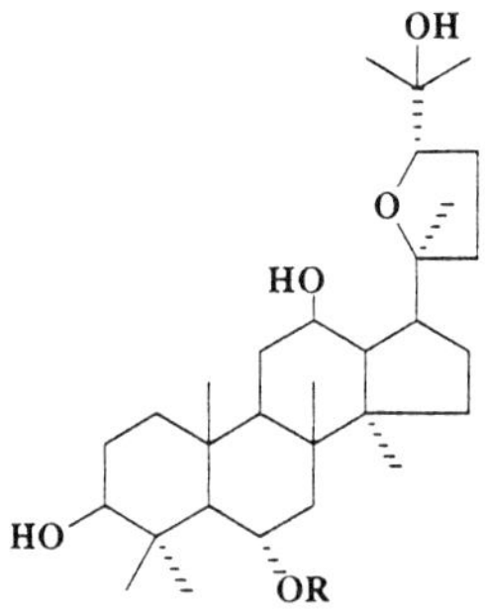

	R
Pseudo-ginsenoside-RT_4:	-glu
Pseudo-ginsenoside-F_{11}:	-glu^2-rham
Vina-ginsenoside-R_1:	-glu^2-rham (^{6}acetate)
Majonoside-R_2:	-glu^2-xyl
Majonoside-R_1:	-glu^2-glu
Majonoside-R_2-monoacetate:	-glu^2-xyl (^{6}acetate)

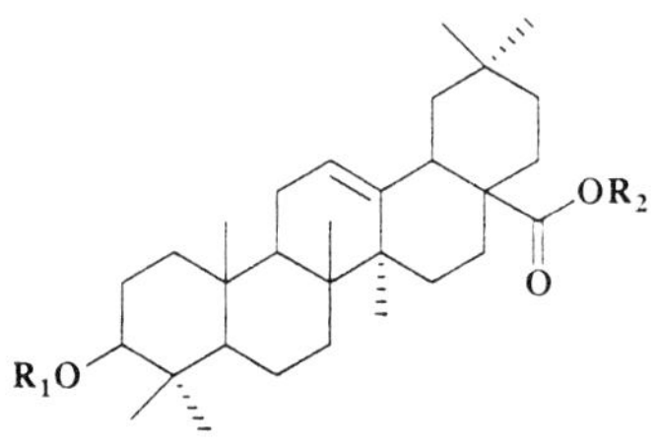

	R_1	R_2
Ginsenoside-Ro:	-gluUA^2-glu	-glu
Hemsloside-Ma_3:	-gluUA^2-glu (^{3}ara(p))	-glu

glu = β-D-glucopyranosyl
rham = α-L-rhamnopyranosyl
gluUA = β-D-glucuronopyranosyl
xyl = β-D-xylopyranosyl
ara(p) = α-L-arabinopyranosyl

Figure 1. Structures of some typical saponins isolated from ginseng (Duc *et al.*, 1993).

both the rate of tumor production and tumor weight were reduced, and that the natural killer cells activity was markedly increased as compared with non-treated mice. It has also been demonstrated by Yang and Yu (1990), that ginseng possesses 'two way' effects on the production of specific antibodies. After immunization with a low dose of an antigen, diphtheria toxoid, ginseng could conspicuously immunopotentiate the primary antibody response, whereas after immunizing with a high dose of the same antigen ginseng had only a slight immunopotentiating effect. This outstanding action pattern of ginseng is the so called 'immunomodulation', that is, it stimulates functions which had been curbed and restrains those which had been excited (Yang and Yu, 1990).

Based on the numerous valuable stimulating and therapeutical properties of the constituents of ginseng, extensive studies in the new field of plant biotechnology on the possibility of cultivating *Panax* plant cells *in vitro* have been made in the past few years. As a result of these efforts technologies for large scale cultivation in bioreactors have been

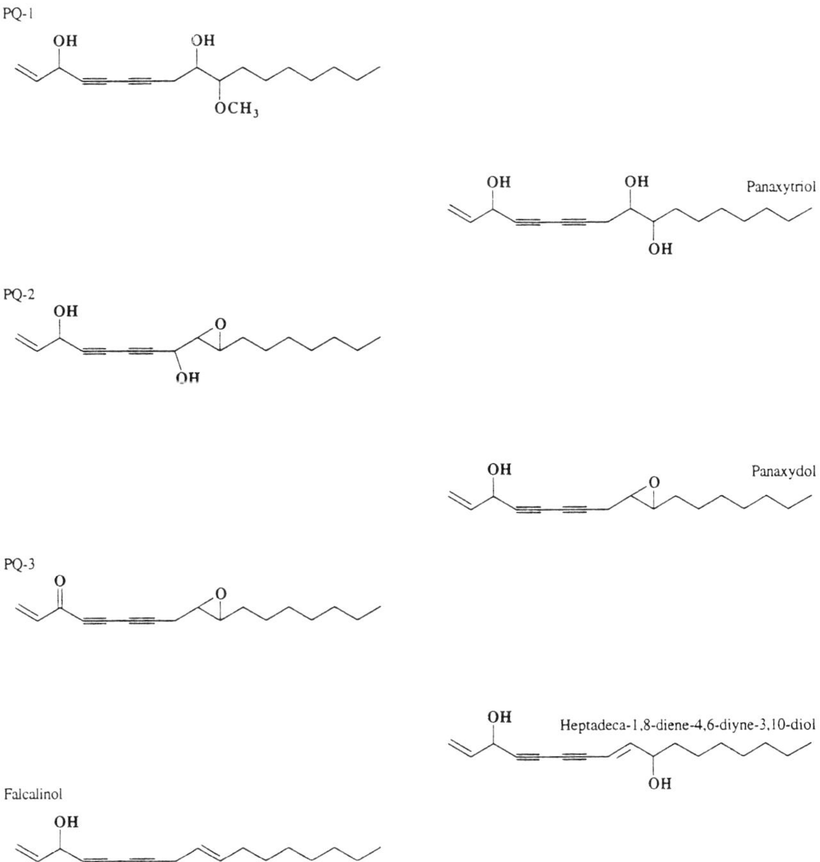

Figure 2. Structures of some unsaturated long chain alcohols from ginseng (Fujimoto *et al.*, 1991).

established with saponin contents similar to those of the intact plants (Zhou and Zheng, 1991; Asaka *et al.*, 1993).

The structures of the main active principles of *Panax* fall into two classes – triterpene glycosides (ginsenosides), and unsaturated long chain alcohols. Representative members are shown in Figures 1 and 2.

Pharmacological activities of *Echinacea*

The most studied species of the genus *Echinacea* for pharmacological activity are *E. purpurea*, *E. angustifolia* and *E. pallida*. The first documented experiments to prove the immunological activity of *Echinacea* extracts were made by Unruh in 1915 (Bauer *et al.*, 1988), who demonstrated that *E. angustifolia* stimulated leucocytes and suggested that it enhanced the intrinsic defense forces of the organism. Since that time, *Echinacea* is one of the most widely used medicinal plants. In the European Pharmacopoeia there are two species listed, *E. angustifolia* and *E. purpurea*, which differ from each other in morphology and chemical composition. Since the results from some of the earlier studies are to a certain degree somewhat contradictory as demonstrated by Schumacher and Friedberg (1991), which is mainly due to the fact that different authors have been using non-standardized *Echinacea* extracts, and that an exact taxonomical determination of the plant material under investigation had not always been made, numerous studies on the immunostimulating activity of *Echinacea* extracts have been reported in the literature in the past several years (Bauer *et al.*, 1988; Wagner *et al*, 1988; Lüttig *et al.*, 1989; Mengs *et al.*, 1991; Roesler *et al.*, 1991a; Roesler *et al.*, 1991b; Schumacher and Friedberg, 1991; Facino *et al.*, 1993; Lersch *et al.*, 1992; Steinmüller *et al.*, 1993).

In a study by Mengs *et al.* (1991) on the toxicity of *E. purpurea* extracts it was demonstrated that after 4 weeks of oral administration to rats in doses amounting to many times the human therapeutic dose, laboratory tests and necropsy findings gave no evidence of any toxic effects. Single oral or i.v. doses of the extract proved also virtually non-toxic to rats and mice. Tests for mutagenicity carried out with microorganisms and mammalian cells *in vitro* and *in vivo* with mice all gave negative results. In the same experiment, after *in vitro* carcinogenicity study *E. purpurea* did not produce malignant transformation in hamster embryo cells.

In another study by Bauer *et al.* (1998) a comparative investigation of immunological activities of extracts from *E. angustifolia*, *E. purpurea* and *E. pallida* was performed. Ethanol extracts from roots of all three species showed significant increase of phagocytosis determined by carbon clearance test compared with the controls. In this study, extracts from *E. purpurea* caused a 3–fold increase of the elimination rate, whereas *E. angustifolia* and *E. pallida* showed a 2–fold increase of the elimination rate, of carbon particles as compared with the controls. It was also demonstrated in the same study that depending on the type of extraction different substances prevailed in the corresponding extracts and hence different pharmacological activities were manifested by the same species.

Table 3. Some of the main types of compounds in roots from *Echinacea* sp. after Bauer *et al.* (1988).

Fraction	*E. purpurea*	*E. angustifolia*	*E. pallida*
Lipophilic	Isobutylamides	Isobutylamides	Polyines
Hydrophilic	Chicoric acid	Echinacoside, Cynarine	Echinacoside

Some of the main types of compounds detected by HPLC are summarized in Table 3 based on extraction with different solvents.

These authors demonstrated that the lipophilic fractions of the ethanolic extracts showed more intensive influence on the phagocytic activity compared with the hydrophilic ones, and therefore isobutylamides, polyines and polyenes have been suggested by the same authors as the main active components.

By means of fast atom bombardment and fast atom bombardment tandem mass spectrometry several caffeoyl conjugates from crude extracts from roots of *E. angustifolia* with antihyaluronidase activity in the ethyl acetate, butyl acetate and chloroform fractions have been detected and identified by Facino *et al.* (1993). Hyaluronidases are enzymes which depolymerize or hydrolyze hyaluronic acid and chondroitin sulfates, the constituents of the amorphous substance of connective tissue. These enzymes are thought to play a key role in the etiology and development of the inflammatory disease, since the destructuring of the extracellular matrix promotes the spreading of chemotactic factors of inflammation. Among the four main caffeoyl esters detected, chicoric acid (2,3-*O*-dicaffeoyltartaric acid) and caftaric acid (2-*O*-caffeoyltartaric acid) (Figure 3) were found to exhibit the greatest antihyaluronidase activity.

Wagner and Jurcic (1991a) tested in a comparative study the activity of phagocytosis by *in vitro* granulocyte test and by *in vivo* carbon-clearance test in mice after application of *E. angustifolia* extract alone and combined with extracts from *Arnica montana*, *Baptisia tinctoria* and *Eupatorium perfoliatum*. In both immune test models, a step by step stimulation of the activity of phagocytosis by the addition of the four plant extracts was demonstrated with an increase in effectiveness of partially over 50%, compared with the pure *E. angustifolia* mono-extract. This extract combination showed also a higher efficiency than two differently composed combined preparations and two *Echinacea* mono-extracts. As possible active principles regarding the stimulation of phagocytosis, acidic amides and phenolic carbon acid esters like chicoric acid have been suggested. It has been found in earlier studies (Wagner and Jurcic, 1991a) that sesquiterpene lactones from *Eupatorium* and *Arnica* also possess immunomodulating activity.

A study by Lersch *et al.* (1992) on outpatients with metastasizing far advanced colorectal cancers receiving combined immunotherapy with low-dose cyclophosphamide, thymostimulin and 'echinacin' (extracts of *E. purpurea*) showed a partial tumor regression in one and a stable disease in six other patients, together with a decrease of the tumor marking carcinoembryonic antigens, and an increase of the survival

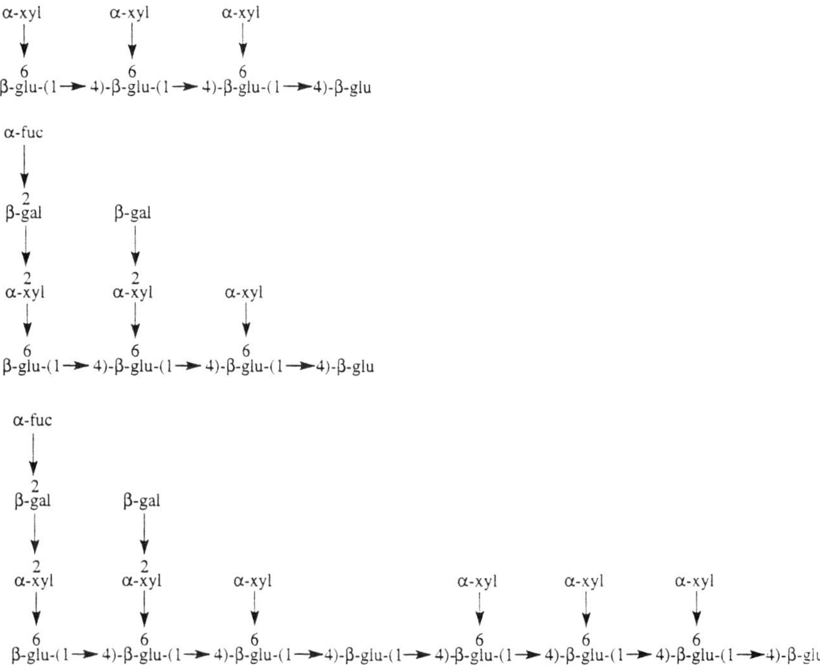

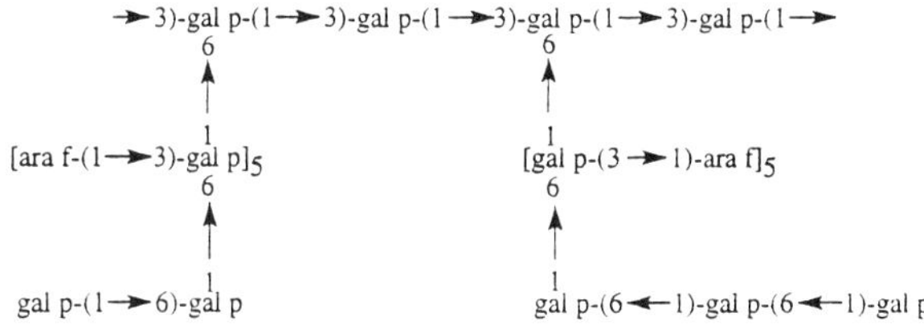

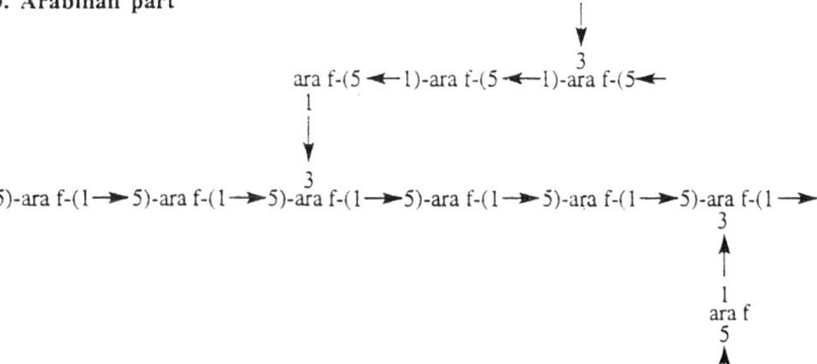

Figure 3. Representative structures of compounds identified from *Echinacea* extracts.

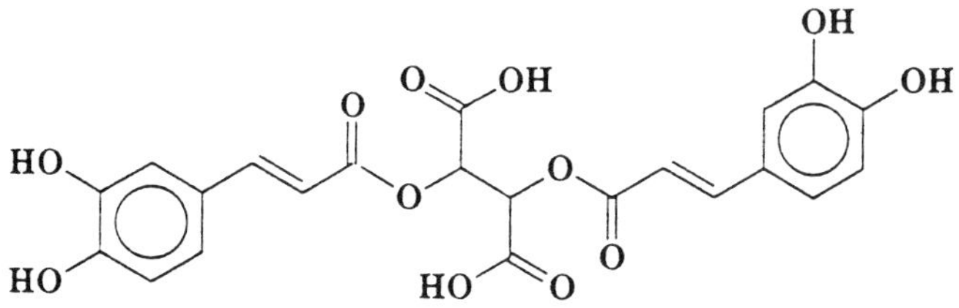

Chicoric acid

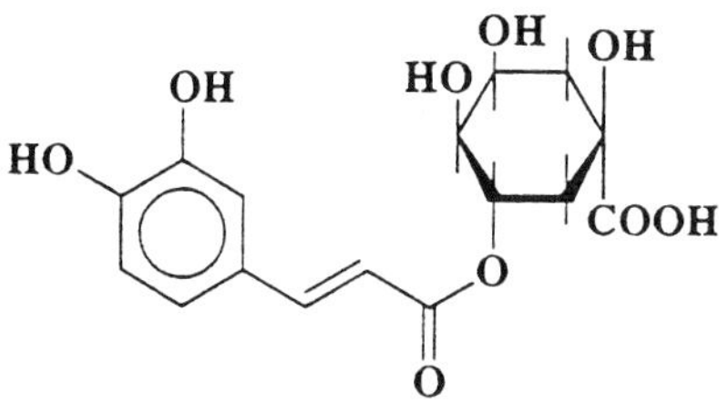

Chlorogenic acid

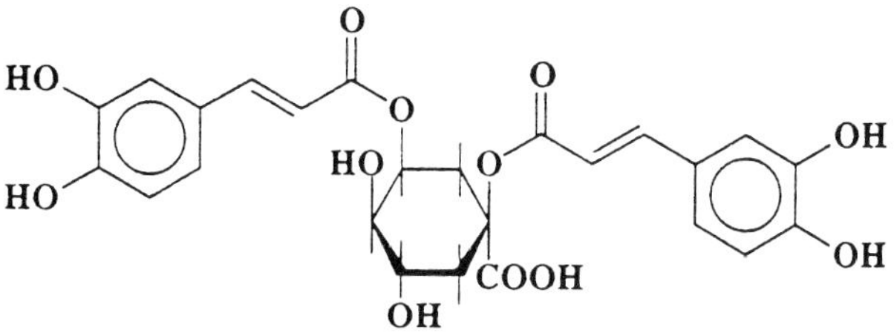

Cynarine

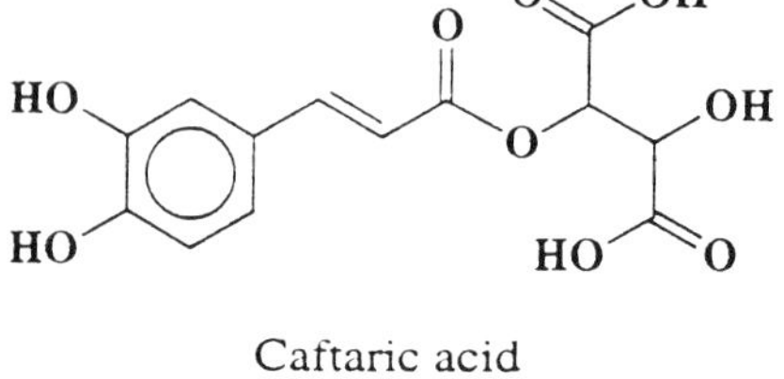

Caftaric acid

Figure 3. (Continued)

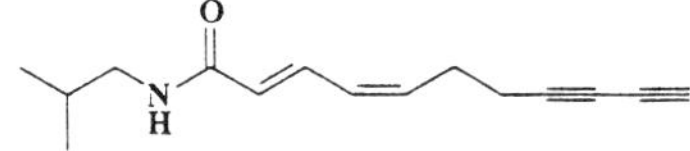

Undeca-2*E*,4*Z*-diene-8,10-diynoic acid isobutylamide

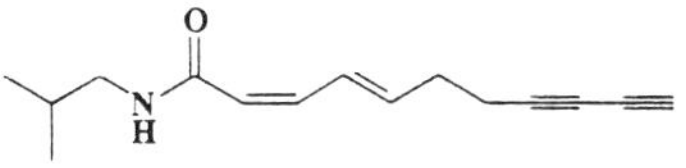

Undeca-2*Z*,4*E*-diene-8,10-diynoic acid isobutylamide

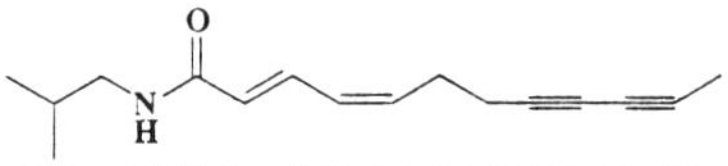

Dodeca-2*E*,4*Z*-diene-8,10-diynoic acid isobutylamide

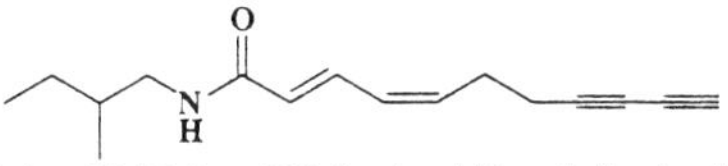

Undeca-2*E*,4*Z*-diene-8,10-diynoic acid 2-methylbutylamide

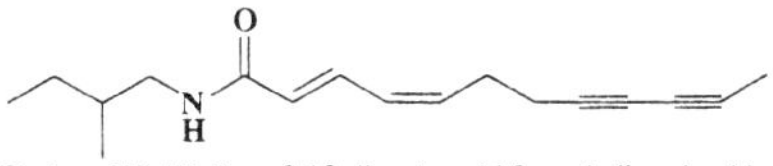

Dodeca-2*E*,4*Z*-diene-8,10-diynoic acid 2-methylbutylamide

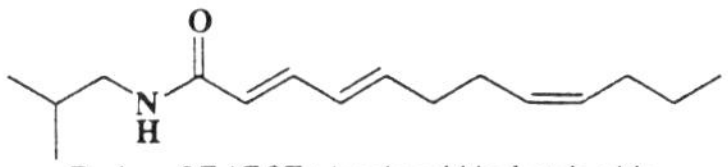

Dodeca-2*E*,4*E*,8*Z*-trienoic acid isobutylamide

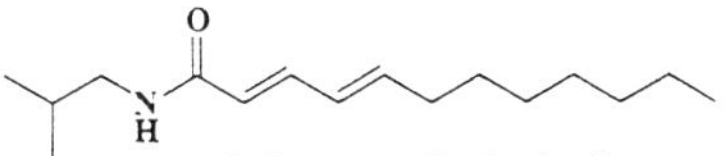

Dodeca-2*E*,4*E*-dienoic acid isobutylamide

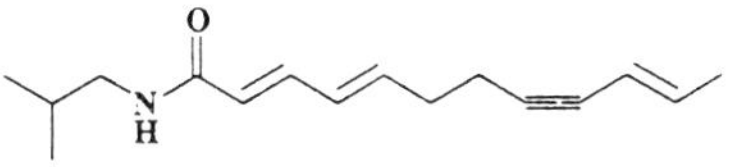

Dodeca-2*E*,4*E*,10*E*-trien-8-ynoic acid isobutylamide

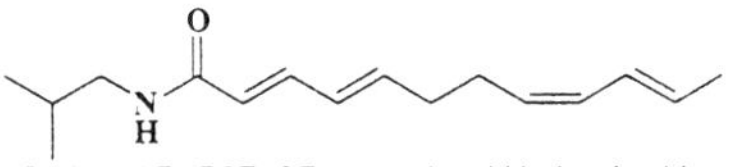

Dodeca-2*E*,4*E*,8*Z*,10*E*-tetraenoic acid isobutylamide

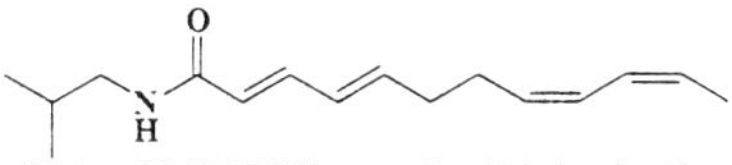

Dodeca-2*E*,4*E*,8*Z*,10*Z*-tetraenoic acid isobutylamide

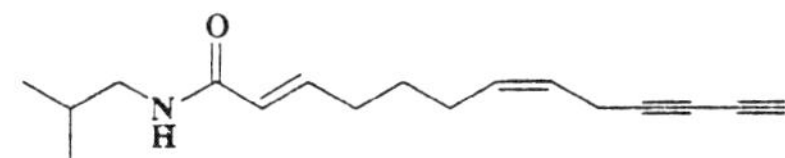

Trideca-2*E*,7*Z*-diene-10,12-diynoic acid isobutylamide

Figure 3. (Continued)

time of some of the patients. This combined immunotherapy was well tolerated by all patients without side effects. Echinacin caused an enhancement of the phagocytic activity of peripheral blood leukocytes and an increase of $CD3^+$ cells and the $CD4^+$ subset after low-dose cyclophosphamide.

Extensive studies on the immunostimulating activity of extracts of *E. purpurea* not only from intact plants, but also from cell cultures of the same species have been reported in the literature during the past five years (Wagner *et al.*, 1988; Lüttig *et al.*, 1989; Roesler *et al.*, 1991a; Roesler *et al.*, 1991b; Steinmüller *et al.*, 1993). Wagner *et al.* (1988) isolated from the medium of cell cultures of *E. purpurea* three homogenous polysaccharides, two fucogalactoxyloglucans with molecular weights 10 000 and 25 000 and an acidic arabinogalactan with molecular weight of 75 000. Tested for immunological activity by *in vitro* granulocyte test and *in vivo* carbon clearance test, the fucogalactoxyloglucan with molecular weight of 25 000 was demonstrated to enhance phagocytosis *in vitro* and *in vivo*. The arabinogalactan specifically stimulated macrophages to excrete tumor necrosis factor.

In another study by Lüttig *et al.* (1989) with this acidic arabinogalactan isolated from the medium of the same plant cell culture, the polysaccharide was demonstrated to be able to strongly activate macrophages to cytotoxicity against tumor cells and microorganisms like *Leishmania enriettii*. The polysaccharide induced macrophages for production of tumor necrosis factor TNF-a, interleukin-1 and interferon-b_2, and induced an increase in T-cell proliferation. The stimulation of the macrophages after intraperitoneal injection can have therapeutic implication in the defense against tumors and infectious diseases.

In other experiments by Roesler *et al.* (1991a) purified polysaccharides of the same *E. purpurea* cell culture were investigated for their ability to enhance the activity of phagocytes regarding non-specific immunity *in vitro* and *in vivo*. Macrophages from different organ origin could be activated to produce interleukin-1, TNF-a and interleukin-6, to produce elevated amounts of reactive oxygen intermediates and to inhibit growth of the microorganism *Candida albicans in vitro*. The polysaccharides could induce increased proliferation of phagocytes in spleen and bone marrow and migration of granulocytes to the peripheral blood in *in vivo* experiment. The demonstrated effects resulted in an excellent protection of mice against lethal infections with a predominantly macrophage dependent pathogen, *Listeria monocytogenes*, and a predominantly granulocyte dependent pathogen, *Candida albicans*. Purified polysaccharides from large-scale plant cell cultures of *E. purpurea* were tested for their ability to activate also human phagocytes *in vitro* and *in vivo* by the same authors (Roesler *et al.*, 1991b). The polysaccharides enhanced the spontaneous motility of polymorphonuclear cells under soft agar and increased the ability of these cells to kill staphylococci. Monocytes were activated to secrete TNF-a, interleukin-6 and interleukin-1. Intravenous application of the polysaccharides to test subjects immediately induced a fall in the number of the polymorphonuclear cells in the peripheral blood, indicating an activation of adherence to endothelial cells. The fall was followed by a leukocytosis due to an increase in the number of the polymorphonuclear cells and a lesser increase of monocytes. The appearance of stab cells and some juvenile forms and even myelocites indicated the migration of cells from the bone marrow into the peripheral blood. The acute phase C-reactive protein was induced, due to activation of monocytes and macrophages, to produce interleukin-6. In another study, the same authors (Steinmüller *et al.*, 1993) investigated the influence of the polysaccharides isolated from a large-scale plant cell culture of *E. purpurea* on the non-specific immunity in immunodeficient mice. The polysaccharides were found to be effective in activating peritoneal macrophages isolated from animals after administration of cyclophosphamide or cyclosporin A. Macrophages treated with the polysaccharides exhibited increased production of TNF-a and an enhanced cytotoxicity against tumor target WEHI 164 as well as against the intracellular parasite *Leishmania enriettii*. After a cyclophosphamide-mediated reduction of leukocytes in the periferal blood, the polysaccharides induced an earlier influx of neutrophil granulocytes as compared to controls treated with phosphate buffer saline. Similar to earlier results reported by Roesler *et al.* (1991a) the polysaccharide treatment of mice, immunosuppressed with cyclophosphamide or cyclosporin A, restored their resistance against lethal infections with the predominantly macrophage-dependent pathogen *Listeria monocytogenes* and predominantly granulocyte-dependent *Candida albicans*.

Compounds which have been detected from *Echinacea* extracts include complex polysaccharides, phenolic acids and long chain unsaturated amides (Figure 3).

Other plants

Typical structures which have been highlighted from other plants for potential immunostimulating effects include a wide range of secondary metabolites such as phenyl flavonoids, steroids and polysaccharides as well as simple quinones (Figure 4).

Screening systems for plant-derived immunostimulants

The most prominent advantages of target-directed screening programs are that they are in most cases specific, sometimes even organ-specific, and that they lend themselves readily to automation (Wagner, 1993). These screening methods also have practical and important implications with regard to operational time, costs and especially medical ethics. The disadvantage of these programs is that the results obtained *in vitro* may have no counterpart *in vivo*. In other words, efficacy *in vivo* often must be proven or confirmed in experimental animal studies.

It should be also kept in mind that due to the complex pathophysiology of many diseases it is not always possible to find and select adequate target models. If the etiology of a disease or the place of a deficiency are unknown, the screening program usually yields only symptomatically acting compounds. This means that even by employing modern screening techniques, causally acting drugs will be found only in some cases. Apart from these limitations, however, each plant screening program opens the possibility of finding lead structures. These might become the starting templates for biotransformational and/or synthetic modifications with

Epimedium sp.:

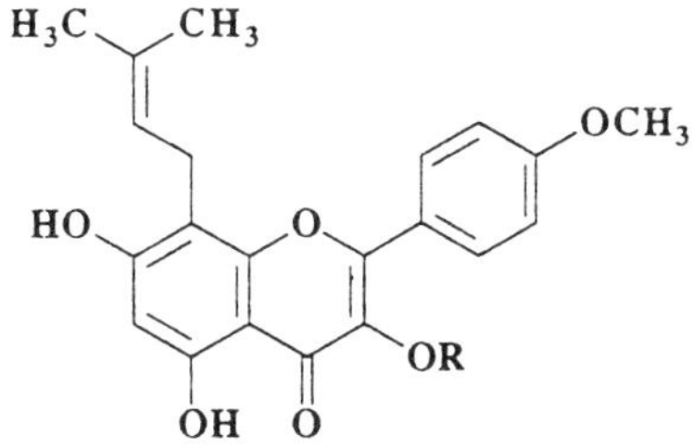

H$_3$C CH$_3$ OCH$_3$ HO O OR OH O

Icaritin, R = H
Icariside-II, R = rham

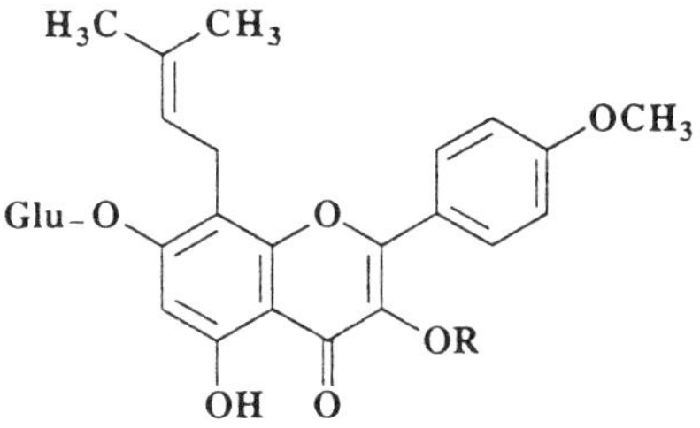

H$_3$C CH$_3$ OCH$_3$ Glu-O O OR OH O

Icariin, R = rham
Icariside-I, R = H
Epimedin-A, R = rham-glu
Epimedin-B, R = rham-xyl
Epimedin-C, R = rham-rham

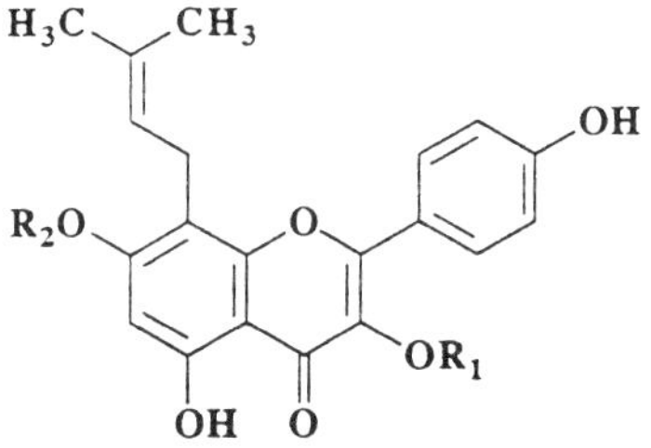

H$_3$C CH$_3$ OH R$_2$O O OR$_1$ OH O

Ruhuoside, R$_1$ = rham (4 → 1) glu; R$_2$ = glu
Epimedoside-A, R$_1$ = rham; R$_2$ = glu
Baohuoside-II, R$_1$ = rham; R$_2$ = H

Figure 4. Plant-derived compounds with immunopotentiating activity.

Cynanchum auriculatum:

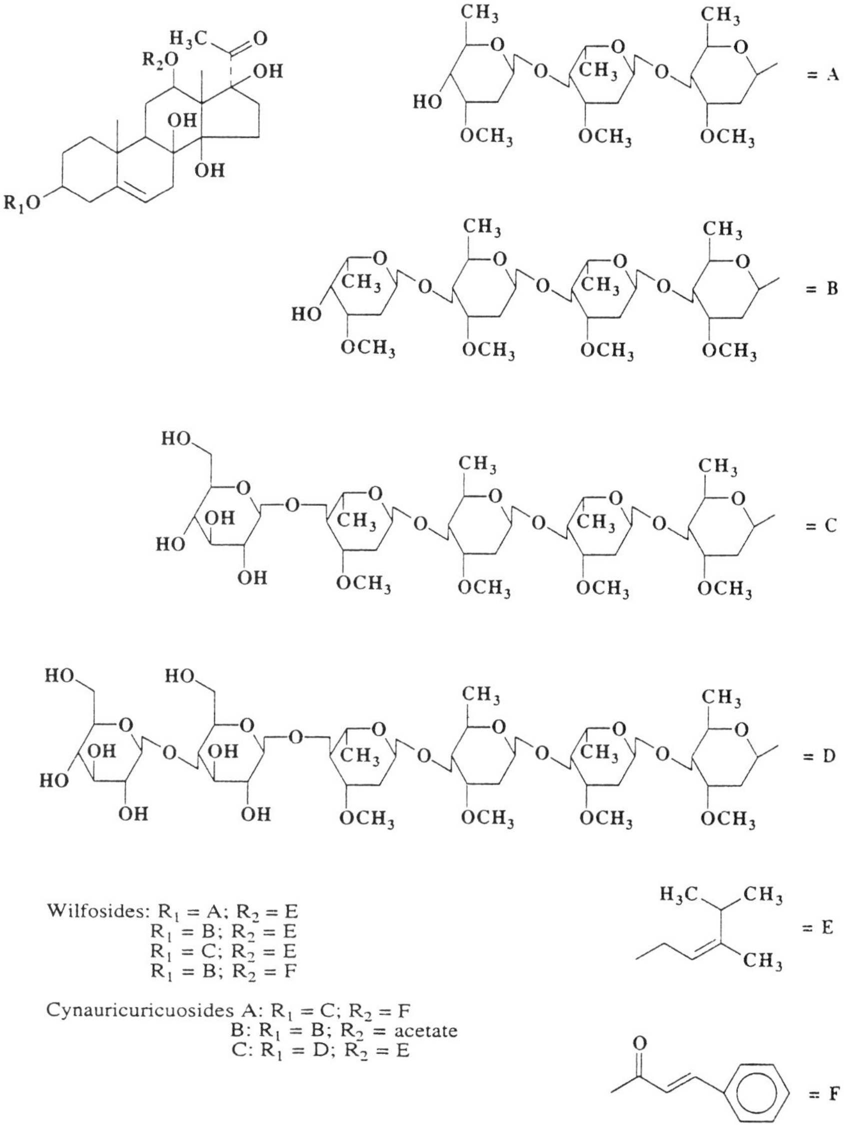
H3C
O
R2O
OH
OH
OH
R1O
CH3
O
HO
OCH3
= A
= B
= C
= D
HO
OH
H3C
CH3
= E
= F
Wilfosides: R1 = A; R2 = E
R1 = B; R2 = E
R1 = C; R2 = E
R1 = B; R2 = F
Cynauricuricuosides A: R1 = C; R2 = F
B: R1 = B; R2 = acetate
C: R1 = D; R2 = E

Figure 4. (Continued)

the aim of optimizing the bioavailability and pharmacokinetics, thus considerably improving the efficiency of plant constituents for therapy.

One of the historical difficulties in screening plant extracts for potential biological activity has been the reliability of supply and reproducibility of resupply. These problems are well addressed by Phytopharmaceuticals Inc.'s program of world-wide plant sourcing which ensures absolute authenticity in supply and resupply by using only internationally recognized taxonomists for collection and identification of plant material and satellite location technology for precise knowledge of the locality of the collection site. This combined strategy allows efficient and precise screening of 15 000 different plant extracts per year for selective pharmacological evaluation against targets related to disorders of the cardiovascular and central nervous systems as well as of the immune functions.

Highly specialized natural products chemistry and the extensive use of relational databases allow the rapid identification of novel molecules as potential lead compounds for the pharmaceutical industry.

Directed screening programs are also available where the source plants may be selected from amongst those plants with a reputation for potential immunostimulating activity, which are usually plants claimed to have anti-infectious, antitumoral, antiviral or parasiticidal activities (Wagner, 1993).

A typical screening program might include four or five simultaneous assays, selected from *in vitro* and *in vivo* targets. The most important cell types which should be used are granulocytes, macrophages and T-lymphocyte populations. An immune-induced cytotoxicity test should be also included in the screening program. Carbon-clearance tests and *in vitro* infection stress tests using the pathogens *Candida albicans*, *Leishmania*, or *Listeria* have been also used frequently (Steinmüller *et al.*, 1993; Wagner, 1993). Following, some of the most frequently used *in vitro* and *in vivo* tests according to Wagner and Jurcic (1991b) for detection and quantification of immunopotentiating activity of plant extracts are highlighted.

GRANULOCYTE PHAGOCYTOSIS ASSAY *IN VITRO*

The polymorphonuclear neutrophils are the most important phagocytic cells against bacterial infections. Phagocytosis may be analysed quantitatively by measuring the rate of ingestion of opsonized microorganisms by phagocytic cells. Different microorganisms can be used to initiate a phagocytic reaction like *Candida albicans*, *Escherichia coli*, *Staphylococcus aureus* or baker's yeast. After certain time of incubation of granulocytes together with the microorganism and the substance(s) to be tested, and after staining the cells, the phagocytosis index (the sum of ingested microorganism's particles / number of granulocytes) is determined by microscopical observation. In addition to granulocytes, peripheral human blood monocytes, peritoneal macrophages from normal mice, or Kupffer cells prepared from mouse liver can be used for phagocytosis test.

CHEMOLUMINESCENCE ASSAY

Chemoluminescence occurs as a consequence of phagocytosis of bacteria or inert particles by phagocytic leukocytes. After opsonization, bacteria can subsequently be killed by oxygen species such as superoxide anion

(O_2^-), hydrogen peroxide (H_2O_2), singlet oxygen ($^1O_2$) and hydroxyl radicals (OH). A flash of light is associated with the release of oxidizing species in combination with a myeloperoxidase. The light emission can be amplified by adding reagents like luminol or lucigenin. A mixture of polysaccharides and zymosan obtained from yeast by cell wall disruption is used as a challenge.

CHEMOTAXIS ASSAY

Optimal host defense against infection is understood as a high capacity of leucocytes to respond by chemotaxis and random migration. Often, chronic and recurrent infections, cancer, atopic dermatitis, rheumatoid arthritis and diabetes are associated with diminished chemotaxis *in vitro*, and this is the reason of the interest in agents which improve or stimulate leukocyte locomotion. *In vitro* assays are performed to determine the extent to which leukocytes can respond to chemotactic stimuli and are most frequently performed on blood neutrophils. Two different techniques for the *in vitro* assessment of leukocyte chemotaxis are available: the micropore filter method, and the 'under agarose' method. In the former method, cells placed on a filter are allowed to migrate into the channels of the filter, and the number of migrating cells or the distance of migration into the filter is then analysed. In the latter method, cells placed in a central well are allowed to migrate under agarose toward peripheral wells containing either medium alone or a chemoattractant. The distance of migration is used to evaluate the response.

CARBON-CLEARANCE ASSAY FOR *IN VIVO* PHAGOCYTOSIS

The rate of removal of injected colloidal carbon particles from the bloodstream is a measure of reticuloendothelial phagocytic activity. When a preparation of colloidal carbon is injected intravenously, the carbon is removed by the sessile intravascular phagocytes in the liver and spleen. The liver's Kupffer cells take up approximately 90% and the splenic macrophages 10% of the carbon preparation. After blood collection, and after lysis of the erythrocytes, the absorbance of each blood sample is measured spectrophotometrically at 650 nm. Optical densities of samples from treated animals are then compared with controls. In general, the carbon-clearance test correlates well with the *in vitro* granulocyte test.

LYMPHOCYTE PROLIFERATION ASSAY

Lymphocytes comprise a population of cells with different immunological properties – functional cells which can mediate a primary response, prefunctional precursors, memory cells and immunologically active cells. They can be subdivided into T- and B-lymphocytes, and the T-lymphocytes further into helper, suppressor, cytotoxic and memory cells. Information of lymphocyte function is provided by measuring their ability to proliferate, produce mediators, induce cytotoxic responses and regulate immune response. Lymphocyte proliferation can be quantitated by determination of increased DNA synthesis in response to a mitogen like phytohaemagglutinin, pokeweed mitogen, or concanavalin A, or any other appropriate stimulator by incubating lymphocytes in a medium containing a radiolabelled precursor of DNA, which usually is tritiated thymidine. Cells incubated without mitogen are used as controls. The amount of radioactivity incorporated into the cells is determined by scintillation counting.

ASSAY OF NATURAL KILLER CELL ACTIVITY

Natural killer cells are defined as effector cells with spontaneous cytotoxicity against various target cells such as malignant cells or normal cells infected with viruses, fungi and parasites. The cytotoxic effect is a principal mechanism of immune response against the metastatic spread of tumor cells and a natural resistance against tumors. Natural killer cells are closely associated with a subpopulation of lymphocytes, identified as large granular lymphocytes, that include about 3% mononuclear cells and 5% peripheral blood lymphocytes. These cells are non-phagocytic, non-adherent and are able to bind antibody-coated target cells and mediate antibody dependent cell-mediated cytotoxicity except for natural cell-mediated cytotoxicity. They act independently of antigens and belong to the unspecific immune system. All cells can be activated by interferons and interleukin-2. Generally, five main stages in interaction between effector and target cells which lead to cytotoxicity are defined: recognition, binding, activation of lytic machinery, lysis of the target cell and its dissolution. *In vitro* responses of effector or target cells can be measured by different techniques. For cell-mediated immunity most used is the ^{51}Cr-release technique, where intracellular components of target cells are labelled by incubation with ^{51}Cr, and cytolysis after addition of natural killer cells is determined by the release of the label from the target cell.

ASSAY FOR TUMOR NECROSIS FACTOR (TNF) PRODUCTION

TNF was first described and defined as a soluble factor found in the sera of animals that have been sequentially treated with a reticuloendothelial stimulator and bacterial cytotoxin. TNF activity in sera is maximal about 2 h after administration of lipopolysaccharide and cannot be induced again with further administrations for at least 5 days. TNF is preferentially toxic for tumor cells. With the exception of cytostatic and cytocidal activities for a variety of tumor cell lines in culture, new studies have shown that TNF stimulates normal cell proliferation, neutrophils to increase superoxide production, antibody dependent cell-mediated cytotoxicity, phagocytosis and lysozyme release. Activated macrophages have been shown to be the principal producer of TNF for non-specific tumoricidal activity. TNF can be assayed by several methods with cultures of macrophages activated with lipopolysaccharide and L 929 fibroblasts pretreated with actinomycin D. The released TNF is toxic to the L 929 cells, and after incubation of supernatants from macrophages activated by lipopolysaccharide with the fibroblasts the toxic effect of TNF contained in the supernatants is scored.

Conclusion

Plants have played an important role in the development of modern medicines. Technology of assay selection, design and operation is now advanced sufficiently to allow the rapid screening of many thousands of plant extracts across many targets. Immunologically-based assays are now sensitive and robust enough to reproducibly detect active metabolites with activities in the low nanomolar range. Past problems associated with plant extract sourcing and resupply have been solved and it is at last possible to contemplate substantial programs of plant screening against a broad target range.

Such studies could lead to the identification of lead substances for the pharmaceutical industry but also the highlighting of potentially useful crop plants, the inclusion of which in animal food in an appropriate way could lead to the enhancement of the health of animals which may thus require less prophylactic and acute therapy with all its concominant advantages.

References

Asaka, I., I. Ii, M. Hirotani, Y. Asada and T. Furuya. 1993. Production of ginsenoside saponins by culturing ginseng (*Panax ginseng*) embryogenic tissues in bioreactors. Biotechnol. Lett. 15:1259.

Bauer, R., K. Jurcic, J. Puhlmann and H. Wagner. 1988. Immunologische *In-vivo-* und *In-vitro*-Untersuchungen mit *Echinacea*-Extrakten. Arzneim.-Forsch. 38:276.

Duc, N.M., N.T. Nham, R. Kasai, A. Ito, K. Yamasaki and O. Tanaka. 1993. Saponins from Vietnamese ginseng, *Panax vietnamensis* Ha *et* Grushv. collected in central Vietnam. I. Chem. Pharm. Bull. 41:2010.

Facino, R.M., M. Carini, G. Aldini, C. Marinello, E. Arlandini, L. Franzoi, M. Colombo, P. Pietta and P. Mauri. 1993. Direct characterization of caffeoyl esters with antihyaluronidase activity in crude extracts from *Echinacea angustifolia* roots by fast atom bombardment tandem mass spectrometry. Il Farmaco 48:1447.

Fujimoto, Y., M. Satoh, N. Takeuchi and M. Kirisawa. 1991. Cytotoxic acetylenes from *Panax quinquefolium*. Chem. Pharm. Bull. 39:521.

Hong, C.Y., L.J. Lai and S.F. Yeh. 1993. Linoleate-rich triacylglycerol in *Panax pseudo-ginseng* improves erythrocyte deformability *in vitro*. Planta Med. 59:323.

Lersch, Ch., M. Zeuner, A. Bauer, M. Siemens, R. Hart, M. Drescher,

U. Fink, H. Dancygier and M. Classen. 1992. Nonspecific immunostimulation with low doses of cyclophosphamide (LDCY), thymostimulin and *Echinacea purpurea* extracts (Echinacin) in patients with far advanced colorectal cancers: preliminary results. Cancer Invest. 10:343.

Liu, C.-X. and P.-G. Xiao. 1992. Recent advances on ginseng research in China. J. Ethnopharmacol. 36:27.

Lüttig, B., Ch. Steinmüller, G.E. Gifford, H. Wagner and M.-L. Lohmann-Matthes. 1989. Macrophage activation by the polysaccharide arabinogalactan isolated from plant cell cultures of *Echinacea purpurea*. J. Natl. Cancer Inst. 81:669.

Mengs, U., C.B. Clare and J.A. Poiley. 1991. Toxicity of *Echinacea purpurea*. Acute, subacute and genotoxicity studies. Arzneim.-Forsch. 41:1076.

Puhlmann, J., U. Knaus, L. Tubaro, W. Schaefer and H. Wagner. 1992. Immunologically active metallic ion-containing polysaccharides of *Achyrocline satureioides*. Phytochem. 31:2617.

Roesler, J., Ch. Steinmüller, A. Kiderlen, A. Emmendörffer, H. Wagner and M-L. Lohmann-Matthes. 1991a. Application of purified polysaccharides from cell cultures of the plant *Echinacea purpurea* to mice mediates protection against systemic infections with *Listeria monocytogenes* and *Candida albicans*. J. Immunopharmac. 13:27.

Roesler, J., A. Emmendörffer, Ch. Steinmüller, B. Lüttig, H. Wagner and M-L. Lohmann-Matthes. 1991b. Application of purified polysaccharides from cell cultures of the plant *Echinacea purpurea* to test subjects mediates activation of the phagocyte system. J. Immunopharmac. 13:931.

Schumacher, A. and K.D. Friedberg. 1991. Untersuchungen zur Wirkung von *Echinacea angustifolia* auf die unspezifische zelluläre Immunantwort der Maus. Arzneim.-Forsch. 41:141.

Steinmüller, Ch., J. Roesler, E. Gröttrup, G. Franke, H. Wagner and M-L. Lohmann-Matthes. 1993. Polysaccharides isolated from plant cell cultures of *Echinacea purpurea* enhance the resistance of immunosuppressed mice against systemic infections with *Candida albicans* and *Listeria monocytogenes*. J. Immunopharmac. 15:605.

Wagner, H., H. Stuppner, W. Schaefer and M. Zenk. 1988. Immunologically active polysaccharides of *Echinacea purpurea* cell cultures. Phytochem. 27:119.

Wagner, H. and K. Jurcic. 1991a. Immunologische Untersuchungen von pflanzlichen Kombinationspräparaten. *In vitro*- und *in vivo*-Studien zur Stimulierung der Phagozytosefähigkeit. Arzneim.-Forsch. 41:1072.

Wagner, H. and K. Jurcic. 1991b. Assays for immunomodulation and effects on mediators of inflammation. Methods in Plant Biochemistry. Academic Press, London 6:195.

Wagner, H. 1993. Leading structures of plant origin for drug development. J. Ethnopharmacol. 38:105.

Xiao, P.-G., S.-T. Xing and L.-W. Wang. 1993. Immunological aspects of Chinese medicinal plants as anti-ageing drugs. J. Ethnopharmacol. 38:167.

Xiao-Yu, L. 1991. Immunomodulating Chinese herbal medicines. Mem. Inst. Oswaldo Cruz, Rio de Janeiro, Suppl. II, 86:159.

Yang, G. and Y. Yu. 1990. Immunopotentiating effect of traditional Chinese drugs – ginsenoside and *Glycyrrhiza* polysaccharide. Proc. Chin. Acad. Med. Sci. and Peking Univ. Med. College 5:188.

Zee-Cheng, R.K.-Y. 1992. Shi-Quan-Da-Bu-Tang (Ten significant tonic decoction), SQT. A potent Chinese biological response modifier in cancer immunotherapy, potentiation and detoxification of anticancer drugs. Meth. Find Exp. Clin. Pharmacol. 14:725.

Zhou, L. and G. Zheng. 1991. A study of the technology of mass cell culture of American ginseng. Chin. J. Biotechnol. 7:191.

A SHOWCASE ON THE DIVERSITY OF AGRICULTURE AROUND THE WORLD

DIETETIC FEEDS AND NUTRITIONAL SUPPLEMENTS

R. WOLTER

Ministry of Agriculture, Nutrition and Feed Department, National Veterinary School, Alfort, France

Introduction

The European Community Council Directive, dated Sept 8, 1993 (no. 8130/93) created the category of dietetic feedstuffs' for animals. This directive defines the term dietetic with a specific meaning in this context which is to identify a sector of feedstuffs intermediate between standard (normal diets) and medicinal feeds. Creation of this category is part of the goal to harmonize the European market, favor free competition, provide the buyer with useful information and all possible guarantees of effectiveness and safety, clearly define indications for use and provide for regulatory control. This led to the publication of an exhaustive list of uses and purposes for dietetic feedstuffs on August 10, 1994 to be put into practice by June 30, 1995.

The area of 'nutritional supplements', previously called 'specific nutritional adaptation supplements' (ANSA) remains in a legal vacuum; however these supplements will be similarly regulated at a later time, at which point many will be found to fall into either dietetic, normal or medicinal categories.

Dietetic feedstuffs

The term dietetic now has an established legal meaning in the context of animal feed. There are strict rules covering the composition and application of these feedstuffs. Dietetic feedstuffs are defined as having a particular purpose or function (e.g. to prevent or fight against a pathological or metabolic condition) because of their specific nutritional composition. As such, dietetic feedstuffs are set apart from standard feeds, medicinal feeds and nutritional supplements.

Dietetic feedstuffs are not intended to meet the nutritional needs for, or affect, physiological functions such as maintenance, reproduction, growth, milk production, muscular work, ageing, etc. Standard feedstuffs are formulated to cover all normal requirements of animals of various species, ages, physiological stages or productivity levels. Standard feeds carry an assurance of palatability, digestibility, of meeting nutritional requirements, and of providing a balanced diet as per current scientific and technical knowledge. These feeds enable animals to fully reach their genetic potential in terms of performance and/or athletic work. Standard feedstuffs also contribute to the condition, health, longevity and well-being of the animal and are formulated with consideration for environmental concerns.

In contrast to medicinal feeds, which contain compounds such as antimicrobials or antiparasitic agents against specific problems, dietetic feedstuffs display an exclusively nutrition-related mode of action. They are in no way enhanced by chemical substances with pharmacological, anti-infectious, or anti-prasitic properties. Additionally, dietetic feeds do not require special authorization to be put on the market. They do not require veterinary prescription nor are they subject to any professional domain. However, as dietetic feeds might be part of a program designed to prevent or ameliorate a pathological or metabolic condition, it is recommended that a veterinarian be consulted in order to clarify the diagnosis, justify use and combine with necessary medical treatment and/or prevent possible complications.

When compared with nutritional supplements, dietetic feedstuffs meet European requirements, according to which every feed mix should represent a significant part of the total ration. On the other hand, concentrated nutritional preparations are developed only for occasional use in critical periods to supplement an already complete ration. Consisting mainly of minerals and vitamins in strong dosages for a very specific use, they can be used as feed supplements provided that the minimum required mineral level in the ration is reached. We differentiate between standard supplements, which are formulated to support physiological functions such as reproduction, lactation peak, growth, hair regrowth, preparation for sporting events, recovery following exertion, etc., and dietetic supplements which are recommended for pathological conditions such as poor digestion, locomotive disorders, heat stress, convalescence, etc.

PURPOSES OF DIETETIC FEEDSTUFFS

Dietetic feedstuffs are distinguished by their special purposes and nutritional characteristics adapted to meet specified aims. These feedstuffs carry additional guarantees which are supported and can be verified and strictly defined conditions of use (Figure 1).

In the interest of free and fair competition, animal health and consumer protection, European regulators are imposing a exclusive and closed list of dietetic feedstuffs for different species. New products can be added to this list at a later time at the request of industry and in keeping with the evolution of knowledge, provided that a dossier outlining health and nutritional or economic benefits with scientific support is submitted and approved.

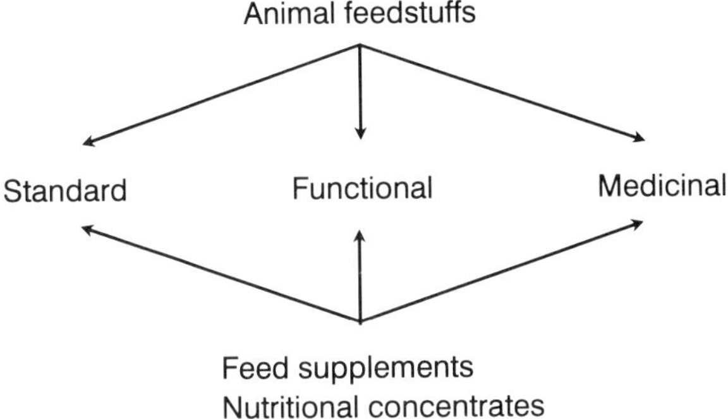

Figure 1. Different ways in which feed and nutritional and feed supplements are used.

Dietetic feedstuffs must have the following objectives and meet the following requirements.

1. Dietetic feedstuffs must deal with a pathological or metabolic condition that either potentially or presently exists (regardless of origin) and can be improved or alleviated, compensated for or better endured through a specially adapted preventive or palliative feeding program. This can complement medical treatment without purporting to substitute for it, as in the cases of infections and parasites.
2. There must be scientific, experimental, epidemiological, or clinical proof of the beneficial effect of the feed additive or nutritional factor in overcoming the digestive or metabolic inadequacy(s) associated with a given ailment.
3. Dietetic feedstuffs must have additional composition guarantees beyond the usual guarantees for any feed mix in order to indicate the value of the particular formulation with respect to its stated goal and in order to provide a means of regulatory control. These guarantees should therefore be useful, meaningful, objective and easily verified.
4. Dietetic feedstuffs should be used in exceptional circumstances (e.g., during critical periods) without competing with standard feeds which are technically and economically more attractive under normal circumstances, i.e., for healthy animals.
5. There must be a real market demand for a proposed dietetic feedstuff. Manufacturers of such feedstuffs must make every effort to improve the health and the well-being of the animals without exposing the market to a multitude of formulas of no practical interest.

LABELING DIETETIC FEEDSTUFFS

The legally intended objectives of all dietetic feedstuffs appear very clearly in the strictly defined product name. These names are used uniformly throughout the European community (with translations in each national language). To the extent possible these names and definitions avoid the use of medically-related terms which might have unintended therapeutic connotations and evoke illegal veterinary practice. In any case product designations must be perfectly intelligible to the general public to whom they are addressed.

The essential nutritional features should reflect the intended purpose and be stated on the label of any dietetic feedstuff. As do the names of dietetic feedstuffs, the label informs any potential buyer regardless of his/her training in useful and easily understood terms. The label will recommend the advice or the consultation of a veterinarian if indicated, though it is understood that the option to do so is left with farmer.

The reason for additional composition guarantees on the label of all dietetic feedstuffs is to validate individual nutritional claims. This implies that the claims must be well-founded and consistent in terms of stated purposes and characteristics. These nutritional guarantees, unique to dietetic feedstuffs, must be based on official or at least scientifically recognized measurements (e.g. total, soluble and insoluble fiber content, individual essential amino acids, essential fatty acids, essential trace elements, and vitamins).

Additionally, labeling feedstuffs in this manner makes adjustments to, and even extends, general feed additive legislation. For example, the clays currently authorized as technological additives (anti-caking clay for meal, expandable clay for pellets) should also be authorized as anti-diarrhetic factors. Chromium, to which anti-stress and glycemia regulatory properties are now being attributed, must be cleared as an additive for this purpose before it can appear on labels.

Correct administration of these feedstuffs on the farm is important as distribution errors can result in ineffectiveness and/or worsening of the ailment while delaying suitable treatment. This is why veterinary consultation is recommended in all ambiguous cases, despite the fact that this consultation is not obligatory for non-medicinal feed use. Correct identification of the cause and the nature of the problem is critical in determining whether it can be solved through nutritional means and in selecting the appropriate dietetic feedstuff. Moreover, in all serious cases it is important to give priority to a suitable medical treatment (if it exists), without overlooking the possibility of applying a specialized diet and to prolong this diet in certain chronic irreversible cases.

However the basic rule is to limit duration of dietetic feedstuff application to the minimum time necessary for optimum effectiveness. Unjustified application over too long a period should be avoided. This has no health benefit for the animal and is poor economic practice. More importantly, it may be harmful for the animal as dietetic feedstuffs are formulated to alleviate a specific condition and do not necessarily ensure an optimally balanced diet over the long term. Additionally, too great a focus on specialized needs leads producers to forget that well-balanced standard feeds are the best option for healthy animals.

Tables 1–3 and Figure 2 summarize the current proposals regarding use of dietetic feedstuffs, their specific functions, essential characteristics and the duration of use and other recommendations.

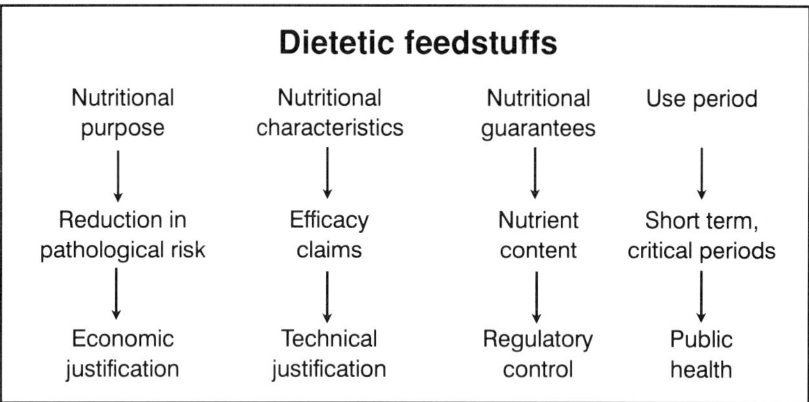

Figure 2. Description and objectives of dietetic feedstuffs.

Table 1. Animal husbandry objectives for dietetic feedstuffs

Hypocalcaemia	Dairy cows
Ketosis	Dairy cows and sheep
Hypomagnesemia	Ruminants
Acidosis	Ruminants
Urinary calculi	Ruminants
Water and electrolyte balance	Calves, piglets, lambs, young goats, foals
Stress response	Pigs
Digestive disorders	Piglets and pigs
Constipation	Sows
Hepatic steatosis	Laying hens
Malabsorption syndrome	Poultry

Table 2. Proposed objectives for functional feedstuffs for horses.

Digestive disorders of the small intestine
Digestive disorders of the large intestine
Stress (reduced reaction to)
Dehydration (electrolyte imbalance)
Chronic renal insufficiency
Feeding during convalescence

Table 3. Objectives of dietetic feedstuffs for carnivores.

Poor digestion	Poor absorption
Feed intolerance	Feeding during convalescence
Dermatosis	Diabetes
Chronic hepatic insufficiency, hepatic Cu build-up	Obesity
Urolithiasis: struvite (prevention and dissolution crystals), oxalate, cystine	Constipation
Chronic renal insufficiency	
Hyperlipemia	Cardiac insufficiency

Nutritional supplements

A wide variety of feed supplements are currently being marketed for different species. With no legal status at present, these supplements do not always provide the necessary effectiveness, safety and viability guarantees. Within the framework of European regulation these nutritional mixes are currently being examined to determine their nature, importance and appropriate use both in the interest of performance benefits and in the interest of public health.

There is a move to define nutritional supplements more precisely. However, they remain heterogeneous and currently cross over into the areas of both standard and dietetic feedstuffs. Additionally, they raise many as yet unanswered questions. Nutritional supplements include mixtures of amino acids, trace elements, vitamins, pH regulators, etc. Their purpose is to supplement inadequate diets on a temporary basis or under certain exceptional conditions. Nutritional supplements are typically administered through the drinking water or as a topdress on feed. 'Exceptional conditions' might have either physiological or accidental causes. Physiological circumstances calling for nutritional supplements would include phases of very high production, prepartum diets, rapid growth, very intense work, dramatic diet changes, etc. These would require the use of standard nutritional supplements. The assumption used in justifying nutritional supplements is that the standard feeds, produced at minimum cost in order to meet severe economical restraints, have narrow safety margins. Because of this standard feeds may be inadequate in cases where there are peaks in nutritive demand. Supplements are technically useful in such cases.

'Accidental' causes of circumstances indicating use of nutritional supplements are unpredictable and include stress caused by weather change, transport, vaccination or anti-parasite treatment. They also include normal and subclinical bacterial, viral or parasitic ailments. Effects of these stressors include:

- Low intake, which can become extreme (anorexia).
- Reduction in digestibility as a result of faster digesta passage time, reduction in enzymatic secretion, changes in the intestinal wall.
- Reduction in metabolic yield because of hormonal imbalance (excess corticoid) which increases the rate of catabolism.
- Increase in nutritional requirements to support hormonal synthesis, erythropoiesis, immunoglobin production and tissue repair.

Consequently, timely nutritional supplementation could be necessary to support resistance to disease, prevent complications and quickly re-establish production and profitability. Use of these supplements enables the animal to adapt immediately to an unpredictable yet critical situation. Because of claims regarding disease risk, nutritional supplements often fall into the dietetic category and therefore manufacturers should consider redefining their products in light of the new European legislation on objectives, nutritional characteristics, additional guarantees and directions for use.

Official authorization of nutritional supplements, which are presently simply tolerated, would mean that their legal status, effectiveness, conditions for use and guarantees of safety should be specified beforehand.

Nutritional supplements, because they do not represent a significant part of the ration, do not qualify as feedstuffs and therefore have no legal status within the present framework of European regulations. These regulations must acknowledge the notion of premixing the active elements (or core ingredients), as long as these premixes respect Community guidelines on feed mixes.

Regarding effectiveness, nutritional supplements, like dietetic feeds, should provide nutritional characteristics, composition guarantees and directions for use and also prove that their stated function can be achieved. For this, manufacturers should refer to available nutritional science for guidance in preparing useful products. For example, the classic vitamin 'cocktail' supplement added to drinking water over a 3–5 day period may not have the desired effect and polyvalence and may cause reduction in appetite, poor digestion and metabolic problems. Where concentrated additives are concerned, the practicalities of high incorporation rates must be considered in relation to the short usage periods in order to achieve efficacy.

Article 12 of the EEC 70/524 directive has previously stipulated that feed supplements should represent at least 5% of the total feed. This means that the concentration of the additives should be less than 1/20 of the dose required for the total ration so as not to exceed the maximum quantity allowed per animal per day. A change to this directive is proposed which would reduce the maximum inclusion rate to 2% of the total ration of monogastrics (excepting horses). This would correspond to a maximum concentration of additives which would be 1/50 of the total ration. For ruminants and horses, this minimum threshold would be set at 2% of the supplemental ration, with an additional requirement not to exceed an antibiotic content of 2000 mg/kg (for ruminants).

Supplements in the form of liquid or solid licks as well as soluble powders destined for mixing with drinking water would not be allowed to contain antibiotics, growth factors, or coccidiostatics because of variations in intake. In cases where the supplements are distributed in the drinking water, it will also be necessary to have the certified equipment which guarantees even distribution.

For the overall protection of animal health and quality of meat or milk as well as consumer and environmental health, use of nutrient supplements must be carefully monitored to guarantee maximum allowed additive inclusion rates (where they exist) per animal per day. In this regard the rules vary in relation to the potential risk of the additive. We differentiate among:

- purely nutritional additives, very harmless (e.g. vitamins E, B, C, sorbitol);
- nutritional additives with toxic effects when overload levels are reached (e.g. selenium, copper, vitamins A and D);
- health additives (antibiotics, anticoccidia factors, growth factors) which are still forbidden in equine feeds.

On the whole, the supplement manufacturer should promote the exclusively nutritional function of these ingredients, bearing in mind regulations pertaining to standard feeds on the one hand, and functional feeds on the other. Leaving aside preparations without true technical or economic value, nutritional supplements can indeed render excellent and practical services.

Other additives

A diverse range of other additives is also being proposed. Many have implied pharmacodynamic properties which would place them in the catagory of medicines and are of varying effectiveness. Use of these products, for example certain herbal medicines and β-agonists, is not allowed in livestock feeds.

Herbal preparations can be divided into two groups. In the first group extracts are derived from plants and are applied in small dosages, therefore the herbal medicine effect seems almost negligible. Such extracts are allowed in feeds for their aromatic properties. The second group of plant products have herbal medicinal properties and their extracts are included at high levels, ostensibly to improve athletic ability. Consequently, these products are medicines and should be considered as such. These include eleutherocoque (neurostimulant), harpagophytum (anti-inflammatory agent) and ginsing (neurotonic) extracts.

β-agonist substances are characterized by properties similar to adrenalin (and other catecholamines). Used fraudulently in meat production, they greatly improve the quality of the carcass by increasing the proportion of muscle to the detriment of surplus fat. Controls should be very specific and wide-ranging as there are many possibilities.

Other fraudulent products such as anabolic steroids (taken orally), erythopoietine, (which is in injectable form, very expensive and of very dubious quality) are also of concern.

Conclusion

Dietetic feedstuffs will be the subject of very rigid and effective European regulation. They offer interesting possibilities for better protection of animal health and maximum athletic potential. In this way, we will be better equipped to fight against stress, dehydration, digestive problems, bone, muscle and hoof problems, disorders which threaten the performance horse in particular.

Supplements could later play a comparable role, if the same rules are imposed regarding use, effectiveness, safety and viability. Preparations with a marketing rather than a technical base should be removed from the market.

THE ANIMAL AND POULTRY WASTE MANAGEMENT CENTER AT NORTH CAROLINA STATE UNIVERSITY: A MULTI-DISCIPLINARY PROGRAM FOR ADDRESSING ENVIRONMENTAL CONCERNS RELATED TO FOOD-ANIMAL INDUSTRIES

C.M. WILLIAMS

Departments of Animal and Poultry Science,
North Carolina State University,
Raleigh, North Carolina, USA

Introduction

In North Carolina, the farm value of poultry and swine is estimated to exceed $3 billion in 1995. Both nationally and in North Carolina, food animal related industries have grown tremendously during the past two decades, and are expected to continue to grow at a rapid rate in coming years. However, the size and magnitude of these industries results in the production of tremendous quantities of manures and litters, hatchery by-products, feathers and hair, animal mortalities, processing offal and processing waters containing compounds of environmental and regulatory concern. Under certain conditions, these by-products can potentially result in air quality problems and the pollution of ground and surface waters by inorganic chemicals and pathogens. Environmental concern has resulted in certain federal, state, and local regulatory agencies establishing requirements that waste minimization, on-site waste recovery and recycling replace land application as the predominant method of waste disposal.

Considering the current regulatory focus on agriculture, there is little doubt that the future growth and economic stability of the agricultural food animal industries will be affected by their waste management and co-product utilization practices. The College of Agriculture and Life Sciences at North Carolina State University has initiated a program to expand on research needs for improved treatment and management procedures for converting manures, litters, and processing wastes into aesthetically acceptable and economically valuable co-products. This program, the Animal and Poultry Waste Management Center, provides the infrastructure (composting facilities, extruders, fluidized-bed dryers, pellet mills, waste handling and conversion equipment, etc.) to evaluate alternative waste management procedures and the development of co-products beyond the laboratory bench on a pilot or commercial scale basis. Interdisciplinary, broad-based, and cooperative research by industry, universities, and government agencies to address short and long term co-product management needs of the food animal agriculture industry is the primary focus of this program.

The animal and poultry waste management center concept

Numerous alternative processing procedures for food animal co-products have been proposed by researchers at many institutions in the US and abroad. However, most such alternatives have not been tested beyond the laboratory on a commercial or full scale by the investigators conducting waste management research and extension work due to the lack of infrastructure to do so. The North Carolina State University (NCSU) Animal and Poultry Waste Management (A&PWM) Center facilities are to be housed on approximately 5 acres at the NCSU Agricultural Field Laboratory, located approximately 8 miles south of the NCSU main campus. Two buildings (a 70ft × 100ft waste processing equipment building and a 36ft × 140ft composting building) will house the equipment necessary for research, development, and demonstration of advanced waste management technologies. Capital waste conversion equipment available for research utilization includes extruders, a fluidized bed dryer, cooler, roaster, mixer, screw press, and pellet mill. In addition, NCSU's poultry, swine and cattle research units, located on adjacent property, will supplement the A&PWM facilities. These units are equipped with grow-out facilities and standard feed mill equipment such as ribbon mixers, pellet mills, scales, bagging equipment, and meat grinders. Collectively, these facilities will provide the infrastructure for the continued evaluation of innovative alternatives for animal waste management.

The A&PWM Center, however, is not a facility as much as it is an organizational concept. In order to efficiently and effectively address the waste management requirements of the food animal industries, a broad based and interdisciplinary participation and input into the A&PWM Center activities is needed. The A&PWM Center has, therefore, established an operational structure in which NCSU, industry, commodity groups, other universities and government agencies may form a partnership to address the agricultural food animal waste management research area.

A&PWM CENTER OBJECTIVES

The A&PWM Center will be addressing all aspects of food animal waste management by utilizing a variety of technological approaches. The specific objectives for the A&PWM Center are:

1. To provide a modern facility and associated equipment (i.e., the infrastructure) for carrying out research and extension educational activities on the management and utilization of food animal waste products; and, for the development of economically and environmentally acceptable procedures for conversion of these wastes into value-added products for the food producing animal industries.
2. To provide personnel to operate the facility and its equipment on a daily basis and to work with the faculty and industry groups in carrying out the research and extension educational activities.

3. To provide the infrastructure that will allow faculty and organizations associated with the A&PWM Center to be successfully competitive for individual and multi-disciplinary research funding on a national basis in the waste management arena.
4. To provide the national and world food animal producing industries with economically feasible and safe alternatives for handling and recycling by-products and wastes produced by these industries in the course of food production.
5. To reduce nutrient output in the waste stream of food-producing animals through dietary manipulations to improve the nutrient utilization of traditional foodstuffs, as well as of any value-added foodstuffs which are developed from animal by-products produced by the Center.
6. To facilitate in-service training in new technologies for waste management for extension agents, agricultural agencies, waste management system operators, agribusiness personnel, and other technology-user groups.

Issues affecting the food-animal agriculture industry

Some key research, development, and education issues facing the food-animal agriculture industry which will be addressed by the A&PWM Center include:

- dietary manipulation to affect digestibility and nutrient concentrations in manure,
- odor control,
- utilization of farm animal mortality as well as other agricultural co-products in feedstuffs,
- nutrient management and utilization of manures as fertilizer,
- food safety and animal welfare,
- education of and communication between industry, academia, farmers, government agencies, and the consumer public regarding co-product utilization applications and the resulting environmental impacts.

Dietary manipulation to affect digestibility and nutrient concentrations in manure

A logical approach to reducing the environmental impact of nitrogen, phosphorus, copper, zinc, and other elements in manure is the improved efficiency of utilization of these nutrients by the animal. More work is needed to 'fine tune' the nutritional requirements of food animals as they relate to environmental issues as well as consumer requirements for meat, eggs, and milk. The efficacy of using exogenous enzymes in animal feedstuffs for improved utilization of oligosaccharides, phytate, and selected proteins needs to be determined. The effect of new feed processing technologies (dryers, extruders, expanders, etc.) on co-product nutrient availability and digestion also needs to be determined.

Odor control

One of the most sensitive environmental issues currently facing the food animal agriculture industry in general, and the pork producers in particular, is odor control. As a result of the ever increasing attention directed to this issue, many producers are considering the use of, or are already expending revenues for the use of, commercial products that are alleged to abate and/or control manure related odor. These products can generally be classified as chemical, enzymatic, or microbial in composition. Some are proposed to be incorporated into the diet. However, very little objective scientific evidence is available regarding effectiveness of most of these products as odor control agents which presents a dilemma for producers considering their use. Consensus protocol(s) by which dietary odor control products may be evaluated under controlled scientific procedures need to be established.

Utilization of farm animal mortality as well as other agricultural co-products in feedstuffs

Most food animal mortalities (especially for poultry and young swine) are currently being buried in pits, incinerated, or are being placed in sanitary landfills. Both practices are likely to be banned in the near future, at least in areas with certain soil types and/or high water tables, so alternatives must be developed to manage and recycle these potentially valuable co-products. Over the past few years, a number of poultry enterprises have begun utilizing the natural biological process of composting to handle their animal mortalities. This process provides a good, relatively low-cost alternative for disposing of animal mortalities in an environmentally safe manner, but still results in a product that must be land applied. Carcasses from large animals (beef, dairy, swine, horses, *etc.*) have for many years been rendered and returned to the animal food chain, but due to the small amount of mortality which generally occurs on a daily basis on poultry operations, it has not generally been considered economically feasible for those operations to transport their mortalities to the renderer. Thus, alternative systems need to be developed, demonstrated, and refined which will allow on-farm preservation and storage of the preserved carcasses so that larger quantities can be collected and transported to rendering, drying and/or extrusion facilities for conversion into animal feed-grade meals.

Offal from food animal processing plants has for many years been sold to rendering plants for conversion into feed-grade meat, bone and blood meals. Feathers, another major co-product of poultry processing plants, have been hydrolyzed and converted into feed-grade feather meals. A number of what would appear to be economically viable and better alternatives such as acid fermentation, fluidized bed drying, extrusion, and treatment with newly developed enzymes, have been introduced over the past few years, but these alternatives need to be scaled up and possibly modified so they can be demonstrated to be economically feasible if they are to become adopted by the animal industries.

Nutrient management and utilization of manures as fertilizer

Nutrient management presents a significant challenge to the food animal producer. Poor management by one or a few producers can have widespread effects on the entire industry regarding perception and potential legislation. Nutrients contained in co-products such as manure slurry or litter can be cycled from crops to feedstuffs to animals to soil and again to crops. Under proper management practices, this cycle can be maintained under economical and environmentally sound conditions. This includes analysis of the manure content for fertilizer value, uniform application rates to the soil at times of the crop growing cycle such that uptake is maximized, and subsequent adjustment of commercial fertilizer rates.

Food safety and animal welfare

The conversion of food animal co-products into feedstuffs draws attention to feed safety and animal welfare issues. Although the quality assurance and quality control evaluations regarding nutritional value and potential health hazards will be the same as for conventional feedstuffs, the risk that this practice may provide a vehicle to support 'agendas' such as diet/health, vegetarianism, and animal welfare is real and must be taken seriously. Some may argue that it is best to simply ignore the special interest groups that bring food animal agriculture under attack. We cannot accept that. To address such groups, cavalier attitudes and a siege mentality must be avoided by all in the food animal industries. A prepared and proactive response regarding the safety and nutritional value of food animal products, based upon knowledge that is unbiased, factual and scientifically based, will likely gain more positive results as well as the respect of the consumer public.

Education of, and communication among industry, academia, farmers, government agencies, and the consumer public regarding co-product utilization applications and the resulting environmental impact

As technology advances the economic and environmental benefits of co-product utilization, it is in our industry's best interest to not only utilize these applications but to 'market' the results. Communication and partnerships among the parties noted above will result in efficient identification of real and emerging issues and subsequent solutions to address issues concerning co-product management. If we pool our resources and talents to meet common goals and objectives to benefit society and the environment, food-animal agriculture and its allied industries will continue to grow and be profitable.

DEER AND DEER FARMING – THE NEW ZEALAND EXPERIENCE

PETER FENNESSY

AgResearch (New Zealand Pastoral Agriculture Research Institute Ltd). Invermay Agricultural Centre, Mosgiel, New Zealand

Introduction

International interest in the utilization of non-traditional animal species is increasing rapidly as reflected in numerous conferences and publications over the last 10–15 years (see Wemmer, 1987; Hudson *et al.*, 1989; Renecker and Hudson, 1991). The interest covers the spectrum from farming to managed utilization of animals in rangeland situations. Deer farming is the most spectacular example, stimulated by the developments in Scotland and New Zealand. However managed utilization of deer, which often involved herding, has been practised for centuries in the Arctic with the reindeer (Hudson *et al.*, 1989).

New Zealand is not alone in deer farming. In many ways the New Zealand developments are very recent in comparison with other cultures, most notably the Chinese who have been farming deer, keeping them in small enclosures and harvesting their velvet antler for 400 years or so. In fact there is reference to Chinese emperors of the Shang dynasty keeping deer, probably for game, over 3000 years ago (Xu, 1992); while 2000 years ago the Romans also kept deer for game (Anderson, 1985). The interest in the use of deer for medicinal purposes in China goes back to around the time of the Qing dynasty (221–207 BC) with references to the use of velvet antler and hard antler in Shen Nung Pentsao Ching of around AD 200 (Kong and But, 1985; Xu, 1992). Thus deer farming itself has a long history and the opening up of China and the USSR through the late 1970s revealed some of the extent and history of deer farming in these countries (Pinney, 1981; K.R. Drew, personal communication).

Deer and history

Deer have fascinated humanity for thousands of years – prehistoric cave drawings and the ancient Greek and Chinese writings all witness to that. In particular the antlers of deer have generated a folklore of their own. This is not surprising for a tissue of such spectacular appearance which grows so rapidly, turns into bone and is then replaced a few months later.

Many of the cultures which had contact with deer, such as the Celts and the Egyptians, found uses with mystical overtones for their antlers, while the Chinese used them for medicines and tonics. In essence it was sex and seasonality as reflected in the antler cycle that generated so much interest. Aristotle was struck by the effect of castration on the antlers – 'if stags are castrated before they are old enough to have horns, these never appear; but if castrated after they have horns their size never varies, nor are they subject to annual change' (after Cresswell, 1862, cited by Chapman, 1975). The Chinese saw deer with its regenerating antler as a lucky animal which brought health and longevity and have been using deer products for more than 2000 years in medicine (Kong and But, 1985). Longevity was also a feature of some of the European mythology surrounding the deer such as the old Gaelic rhyme '. . . thrice the age of a man is that of a deer . . .' (cited by Blaxter, 1979). In their 16th century writings it was the change in the male's demeanour and the dramatic change in appearance and the lack of appetite that so fascinated du Fouilloux and Tubervile (cited by Short, 1985) as 'during the time of their Rut they lyve with small sustenance . . . which helpeth well to make them pysse their greace'.

With all this interest and contact with deer through history, it is perplexing as to why deer, with the exception of the reindeer, were not domesticated. Some species, notably red and fallow deer, would seem to possess the first prerequisite in that they are social species. The practice of deer husbandry probably has a very long history and may even go back to pre-Neolithic times (Hudson *et al.*, 1989). However there is some evidence that with the development of agriculture in the Neolithic period, deer species gave way to other species such as sheep and goats (e.g. the Natufian culture of Palestine – Cole, 1963; Henry, 1985). Since domestication was almost certainly attempted thousands of years ago it is quite possible that it was seasonality and the associated difficulties of handling which led to failure. In this respect it is seasonality which has provided the challenge in our recent experiences in farming deer.

Deer farming in New Zealand

The New Zealand deer industry arose out of demand for venison in Germany. The wild red deer (*Cervus elaphus*) population of NZ had increased rapidly since the first introductions in 1851, such that by the 1950s they were regarded as a serious pest. In the 1960s large quantities of wild shot venison were exported and concerns about the future of this industry led to legislation allowing deer farming in 1969 (Drew, 1976a; Fennessy and Drew, 1983). From the establishment of the first deer farm in 1970, the industry grew so that by June 1994 there were 1.2 million farmed deer. The basis of the industry is venison, but there is a substantial income from velvet antler. Over 90% of the deer are red deer (mostly of *C.e. scoticus* background) and their hybrids with European strains of red deer and North American wapiti (both *C. elaphus* ssp). The remainder are virtually all European fallow deer (*Dama dama*).

For the first 10–15 years, handling, and management to deal with the seasonality of the deer, dominated the concerns of deer farmers. While farmers may not always recognize it as such, it is their seasonality which makes deer what they are as a farm animal and provides the challenge. Seasonality is most apparent in the reproductive cycles but this is not uncommon in domestic species. However, seasonality in deer exhibits many other facets, especially in males – the antler cycle, seasonal patterns of live weight change, seasonal changes in the composition of the animal body, especially the fat component, and of course, behaviour.

Biology of deer and farming

The interest in wild deer and their management throughout the world had generated a considerable amount of research by the late 1960s, especially with white-tailed deer in the US and red deer in Britain. However, the interest in red deer as a potential farmed species spawned research efforts in both Scotland in 1970 (Blaxter *et al.*, 1974) and in New Zealand in 1968 (Coop and Lamming, 1976), with New Zealand's second wave coming with the Invermay programme starting in 1973 (Drew, 1976b). These programmes marked a new phase of intensive research with the emphasis on the farming of deer. The publication by Blaxter *et al.*, (1974) of their early research provided a wealth of preliminary data. The Rowett group hand-reared deer calves to ensure they were amenable to handling although they noted that they were not sure this would be necessary in the future (Blaxter *et al.*, 1974). In contrast, the New Zealand approach involved development of management systems for grazing deer on pasture – the system had to deal with animals of all ages, many of which had been captured from the wild. The increasing sophistication of management came later with the offspring of these wild deer as they became accustomed to their human handlers. By the mid-1980s, the development of yarding and handling systems, including the use of body crushes, allowed deer to be handled like any other farm animal.

Venison and marketing

The ongoing development of export markets for the high quality farm-raised venison is the key to the future of deer farming in New Zealand. Venison accounted for over 70% of the total export receipts of nearly $200m from the deer industry in the 1994 year.

Quality is the key marketing feature of New Zealand venison. The New Zealand Game Industry Board (GIB) is the industry-funded body responsible for promoting and assisting in the orderly development of the deer industry and products derived from deer. Over the last 5 years the GIB has lead the move towards a repositioning of New Zealand's farm-raised venison in the international market. The accreditation of processing plants to ISO 9000 quality standards has been an integral part of this process. Transport operators and farmers are increasingly becoming involved in the accreditation process to ensure a total farm-to-market quality approach. An integral part of the total repositioning package is the industry brand, Cervena®, which was first launched in the USA and New Zealand in April 1993. The total quality approach is also a feature in other markets which operate under the Zeal® quality mark.

A basic component of the total quality approach is the leanness of the product. The rapidly growing young animal is very lean and the highly seasonal pattern of growth in these young animals indicates the obvious biological optima for slaughter of young stags – that is at the end of the spring–summer periods of rapid growth. Even though such stags will have reached 50–55% (at 15 months) or 65–70% (at 27 months) of their mature body weight, they are still very lean when compared with traditional farmed livestock such as sheep and cattle (red deer stags at 15 months will have only about 9% body fat). However market requirements (as reflected in prices paid to farmers) often dictate other slaughter times and as with all farm operations it is the balance between biological and financial considerations which determines the operation of the farming business. Clearly, seasonality has interesting implications for the venison industry.

Seasonality and nutrition

The seasonal pattern of body weight change has been a well-reported phenomenon in the cervid literature (French *et al.*, 1956; Wood *et al.*, 1962). The observations were repeated when measurements were taken of farmed deer (Blaxter *et al.*, 1974; Moore and Brown, 1977) and such growth patterns occurred even when young male and female deer were fed indoors on high quality diets *ad libitum* (Drew, unpublished data; Blaxter *et al.*, 1974). Such seasonal patterns lead to the conclusion by farmers that since the animals barely grew over winter and the stags ate virtually nothing over the rut, they did not require very much feed nor did they require high quality feed over the autumn–winter period. Unfortunately this overlooked the reality of fat mobilization during the rut and winter in stags, and maintenance energy costs, especially the impact of adverse climatic changes on energy demand during winter in both sexes, but especially in males and in young animals.

Nutritional experiments in the late 1970s resulted in strong messages to New Zealand farmers on winter feed requirements for deer. However, the publication of the feed requirement recommendations in 1981 (Fennessy *et al.*, 1981; Fennessy, 1981) had a major impact. The feed requirements emphasized the very high maintenance costs of deer, especially stags in winter (readily apparent when stags outdoors were compared with those indoors) and lactating hinds in summer. The demonstration of the benefits of improved nutrition (Harbord, 1982; Milligan, 1984) was a key to the uptake of this technology.

The improved nutrition of stags had some immediate results. The incidence of 'winter death' (McAllum, 1980) (essentially death resulting from a stag's inability to consume sufficient energy in the absence of readily mobilizable body fat reserves) and deaths from malignant catarrhal fever, a fatal viraemia caused by a bovid gammaherpes virus declined dramatically (van Reenen and Innes, 1985). As well, improved winter nutrition increased velvet antler yields in the following spring (Muir and Sykes, 1988; Fennessy, 1989). Essentially the adult red deer stag in winter is very susceptible to adverse environmental changes, as insulation is relatively poor and as well, the stag is left with very low fat reserves following the rut. The improved feeding reduced weight losses in early winter and in some cases even promoted body weight gain in late winter (Fennessy *et al.*, 1981). The critical management techniques to enable stags to weather southerly storms were the provision of shelter and adequate high quality (i.e. high energy) feed. A further management technique involved reducing the impact of the self-induced rut fast by providing stags with more paddock space during the rut. While the stags still reduced their feed intakes, weight losses tended to be lower and they were more settled and fought less. Special attention to the nutrition and space requirement of breeding stags after the rut also helped by minimizing further loss of condition over the winter (e.g. van Reenen and Innes, 1985).

The rut

The changes in behaviour and body condition of the red deer stag around the time of the rut (mating season) are striking. The rut inappetance enables the red stag to spend time in gathering and protecting his harem of hinds which increases the chances of mating and thus contributing his genes to the next generation (see Clutton-Brock *et al.*, 1982). The basics of the pattern of changes in body composition of stags around the rut have been known for hundreds of years as recorded by Tubervile (see Short, 1985). More recently Drew (1985) quantified this (Tables 1 and 2) in comparative slaughter experiments and then Jopson *et al.* (1993) used a whole body CAT scanner to follow the patterns of tissue mobilization in male fallow deer during the rut. They found that castrated fallow bucks mobilized both muscle and fat when they were fed at restricted levels over the 'rut' period in marked contrast to their rutting male contemporaries who 'naturally' restricted their feed intake but preferentially mobilized body fat, thus helping to conserve their lean body mass. However when entire males were feed-restricted in the spring (i.e. when testosterone levels were very low or undetectable) they too

Table 1. Composition (dissected components) of carcasses of adult male red deer stags slaughtered pre- and post-rut (Drew, 1985).

	Pre-rut	Post-rut
Carcass weight (CW, kg)	120	87
Composition (% CW)		
Fat	20.8	1.3
Lean	66.0	83.2
Bone	12.9	15.5

Table 2. Composition (chemical components) of the carcasses of adult red deer stags through the year (Drew, 1985).

Season	Month	Weight, kg		Components, kg			Water:Protein Ratio
		Body	Carcass	Water	Protein	Fat	
Autumn	March	203	122	68	23	25	2.91
	May	151	87	58	20	3	2.88
Winter	July	146	82	56	17	2	3.26
Spring	September	134	78	54	17	2	3.09
	November	168	96	53	19	19	2.77
Summer	December	181	111	62	23	19	2.64
	February	196	112	62	22	23	2.86
Autumn	March	203	122	68	23	25	2.91

Table 3. Rates of carcass lean tissue mobilization in castrated fallow deer during the rut and entire fallow deer in the spring compared with entire males over the rut. All deer were fed at the same level as the entire males during the rut. The predicted rates of tissue mobilization are those for the entire males during the rut from the negative exponential relationship between lean tissue mobilization (y) and initial fat content (x) where y=0.904 (1+0.604x); F2,8=19.6* (Jopson *et al.*, 1993).**

	Rate of lean tissue mobilization (kg/kg empty body weight)	
	Actual ± s.e.	Predicted
Castrate males in rut	0.26 ± 0.038	0.12
Entire males in spring	0.35 ± 0.042	0.18

mobilized muscle and fat tissue at a similar rate to the castrates (i.e. at around twice the rate of entire males during the rut, Table 3). Thus these studies provided convincing evidence that it was some factor associated with the high testosterone environment during the rut which 'protected' the stag's musculature while preferentially drawing upon fat reserves to meet the energy deficit. Interestingly, fallow bucks which had not had the opportunity to build up fat reserves before the rut, maintained a substantially higher food intake during the rut than their contemporaries which had accumulated large fat reserves (Jopson, 1993).

While the cycle is fascinating, there is arguably an even more fascinating phenomenon associated with the rut – the accumulation of water in the musculature. This was first observed in comparisons of young castrated and entire red stags where the water to protein ratios were markedly different, especially in the androgen-responsive neck muscles (Tan and Fennessy, 1981; Table 4). However subsequent observations of adult stags held indoors revealed extraordinary body weight gains in the pre-rut period while food intake was declining rapidly. The peculiarities

Table 4. Composition of selected neck muscles in 2-year-old castrate and entire red deer stags around the time of the rut (the ratio of water to protein in the complete half carcasses were 2.95 and 3.14 for castrates and entires respectively, $P<0.01$) (Tan and Fennessy, 1981).

Muscle	Group	Muscle weight,g	Components, %			Water:Protein ratio
			Water	Protein	Fat	
Rhomboideus	Castrate	145	74.6	21.5	2.6	3.47
	Entire	229	77.1*	19.4*	2.4	3.98*
Splenius	Castrate	55	75.1	21.5	2.3	3.50
	Entire	120	78.0**	18.9*	1.8	4.13*

of the change in the water to protein ratio are also readily apparent in the compositional data in Table 2.

The rut is not an 'all or nothing' phenomenon. The severity of the rut, as defined by bodyweight loss, is highly variable among stags. For example, the food intake depression in young red deer stags during the first rut at around 16 months of age is small, but as the stag ages, the severity of the rut also increases. The impact of these changes with age and the variability between animals in the live weight change are apparent in Table 5. These data are from an indoor study where stags were fed high quality diets *ad libitum* – the very high standard deviations for the live weight change over the rut indicate this variability.

While the rut is a spectacular manifestation of seasonality, stags actually exhibit two seasonal cycles. The underlying cycle is under photoperiodic control and follows the changing pattern of daylength with a lag of around 6–10 weeks. This cycle occurs in both males and females and is marked by a rising food intake in the spring and a declining food intake in the autumn. Table 6 presents some data from a series of indoor experiments which show the amplitude of the cycle and the lag phase. In essence, the annual rut depression in food intake is superimposed on this annual cycle in the stag. It is the interaction of these two cycles of food intake which results in the stag being so vulnerable to the exigencies of the environment over winter.

Table 5. Mean liveweight change (kg ± sd) over the rut (pre-rut maximum to rut minimum) for two groups of red deer stags fed indoors on high quality diets *ad libitum* (Fennessy *et al.*, 1991a).

Rut	Live weight change (kg ± sd)	
	Experiment 1	Experiment 2
Number of animals	6	10
Yearling	−4.7 ± 2.3	+1.3 ± 3.0
2 year	−9.6 ± 3.8	−17.3 ± 13.5
3 year	−28.2 ± 10.7	−29.4 ± 15.7

Table 6. Parameters of food intake in red deer stags and hinds (Fennessy *et al.*, 1991a)

	Stags	Hinds
Food intake at maturity (C, MJ ME/week)*	213	144
Oscillation in food intake (d,MJ ME/week)	47	25
Proportional oscillation (d/C)	0.22	0.18
Phase delay in food intake from photoperiodic oscillation (weeks)	6.4	9.9

*Food intake at maturity is effectively the feed requirement for maintenance.

Mobilization of energy reserves to service the energy deficit over the rut is one thing, but then to be followed by a depressed voluntary intake over the winter adds to the problem. This dual cycle and its ultimate impact on survival, especially in the wild (see Clutton-Brock *et al.*, 1982), raises some intriguing evolutionary questions (Fennessy *et al.*, 1991a). In this respect, there is evidence that the intake depression over the rut is subject to behavioural modification. For example, observations show that stags who have been displaced or defeated in an encounter with another stag for access to hinds will increase their food intake and thus suffer a lesser weight loss (e.g. Rapley, 1985). At least this will increase their chances of survival in that year and thus provide them with another opportunity to spread their genes the following year. A further intriguing observation from indoor feeding studies was that the presence of older stags in close proximity to younger stags resulted in a much smaller than expected depression in food intake in the latter (Fennessy *et al.*, 1991a). The mechanism is the fascinating question.

The antler cycle

The annual antler and reproductive cycles are synchronized in the temperate species of deer (see Goss, 1983). The antler, which is deciduous, grows from the pedicles, which are permanent boney outgrowths of the frontal bone. In the young male deer, the pedicle grows in response to the rise in testosterone associated with the first phase of puberty (Suttie *et al.*, 1984). The initiation of puberty is a function mainly of the stag's bodyweight and in farmed red deer in New Zealand occurs in late winter/early spring when the stags are 6–9 months old (Fennessy, 1982; Meikle *et al.*, 1992).

The first antler can be clearly seen when the pedicle is about 5 cm long and thereafter the first antler grows rapidly. As the antler is laid down the mesenchymal cells at the tip differentiate into cartilaginous tissue (see Suttie and Fennessy, 1991). The subsequent bone formation and stripping of the velvet or skin as the antler reaches its final stage of maturation are under the control of testosterone. In the red deer, the old hard antlers are cast when the circulating testosterone declines to very low or undetectable levels in the spring (Lincoln, 1971; Fennessy *et al.*, 1988).

THE GROWING ANTLER

The antler grows at an extraordinary rate especially in the larger subspecies of *Cervus elaphus* where a 500 kg stag can be expected to grow an antler weighing around 21 kg (Huxley, 1931). For example even in a young stag with a final antler weight of around 2 kg the rate of linear growth of the main beam during the main phase of growth is around 5 cm per week (Fennessy *et al.*, 1991b). This represents a considerable deposition of nutrients, especially calcium and phosphorus (see Hyvarinen *et al.*, 1977). While the timing of the antler cycle in temperate species is regulated by the reproductive hormones, there is a dearth of knowledge as to what controls or influences the rate of growth and hence the ultimate antler size. However there is evidence that IGF-I may have a role in that the rate of antler growth in stags is correlated with the plasma IGF-I concentration (Suttie *et al.*, 1985; 1989). In a further fascinating experiment, this group (Suttie *et al.*, 1988) removed the pedicles and antlers from three stags and compared the plasma IGF-I levels with those in intact stags over the antler growth period in spring. Those without antlers had IGF-I levels which were up to 60% higher. Significant IGF-I binding in the antler tip (in the absence of growth hormone binding) provides more evidence for a trophic role for IGF-I (Elliott *et al.*, 1992). Antler tip cells *in vitro* are highly responsive to both IGF-I and IGF-II (Sadighi *et al.*, 1994) but more recent evidence would suggest that both operate through the Type I receptor (M. Sadighi and J.M. Suttie, unpublished data). The antler tip promises to be a particularly rewarding area. For example, a feature of these cells is the relative concentration of selenium compared with the more differentiated cell types further down the antler (Suttie and Fennessy, 1991). The concentration in the growing tips is five- to ten-fold higher than any other part of the antler. However, virtually none of this selenium can be associated with glutathione peroxidase (GSH-Px) as the levels of GSH-Px (and other antioxidants such as superoxide dismutase, SOD) in the antler tip are very low (Table 7). The actual form of the selenium is not known at this stage. The overall situation with the antioxidant enzymes in the growing antler tip is similar to some tumours which exhibit low levels of antioxidants presumably because they

Table 7. Selenium content, glutathione peroxidase (GSH-Px) and superoxide dismutase (SOD) activity in the rapidly growing antler tip (*n*=3 antlers) (Suttie and Fennessy, 1991) compared with rat liver.

Distance from the antler tip (cm)	Selenium (mg/kg DM)	GSH-Px (IU $\times^{-6}$)	SOD μ
0–0.5	0.90	20.3	3.03
1.0–1.5	0.60	13.7	2.43
2.0–2.5	0.47	14.7	2.87
SED	0.07	1.8	0.24
Rat liver	3.63	12.9	20.6

rely largely on glycolysis rather than oxidative phosphorylation for their energy source (Suttie and Fennessy, 1991). The parallels with tumours have been a feature of the literature on antlers for more than 60 years (Steinwedel, 1929).

NUTRITION AND ANTLER GROWTH

The putative role for IGF-I as trophic for antler growth and the frequently described effects of nutrition on circulating IGF-I stimulated further interest in the effects of nutrition on antler growth. While the level of nutrition does influence antler size in that stags fed at levels well below their *ad libitum* intakes grow smaller antlers (Table 8), the effect is relatively minor. As well, the level of energy intake in the winter while the stags are still in hard antler (i.e. before the antler growth period) can affect subsequent antler size (Table 9). However, in stags which are well-fed over winter any nutritional effects on antler growth are negligible, although the costs of high energy supplements (e.g. grains) in order to ensure that stags are well-fed are probably economic. Some early studies at Invermay did suggest that the level of bypass protein in the diet (and hence protein: energy ratio at the duodenum) during the winter (before antler casting) influenced the rate of antler growth in the subsequent spring. However, while a hypothesis relating

Table 8. Effect of restricting feed intake on antler weight; intake of the restricted group was 80% of the *ad libitum* group for 65 days from hard antler casting (n=5 individually-fed 4-year-old stags per group; one antler was removed at 65 days and the other within 7 days after stripping of the velvet).

	Weight of one antler (kg)	
Level of nutrition	At 65 days	At maturity (kg)
Ad libitum	1.10	1.59
Restricted	0.94$^{(*)}$	1.48^{NS}

Table 9. Comparison of the effect of food intake over winter on subsequent velvet antler weight (kg) on four farms (A and D – Muir and Sykes, 1988; B and C – Fennessy 1989).

	Velvet antler weight (kg)		
	Hay	Hay + restricted supplement	Hay + *ad libitum* supplement
Property A	1.22	1.38	1.46
Property B	1.37	1.61	1.73
Property C	1.77	1.94	1.77
Property D	2.28	2.30	2.17

to a modification of the endocrine environment was highly attractive, subsequent experiments have not been able to repeat the early results (Fennessy, 1989). Thus the search for a practical method to increase the rate of antler growth goes on, but to date it is in genetics that the greatest impact has been made.

GENETICS AND ANTLER GROWTH

New Zealand livestock farmers have a strong interest in genetic improvement and this has extended to deer especially for antler growth. The possibilities for improved genetics based on the local herd were being explored during the late 1970s and early 1980s (e.g. Fennessy, 1982). At this time the emphasis was on the different strains of red deer derived from the various importations a century or so earlier. As well as the variation between strains, that within strains also offers the potential as a source of superior animals. While there is considerable variability in velvet antler weight within a group of stags cut at the same stage of growth, the major question relates to the genetic component of this variability. There are few data available but Table 10 summarizes one progeny test of five selected high-performing sire stags with the progeny of the top sire (E) around 7% above average for cumulative velvet antler weight over 4 years. In a more extensive analysis involving the same data set plus a further five sires, there was a 30% range in velvet antler weight between progeny groups (Ball *et al.*, 1994). Clearly there is considerable potential for selection and breeding even within strains considering that there was little difference in live weight (and hence estimated feed costs) between the different sire groups.

In the early 1980s, hybridization became an option. It was the natural hybridization between the larger North American wapiti (*C.e. nelsoni*) and the smaller red deer in the Fiordland National Park (Smith, 1974) which alerted deer farmers to the possibility of such hybridization to increase body size and antler size in the farmed deer. Other examples of hybridization including that where populations of red deer and the smaller sika deer have hybridized (Harrington, 1982) and that between

Table 10. Progeny test of five red sires: comparison of the mean cumulative velvet antler production (2–5 years of age) and the 3-year-old winter lean liveweight.

Sire	Number of male progeny n=301)	Deviations from average, kg	
		Cumulative velvet antler weight (mean 9.11 kg)	3-year-old winter lean liveweight (mean 126 kg)
A	29	−0.46	+4.6
B	32	−0.45	−1.4
C	22	−0.27	−0.7
D	17	+0.37	+4.6
E	35	+0.61	+2.0

malu and meihualu in China (Xu, 1992) also provided a stimulus. It is the remarkable capacity for hybridization among the greater red deer–wapiti family which offers so much flexibility to the deer producer to produce the appropriate products (both venison and velvet antler) for the market.

Investigation of the possibilities of new genotypes indicated that the red deer originating from mainland Europe were apparently genetically larger than the Scottish red deer which formed the basis of the New Zealand farmed deer industry. As well, there was the well-known relationship between antler size and body size where hard antler weight increased at a rate about 1.6 times faster than that of body weight (data from over 500 red deer stags from around Europe, Huxley, 1931. The selection for antler size (and often numbers of antler points) in the European deer parks also added to the possibilities for New Zealand deer farmers searching for new genetic material. The large wapiti subspecies were also available from Canada. Consequently there were large numbers of red deer from Europe and wapiti from North America imported during the 1980s. The impact of these importations together with the selection within herds is now becoming very apparent in the quality of velvet antler being produced.

Antlers in traditional medicine

No discussion of antlers would be complete without a brief mention of the use of antlers in traditional medicine and the increasing international interest in this area. The emphasis in traditional Chinese medicine on the promotion of health/prevention of illness is in marked contrast to that largely practised in Western medicine, which is more concerned with the treatment of ill health (Fulder, 1980). It is in the 'health tonic' area where (velvet) antler has traditionally found a niche although it is also used in the treatment of a number of specific conditions (Kong and But, 1985; Yoon, 1989). The use of velvet antler as a tonic and the importance of sexual wellbeing in the Oriental tradition, have resulted in velvet antler generally being regarded by Western commentators as an aphrodisiac. This is unfortunate and has resulted in velvet antler being disregarded as a serious candidate for pharmacological activity or application. It is therefore somewhat ironic that Yoon (1989) notes that about 70% of the velvet antler users in his Korean clinic are children.

At the stage when the growing antler is harvested as velvet antler for use as a high quality product in traditional Oriental medicines, it is an actively growing cartilage-type tissue and is not of uniform composition. The degree of mineralization or calcification of the velvet antler is generally regarded as an indicator of the likely pharmacological activity, with the more calcified tissues being downgraded. Traditionally, the different parts of the antler have different uses. The upper portions are often used as preventative medicines or tonics in children and young people, the middle portion for treatment of arthritis and other bone- and joint-related conditions and the lower portion for older people (see Yoon, 1989; Young, 1990).

There is increasing evidence in the scientific literature for a number of biologically active features of products derived from antler. For example, studies have shown improved stamina in both mice and humans (Yudin and Dobryakov, 1974) while a number of animal studies have also identified gonadotrophic (Kong and But, 1985) and haematopoietic effects (Yong, 1964). Much of this literature has been summarized by Fennessy (1991). The New Zealand studies now involve the development of *in vitro* cell line procedures to assess the biological activity of various antler fractions or extracts.

Clearly a more detailed scientific understanding of the bioactive components of velvet antler is necessary to define the nature of the compounds and their effects in animal systems. This search may yield new bioactive molecules with pharmacological activity but which could also provide new insights into the regulation of differentiation, growth and metabolism.

Deer opportunities

The opportunities highlighted by our studies of deer are numerous. In particular, genetic, endocrinological and nutritional possibilities abound. There are opportunities for new types of 'functional foods' in addition to the yeasts and organic minerals. Natural supplements with an endocrine effect such as those which modify hormonal patterns which then have longer term effects are one possibility. For example, they could well have a role in influencing antler growth or food intake post-rut. Supplements which modify the endocrine environment of the pregnant female and thus influence the fetus and subsequent post-natal growth are a real possibility.

Working with deer has provided numerous insights into aspects of biology which are relevant to our farming and management of more traditional species. This is especially so in relation to seasonality with respect to reproduction, food intake and tissue mobilization while the antler has opened up whole new areas. Undoubtedly our studies of deer have provoked us in to thinking about animal management in terms of the evolutionary biology of the other species we also farm.

References

Anderson, J.K. 1985. Hunting in the Ancient World. University of California Press, Berkeley.

Ball, A.J., J.M. Thompson and P.F. Fennessy. 1994. Relationship between velvet antler weight and liveweight in red deer (*Cervus elaphus*). N. Z. J. Agric. Res. 37:153.

Blaxter, K.L. 1979. *Cervinus annos vivre*: an account of opinion about the length of life of the deer. Br. Vet. J. 135: 591.

Blaxter, K.L., R.N.B. Kay, G.A.M. Sharman, J.M.M. Cunningham and W.J. Hamilton. 1974. Farming the Red Deer. Dept of Agriculture & Fisheries for Scotland.

Chapman, D.I. 1975. Antlers – bones of contention. Mam. Rev. 5:121.

Clutton-Brock, T.H., F.E. Guinness and S.D. Albon. 1982. Red deer: behaviour and ecology of two sexes. University of Chicago Press, Chicago.

Cole, S. 1963. The Neolithic Revolution. 3rd edn. British Museum (Natural History), London.

Coop, I.E. and R. Lamming, 1976. Observations from the Lincoln College Deer Farm. In: Deer Farming in New Zealand, Progress and Prospects, N.Z. Soc. Anim. Prod. Occ. Publ. 5. K.R. Drew and M.F. McDonald (Eds), p.32.

Drew, K.R. 1976a. The farming of red deer in New Zealand. World Rev. Anim. Prod. 12:49.

Drew, K.R. 1976b. In: Deer Farming in New Zealand, Progress and Prospects, N. Z. Soc. Anim. Prod. Occ. Publ. 5. K.R.Drew and M.F. McDonald (Eds), p.1.

Drew, K.R. 1985. Meat production from farmed deer. In: Biology of Deer Production, The Royal Society of N. Z. Bulletin 22. P.F. Fennessy and K.R. Drew (Eds), p.285.

Elliott, J.L., J.L. Oldham, G.R. Ambler, J.J. Bass, G.S.G. Spencer, S.C. Hodgkinson, B.H. Breier, P.D. Gluckman and J.M. Suttie. 1992. Presence of insulin-like growth factor-I receptors and absence of growth hormone receptor in the antler tip. Endocr. 130:2513.

Fennessy, P.F. 1981. Nutrition of red deer. In: Proc. of a Deer Seminar for Veterinarians (N.Z. Veterinary Assn), p.8.

Fennessy, P.F. 1982. Growth and nutrition. In: The Farming of Deer: World Trends and Modern Techniques. D. Yerex (Ed.), Agricultural Promotion Associates, Wellington, p.105.

Fennessy, P.F. 1989. High returns from velvet antler. In: Proc. Ruakura Deer Industry Conf., p.21.

Fennessy, P.F. 1991. Velvet antler: the product and pharmacology. Proc. Deer Course for Veterinarians (Deer Branch NZ Vet. Assn) 8:169

Fennessy, P.F. and K.R. Drew. 1983. Development of deer farming in New Zealand. Phil. J. Vet. Animal. Sci. 9:197.

Fennessy, P.F. and J.M. Suttie. 1985. Antler growth: nutritional and endocrine factors. In: Biology of Deer Production, The Royal Society of N. Z. Bulletin 22. P.F. Fennessy and K.R. Drew (Eds), p.239.

Fennessy, P.F., G.H. Moore and I.D. Corson. 1981. Energy requirements of red deer. Proc. N. Z. Soc. Anim. Prod. 41:167.

Fennessy, P.F., J.M. Suttie, S.F. Crosbie, I.D. Corson, H.J. Elgar and K.R. Lapwood. 1988. Plasma LH and testosterone responses to gonadotrophin-releasing hormone in adult red deer (*Cervus elaphus*) stags during the annual antler cycle. J. Endocrinol. 117: 35.

Fennessy, P.F., J.M. Thompson and J.M. Suttie. 1991a. Season and growth strategy in red deer: evolutionary implications and nutritional management. In: Wildlife Production: Conservation and Sustainable Development. L.A. Renecker and R.J. Hudson (Eds), Univ. of Alaska, Fairbanks, AFES misc. publ. 91–6: 495.

Fennessy, P.F., I.D. Corson, J.M. Suttie and R.P. Littlejohn 1991b. Antler growth patterns in young red deer stags. In: Biology of the Deer. R.D. Brown (Ed.), Springer-Verlag, p.487.

French, C.E., L.C. McEwan, N.D. Magruder, R.H. Ingram and R.W. Swift. 1956. Nutrient requirements for growth and antler development in the whitetailed deer. J. Wildl. Manage. 20:221.

Fulder, S. 1980. The hammer and the pestle. New Scientist 87 (1209):120.

Goss, R.J. 1983. Deer Antlers. Regeneration, Function and Evolution. Academic Press, New York.

Harbord, M. 1982. An approach to advisory work with deer farmers in Southland. Proc. N. Z. Soc. Anim. Prod. 42:153.

Harrington, R. 1982. The hybridisation of red deer (*Cervus elaphus* L 1758) and Japanese sika deer (*C. nippon* Temminck 1838). In Transactions Fourteenth Int. Congr. Game Biologists. F. O'Gorman and J. Rochford (Eds). Irish Wildlife Publ., Dublin, p. 559.

Henry, D.O. 1985. Preagricultural sedentism: The Natufian example. In: Prehistoric Hunter-Gatherers: The Emergence of Cultural Complexity. T.D. Price and J.A. Brown (Eds). Academic Press, New York, p. 365.

Hudson, R.J., K.R. Drew and L.M. Baskin (Eds). 1989. Wildlife Production Systems: Economic Utilisation of Wild Ungulates, Cambridge University Press, Cambridge.

Huxley, J.S. 1931. The relative size of the antlers of the red deer (*Cervus elaphus*). Proc. Zool. Soc. Lond. 72:819

Hyvarinen, H., R.N.B. Kay and W.J. Hamilton. 1977. Variation in the weight, specific gravity and composition of the antlers of red deer (*Cervus elaphus*). Br. J. Nutr. 38:301

Jopson, N.B. 1993. Physiological adaptations in two seasonal cervids. PhD. thesis, Univ. of New England, Armidale, Australia.

Jopson, N.B., J.M. Thompson, and P.F. Fennessy. 1993. Body compositional changes during fasting periods in fallow deer bucks. In: Proc. First World Forum on Fallow Deer Farming. G.W. Asher (Ed.), p. 199.

Kong, Y.C. and P.H. But. 1985. Deer – the ultimate medicinal animal. In: Biology of Deer Production, The Royal Society of N. Z. Bulletin 22. P.F. Fennessy and K.R. Drew (Eds), p. 311.

Lincoln, G.A. 1971. Puberty in a seasonally breeding male, the red deer stag (*Cervus elaphus* L.). J. Reprod. Fert. 25:41.

McAllum, H.J.F. 1980. Winter death syndrome in deer: exposure and starvation. AgLink AST92, NZ Ministry of Agriculture and Fisheries, Wellington.

Meikle, L.M., P.F. Fennessy, M.W. Fisher, and H.J. Patene. 1992. Advancing calving in red deer: the effects on growth and sexual development. Proc. N. Z. Soc. Anim. Prod. 52:187

Milligan, K.E. 1984. Deer nutrition: feed demands and how to meet them. Proc. Deer Course for Veterinarians (Deer Branch NZ Veterinary Assn) 1:46.

Moore, G.H. and C.G. Brown. 1977. Growth performance in farmed red deer. N. Z. Agric. Sci. 11(4):175.

Muir, P.D. and Sykes, A.R. 1988. Effect of winter nutrition on antler development in red deer (*Cervus elaphus*): a field study. N. Z. J. Agric. Res. 31(2):145.

Pinney, B. 1981. Delegation to China. The Deer Farmer (Spring): 22–35.

Rapley, M.D. 1985. Behaviour of wapiti during the rut. In: Biology of Deer Production, The Royal Society of N.Z. Bulletin 22. P.F. Fennessy and K.R. Drew (Eds), p. 357.

Renecker, L.A. and R.J. Hudson (eds). 1991. Wildlife Production: Conservation and Sustainable Development. AFES misc. pub. 91–6. University of Alaska, Fairbanks.

Sadighi, M., S.R. Haines, A. Skottner, A.J. Harris and J.M. Suttie. 1994. Effects of insulin-like growth factor-I (IGF-I) and IGF-II on the growth of antler cells *in vitro*. J. Endocrinol. 143:461.

Short, R.V. 1985. Deer: yesterday, today, and tomorrow. In: Biology of Deer Production, The Royal Society of N. Z. Bulletin 22. P.F. Fennessy and K.R. Drew (Eds), p. 461.

Smith, M.C.T. 1974. Biology and Management of the Wapiti (*Cervus elaphus nelsoni*) of Fiordland, New Zealand. NZ Deerstalkers Assn. Inc., Wellington, New Zealand.

Steinwedel, M. 1929. Histologische und histogentische Studien uber das Geweih von *Cervus elaphus*, von *Cervus capreolus* und *Rangifer tarandus*; zugleich ein Beitrag zur Kenntnis des Verknocherungsvorganges. Inaugural dissertation, Doctor medicinae veterinariae der Tierarztlichen Hochschule, Berlin.

Suttie, J.M. and P.F. Fennessy. 1991. Recent advances in the physiological control of velvet antler growth. In: The Biology of Deer. R.D. Brown (Ed.), Springer-Verlag, p.471.

Suttie, J.M., G.A. Lincoln and R.N.B. Kay. 1984. Endocrine control of antler growth in red deer stags. J. Reprod. Fert. 71:7.

Suttie, J.M., P.D. Gluckman, J.H. Butler, P.F. Fennessy, I.D. Corson and F.J. Laas. 1985. Insulin-like growth factor 1: antler stimulating hormone? Endocr. 116:846.

Suttie, J.M., P.F. Fennessy, P.D. Gluckman, and I.D. Corson. 1988. Elevated plasma IGF-1 levels in stags prevented from growing antlers. Endocr. 122:3005.

Suttie, J.M., P.F. Fennessy, I.D. Corson, F.J. Laas, S.F. Crosbie, J.H. Butler, and P.D. Gluckman. 1989. Pulsatile growth hormone, insulin-like growth factor and antler development in red deer (*Cervus elaphus scoticus*) stags. J. Endocrinol. 121:351.

Tan, G.Y. and P.F. Fennessy. 1981. The effect of castration on some muscles of red deer (*Cervus elaphus* L.). N.Z. J. Agric. Res. 24:1

Yoon, P. 1989. Velvet pharmacology. In: Proc 14th Deer Farmers Conf., p.11.

Yong, Y.I. 1964. The effect of deer horn on the experimental anaemia of rabbits. J. Pharmaceutical Society Korea 8:6.

Young, L.M. 1990. Health plan: do you take deer horn with any understanding of it? In: translation of article from Korean Airlines Magazine, p. 106.

Yudin, A.M. and Y.I. Dobryakov. 1974. A guide for the preparation and storage of uncalcified male antlers as a medicinal raw material. In: Reindeer Antlers. (Translated by M. Chapin). I.I. Brekhman (Ed.). Academy of Sciences of the USSR Far East Science Centre, Vladivostock.

van Reenen, G.M. and J.I.S. Innes. 1985. Preventive veterinary medicine in a commercial red deer herd. In: Biology of Deer Production, The Royal Society of N. Z. Bulletin 22. P.F. Fennessy and K.R. Drew (Eds), p. 87.

Wemmer, C.M. (Ed.). 1987. Biology and Management of the Cervidae. Smithsonian Institution Press, Washington.

Wood, A.J., I.McT. Cowan, and H.C. Nordan. 1962. Periodicity of growth in ungulates as shown by the genus *Odocoileus*. Can. J. Zool. 40:593.

Xu, H. 1992. Deer farming and management. In: The Deer in China. H. Sheng (Ed.), East China Normal University Press, p. 267.

OSTRICH PRODUCTION – A SOUTH AFRICAN PERSPECTIVE

W.A. SMITH[1], S.C. CILLIERS[2], F.D. MELLETT[1] and S.J. VAN SCHALKWYK[2]

[1]*Department of Animal Sciences, University of Stellenbosch, Stellenbosch 7600, South Africa*
[2]*Little Karoo Agricultural Development Center, Oudtshoorn Experimental Farm, P.O. Box 313, Oudtshoorn 6620, South Africa*

Introduction

Although the first ostrich farm in South Africa was established in 1838 to supply feathers (Sauer *et al.*, 1972), commercial ostrich farming in the Little Karoo of the Cape Province was only recognized between 1857 and 1864 as a new branch of the agricultural industry (Smit, 1963). The ostrich feather trade, however, is several thousand years old, dating back to the period of the early civilizations of Egypt, Assyria and Babylonia.

In 1913 ostrich feathers was the fourth largest export product of South Africa (Smit, 1963). Only gold, diamonds and wool earned more international money. The world market for ostrich feathers collapsed completely with the 1914 world depression. Ostrich numbers in South Africa decreased from 776 313 in 1913 to 32 500 in 1930 (Smit, 1963). By 1945 most ostrich farmers in South Africa had either slaughtered or sold all their ostriches. However, a few farmers in the Little Karoo kept only their 'highest quality' birds hoping that the ostrich industry would revive. In 1946 there was a renewed interest in ostrich feathers with the revival of the world economy and ostrich numbers increased steadily to 100 000 in 1983. The South African ostrich industry produced 51 500 hides, 117 tons of feathers (mainly for the duster industry) and 1500 tons of meat during 1983. Of the total income 58% was from hides, 25.9% from feathers and 16.1% from the meat (Swart, 1985). During 1993 there were approximately 200 000 ostriches in South Africa, while the total export of ostrich products reached R190 million (approx. US$54 million). Of the total income 76% was from hides, 7.5% from feathers and 16.5% from meat (Anon, 1994).

The aim of this chapter is to provide a short overview of only the most recent South African developments in the areas of:

- Genetics and reproduction
- Layer bird management and egg hatching
- Nutrient requirements and feedstuffs values
- Diseases and mortality
- Ostrich products
- The future of the ostrich industry

Genetics and reproduction

No breeding structures for ostriches existed in South Africa until the first breeding society was founded in 1994.

Ostriches can live a long time and layer birds with an age of 25 years or more are quite common. Male and female birds are ready for breeding at the age of 3½ and 2½ years, respectively. The average age of layer birds is therefore a major constraint in achieving genetic progress.

The heritability (h^2) of live body weight is relatively high for most animal species. Therefore it can be expected that fast results can be achieved when selecting for body weight.

The repeatability of several characteristics has been established for female birds at the Oudtshoorn Experimental farm. Repeatability is an indication of the ability of the female bird to repeat a specific level of production in consecutive years. The repeatability of a characteristic is normally higher than its heritability and the repeatability value can be regarded as a possible maximum heritability value for the specific characteristic.

The repeatability (%) of various production parameters for ostriches are as follows (Cilliers and Van Schalkwyk, 1994):

Egg production (per year)	20
Egg weight	74
Percentage infertile eggs	27
Percentage embryonic deaths	10
Brooding time	19
Average daily gain	39
Percentage eggs that hatch	14

Averages for reproduction performance parameters of ostriches are as follows (Cilliers and Van Schalkwyk, 1994):

Percentage hatchability of eggs	50 ± 30
Laying days per year	120 ± 20
Egg production per season (Jun.–Feb.)	50 ± 20
Days from mating till nesting	14 ± 7
Breeding cycles per breeding pair per season (natural)	2 ± 1
Eggs per nest (natural)	12 ± 6

The average chick mortality percentages are as follows (Cilliers and Van Schalkwyk, 1994):

Day old till 3 months	50 ± 30%
3–6 months	10 ± 8%
6–14 months	3 ± 2%

Cilliers and Van Schalkwyk (1994) made the following practical comments:

- There was no significant difference in the live weight of good and bad layers.
- Male body weight had a negative correlation with egg production.

- There was a tendency for bad layers to have a larger chest circumference.
- The best layers had longer legs with visually lighter thighs.
- Most reproduction parameters had a relative high repeatability value and should be included in a selection program.
- Best hatchability percentages were achieved with eggs of average weight.
- Poor hatchability percentages were achieved with heavy eggs.
- Egg weight responded positively on selection.

Layer bird management and egg hatching

The breeding season of the ostrich is synchronized photoperiodically starting during periods of decreasing daylight (fall) and lasting well into spring. Approximately 10 days after mating the female bird begins laying eggs at the rate of one egg every two days until 12–15 eggs have been laid. Eggs must be collected on a daily basis.

The practical egg hatching program entails daily collection, fumigation (80 g potassium permanganate in 130 ml 40% formaldehyde solution per m^2 for 20 min), washing or spraying eggs with a commercial sterilizing agent (0.05 or 0.1% solution), storing eggs for a maximum of 7 days in an upright position at 15–20°C and 75–80% relative humidity. Eggs must be turned once every 24 h while being stored.

Before incubation eggs should be preheated to 25°C for a minimum period of 12 h. Within incubators eggs are placed in a horizontal position for the first 2–3 weeks. For the first 3 days eggs are turned (180°) three times daily. For the rest of the hatching period eggs are kept in a vertical position with the air-sac at the top. In electronic incubators eggs are placed at a 45° angle and turned 24 times a day automatically. The incubator temperature is kept at 36°C and the relative humidity at 28–34%.

Infertility of eggs and embryonic deaths are the major problems in ostrich production in South Africa. Of the more than one million eggs that are laid yearly in the Little Karoo area, only 45% hatch (Cilliers and Van Schalkwyk, 1994). This is much lower than the generally accepted 85% for turkeys and 93% for broilers and layers. On average 25% of all eggs collected are infertile while embryonic deaths account for 20% of unhatched eggs (Van Schalkwyk, 1995).

Infertile eggs have several causes. Infertility in male birds (or low sperm count) is a major factor. Ostrich producers in South Africa have just started semen evaluation of male birds. Overly fat females also tend to lay more infertile eggs. Experimental results from the Oudtshoorn Research Farm showed that layer birds which were fed *ad libitum* laid 24.5% infertile eggs, while birds that received only 2 kg dry matter per bird per day of the layer diet laid 11.5% infertile eggs (Cilliers and Van Schalkwyk, 1994). A further complicating factor is that male birds are over-consuming high calcium layer diets, causing overly fat males with poor sperm counts, mainly because of poor zinc availability in the layer diet (high calcium). Du Preez (1995 – personal communication) recommends that male birds should be kept in adjacent paddocks on maintenance diets and be let in with female birds to mate for several hours every second day only after the female birds have consumed most of their daily ration.

Embryonic deaths in the Little Karoo region are mainly caused by:

- Poor heat distribution in incubators.
- Inability to dispose of excess moisture in incubators, especially during summer months.
- Poor ventilation with CO_2 levels exceeding 0.5%.
- Incorrect egg turning procedures (turning frequency, axis of rotation and axis of setting).
- Poor storage facilities for eggs before incubation.

A combination of these factors often cause embryonic deaths exceeding 30%. Van Schalkwyk *et al.* (1993) compared the traditional horizontal egg position for the total 6-week hatching period with five treatments in which eggs were kept in the horizontal position with 1-week increments before being turned vertical. A seventh treatment in which eggs were kept vertical for the total 6-week hatching period was also compared. No assistance was given to any hatching chicks. Turning eggs vertical after 2–3 weeks in a horizontal position increased the hatchability percentage by 30% when compared with eggs that were kept in the horizontal or vertical position for the total 6-week hatching period (Figure 1).

According to Van Schalkwyk (1995) less than 5% of the eggs hatched in the Little Karoo are from automatic, electronic incubators. Most producers still use wooden incubators in which eggs must be turned

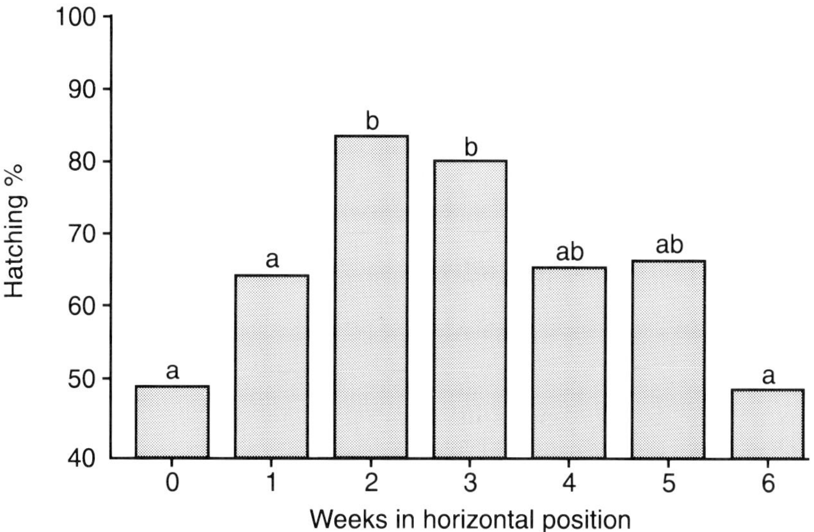

Figure 1. Influence of egg position on the percentage of eggs that hatched without help relative to fertile eggs) (Van Schalkwyk *et al.*, 1993).

by hand. Wooden incubators do not always cool sufficiently and the optimal temperature of 36°C is often exceeded. Increasing incubator temperature to 37.3°C decreased the hatching percentage from 73.1 to 44.0% ($P \leq 0.01$). Van Schalkwyk (1995) concluded that eggs should first be placed in the warmer areas of wooden incubators and then gradually moved to the cooler spots, as embryos generate a lot of heat from day 27 until hatching.

Van Schalkwyk *et al.* (1994) also studied the pattern of gas exchange of the developing ostrich embryo. The O_2 consumption and CO_2 production inclined exponentially during the first part of incubation (before 35 days) followed by a decline of about 25% in O_2 consumption from day 35 to 41 (Figure 2). The O_2 consumption of ostrich eggs increased sharply just before pipping. It was concluded that CO_2 concentrations be maintained below 0.5%. Therefore a 1000 egg incubator requires an airflow of 45 litres/h for proper ventilation.

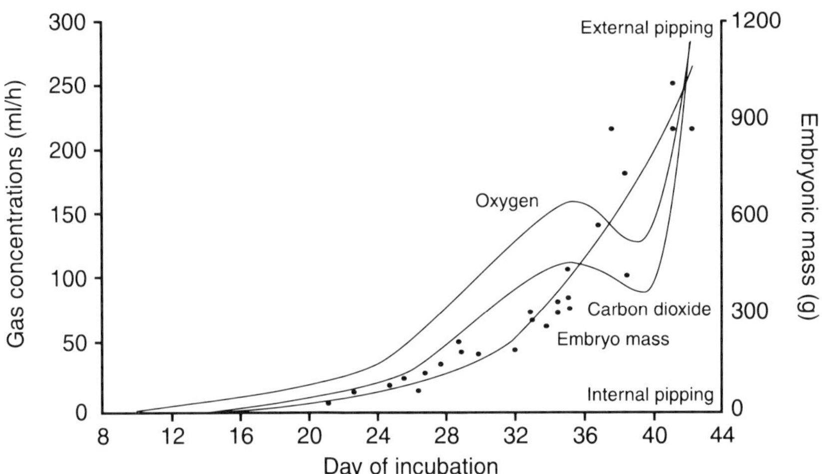

Figure 2. The gas exchange and mass increase of embryos in relation to day of incubation (Van Schalkwyk *et al.*, 1994).

Nutrient requirements and feedstuff values

GROWTH CURVES FOR GROWING OSTRICHES

According to Emmans and Fisher (1986) the genetic potential of a bird may be described in terms of its characteristic growth curve.

Du Preez *et al.*(1992) published results on fitting a Gompertz model to growth performance data of ostriches from different localities in Southern Africa, including Oudtshoorn birds. Cilliers (1995) verified the results using only Oudtshoorn birds (43 males and females), and from the results suggested minor alterations to the growth parameters published by Du Preez *et al.*(1992) for Oudtshoorn stock. The dietary

Table 1. Feedstuffs and chemical composition of diets offered to ostriches from day old to maturity (Cilliers *et al.*, 1994).

	0–2 months Pre-starter	2–4 months Starter	4–6 months Grower	6–10 months Finisher	10–14 months Post Finisher	14 months onwards Maintenance
Ingredients (g/kg on a 90% DM basis)						
Alfalfa	22.7	260.0	428.0	812.0	884.6	420.0
Yellow maize	577.0	501.3	463.5	172.7	100.0	0
Maize oil	20.0	20.0	0	0	0	0
Soybean oilcake meal	232.0	86.0	30.0	0	0	0
Fish meal	120.0	106.0	59.0	0	0	8.9
Dicalcium phosphate	5.3	7.2	11.0	11.0	11.2	15.0
Limestone	17.0	12.3	3.0	0	0	0
Methionine	1.0	2.23	1.0	1.8	1.7	1.6
Ostrich vitamin/mineral premix	4.5	4.5	4.5	2.5	2.5	2.5
Zincbacitracin 10%	0.5	0.5	0	0	0	0
Lucerne straw	0	0	0	0	0	552.0
Nutrient content						
AME_n MJ/kg (ostrich)	12.5	11.5	10.5	9.2	8.5	7.0
Protein g/kg	230.0	190.0	155.0	140.0	120.0	100.0
ARG	12.0	9.5	7.0	5.5	4.2	3.2
LYS	12.8	11.0	7.8	5.8	4.5	3.0
MET	5.0	4.5	3.5	3.0	2.5	1.1
TSA*	9.0	7.0	5.5	4.5	3.5	1.6
THR	7.5	6.0	4.2	3.1	2.3	1.7
Calcium	14.0	14.0	12.0	12.0	10.0	10.0
Phosphorus (available)	4.3	4.3	3.8	3.8	3.5	3.2

*Total sulphur containing amino acids (methionine + cystine)

components and the chemical composition of the diets used in Cilliers' growth study are presented in Table 1. The estimated mean mature (14 months) body weights were 119.2 kg for males and 122.3 kg for females. None of the estimated parameters differed significantly between sexes. The combined growth curves for males and females are shown in Figure 3. The Gompertz growth curve has multiple uses in production and research as demonstrated by Emmans (1989). It may be used as a tool to measure the standard of management and feeding compared to the potential growth of the ostriches.

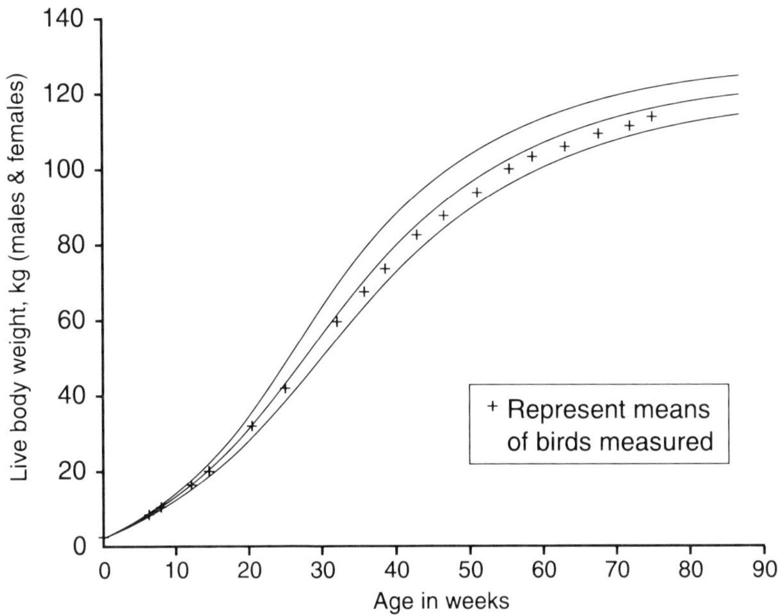

Figure 3. Growth curve of Oudtshoorn ostriches and estimated confidence (1%) limits (Cilliers, 1995).

ENERGY REQUIREMENTS

Commercial farming of ostriches in feedlots on total mixed diets has for many years lacked estimations of specific nutrient requirements for ostriches to achieve maximum growth rates on minimum balanced inputs. Although the nutrition of ostriches has received some interest in recent years (Du Preez, 1991; Vohra, 1992) most of the information was based on poultry data. The world-wide general practice of using nutrient values for feedstuffs and nutrient requirements suggested for poultry in diet formulation, probably resulted in less profitable ostrich feeding systems (Swart *et al.*, 1993). In one of the first scientific studies on nutrient utilization by ostriches, Swart (1988) demonstrated the production of volatile fatty acids, specifically acetate, in the colon of immature growing birds. Digestibility coefficients of 0.66 for hemicellulose and 0.38 for cellulose in a diet containing 340 g alfalfa meal per kg were found. Swart (1988) concluded that the end products of fibre fermentation could contribute as much as 76% of the metabolizable energy (ME) requirements for maintenance of the growing ostrich and that the use of ME values for poultry or pigs in diet formulation for ostriches will result in an underestimation of ME contents of feedstuffs. This was confirmed by Cilliers (1995) who conducted various comparative studies with poultry and ostriches regarding the true metabolizable energy contents corrected for nitrogen retention (TME_n) of feedstuffs generally used in ostrich diets. Apparent metabolizable energy values, corrected for nitrogen retention (AME_n), were determined according to the replacement method (Hill *et al.*, 1960) while TME_n values were calculated according to the regression method of McNab (1990). In Table 2 the TME_n values of various feedstuffs generally used in South African ostrich diets are given as determined with ostriches and roosters (Cilliers, 1995; Cilliers *et al.*, 1994).

The difference in TME_n values between roosters and ostriches indicate that the ostrich is capable of also digesting the fibrous parts of feedstuffs such as cellulose and hemicellulose, which was also shown by Swart (1988) and Swart *et al.*(1993).

The additivity and accuracy of the values in Table 2 in practice were evaluated by feeding an experimental diet, comprising a number of the feedstuffs, to ostriches and roosters. The TME_n value of the test diet was determined by the balance method of continuous feeding 36, 7-month-old ostriches for 5 days while results in roosters were determined

Table 2. TME_n values of various feedstuffs as determined with ostriches and roosters on a 90% DM basis (Cilliers, 1995).

Ingredients	Ostriches	Roosters
	Mean ± sd	Mean ± sd
Yellow maize	15.06 ± 0.228	14.42 ± 0.056
Lucerne hay†	8.91 ± 0.119	4.03 ± 0.118
Malting barley‡	13.93 ± 0.251	11.33 ± 0.212
Oats	12.27 ± 0.291	10.63 ± 0.783
Triticale	13.21 ± 0.241	11.82 ± 0.224
Wheat bran	11.91 ± 0.221	8.55 ± 0.375
Sunflower oilcake meal	10.79 ± 0.278	8.89 ± 0.494
Soybean oilcake meal	13.44 ± 0.173	9.04 ± 0.165
Saltbush hay (*Atriplex nummularia*)	7.09 ± 0.238	4.50 ± 0.271
Common reed (*Phragmites australis*)	8.67 ± 0.337	2.79 ± 0.147
Sweet white *Lupinus albus* (cv Buttercup)	14.61 ± 0.340	9.40 ± 0.642
Ostrich meat and bone meal	12.81 ± 0.203	8.34 ± 0.126
Fish meal	15.13 ± 0.315	13.95 ± 0.190

*Mean of 2 measurements in ostriches and 6 in roosters
†Mean of 10 measurements in ostriches
‡Mean of 2 measurements in ostriches
The remainder were values determined in one balance study

according to the DSQ-method (Du Preez *et al.*, 1986), where the test diet was offered for 3 days after an adaptation period of 24 hours. Theoretical values for the test diets of 11.69 ± 0.189 and 8.28 ± 0.181 MJ TME_n per kg feed (90% DM) for ostriches and roosters respectively, compared favorably to the practically determined values of 11.25 ± 0.0724 and 8.02 ± 0.445 MJ TME_n per kg feed (90% DM). Cilliers (1995) concluded that more reliable TME_n of feedstuffs for ostriches were now available for the establishment of energy requirements and diet formulation.

AMINO ACID AVAILABILITY AND REQUIREMENTS

Amino acid availability is defined as the digested dietary fraction available for normal metabolic processes (Low, 1990). Accurate data on the available amino acid content of feedstuffs are required for least cost formulation of diets. It is a general practice to extrapolate amino acid availability data from poultry to establish values for feeding ostriches. Cilliers (1995) compared apparent and true amino acid availabilities in roosters and ostriches fed a high protein (21% on a 90% DM basis) diet comprising seven feedstuffs (Table 3). Improved digestion of dietary amino acids in ostriches indicates that values derived from poultry will underestimate the true availability of amino acids for ostriches. According to Cilliers (1995) it is possible that the higher availability values in ostriches could possibly be ascribed to the production of *de novo* amino acids although no results are available to prove this theory.

Table 3. Apparent and true availability of amino acids as estimated for an experimental diet in ostriches and roosters (Cilliers, 1995).

	Ostriches		Roosters	
Amino acid	Apparent availability	True availability	Apparent availability	True availability
THR	0.861 ± 0.00299	0.831 ± 0.00263	0.774 ± 0.0104	0.804 ± 0.0273
SER	0.874 ± 0.00327	0.849 ± 0.00272	0.814 ± 0.0109	0.823 ± 0.0240
ALA	0.942 ± 0.00270	0.862 ± 0.00282	0.797 ± 0.0115	0.810 ± 0.0344
MET	0.837 ± 0.00319	0.816 ± 0.00300	0.782 ± 0.0162	0.776 ± 0.0444
PHE	0.815 ± 0.00544	0.809 ± 0.000527	0.748 ± 0.00922	0.723 ± 0.0225
HIS	0.875 ± 0.00502	0.854 ± 0.00362	0.777 ± 0.0637	0.806 ± 0.0210
LYS	0.836 ± 0.00350	0.832 ± 0.00386	0.768 ± 0.00726	0.755 ± 0.0187
ILE	0.842 ± 0.00268	0.829 ± 0.00239	0.814 ± 0.0103	0.817 ± 0.0288
TYR	0.831 ± 0.00399	0.816 ± 0.0374	0.753 ± 0.0106	0.764 ± 0.0257
ARG	0.612 ± 0.00626	0.780 ± 0.00609	0.756 ± 0.00427	0.736 ± 0.0102
CYS	0.814 ± 0.00984	0.806 ± 0.0188	0.766 ± 0.00378	0.781 ± 0.00960
LEU	0.871 ± 0.00240	0.859 ± 0.00206	0.836 ± 0.00910	0.825 ± 0.0246
Protein	0.653 ± 0.0133	0.646 ± 0.0114	0.568 ± 0.0333	0.609 ± 0.0643
Lipid	0.857 ± 0.00464	0.870 ± 0.0041	0.862 ± 0.00543	0.892 ± 0.0174

Values are mean ± sd

Table 4. Estimated dry matter intake (DMI), energy, protein and amino acid requirements for maintenance and growth of ostriches (Cillers, 1995).

AGE (Days)	LW kg	EBLW kg	ADG g/b/d	DMI g/b/d	TME_n MJ/kg DMI	EE MJ/kg DMI	PROT g/kg DMI	LYS g/kg DMI	MET g/kg DMI	CYS g/kg DMI	MET + CYS g/kg DMI	ARG g/kg DMI	THR g/kg DMI	VAL g/kg DMI	ISL g/kg DMI	LEU g/kg DMI	HIS g/kg DMI	PHE g/kg DMI	TYR g/kg DMI
Day old	0.85	0.80																	
30	4.0	3.3	105	220	15.2*	14.7*	239	10.6	3.1	2.8	5.9	9.8	6.5	7.9	8.7	14.5	3.6	8.5	4.4
60	11.0	9.1	233	440	17.5*	15.6*	272	12.5	3.6	3.3	6.9	11.5	7.6	9.3	10.3	17.0	4.3	10.0	5.1
90	19.5	16.6	283	680	15.3*	13.6*	224	10.8	3.2	2.8	6.0	10.1	6.6	8.2	9.0	14.7	3.8	8.7	4.5
120	28.5	25.0	300	820	14.9*	13.3*	207	10.6	3.2	2.7	5.9	9.9	6.4	8.1	8.8	14.3	3.8	8.5	4.5
150	39.5	36.2	367	1220	12.5*	11.5*	174	9.1	2.7	2.3	5.0	8.5	5.5	7.0	7.6	12.3	3.3	7.3	3.9
180	52.1	47.9	420	1490	12.2*	11.4*	168	9.0	2.7	2.3	5.0	8.5	5.5	6.9	7.6	12.2	3.3	7.2	3.9
210	63.4	58.3	375	1630	11.3	10.8	148	8.5	2.6	2.1	4.7	8.0	5.1	6.6	7.2	11.4	3.1	6.8	3.7
240	73.3	67.4	330	1710	10.8	10.5	135	8.2	2.5	2.0	4.5	7.8	5.0	6.4	7.0	11.0	3.1	6.6	3.6
270	82.4	75.8	305	1760	10.7	10.6	130	8.3	2.6	2.0	4.6	7.9	5.0	6.5	7.1	11.1	3.1	6.6	3.7
300	91.0	83.7	287	1800	10.8	10.8	128	8.4	2.6	2.0	4.6	8.1	5.1	6.7	7.2	11.2	3.2	6.7	3.8
330	96.3	88.6	177	2160	8.0	8.1	85	6.3	2.0	1.5	3.5	6.1	3.8	5.1	5.4	8.4	2.4	5.0	2.9
360	99.9	91.9	120	2210	7.4	7.6	74	5.9	1.9	1.3	3.2	5.7	3.5	4.8	5.1	7.8	2.3	4.7	2.7
390	103.5	95.2	120	2250	7.4	7.7	74	5.9	1.9	1.4	3.3	5.8	3.6	4.8	5.2	7.9	3.3	4.7	2.7
420	107.0	98.4	117	2250	7.5	7.8	75	6.1	2.0	1.4	3.4	5.9	3.7	4.9	5.3	8.1	2.4	4.8	2.8
450	110.0	101.2	100	2250	7.5	7.9	73	6.1	2.0	1.4	3.4	5.9	3.7	5.0	5.3	8.0	2.4	4.8	2.8
480	112.3	103.3	77	2250	7.3	7.8	69	6.0	1.9	1.3	3.2	5.9	3.6	4.9	5.2	7.9	2.4	4.8	2.8
510	114.2	105.1	63	2250	7.3	7.7	67	6.0	1.9	1.3	3.2	5.9	3.6	4.9	5.2	7.9	2.4	4.7	2.8
540	116.0	106.7	60	2250	7.3	7.8	67	6.0	2.0	1.3	3.3	5.9	3.6	5.0	5.3	8.0	2.4	4.8	2.8
570	118.6	109.1	87	2250	7.7	8.3	74	6.4	2.1	1.4	3.5	6.2	3.8	5.2	5.6	8.4	2.5	5.1	3.0
600	120.3	110.7	57	2250	7.5	8.1	68	6.2	2.0	1.4	3.4	6.1	3.7	5.1	5.4	8.2	2.5	4.9	2.9

Total energy retention was measured in empty body weight (EBW). Maintenance was calculated according to 0.678 g/kg EBW. Energy retention as lipid and protein retention was separately calculated. Amino acid retention in feathers and body was calculated separately and requirements for maintenance were calculated using mature defeathered amino acid weight and defeathered tyrosine weight.

* In calculating TME_n and EE requirements from results obtained for 7-month-old birds similar energy contents were assumed for younger birds. This assumption is incorrect, resulting in an overestimation of dietary energy requirements.

LW, live weight; EBLW, empty body live weight; ADG, average daily gain; EE, effective energy.

FEED CONVERSION RATIOS (FCRs)

It is well known that feed conversion ratios for ostriches are higher than that of the different poultry species. According to Cilliers (1995), the following FCRs can be expected for ostriches:

Birth to 2 months	2 : 1
2–4 months	2 : 1
4–6 months	3.8 : 1
6–10 months	5.5 : 1
10–14 months	10 : 1

(FCR will vary according to source of dietary supply and the extent to which the diet meets the requirements.)

The main reason for the high feed conversion ratios for older birds is probably the poor utilization of metabolizable energy (ME) by ostriches. Cilliers (1995) reported ME-utilization values of 0.414, 0.426 and 0.443 for three groups of ostriches weighing between 70 and 75 kg live weight. The values are markedly lower than values reported in the literature for other species viz. 0.72 for broilers (Chwalibog, 1991), 0.82 for piglets (Huang *et al.*, 1981) and 0.62 for calves (Thorbek, 1980). Chwalibog (1991) reported a ME utilization value of 0.52 for ostrich chicks with live weights between 0.3 and 1.2 kg which is in agreement with the values Cilliers (1995) reported for older birds.

Cilliers (1995) determined the TME_n, effective energy (EE), protein and amino acid requirements for maintenance and growth for ostriches in a response study with 7-month-old birds. The results were used to estimate dietary requirements for these birds (day 210 to 240 of age) and these results were then extrapolated to estimate requirements from day old to 20 months of age (Table 4). In extrapolating, it was assumed that the caloric value for younger birds would be the same as for the experimental birds, resulting in an overestimation of energy requirements for birds under 6 months of age.

PRACTICAL DIET SPECIFICATIONS AND MANAGEMENT

The chest circumference of ostriches was found to be an excellent prediction of both the live weight and skin surface (Table 5). It is suggested that the chest circumference of at least 10 birds of a group should be measured and the average computed before the corresponding live weight and skin surface for the total group can be read from Table 5.

Cilliers (1995) suggested that ostriches should be fed five different diets from birth to slaughter at 14 months of age. Chest circumference (weight) should be used to determine when to change between the five diets (Cilliers, 1995). ME (ostriches) and amino acid specifications for the five different diets, a maintenance and layer diet are given in Table 6.

Cilliers and Van Schalkwyk (1994) suggested the Ca, P and Na

Table 5. The chest circumference and corresponding average live weight and skin surface of ostriches (Cilliers and Van Schalkwyk, 1994).

Chest circumference (cm)	Live weight (kg)	Skin surface (dm²)	Age (months)	Chest circumference (cm)	Live weight (kg)	Skin surface (dm²)	Age (months)	Chest circumference (cm)	Live weight (kg)	Skin surface (dm²)	Age (months)
18	0.8	–	–	70	19.4	–	–	105	–	123.2	–
20	1.1	–	–	72	21.1	–	3	106	76.4	124.8	–
22	1.3	–	–	74	23.0	–	–	107	–	126.4	–
24	1.5	–	–	76	25.0	–	–	108	81.4	128.0	12
26	1.8	–	–	78	27.1	–	–	109	–	129.6	–
28	2.1	–	–	80	29.4	–	–	110	86.5	131.3	13
30	2.4	–	–	82	31.9	–	4	111	–	132.9	–
32	2.8	–	–	84	34.5	–	–	112	91.8	134.5	–
34	3.2	–	–	86	37.4	–	–	113	–	136.1	14
36	3.6	–	–	88	40.4	–	5	114	97.2	137.7	–
38	4.1	–	–	90	43.6	99.1	–	115	–	139.3	–
40	4.5	–	–	91	–	100.7	6	116	102.6	140.9	–
42	5.1	–	1	92	47.0	102.3	–	117	–	142.5	–
44	5.7	–	–	93	–	103.9	–	118	108.2	144.1	–
46	6.3	–	–	94	50.5	105.5	7	119	–	145.7	–
48	7.0	–	–	95	–	107.1	–	120	113.7	147.3	–
50	7.7	–	–	96	54.4	108.8	–	121	–	148.9	–
52	8.5	–	–	97	–	110.4	8	122	119.2	150.5	–
54	9.4	–	–	98	57.4	112.0	–	123	–	152.1	–
56	10.9	–	–	99	–	113.6	–	124	–	153.8	–
58	11.3	–	–	100	62.6	115.2	9	125	–	155.4	–
60	12.4	–	–	101	–	116.8	–	126	–	157.0	–
62	13.6	–	2	102	67	118.4	10	127	–	158.6	–
64	14.9	–	–	103	–	120.0	–	128	–	160.2	–
66	16.3	–	–	104	71.6	121.6	11	129	–	161.8	–
68	17.8	–	–					130	–	163.4	–

Table 6. Practical diet specifications for ostriches in various stages of production (Cilliers and Van Schalkwyk, 1994).

	TME_n (MJ/kg DM)	PROTEIN (g/kg DM)	LYS (g/kg DM)	METH (g/kg DM)	TSAA* (g/kg DM)	ARG (g/kg DM)	THR (g/kg DM)	ILE (g/kg DM)	LEU (g/kg DM)
PRE STARTER DIET (From 0.8 to 11 kg live body weight or from day old to 2 months of age, 18 cm chest)									
	13.2	255.0	12.5	3.6	6.9	11.5	7.6	10.3	17.0
STARTER DIET (From 11 kg to 28 kg live body weight or 2 to 4 months of age, 56 cm chest)									
	12.8	215.0	10.7	3.2	6.0	10.0	6.5	8.9	14.8
GROWER DIET (From 28 to 52 kg live body weight or from 4 to 6 months of age, 79 cm chest)									
	12.2	171.0	9.0	2.7	5.0	8.5	5.5	7.6	12.2
FINISHER DIET (From 52 to 91 kg live body weight or from 6 to 10 months of age, 95 cm chest)									
	10.9	135.0	8.4	2.6	4.6	8.1	5.1	7.2	11.2
POST FINISHER (From 91 to 107 kg live body weight or from 10 months to 20 months of age, 112 cm chest)									
	8.0	85.0	6.3	2.0	3.5	6.1	3.8	5.4	8.4
MAINTENANCE DIET AFTER MATURE BODY WEIGHT IS ATTAINED									
	6.5	80.0	2.7	1.1	2.1	3.2	1.7	1.6	3.3
LAYER DIET IN MATING SEASON									
	9.2	140.0	6.8	3.2	5.3	7.0	5.3	5.1	8.8

*TSAA, total sulphur amino acids

Table 7. Ca, P and Na specifications for grower, maintenance and layer diets for ostriches (Cilliers and Van Schalkwyk, 1994).

	Total Ca (%)	Available P (%)	Total Na (%)
Pre-starter, starter and grower diets	1.2–1.5	0.4–0.45	0.20–0.25
Finisher and post finisher diets	0.9–1.0	0.32–0.36	0.15–0.30
Maintenance diet	0.9–1.0	0.32–0.36	0.15–0.30
Layer diet	2.0–2.5	0.35–0.40	0.15–0.25

specifications in Table 7, for grower, finisher, layer and maintenance diets.

The suggested trace element requirements and vitamin supplementation per ton of total diet are presented in Table 8 (Cilliers and Van Schalkwyk, 1994).

NUTRIENT REQUIREMENTS FOR LAYERS

After 14 months of age the accretion of body substance in female birds is relatively small and nutrient requirements can be considered equal to

Table 8. Trace element and vitamin supplementation in total diets for ostriches (amounts per tonne of feed, Cilliers and Van Schalkwyk, 1994).

		Grower diets Birth to 6 months	Grower and finisher diets 6 months to slaughter	Layer diets
Vit. A	IU	12 000 000	9 000 000	15 000 000
Vit. D_3	IU	3 000 000	2 000 000	25 000 000
Vit. E	IU	40 000	10 000	30 000
Vit. K_3	g	3	2	3
Vit. B_1	g	3	1	2
Vit. B_2	g	8	5	8
Niacin	g	60	50	45
Calc. panth. A	g	14	8	18
Vit B_{12}	mg	100	10	100
Vit. B_6	g	4	3	4
Choline chloride	g	500	150	500
Folic acid	g	2	1	1
Biotin	mg	200	10	100
Endox[R*]	g	100	–	100
Magnesium	g	50	–	40
Manganese	g	120	80	120
Zinc	g	80	50	90
Copper	g	15	15	15
Iodine	g	0.5	1	1
Cobalt	g	0.1	0.3	0.1
Iron	g	35	20	35
Selenium	g	0.3	0.15	0.3

*Endox is a registered trade name

maintenance until sexual maturity is reached at 24 months (Du Preez, 1991). It is important however to anticipate increased requirements for minerals (calcium and phosphorous), amino acids, vitamins and energy before the first egg is formed. The period of follicular growth in the domestic fowl is 7–8 days (King, 1972) and probably 16 days for ostriches (Du Preez, 1991) in which case the demand for additional nutrients would start 18 days before the first egg is laid. The demand for extra nutrients would increase in a sigmoidal pattern reaching a maximum approximately 8 days before the first egg is laid. From this point onward the nutrient requirements for egg production would remain at the plateau level and are only dependent on the quantity of nutrients deposited in each egg, which amounts to half an egg per day, since ostriches lay one egg every second day (Du Preez, 1991). Amino acid and energy requirements for maintenance and egg production for ostriches were calculated by Du Preez (1991). These are presented in Tables 9 and 10. Diet specifications for breeding diets are also given in Tables 6, 7 and 8.

The breeding diet is totally inappropriate for male birds, but it is very difficult to find practical ways around this dilemma. The over consumption of calcium suppresses the absorption of zinc which is essential for sperm production (see section on **layer bird management and egg hatching** for recommendations).

Table 9. Requirements for some amino acids of breeding ostriches in production (Du Preez, 1991).

	For maintenance of body mass (kg)			For egg production (kg) including shell		
	100	105	110	1.2	1.4	1.6
Protein, g	67	69	72	119	138	158
Amino acid, (g*)						
Arginine	5.70	5.87	6.12	3.56	4.15	4.74
Lysine	5.78	5.95	6.21	6.41	7.48	8.55
Methionine	1.86	1.90	2.00	2.67	3.10	3.56
Histidine	2.54	2.61	2.73	1.91	2.20	2.50
Threonine	3.54	3.64	3.80	6.85	8.00	9.13
Valine	4.32	4.46	4.65	5.50	6.40	7.30
Isoleucine	3.50	3.60	3.76	4.55	5.30	6.10
Leucine	6.90	7.14	7.45	9.00	10.50	12.00
Tyrosine	2.33	2.40	2.50	3.70	4.30	4.90
Phenylalanine	3.82	3.90	4.10	4.06	4.67	5.30
Cystine	0.89	0.92	0.96			
Tryptophan	0.73	0.75	0.78			

*Daily requirements of dietary amino acids per gram of egg produced. A female weighing 110 kg producing a 1.2 kg egg each 2nd day will require:

$$\frac{(6.21 + 6.41)}{2000} \times \frac{100}{1} = 0.63\% \text{ lysine in the diet.}$$

Table 10. Estimated energy requirement (MJ per day) for egg production of ostriches in 0.25 hectare breeding pens (Du Preez, 1991).

	Energy requirements (MJ ME) for maintenance and activity Body mass kg			Energy MJ requirements (MJ ME) for egg production Egg mass (kg)		
	100	105	110	1.2	1.4	1.6
Maintenance*	13.64	14.12	14.60			
Activity	1.37	1.41	1.46			
Egg lipid				2.30	2.68	3.07
Egg protein				3.58	4.18	4.77
Shell† (18 % of egg mass)				0.26	0.30	0.35
Total	15.01	15.53	16.06	6.14	7.16	8.19

*1.63 $Pm^{0.73}$ where Pm = protein mass in body at maturity (Emmans and Fisher, 1986)
†1.2 MJ per kg shell formed

REARING THE NEWLY HATCHED OSTRICH

Newly hatched ostriches are equipped with reservoirs of food in the form of a yolk sac which is taken up into the abdominal cavity before hatching and on which it can survive for a few days without ingesting food (Mellett, 1993).

Hatched ostrich chicks can be reared extensively with foster parents (25–30 chicks per adult pair) without any housing; semi-extensively with foster parents (approx. 100 chicks per adult pair) with housing during night time only; semi-intensively (moving outdoors on pastures during daytime from the age of 3 days); or intensively. The intensive rearing of young ostriches is a well established practice, but requires excellent hygiene and management skills. Chicks are kept in a building for at least one week. The rearing house must be sterile with under-floor heating while the floor (concrete) should be coated with a rubberizing agent. Daily cleaning of the rearing house is essential. Rearing houses should be temperature controlled (22–26°C) with adequate ventilation but avoiding any drafts (Cilliers and Van Schalkwyk, 1994). After one week chicks can be let out on pastures (alfalfa) during daytime. Care should be taken that no sticks or pieces of wire are lying around. When available, chicks will ingest these objects which will penetrate the proventriculus resulting in death.

Diseases and mortality in ostriches

Allwright (1994) listed the majority of conditions encountered in ostriches into the following four groups:

- Respiratory
- Gastrointestinal
- Musculoskeletal/nervous
- Miscellaneous

The most common respiratory syndromes are rhinotracheitis and airsacculitis, with rhinotracheitis often leading to airsacculitis. The main causing factors of respiratory diseases are:

Influenza	Bordetella
Pseudomonas	Aspergillus
E. coli	Mycoplasma
Pasteurella	Environmental
Klebsiella	Aspiration
Staphylococcus aureus	

Stress, dust, high ammonia levels (20 ppm +), incorrect environmental temperatures, poor ventilation, drafts and lack of shelter will predispose ostriches to respiratory diseases. According to Mellett (1993), severe cold or defective heating equipment cause respiratory diseases and are the major causes of mortality in young ostrich chicks. Although some chicks may survive these conditions, their metabolism is disrupted causing poor feed intake resulting in death within 3–10 days.

Gastrointestinal conditions can be divided into those affecting the stomach and those affecting the intestines. The main causing factors are:

GASTRIC	ENTERIC	
Megabacteria	Coronavirus	Salmonella
Candida spp.	Paramyxvirus	*Campylobacter jejuni*
Mucor spp.	Reovirus	*Balantidium coli*
Libyostrongylus	Other viruses	Cryptosporidum
Foreign body/impaction	*Clostridium perfringens*	Coccidia
Stasis	*E. coli*	Houttuynia
	Pseudomonas	Codiostomum

Ingestion of foreign material such as woodshavings, large bones and sharp sticks occurs mainly when the diet is not balanced or insufficient. The substances may penetrate the proventriculus and result in the death of birds up to the age of 6 months. Balanced diets, adequate roughage and correct sized stones are essential to prevent ingestion of foreign material (Mellett, 1993).

Bacterial enteritis infection is often precipitated by deworming, overeating of lush alfalfa, parasitic infestations or viral infections (Allwright, 1994). The *C. perfringens* type B 'lamb dysentery' vaccine has given good results when administered at ½ dose at 10 weeks and 16–24 weeks.

The following were listed as possible causes of musculoskeletal/nervous conditions (Allwright, 1994):

Newcastle	Heavy metals
Encephalopathy	Leg deformities
Botulism	Fractures
Pesticides	Myopathy
Furanzolidone	Hypoglycaemia
Ionophores	Airsacculitis
Nitrates	Gastrointestinal disease
Avocado	

Fast-growing chicks may develop porosis of the ankle or stifle joints at the age of 4–8 weeks and even up to 4 months with impairment of growth of the long bones and slippage of the Achilles tendon. This is also associated with a shortage of dietary manganese or too high levels of dietary calcium (Mellett, 1993).

Causes of miscellaneous conditions are divided into three groups:

DERMATOLOGICAL	HEPATITIS	REPRODUCTIVE
Avian pox	Influenza	Infertility
Dermatophytes	Herpes virus	Embryonic death
Ectoparasites	Other viruses	Chick quality
Nutrition	Sporidesmin	

Ostrich products

FEATHERS

During the early 1900s a lot of attention was focused on the ostrich since the feathers were high fashion amongst the wealthy. Swart (1979) and Swart and Heydenrych (1982) quantified the quality characteristics of ostrich feathers relative to their economic value. It is unlikely that ostrich feathers will return as a high fashion item and feathers will probably continue to make only a small contribution to the total income from ostriches.

LEATHER

Well fed and managed ostriches, slaughtered at 14 months with a live weight of at least 75 kg, should have a well developed skin with a minimum size of 120 dm^2. The industry have set standards for top quality ostrich leather. Apart from a minimum size and minimum damage specification, well developed and rounded follicles are important criteria for quality ostrich leather. In ostrich studies there is a strong case to be made for time related studies rather than mass related studies, since development of tissue types, such as the skin, is also of great importance to the profitability of the industry (Mellett, 1991).

MEAT

The ostrich carcass yields approximately 35 kg meat at the age of 14 months. Scientific muscle names, common names and muscle application are given in Table 11.

According to Sales (1994), ostrich meat can be identified relative to beef as an intermediate type red meat (pH_f between 5.8 and 6.2) between normal and extreme DFD (Dark Firm Dry). Mellett (1985) illustrated the rapid decline of the pH of ostrich meat to just below 6.0 after slaughter (Figure 4). Ageing of ostrich meat for more than 4 days is not

Table 11. Average muscle mass (sum of left and right hand side, kg) of the hindquarters and certain hip and thigh muscles of adult ostriches and their possible application (Mellett, 1993).

Body part or muscle	Mass (kg)	Common name	Application
Hindquarters (total)	39.800	–	
Pre-acetabular muscles			
1. *M. iliotibialis cranialis*	1.530	–	processing*
2. *M. ambiens*	0.540	tornedo fillet	whole muscle, frying
3. *M. pectineus*	0.308	–	processing
Acetabular muscles			
4. *M. iliofemoralis externus*	1.150	steak	whole muscle, frying
5. *M. iliofemoralis internus*	0.128	–	processing
6. *M. iliotrochantericus caudalis*	0.092	–	processing
7. *M. iliotrochantericus cranialis*	0.148	–	processing
Post-acetabular muscles			
8. *M. iliotibialis lateralis*	3.280	steak	whole muscle, or products, frying
9. *M. iliofibularis*	3.400	fillet	whole muscle, or products, frying
10. *M. iliofemoralis*	1.160	mongular fillet	whole muscle, frying
11. *M. flexor cruris lateralis*	1.170	–	processing
12. *M. flexor cruris medialis*	0.375	–	processing*
13. *M. pubo-ischio-femoralis*	0.387	–	processing
14. *M. ischiofemoralis*	0.131	–	processing
15. *M. obturatorius medialis*	1.710	small leg	processing*
16. *M. obturatorius lateralis*	–	–	–
Femoral muscles			
17. *M. femorotibialis medius*	1.660	steak	whole muscle, or products, frying
18. *M. femorotibialis accessorius*	1.280	steak	whole muscle, or products, frying
19. *M. femorotibialis externus*	0.368	steak	whole muscle, or products, frying
20. *M. femorotibialis internus*	0.300	steak	whole muscle, or products, frying

*Potential for upgrading for use as whole muscle

recommended (due to the high ultimate pH) and electrical stimulation will probably have very little effect on the quality of ostrich meat since the pH decline is rapid enough and cold shortening is therefore not anticipated (Sales, 1994).

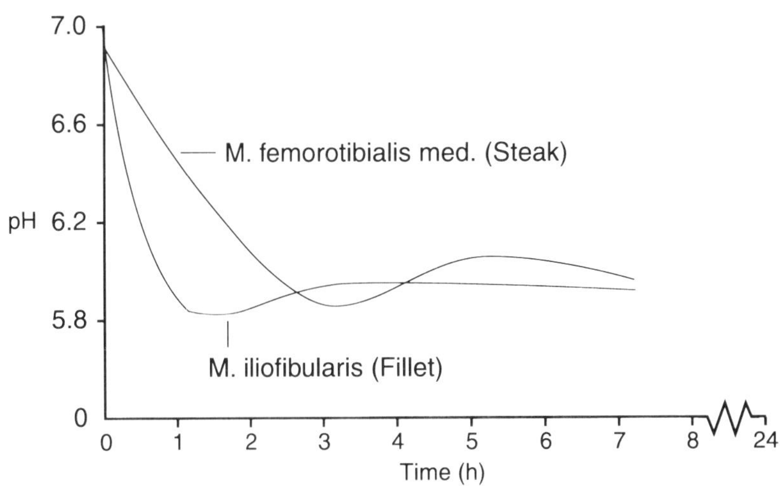

Figure 4. Post mortem pH pattern of two different muscles of the ostrich (Mellett, 1985).

Sales (1994) showed by microscopic investigation that in slaughtered ostriches muscles are not contracted to such an extent that cold shortening could be a cause for concern under standard chilling procedures. Warner-Bratzler shear force values compare well with that of tender beef cuts. Visually the meat texture appears very loose and coarse (muscle fiber **bundle** characteristic), but microscopically the fiber diameter (**single** muscle fiber characteristic) is less than that of beef.

The inherent dark color of ostrich meat may create a marketing problem since the appearance resembles that of raw liver. In combination with the high ultimate pH, this dark color may be anticipated. The high ultimate pH has other advantages, however. High pH meat has a naturally high water binding capacity which makes it ideally suited for the manufacturing of high priced meat products.

The approximate ratio of saturated fatty acids : monounsaturated fatty acids : polyunsaturated fatty acids in ostrich meat is 1 : 1 : 1 (Sales, 1994) which makes ostrich meat outstanding in terms of health characteristics. The ratios of polyunsaturated to saturated fatty acids for poultry and beef are 0.77 (chicken) and 0.22, respectively. Given that a cholesterol content of 57 mg/100 g for trout is considered low compared with that of beef at 68 mg/100 g, then the cholesterol content of 62 mg/100 g of ostrich meat may also be described as 'low'.

The protein content of ostrich meat is 21% which compares well with the meat of other animal species, while the fat content is less than 0.5% (Sales, 1994).

The larger muscles (e.g. *M. iliofibularis*) may be processed into single muscle products such as bacons or other cold smoked products. Medium sized muscles (e.g. *M. femorotibialis medius*) may be successfully processed into combined muscle products such as pressed hams, or single muscle products such as pastrami. The smaller muscles (e.g. *M. iliofemorales*) are suitable for processing only, preferably for minced or emulsion products. A combination of ostrich and pork will result into more intensely colored meat products than the traditional combination of pork and beef.

The future of the ostrich industry

According to Mellett (1993) it is not unlikely that ostrich feathers may return as a fashion item in the foreseeable future, but the ostrich industry will have to provide scientific proof that the production and harvesting of feathers is a painless or natural process. However, if it does not return as a fashion item, the income from feathers will probably be between 5 and 10% of the total income from all ostrich products.

Ostrich meat contains considerably less fat (0.5%) and less cholesterol than other meats. The health aspect of ostrich meat in human nutrition should be exploited much more by the ostrich industry. Due to the unfavorable feed conversion of ostriches up to the age of 14 months, when the hide is mature, ostrich meat has to be very expensive. The development of a special ostrich line for meat production is however a strong possibility. Selected birds at the Oudtshoorn experimental farm have produced 35–38 kg usable meat at the age of 6–7 months with a feed conversion ratio of 4:1 (Cilliers, personal communication). If specialized products could be produced from young ostrich leather the development of a meat ostrich will probably be a reality. The 15–20% income from ostrich meat (as a percentage of all ostrich products) should be increased by efforts to market the health aspects of ostrich meat to consumers who can afford it.

Ostrich leather is in demand for fashion clothing and footwear worldwide. The South African ostrich industry depends heavily on this demand as approximately 75% of the total income from ostrich products are from hides. Oversupply of hides will have negative implications and could cause the total industry to collapse.

The efficiency of ostrich production can still be improved. It is estimated that 25% of layer birds lay 75% of the total eggs that hatch. Only 20% of all eggs laid will produce mature ostriches at the age of 14 months. Selection for better producing birds and better management practices can still increase efficiency as very little science has been implemented in the ostrich industry up to now.

Finally it should be understood that uncontrolled expansion will result in the overproduction of ostrich products and will probably lead to the total collapse of the industry.

References

Allwright, D.M. 1994. Ostrich diseases. Technical pamphlet. Regional Veterinary Laboratory, Private Bag X5020, Stellenbosch, 7599, South Africa.

Anon, 1994. Annual Report of the Little Karoo Agricultural Cooperation Ltd. Oudtshoorn, South Africa.

Chwalibog, A. 1991. Energetics in animal production. Acta Agric. Scand. 41:147.

Cilliers, S.C. 1995. Feedstuffs evaluation in ostriches (*Struthio camelus*). Ph.D. thesis, University of Stellenbosch, South Africa.

Cilliers, S.C., J.P. Hayes, J.S. Maritz, A. Chwalibog and J.J. Du Preez. 1994. True and apparent metabolizable energy values of lucerne and yellow maize in adult roosters and mature ostriches (*Struthio camelus*). Anim. Prod. 59:309.

Cilliers, S.C. and S.J. Van Schalkwyk. 1994. Volstruisproduksie [Ostrich Production]. Technical booklet. Little Karoo Agricultural Development Center, Oudtshoorn Experimental Farm, P.O. Box 313, Oudtshoorn 6620, South Africa.

Du Preez, J.J. 1991. Ostrich nutrition and management. In: Recent Advances in Animal Nutrition in Australia. D.J. Farrell (Ed.) University of New England, Armidale, Australia, 278.

Du Preez, J.J., J.S. Duckitt and M.J. Paulse. 1986. A rapid method to evaluate metabolizable energy and availability of amino acids without fasting and force feeding experimental animals. S. Afr. J. Anim. Sci. 16:47.

Du Preez, J.J., M.F.F. Jarvis, D. Capatos and J. de Kock. 1992. A note on growth curves for the ostrich (*Struthio camelus*). Anim. Prod. 54:150.

Emmans, G.C. 1989. The growth of turkeys. In: Recent Advances in Turkey Science. D. Nixen and T.C. Grey (eds.) Butterworths, London, England, 135.

Emmans, C.G. and C. Fisher. 1986. Problems of nutritional theory. In: Nutritional Requirements of Poultry and Nutritional Research. C. Fisher and K.N. Boorman (Eds.) Butterworths, London, England, 9.

Hill, F.W., D.L. Anderson, R. Remer and L.B. Carew. 1960. Studies of the metabolizable energy of grain and grain products for chickens. Poultry Sci. 39:573.

King, J.R. 1972. In: Breeding Biology of Birds. D.S. Farmer (ed.) National Academy of Science, Washington, D.C., 91.

Huang, Z., G. Thorbek, A. Chwalibog and B.O. Eggum. 1981. Digestibility, nitrogen balances and energy metabolism in piglets raised on soyaprotein concentrate. Z. Tierphysiol., Tierernährg. U. Futtermittelkde 46:102.

Low, A.G. 1990. Protein evaluation in pigs and poultry. In: Feedstuff Evaluation. J. Wiseman and D.J.A. Cole (Eds.) Butterworths, London, England, 91.

McNab, J.M. 1990. Apparent and true metabolizable energy of poultry diets. In: Feedstuff Evaluation. J. Wiseman and D.J.A. Cole (Eds.) Butterworths, London, England, 41.

Mellett, F.D. 1985. The ostrich as meat animal – anatomical and muscle characteristics. M.Sc. thesis. University of Stellenbosch, Stellenbosch, South Africa.

Mellett, F.D. 1991. Die volstruis as slagdier: aspekte van groei [The ostrich as slaughter animal: aspects of growth.] Ph.D. thesis, University of Stellenbosch, South Africa.

Mellett, F.D. 1993. Ostrich production and products. In: Livestock Production Systems. C. Maree and N.H. Casey (Eds.) Agric-Development Foundation, Brooklyn, South Africa, 187.

Sales, J. 1994. Die identifisering en verbetering van kwaliteitseienskappe van volstruisvleis. [The identification and improvement of quality characteristics in ostrich meat.] Ph.D. thesis. University of Stellenbosch, Stellenbosch, South Africa.

Sauer, F., E. Sauer and B. Grzimek. 1972. Ostriches. In: Grzimek's Animal Encyclopedia, Vol. 7: Birds I. Van Nostrand Reinhold, 91.

Smit, D.J. v Z. 1963. Volstruisboerdery in die Klein-Karoo. [Ostrich farming in the Little Karoo.] Pamphlet 358. Department of Agricultural Technical Services, Pretoria, South Africa.

Swart, D. 1979. Die kwantifisering van kwaliteitseienskappe van volstruisvere en die relatiewe ekonomiese belangrikheid daarvan. [The quantifying of quality characteristics of ostrich feathers and its relative economic importance.] M.Sc. thesis, University of Stellenbosch, Stellenbosch, South Africa.

Swart, D. 1985. Volstruisproduksie: 'n bydrae tot diereproduksie in die winter reënstreek. [Ostrich production: a contribution to animal production in the winter rainfall region.] In: Proceedings of a Western

Cape Society of Animal Science Conference, Stellenbosch, South Africa.

Swart, D. 1988. Studies on the hatching, growth and energy metabolism of ostrich chicks (*Struthio camelus* var. *domesticus*). Ph.D. thesis. University of Stellenbosch, South Africa.

Swart, D., R.I. Mackie, and J.P. Hayes. 1993. Fermentative digestion in the ostrich (*Struthio camelus* var. *domesticus*), a large avian species that utilizes cellulose. S. Afr. J. Anim. Sci. 23:119.

Swart, D. and H.J. Heydenrych. 1982. The quantifying of flue quality in ostrich plumes with special reference to the fat content and cuticular structure of the barbules. S. Afr. J. Anim. Sci. 12:65.

Thorbek, G. 1980. Studies on protein and energy metabolism in growing calves. Beretning 49, Statens Husdyrbrugsforsog: 104.

Van Schalkwyk, S.J. 1995. Fertility, incubation and embryonic development in ostriches in relation to production and hatchability in the Little Karoo region. M.Sc. thesis. Rhodes University, Grahamstown, South Africa.

Van Schalkwyk, S.J., C.R. Brown, S.W.P. Cloete and J.A. de Kock. 1993. Die invloed van pakmetode op die uitbroeibaarheid van volstruiseiers onder kunsmatige broeikondisies. [The influence of egg position on the hatchability of ostrich eggs under artificial hatching conditions.] In: Proceedings of the 32nd Congress of the South African Society of Animal Science, Wild Coast Sun, Republic of Transkei, p. 27.

Van Schalkwyk, S.J., C.R. Brown, S.W.P. Cloete and G. van Wyk. 1994. The gaseous environment of the ostrich embryo in relation to its development and incubator design. In: Proceedings of the 33rd Annual Congress of the South African Society of Animal Science, Warmbad, South Africa, p. 97.

Vohra, P. 1992. Information on ostrich nutritional needs still limited. Feedstuffs 64:16.

THE CHALLENGES INVOLVED IN POULTRY PRODUCTION IN BRAZIL: THE NUTRITION PERSPECTIVE

ANTÔNIO MÁRIO PENZ, Jr.

Departamento de Zootecnia, Universidade Federal do Rio Grande do Sul, Porto Alegre, Brazil

Introduction

Many factors contributed to development of the poultry industry in Brazil at the end of the 1960s; however, the main factors were:

- interest of the Brazilian government in developing the industry,
- the desire of certain poultry farmers to improve the business,
- available manpower to fill the needs of the growing industry,
- huge land resources to produce the necessary feedstuffs,
- a growing population ready to appreciate the benefits of poultry meat.

The main limitation at that time was technical knowledge needed to put various aspects of the business together and begin operation. After three decades, Brazil has consolidated its position as one of the main broiler producers in the world and it has a modern industry, ready to keep growing at the same pace at which it has grown to date.

Evolution of the Brazilian poultry industry

Investment in the poultry industry began in the late 1960s and early 1970s. The Brazilian government, convinced that the poultry sector could grow quickly, generate jobs, offer more food to Brazilian citizens and be able to export the production surplus; decided to give all possible support to the business people investing in this industry. Initial needs were on-farm construction technology, hatcheries and slaughter plants. At the same time, new commercial broiler strains were imported and new strains were substituted for lower performing older breeds and local strains. Production technology was also imported. These first steps were all needed to create a modern broiler industry, demonstrate its potential and to motivate investors. Government participation at that time was critical as it directed technical efforts and organized resources to speed progress. In the years following, the Brazilian position in the poultry world consolidated, exports grew, and in 1994 Brazil was the

Table 1. The Brazilian broiler exportation and the average export price/ton

Year	Tons exported*	$USD/ton	Year	Tons exported	$USD/ton
1975	3 469	948	1985	273 010	874
1976	19 636	996	1986	224 652	981
1977	32 829	961	1987	215 163	990
1978	50 805	922	1988	236 302	951
1979	81 096	1 000	1989	243 891	1 079
1980	168 713	1 225	1990	299 218	1 069
1981	293 933	1 205	1991	321 700	1 221
1982	301 793	945	1992	371 719	1 157
1983	289 301	838	1993	433 498	1 105
1984	287 494	936	1994	480 905	1 224

*×1000 tons
Associação Brasileira de Produtores e Exportadores de Frangos, 1994

second largest poultry exporter in the world behind the United States (Table 1).

The poultry industry exports only 10 to 15% of the total Brazilian production. The other 85 to 90% is consumed inside the country. Per capita consumption has increased strongly in recent years as can be seen in Table 2. The growth in broiler consumption has reinforced the Brazilian poultry industry, however the international market has always been important to Brazil. Exports brought money back to the country and forced the industry to become modern and sufficiently updated to compete abroad.

Table 2. Brazilian per capita consumption (kg) of broiler, pig and beef meat.

Year	Broiler	Pig	Beef
1976	5.8	7.2	18.9
1977	6.0	7.4	20.6
1978	7.1	7.5	20.6
1979	8.7	7.7	18.5
1980	8.9	8.2	17.2
1981	8.9	8.0	15.4
1982	8.5	7.7	16.0
1983	9.3	7.4	14.7
1984	8.1	7.1	12.2
1985	8.9	6.9	12.1
1986	10.0	7.3	12.0
1987	12.4	8.0	11.8
1988	11.8	7.0	12.0
1989	12.4	6.6	12.0
1990	13.3	7.0	12.5
1991	15.0	7.0	13.0
1992	16.0	7.3	14.5
1993	17.0	7.6	14.2
1994	18.1	8.0	15.0

União Brasilera de Avicultura, 1994

Last year, due to some changes in Brazilian economic trends, the poultry industry arrived at a very interesting point. Total broiler production was above 3 500 thousand tons, with an average above 200 000 broilers produced each month. This meant an average increase of 9% in this sector. To get to those figures, utilization of available farm space had to be nearly 100% throughout the year. So, the poultry industry finished the year with very optimistic figures and the expectations for 1995 are very good.

The role of poultry nutrition in progress of the Brazilian poultry industry

Many technical areas were important in the development of the Brazilian poultry industry. Naturally, the sector that deals with equipment and infrastructure was crucial to the development process. Many local industries grew and improved their knowledge of hatcheries and slaughter plants. Also, these local industries were vital in developing knowledge of farm construction and equipment. Knowledge of pathology and poultry medicine was very important, also. Many national and international companies developed products in Brazil, helping to improve competitive ability and better meet needs of the birds. Management knowledge evolved; and with updated information farmers were better able to understand production systems and how to get more from the system that they were using.

However, nutrition and feed management knowledge were critical for overall success. Since feed costs are a major portion of final costs (>60% of total production cost), progress in this important aspect of production allowed Brazilian broiler meat to compete both internally and abroad. In the past three decades nutrition knowledge has grown tremendously and has been largely responsible for the success of the Brazilian poultry industry. To understand the participation of nutrition, these 30 years will be divided into decades.

POULTRY NUTRITION IN THE 1970s

In the early stages of the Brazilian poultry industry, nutrition knowledge was weak. Local information was almost nil. International feed companies came to Brazil and built many feed mills to give support to the broiler companies. However, it is important to note that internationally knowledge of poultry nutrition was not that strong, either. In the United States bulletins on poultry nutrition from the National Research Council (NRC) were published in 1960, 1962, 1966 and 1971; and the information in them was far behind what the industry needed at that time. The tables suggested the use of only one diet during the growing phase, from 1 to 56 days. It was the 1977 NRC publication that served as a milestone for use of scientific data in poultry nutrition.

The first big problem faced by poultry nutritionists was the large variation in feedstuff composition. Suppliers did not know important nutritional considerations of their products and buyers did not know exactly what was required. This variation brought about widespread losses in production and still does if not recognized and countered. The first very significant improvement in nutrition was construction of analytical laboratories in which to conduct proximal analysis of feedstuffs. So, the nutrition progress during the first half of the decade was almost entirely due to imported technology applied to our conditions without too much for comparison.

In the second half of the 1970s the poultry industry had become well-established. The goal was to improve it internally, motivate the population to increase chicken consumption, and to go abroad with the Brazilian product to start the exportation process. As shown in Table 1, during those years Brazil had product to export and could compete around the world. The steady increase in exports each year was impressive.

From the beginning, some companies in the south of Brazil introduced the concept of contract growing. These programs became convenient because poultry production in this region was mainly on small farms where producers had few alternatives and land allotments were very small. During the late 1970s many integrators from the South left partnerships with independent feed mills and started producing their own diets. That was needed once the contract grower system was operating well and diet production started to be mandatory to integrators to sustain the system.

Lots of investment in feed mills, silos and feedstuff laboratories began during this period. Instead of waiting for information, technical people from different companies left the country and started visiting competitor markets to learn what they were doing with regard to feed production. It was a timely move as many companies across the world were growing at that time and the international public research institutes had developed significant scientific information about poultry.

One change encouraged by scientific travel was the switch from a one diet to a two diet program. The starter diet was fed through 30 days of age and the finisher diet fed until slaughter, around 56 days. About that same time the second main event in Brazilian poultry nutrition, the computer era, began. Nutritionists had quick access to that technology. The equipment was very expensive and most nutritionists were sharing computer time with people doing administrative work in their respective companies. Feed formulation using this equipment brought a better economic and technical response from the birds.

Synthetic DL-methionine was introduced to the Brazilian market at a reasonable price in the late 1970s. This had a major impact. Dietary protein levels could be re-evaluated once it was possible to compensate total protein reduction with added methionine. Additionally, the first vitamin and mineral supplement companies started offering products and the main feed mills started producing their own supplements.

Nutritionists already knew how to formulate for broilers, heavy breeders and laying hens based on most of the nutrients still considered today, i.e., metabolizable energy, crude protein, calcium, total phosphorus, methionine and cysteine. Still, the National Research Council bulletins together with the tables proposed by the Agricultural Research Council from England formed the basis of international information in use in Brazil. The more capable nutritionists, with the support of the feedstuff laboratory, started formulating the diets based on lab results.

The third big change in poultry nutrition in the 1970s was introduction of diets with high metabolizable energy content. The added energy in US diets was obtained at that time (and is still obtained) from animal and vegetable oil blends, mainly from fast food restaurants. In Brazil, the main source of high energy feedstuff was soybean oil and some poultry oil or lard. That change in feed formulation procedure promoted a significant improvement in field results and also started the era of the grease chicken. The difference in diets before and after use of high energy sources is illustrated in Table 3.

Table 3. Nutritional composition of the diets used in the mid 1970s and later.

Nutrient	Starter		Finisher	
	Mid 1970s	Later	Mid 1970s	Later
Metabolizable energy, kcal	2780	3100	2850	3200
Crude protein, %	23.0	21.0	20.5	19.0
Calcium, %	1.00	1.00	0.95	0.95
Available phosphorus, %	0.45	0.45	0.42	0.40
Methionine, %	0.53	0.53	0.47	0.47
MET + CYS, %	0.91	0.91	0.72	0.82
Lysine, %		1.24		1.10
Metabolizable energy:crude protein ratio	121:1	148:1	139:1	168:1

Academic progress in Brazil was also quite significant during the 1970s. The first graduate courses in animal sciences had begun in the late 1960s and early 1970s and now offered to the market master of science graduates with expertise in poultry nutrition. The universities were important in developing the first research work in poultry science and gave to the professionals the tools of research, especially an understanding of scientific methodology. In the same period of time EMBRAPA (Empresa Brasileira de Pesquisa Agropecuária) was founded. One of its research stations, in Concórdia, Santa Catarina, had pig and poultry production as its main research purpose. This institution, together with the universities, contributed deeply to the field of poultry production. By the late 1970s some of the master of science graduates began being hired to work in the poultry industry in research and technological development.

POULTRY NUTRITION IN THE 1980s

The decade between 1980 and 1990 was a fantastic decade in poultry nutrition. The poultry industry was very much consolidated in Brazil with export levels high and internal consumption continuing to grow. With export figures steadily increasing, the US and France began to recognize the potential of Brazil in poultry production. At that time subsidy programs in the US and France were introduced to restrain Brazilian poultry progress. In early 1980s, Brazil crossed into the second era of its poultry production, the 'era of exportation'. That period of time was important because all the technical people understood that the third era of production should start, an era I have termed one 'of technical professionals and technology'. During this era productivity was the main objective. To get to the desired level of productivity more technical people were hired by the industry which also served to strengthen veterinary, agronomic and animal science educational programs. The safety margins in feed formulation began being reduced by those more progressive nutritionists.

There were many changes in nutritional programs during this decade. It began with a strong introduction of pelleted diets. There were lots of discussions on advantages and disadvantages of that technology. As a matter of fact those disagreements are still ongoing. The benefits of feeding heavy breeder males and females separately was proven. That program was so consistently effective that almost all breeder farms use sex-different feeding programs today. At the same time the broiler finisher diet, offered the last week before slaughter, was introduced. That diet had the advantage of being cheaper as it did not contain the growth promotant additives present in the grower. In 1984 the National Research Council updated its nutrient recommendations for poultry and suggested three diets for broilers over the growing period.

During the same period, many papers were published by Brazilian scientists suggesting that many alternative feedstuffs could substitute for corn and soybean meal. The most relevant ones were toasted soybeans, sorghum, rice bran, defatted rice bran and poultry by-product. Those new alternatives had a significant impact on cost of feed.

Companies in the 1980s encouraged international visits and congress participation by their technical people. However, a very important change was the beginning of internal research and development programs by the poultry industry. Companies started operating research farms with the goal of answering local questions internally.

Separate production of broiler males and females was introduced in the 1980s. This procedure is fairly common today. However, use of different diets for males and females still is not used as it should. In many cases similar diets for males and females are produced as a convenience to avoid complication at the feed mill.

The decade finished with broilers being slaughtered at least 20 days earlier than broilers in the late 1960s. In contrast, little progress in nutrition of breeders and laying hens was made. Traditional concepts prevailed.

POULTRY NUTRITION IN THE 1990s

We still have many years to go in this decade. The main change to date is the introduction of a fourth diet in the feeding program of broilers. The pre-starter diet is offered from 1 to 7 days of age. The main change in this diet is the reduction of energy as birds at this age do not efficiently use lipids. This diet has more crude protein, vitamins and minerals. Also at the beginning of the decade the use of synthetic lysine became economically feasible.

However, it is still expected that during this decade separate feeding of males and females will become more firmly established. The levels of dietary crude protein will probably be reduced as methionine and lysine are already available and certainly other new amino acids such as threonine and tryptophan will become economical. Protein reduction will reduce diet cost and will also be a welcome way of reducing environmental pollution. Studies on digestible amino acid requirements are needed in conjunction with this reduction in protein. The concept of ideal protein will be particularly studied.

Biotechnology will contribute strongly with new nutrients and additives to stimulate performance of poultry. Enzymes, probiotics, yeast cultures, organic acids and mineral proteinates will become better known. In the field of enzymes, it is very important to find a way to use organic phosphorus more efficiently. Phosphorus pollution could be reduced if phytin phosphorus could be better used by non-ruminant animals. Also it would not be possible to begin a new century without better utilization of soybean carbohydrates. It is very expensive for soybean carbohydrates to be used so inefficiently, especially by broilers. Probiotics can also be important for the poultry industry. Many probiotics studies are inconclusive, however, part of the problem with these results is related to the research methodology applied in these studies. Organic acids represent an opportunity to reduce enterobacteria problems. Also, acidification can reduce problems with regard to fungal contamination of feed ingredients in poultry rations.

Another big objective for this decade is reduction of abdominal fat. In order to reduce body fat of birds significantly, some old concepts about relationships between energy and protein must be reviewed. Also, more work must be done with saturated and non-saturated fatty acids and the relationship between essential and non-essential amino acids. Consumers will be increasingly concerned about fat in the human diet and will leave the fatty bird alternative behind them. So, studies on factors affecting carcass composition will be of interest during this decade. Pellet quality will continue to be part of academic and practical discussions.

Alternative feedstuffs such as extruded soybeans will be of increasing interest. Also, triticale, both sunflower seed and meal along with canola meal are potentially useful as alternative feed ingredients.

It is also important that we be aware of environmental and ecological movements which will increase in strength as the decade progresses. These forces are expected to bring about some changes in poultry research, production and even in slaughtering processes.

The technology in poultry nutrition will continue to expand with

Table 4. Evolution of broiler performance in Brazil.

Year	Slaughter age (days)	Average body weight (g)	Average daily gain (g)	Feed conversion (g/g)
1985	50.7	2094	41.3	2.22
1986	51.6	1935	37.5	2.21
1987	49.6	1999	40.3	2.17
1988	47.5	2064	43.4	2.11
1989	45.9	2059	44.8	2.06
1990	44.3	1983	44.8	2.04
1991	44.3	2054	46.4	2.01
1992	44.4	2067	46.6	2.02
1993	43.3	2061	47.6	1.96
1994	43.0	2083	48.4	1.96

increasing numbers of technical people employed by the industry and by strong local research programs.

The analytical area will gain in importance as this decade progresses. Sophisticated equipment such as atomic absorption spectrophotometers, high precision liquid chromatography and near infra-red spectrophotometers will become commonly used. This new equipment will encourage nutritionists to formulate diets based on specific ingredient characteristics and not 'book' values.

The new technologies and nutritional changes in the 1980s and the early 1990s had a profound impact on broiler performance during that period, as can be seen in Table 4.

The future of poultry nutrition in Brazil

Nutritionists will need to expand their knowledge about mathematical modeling to better understand bird growth in the future. Also, they will need to get rid of animal development parameters as the most important data in the poultry industry. The technical people will need to be better trained to understand the concept of cost–benefit analysis for all processes. Nutrition departments in the industry will need to be technically prepared to give support to feed formulation, mill operation, analytical laboratories, research facilities and be closely involved with the sector that buys commodities and other dietary components. So, the nutritionist in the future will need wide-ranging knowledge.

However, the Brazilian government also has an interest in maintaining technical growth of professionals. This means that research programs at universities and institutions should be supported to aid in developing technology for the future and training people for a competitive industry.

NATURAL ANTIOXIDANTS: A NEW PERSPECTIVE FOR THE PROBLEM OF OXIDATIVE RANCIDITY OF LIPIDS

ALFONSO VALENZUELA B.

Unidad de Bioquímica Farmacológica y Lípidos,
Instituto de Nutrición y Tecnología de Alimentos,
Universidad de Chile, Santiago, Chile

Introduction

Lipid oxidation is a normal biological process by which we obtain energy from fats. However, lipid oxidation is also often the decisive factor determining the useful storage life of food products, even those with fat content as low as 0.5% or 1%. Oxidation produces undesirable changes in color, flavor, aroma, and other quality factors of a food, and causes the rancid taste and odor that develop in unprotected fats and oils. The nutritional value of a product is impaired and toxicity may be induced. Texture may also change as a result of side reactions between proteins and the products of fat oxidation. In short, oxidative deterioration of lipids may be considered a spoilage factor affecting all aspects of food acceptability. Much research has been conducted to better understand the mechanism of oxidative rancidity in polyunsaturated lipids and effects of decomposition products of lipid oxidation on the development of rancidity.

Antioxidants are organic molecules of either synthetic or natural origin which can prevent or delay development or progress of oxidative rancidity. Much research has been done on antioxidants. Synthetic antioxidants have been a concern during the last two decades because some deleterious and potentially dangerous effects have been observed. Natural antioxidants offer an alternative to the questioned synthetic antioxidants. However, there are also drawbacks to natural antioxidants when compared with the synthetic products.

OXIDATIVE RANCIDITY: MECHANISMS AND PRODUCTS

The classical mechanism for oxidative rancidity is via oxygen free radical attack, and the initial substrates for oxidation are polyunsaturated fatty acids in fats and oils (Labuza, 1971). The higher the polyunsaturation of a lipid, the greater its susceptibility to oxidative rancidity. It has been proposed that the number of pentadienic units present in the structure of a polyunsaturated fatty acid is correlated with its susceptibility to oxidation (Nawar and Ultin, 1988). A pentadienic unit is defined as a structure formed by five carbon atoms bearing two double bonds separated by a methylene group.

Oxidative rancidity of unsaturated lipids has been well documented and unless mediated by other oxidants or enzyme systems, proceeds through a free radical chain mechanism involving initiation, propagation and termination steps. The free radical chain mechanism has been generally accepted as the only process involved in oxidative rancidity. The initiation step has been and still remains the subject of much research and general uncertainty (Gray and Monahan, 1992). The direct reaction of polyunsaturated fatty acids with molecular oxygen is thermodynamically unfavorable. However, the spin restriction prohibiting the interaction of ground state oxygen with a polyunsaturated fatty acid can be overcome by a number of initiating mechanisms including singlet oxygen, partially reduced or activated oxygen species such as hydrogen peroxide, superoxide anion or hydroxyl radicals; active oxygen iron complexes, and iron mediated homolytic cleavage of the hydroperoxides which generate organic free radicals at the polyunsaturated fatty acid molecule.

The resulting fatty acid free radical reacts with molecular oxygen to form peroxy radicals. This is the propagation step. In this process the peroxy radicals react with more polyunsaturated fatty acid molecules to form fatty acid hydroperoxides, which are the fundamental primary products of oxidative rancidity. Decomposition of fatty acid hydroperoxides constitutes a very complicated process and produces a multitude of materials that may have biological effects and cause flavor deterioration in fat-containing foods. This decomposition proceeds by homolytic cleavage of the fatty acid hydroperoxides to form alkoxy radicals. These radicals undergo carbon-carbon cleavage to form breakdown products including aldehydes, ketones, alcohols, hydrocarbons, esters, furans and lactones. Fatty acid hydroperoxides can react again with molecular oxygen to form secondary products such as epoxyhydroperoxides, ketohydroperoxides, dihydroperoxides, cyclic peroxides and bicyclic endoperoxides. These secondary products can in turn decompose like monohydroperoxides to form volatile breakdown products. Lipid hydroperoxides can also condense into dimers and polymers that can break down and produce volatile material generating the typical 'rancid odor' of the oxidized material (Frankel, 1984). The formation of such a diversity of organic products constitutes the termination step.

Finally, fatty acid hydroperoxides and some of their bifuctional breakdown products can interact with proteins, membranes, and enzymes. These reactions with biological components are of most concern to biochemists and cellular physiologists because they can affect vital cell functions and structures (Logani and Davies, 1980; Kubow, 1993). The development of fat oxidative rancidity in complex food systems is also greatly affected by the interaction of proteins and amino acids with the oxidation products. Complex high molecular weight interaction products are formed during processing and cooking of foods, and their further degradation into volatile compounds is a very complex chemical process not well understood. Figure 1 shows a simplified scheme of the

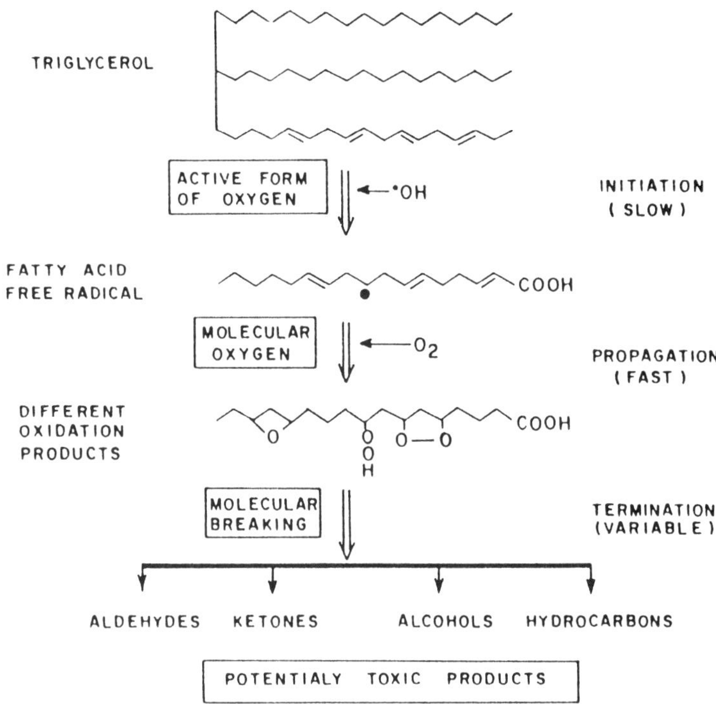

Figure 1. Steps of oxidative rancidity.

various steps of the oxidative rancidity of a polyunsaturated fatty acid (Kubow, 1992).

Synthetic and natural antioxidants: quality protectors

Oxidative rancidity may be avoided or retarded through use of antioxidants. In theory, a substance may act as an antioxidant in a variety of ways, e.g. competitive binding of oxygen, retardation of the initiation step, blocking of propagation by destroying or binding free radicals, inhibition of catalysts, stabilization of hydroperoxides, etc. Antioxidants can scavenge the active forms of oxygen involved in the initiation step of oxidation, or can break the oxidative chain reaction by reacting with the fatty acid peroxy radicals to form stable antioxidant radicals which are either too unreactive for further reactions or form non-radical products.

Antioxidants may be classified as synthetic or natural. The synthetic antioxidants have been generally considered to be relatively 'safe' and are widely applied in a number of manufactured products including pharmaceuticals, cosmetics, human foods and animal feeds. The most popular are those derived from phenolic structures such as butylated hydroxanisole (BHA), butylated hydroxytoluene (BHT), propyl (PG), octyl or dodecyl galates, and terbutyl hydroxyquinone (TBHQ). Another antioxidant usually utilized but only for animal feed is the non-phenolic compound ethoxyquin (ETOX). However, during the past two decades, both consumers and regulators have become suspicious of synthetic antioxidants. The 'safe' status for most of the synthetic antioxidants is now a matter of concern because of their potential toxicity in some biological models (Anonymous, 1986; Federal Register, 1990). Figure 2 shows the chemical structure of the most common synthetic antioxidants.

Although no definitive conclusions may be drawn from the experimental research about synthetic antioxidants, and because other researchers claim positive effects for synthetic antioxidants under different pathological experimental conditions, the public has become aware of questions about synthetic products. In this context natural products appear more healthy and safe than synthetic antioxidants.

The empirical use of natural compounds as antioxidants is very old. The popularity of smoking and spicing in the home for the preservation of meat, fish and cheese and other fat-rich foods may be due, at least in part, to the recognition of the rancidity-retarding effect of these treatments. A number of products, extracted mainly from vegetables, have been

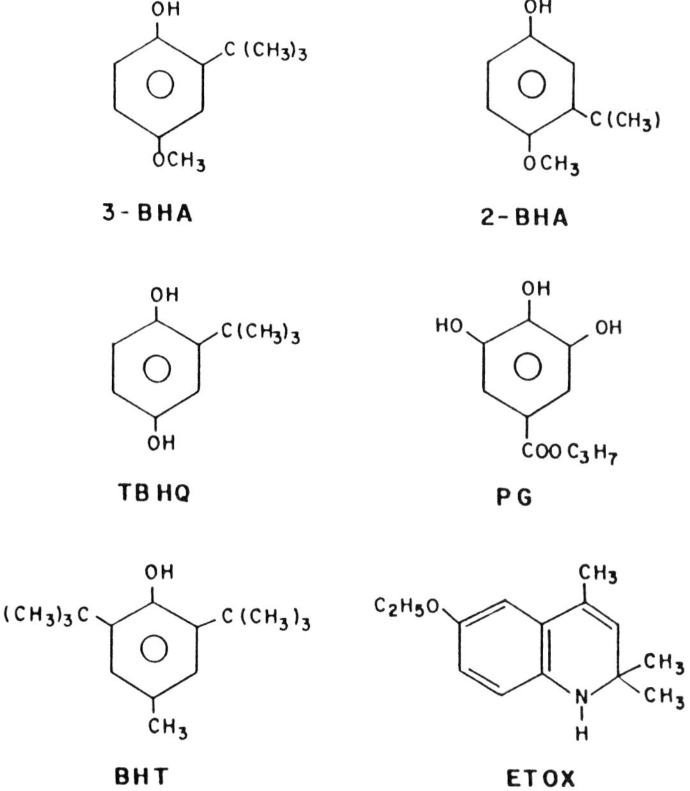

Figure 2. Synthetic antioxidants.

claimed to have antioxidant properties. Products extracted from sesame oil (sesamol), from the seeds of rice (orizanol), from different spices such as sage, mace, black pepper, oregano, and allspice have been demonstrated to be effective antioxidants when assayed under various experimental models (Namiki, 1990). Specific types of molecules such as phenolic flavonoids, tannins, porphyrin-related substances and Maillard reaction products also show antioxidant properties. However, only a few of them have proved effective in retarding or inhibiting oxidative rancidity of oils, fats or more complex products. The most important proven natural antioxidants are the tocopherols, some flavonoids, the extracts from rosemary and boldine, a new natural antioxidant described recently.

TOCOPHEROL ISOMERS (VITAMIN E) AS ANTIOXIDANTS

Among the natural antioxidants, tocopherols deserve special attention. Tocopherols occur as minor constituents in all vegetable oils and are the best known and most widely used antioxidants. The alpha, beta, gamma and delta isomers of tocopherol differ in the degree of methylation of the dihydrochromanol ring. The natural tocopherol isomers (also named tocols) and the tocotrienol isomers (analogs to tocopherol but having a saturated side chain) are collectively named vitamin E. This vitamin is the most important dietary component contributing to antioxidant defenses in tissue (Bieri, 1984). Tocopherols work as antioxidants by donating the hydrogen of the hydroxyl group (Figure 3) to a fatty acid radical. The resulting tocopherol radical is very stable and does not continue the chain reaction of producing new fatty acid radicals. The stability of the tocopherol radical is due to its aromatic structure and its ethereal oxygen. The one paired electron is distributed around the aromatic ring to the ethereal oxygen and on the side chain.

The antioxidant activity of tocopherols depends very much on the food to which they are added, the concentration used, the availability of oxygen, and the presence of heavy metals and various synergists. At high concentrations and in the presence of iron or copper salts tocopherols may act as prooxidants. The prooxidant effect of alpha tocopherol can often be observed at relatively low concentrations, an effect not observed with gamma and delta tocopherols. This leads to the conclusion that for antioxidant properties in foods the gamma and delta tocopherols should be used instead of alpha tocopherol. This means natural vitamin E should be chosen in preference to the synthetic compound which contains mainly alpha isomer. Satisfactory antioxidant activity is usually achieved only when tocopherols are used in combination with synergists such as ascorbic acid, citric acid, some synthetic antioxidants, or various chelating agents. The relative antioxidant activity of tocopherols ranges in descending order: delta isomer> gamma isomer> beta isomer> alpha isomer (the less active) (Pokorny, 1991). Table 1 shows relative effectiveness of tocopherol isomers as antioxidants.

Natural vitamin E increases the stability of animal fats, vegetable oils and processed foods. Vegetable oil naturally contains vitamin E, but it

Table 1. Antioxidant activity of tocopherols compared with biological activity.

	Antioxidant activity	Biological activity
d – Alpha tocopherol*	100	1.49
d, l – Alpha tocopherol†	100	1.10
d – Gamma tocopherol	200	0.15
d – Delta tocopherol	400	0.05

* International units/mg
† Synthetic analog

is lost during the purification process. Therefore, it is useful to add vitamin E at the final stage of product processing. Animal fats contain little or no antioxidants such as vitamin E. To obtain greater stability, therefore, it is also useful to add vitamin E. Natural vitamin E is very stable with respect to heat. It is very tolerant to heat, it evaporates only very slightly even at high temperatures and it has excellent carry-through effect. These characteristics make it particularly useful as an antioxidant for frying oils. In technology, the natural vitamin E content of many oils is important for their stability. However, on an equal weight basis, vitamin E is less effective than synthetic phenolic antioxidants and much more expensive.

Synthetic *dl*-alpha tocopherol or its acetate derivative is frequently added to fats and oil blends, but the addition of natural tocopherols is preferred because the synthetic preparation contains biologically inactive stereoisomers. Tocopherol oxidation products also have some antioxidant activity, and the structurally related alpha, beta, gamma and delta tocotrienols present in some vegetable oils such as palm oil or wheat germ oil have antioxidant activities similar to those of the respective tocopherols. Figure 3 shows the chemical structure of the four tocopherol isomers.

FLAVONOIDS: A SPECIAL TYPE OF NATURAL ANTIOXIDANT

Flavonoids constitute a large group of naturally-occurring plant products that are widely distributed in the vegetable kingdom. All of them are structurally derived from the parent compound flavone (2–phenylcromone or 2–phenylbenzopirone) and are characterized by two benzene rings joined by a C3 structure which is condensed as a six-membered ring and changes with the nature of the flavonoid. Some of the carbons, at any of the three rings, are often hydroxylated and some of these hydroxy groups may have transformed to methoxy groups (Havsteen, 1983). Flavonoids are ubiquitous in photosynthesizing cells, seeds, fruits and flowers. More than 500 different types of flavonoids are now known. Flavonoids, when incorporated into the alimentary chain, may also be present in insects, mollusks, reptiles and even mammals (Middleton, 1984). It has been estimated that the average daily western diet many contain up to 1 g of mixed flavonoids. For centuries, a number of different therapeutic and curative properties have also been ascribed

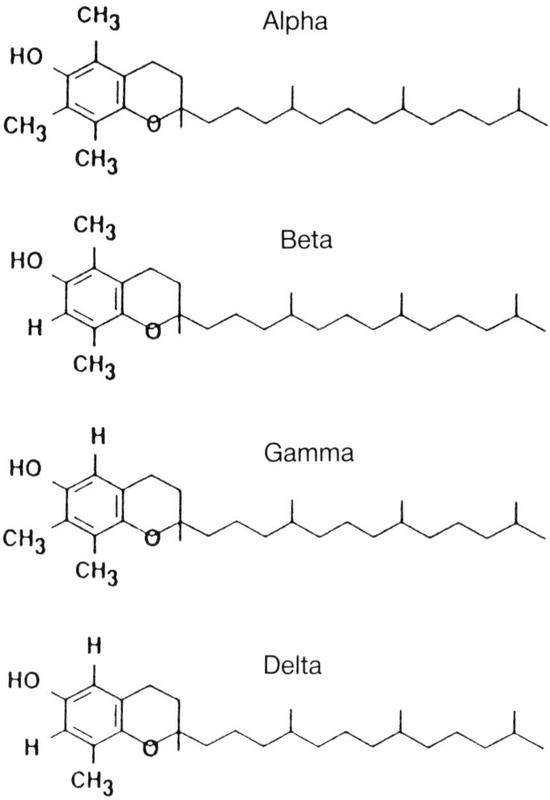

Figure 3. Tocopherol (isomers).

to flavonoids and many of them have been incorporated into popular folk medicine. Flavonoids such as quercetin (Beretz *et al.*, 1982), taxifolin (Vladutiu and Middleton, 1986) and silymarin (Valenzuela and Garrido, 1994) have been used as pharmacological principles, either as such, or mixed in several chemically complex preparations.

Among other properties, flavonoids exhibit high binding affinity to biological polymers and heavy metal ions. They catalyze electron transport reactions and may be active in scavenging free radicals (Robak and Gryglewski, 1988). In view of these properties, their known liposolubility and their hydroperoxide reducing properties, the possible antioxidant activity of flavonoids was investigated by our group. We demonstrated that some commercial flavonoids (such as catechin, morin and quercetin) as well as a group of flavonoids extracted and purified from Chilean native plants, show potent antioxidant activity when assayed against the temperature and metal-induced oxidation of fish oil (Nieto *et al.*, 1993). Although the antioxidant capacity of some flavonoids has been demonstrated both in *in vivo* and in *in vitro* models, their mechanism of action is still a matter of speculation. All flavonoids assayed as antioxidants have in common one or two hydroxyl groups (free or substituted) attached to the B ring of their structure (Figure 4), and their potency appears to be affected by flavonoids having free hydroxyl groups at the ortho position (3′ 4′), should exhibit a reduced antioxidative effect, whereas those containing hydroxyl groups in the para position (2′ 5′) are the most favored (Das and Pereira, 1990).

Flavonoids have been defined as 'high-level' antioxidants (Robak and Gryglewski, 1988). That is, they act by scavenging those free radicals or excited forms of oxygen involved in the first steps of lipid oxidation such as singlet oxygen, the superoxide free radical or the hydroxyl free radical (Fraga *et al.*, 1987). On the other hand, tocopherols defined as 'low-level' antioxidants act at the later steps of oxidation by stabilizing those free radicals formed at the structure of the polyunsaturated fatty acid (methylene, alcohoxy or peroxy free radicals). These different sites of action in the oxidation chain may relate to the synergistic effect noted when flavonoids are assayed in mixture with tocopherols (mainly *dl*-alpha tocopherol) (Nieto, *et al.*, 1993). Additional antioxidative action of most flavonoids may result from their metal-chelating properties. As a matter of fact, quercetin may form strong binding complexes, particularly with copper and iron (Thompson *et al.*, 1976). Results indicate that some flavonoids could be used as natural antioxidants and might substitute for synthetic antioxidants whose use has been questioned due to potential undesirable secondary effects. However, further physiochemical and toxicological evaluations are required to assess the effectiveness and the future of flavonoids such as antioxidants for fish oils or other oils rich in polyunsaturated fatty acids. Figure 4 shows the structure of some antioxidant flavonoids.

QUERCETIN

MORIN

RUTIN

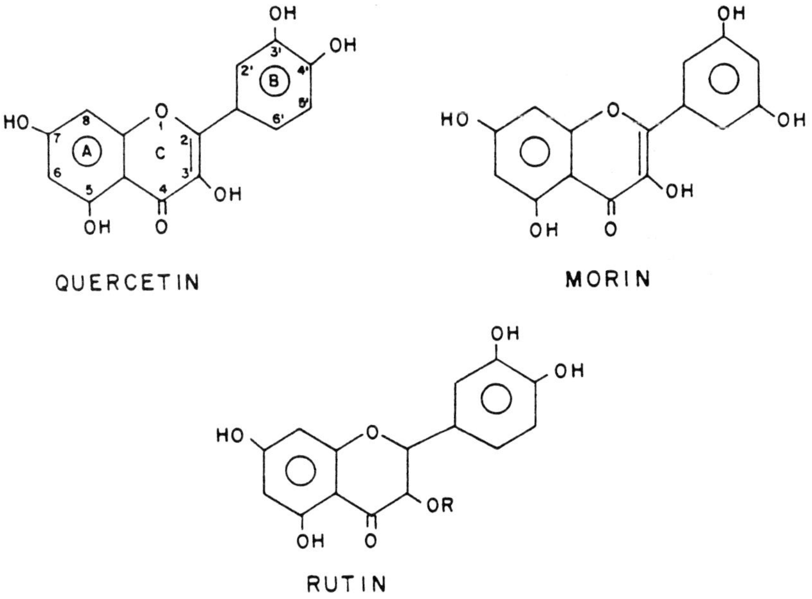

Figure 4. Some antioxidant flavonoids.

ROSEMARY, A SPICE APPROVED AS A NATURAL ANTIOXIDANT

The hexane extract of the leaves of rosemary (*Rosmarinus officinalis* L.) contains four effective antioxidants: carnosol, rosmanol, isorosmanol and rosmaridiphenol. These four are odorless and tasteless diterpenelactones having an *O*-diphenolic group on the C ring which is responsible for antioxidative activity. Among these, rosmaridiphenol and rosmanol show stronger antioxidant activity than BHA when assayed in lard as well as in a water–oil emulsion system. Rosemary is the only spice commercially available for use as an antioxidant in the United States and Europe (were rosemary products reportedly represent about 40 to 50% of the natural antioxidant market). However, because of their primary use as flavoring agents, rosemary extract products are not technically listed as natural preservatives or antioxidants.

Rosemary extracts have been used commercially in processed foods for over 30 years. Dr. S. Chang and colleagues developed and patented a process to concentrate natural antioxidants from rosemary by solvent extraction and subsequent steam vacuum distillation of the extract. This procedure provides an edible oil without a bitter flavor or rosemary odor in the finished product (Chang *et al.*, 1984). Dr. Chang's work subsequently showed the efficacy of these deflavored extracts in retarding soybean oil reversion and in reducing hydroperoxide development in lard as well as in potato chips fried in the oils containing the spice extracts. Figure 5 shows the chemical structure of the four antioxidant active isomers of rosemary extract.

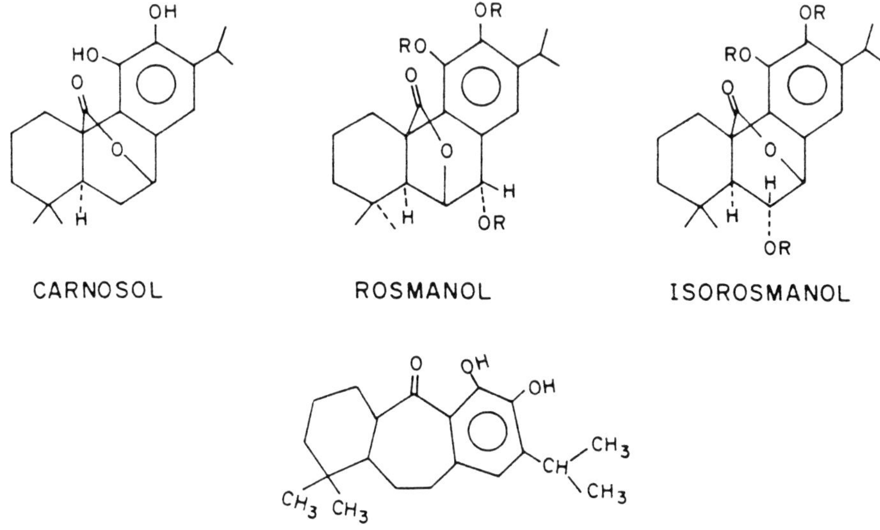

Figure 5. Isomers of rosemary extract.

Kalsec, Inc., which holds an exclusive license to use Dr. Chang's patent, modified procedures to reduce color in its rosemary extract which is said to effectively retard oxidative rancidity, warmed-over flavors and color degradation in snack foods, mayonnaise, salad dressing, citrus oils, processed meats, poultry and seafoods. Drawbacks to using rosemary, however, include its high cost relative to the synthetic antioxidants and the larger quantities that often must be used to obtain the same effect. For instance, marine oils, which are highly unsaturated, require addition of a high level of rosemary extract. Rosemary extracts also must be used at higher levels than TBHQ to obtain the same antioxidant protection in vegetable oils. Even so, demand for rosemary extract antioxidants continues to grow (Fitch-Haumann, 1990).

BOLDINE: NEW PERSPECTIVES ON AN OLD NATURAL PRODUCT

Boldine [(S)-2, 9–dihydroxy-1, 10–dimethoxyaporphine] is the most abundant alkaloid present in the bark and the leaves of boldo (*Peumus boldus* Mol.), a widely distributed native tree from the south region of Chile. Mature boldo trees are usually between 6 and 12 m tall, although some may attain heights up to 20 m. Aqueous infusions of boldo leaves containing boldine have been used for hundreds of years by the natives of Chile for the treatment of gastric and hepatic diseases. The purported choleretic, diuretic, sedative and digestive stimulant properties of boldo extracts and of boldine have been thoroughly described by the pharmacopoeias of natural medicinal products (Speisky and Cassels, 1994). As boldo only grows in abundance in Chile, this country has for decades been the sole original source of the leaves and bark and currently exports about 800 tons of dried boldo leaves per annum mainly to Argentina, Brazil, Italy, France, and Germany.

A few years ago it was demonstrated that boldine in pure form, extracted and purified mainly from the bark of boldo, had a potent antioxidant activity when assayed in both abiotic and biological systems. The aporphine had strong antioxidant activity when tested against either spontaneous or chemically-induced oxidation of biological systems undergoing peroxidative free radical-mediated damage. In studies designed to elucidate the mechanism of the antioxidative action of boldine it has recently been established that the boldine molecule acts as an efficient hydroxyl free radical scavenger (Cederbaum *et al.*, 1992). The latter radicals, recognized as the most reactive oxygen species generated by biological systems, are trapped by boldine with even greater efficiency than that exhibited by dimethyl sulphoxide, a compound used experimentally as a paradigmatic hydroxyl free radical scavenger.

Boldine also inhibits the metal or temperature-induced oxidative rancidity of fish oil, its antioxidative action being three to four times greater than that observed for *dl*-alpha tocopherol and conventional synthetic antioxidants such as BHA, BHT or mixtures of these antioxidants (Valenzuela *et al.*, 1991). In addition, boldine shows a potent synergistic action when mixed with the flavonoid quercetin. Boldine is also effective for stabilizing highly polyunsaturated *n*-3 fatty acid concentrates (up to 80% concentration) (Valenzuela *et al.*, 1994). Although the oil stabilizing activity of boldine resembles that of the natural flavonoid quercetin, the demonstration that the latter compound is mutagenic *in vitro* (Namiki, 1990) is likely to curtail its technological development as a food antioxidant.

Recent studies addressing structure-activity relationships for benzylisoquinolides suggest that all the known boldo alkaloids are likely to exhibit at least some antioxidative activity. Interestingly, in the case of aporphine alkaloids, the presence of phenolic groups is clearly not essential for these molecules to display their activity. In the case of boldine, *O*-methylation of the two phenol functions to produce glaucine was not associated with a loss of potency. Subsequent *N*-methylation led to a derivative shown to be virtually devoid of antioxidative activity. Thus, the available evidence suggests that in the case of aporphine alkaloids, in addition to the expected contribution of phenolic functions to their radical-scavenging ability, the hydrogen atom bonded to the benzylic carbon next to the basic nitrogen atom may be a key to the antioxidative activity displayed by the non-phenolic boldo alkaloids (Cassels *et al.*, 1994). Figure 6 shows the chemical structure of boldine and the alternative sites for electron delocalization in the boldine free radicals.

Concluding remarks

Natural food additives, including antioxidants, are generally preferred by consumers and more easily gain legislative approval than synthetic additives. This is true although most antioxidants are polysubstituted

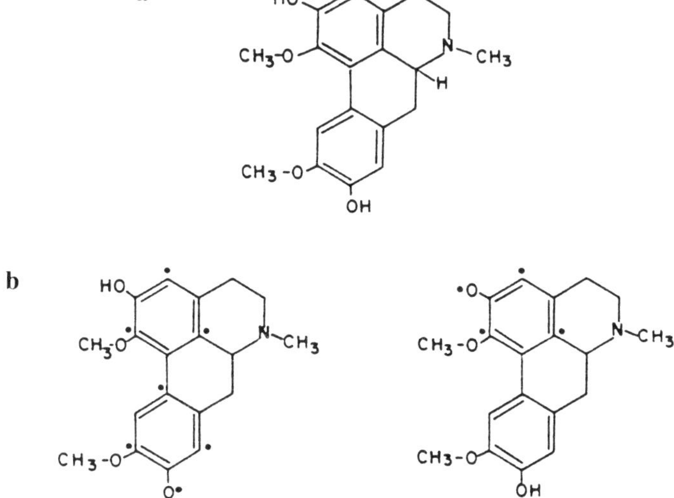

Figure 6. (**a**) Chemical structure of boldine. (**b**) Alternative sites for electron delocalization.

phenolic compounds whether natural or synthetic. Natural food components do not cause acute toxicity, as highly toxic substances would have been eliminated from traditional use as foods. However, the fact that a substance is natural and commonly found in food is not a guarantee that it is entirely non-toxic. It may have low toxicity, or be carcinogenic or mutagenic. Synthetic antioxidants are tested for such effects, but many natural food components have not yet tested.

The advantages and disadvantages of natural and synthetic antioxidants are summarized in Table 2. No rational scientific and technical argument can be given for preferences for natural antioxidants. They are more acceptable to consumers mainly on emotional grounds. While it is important for manufacturers to meet the requirements of consumers, it is imperative that the safety of additives that are not 'generally recognized as safe' (GRAS) be tested before use. Perhaps the safest approach is to avoid the use of both synthetic and natural antioxidants by appropriate packaging and storage methods or by avoiding the use of ingredients that are readily oxidized. However, the benefits of using antioxidants outweigh the risks. Without antioxidants in foods oxidation products are created and cause a greater risk to health than possible hazardous effects of antioxidants. Therefore, antioxidants may be considered as 'quality protectors', but if they are of natural origin we must consider them as 'healthy quality protectors'.

Table 2. Advantages and disadvantages of natural antioxidants, as compared with synthetic antioxidants.

Advantages	Disadvantages
Readily accepted by the consumer, considered to be safe and not a 'chemical'.	Usually more expensive if purified, and less efficient if not purified.
No safety test required by legislation if a component of a food is 'generally recognized as safe'.	Properties of different preparations vary if not purified. Safety often not known.
	May impart color, after taste or off-flavor to the product.

Acknowledgments

Support for the work of the author was provided by FONDECYT (projects 1930808 and 1940422).

References

Anonymous. (1986). Antioxidants' safety examined. J. Am. Oil Chem. Soc. 63:724–727.

Beretz, A., J. Cazenave and R. Anton. 1982. Inhibition of aggregation and secretion of human platelets by quercetin and other flavonoids. Structure-activity relationships. Agents Actions 12:382–387.

Bieri, J. 1984. Sources and consumption of antioxidants in the diet. J. Am. Oil Chem. Soc. 61:1917–1918.

Cassels, B.K., M. Asencio, H. Speisky, L.A. Videla and E. Lissi. 1994. Structure-antioxidative activity relationships in benzyl-isoquinilide alkaloids. Pharmacol. Res. (Submitted).

Cederbaum, A., E. Kukielka and H. Speisky. 1992. Inhibition of rat liver microsomal lipid peroxidation by boldine. Biochem. Pharmacol. 44:1765–1772.

Chang, S., H. Chi-Tang and C. Houlihan. 1984. Elucidation of the chemical structure of a novel antioxidant, rosmaridiphenol, extracted from rosemary. J. Am. Oil Chem. Soc. 61:1036–1039.

Das, N.P., and T.A. Pereira. 1990. Effects of flavonoids on thermal autoxidation of palm oil: structure-activity relationship. J. Am. Oil Chem. Soc. 67:255–258.

Federal Register. 1990. Nov. 29, Vol. 49, pp. 576–577.

Fitch-Haumann, B. 1990. Antioxidants: Firms seeking products they can label as 'natural'. INFORM 1:1002–1013.

Fraga, C., V. Martino, G. Ferraro, J. Coussio and A. Boveris. 1987. Flavonoids as antioxidants evaluated by *in vitro* and *in situ* chemiluminescence. Biochem. Pharmacol. 36:717–721.

Frankel, E.N. 1984. Lipid oxidation: mechanisms, products and biological significance. J. Am. Oil Chem. Soc. 61:1908–1916.

Gray, J.I. and F.J. Monahan. 1992. Measurement of lipid oxidation in meat and meat products. Trends Food Sci. Technol. 3:315–319.

Havsteen, B. 1983. Flavonoids, a class of natural products of high pharmacological potency. Biochem. Pharmacol. 32:1141–1148.

Kubow, S. 1992. Routes of formation and toxic consequences of lipid oxidation products in foods. Free Rad. Biol. Med. 12:63–81.

Kubow, S. 1993. Lipid oxidation products in food and atherogenesis. Nutr. Rev. 51:33–40.

Labuza, P.T. 1971. Kinetics of lipid oxidation in foods. CRC Crit. Rev. Food Technol. 2:335–405.

Logani, M.K. and R.E. Davies. 1980. Lipid oxidation: biological effects and antioxidants, a review. Lipids 15:485–495.

Middleton, E. 1984. The flavonoids. Trends Pharmacol. Sci. 5:335–338.

Namiki, M. 1990. Antioxidants/antimutagens in Food. Food Sci. Nutr. 29:273–300.

Nawar, W.W. and H.O. Ultin. 1988. Stability of fish oils. Omega-3 News III, 1–4.

Nieto, S., A. Garrido, J. Sanhueza, L. Loyola, G. Morales, F. Leighton and A. Valenzuela. 1993. Flavonoids as stabilizers of fish oil: an alternative to the use of synthetic antioxidants. J. Am. Oil Chem. Soc. 70:773–778.

Pokorny, J. 1991. Natural antioxidants for food use. Trends Food Sci. Technol. Sept: 223–226.

Robak, J. and R. Gryglewski. 1988. Flavonoids are scavengers of superoxide radicals. Biochem. Pharmacol. 73:837–841.

Speisky, H. and B.K. Cassels. 1994. Boldo and boldine: an emerging case of natural drug development. Pharmacol. Res. 29:1–12.

Thompson, M., C.R. Williams and G.E. Elliot. 1976. Stability of

flavonoids complexes of copper (II) and flavonoid antioxidant activity. Analyt. Chem. Acta 85:375–381.
Valenzuela, A. and A. Garrido. 1994. Biochemical basis of the pharmacological action of the flavonoid silymarin and of its structural isomer silibinin. Biol. Res. 27:105–112.
Valenzuela, A., S. Nieto, B. Cassels and H. Speisky. 1991. Inhibitory effect of boldine on fish oil oxidation. J. Am. Oil Chem. Soc. 68:935–937.
Valenzuela, A., S. Nieto, A. Ganga and J. Sanhueza. 1994. Inhibition of the thermally-induced oxidation of n-3 polyunsaturated fatty acid concentrates by boldine: a new natural antioxidant. Grasas y Aceites, (Submitted).
Vladutiu, G.D. and E. Middleton. 1986. Effects of flavonoids on enzyme secretion and endocytosis in normal and mucolipidosis II fibroblasts. Life Sci. 38:717–726.

INDIA: A DEVELOPING MARKET FOR ANIMAL PRODUCTION

BHARAT TANDON

VetCare, Bangalore, India

Introduction: Asia action in the marketplace

Considerable attention world-wide has recently focused on the emerging markets in the Asia Pacific region, more specifically the gigantic markets of China and India. While there are many reasons that China is a preferred – in fact more fashionable – destination for American business, opportunities in the Indian market are worth serious consideration.

DEMOGRAPHIC ASPECTS

India's demographic aspects alone point up many opportunities (Figure 1). With the seventh largest land mass (3.3 million square kilometers), India's population is second in size only to China. At last census there were 880 million people. Over 100 million of the population can be considered middle class economic consumers; of which 35 million live in urban areas and 65 million are rural and own more than 4 acres of land.

India has more than 100 universities and advanced management and scientific institutes. English is widely spoken and studied. Economic policies also foster growth. In 1991 the Rupee was devalued by 22%. The present exchange rate, approximately Rs 31.5 per USD, creates competitive cost advantages. The breakdown of socialist controls has brought about a more liberal and open business environment.

Agriculture makes a significant contribution to the gross national product with the value of raw materials at over Rs 800 billion (Table 1). The feed industry contributes 17% of gross national output. All major raw materials for animal feed processing are locally available.

Several factors contribute to the expanding domestic market in India including urbanization, women joining the work-force, household labor shortages and greater incentives for saving.

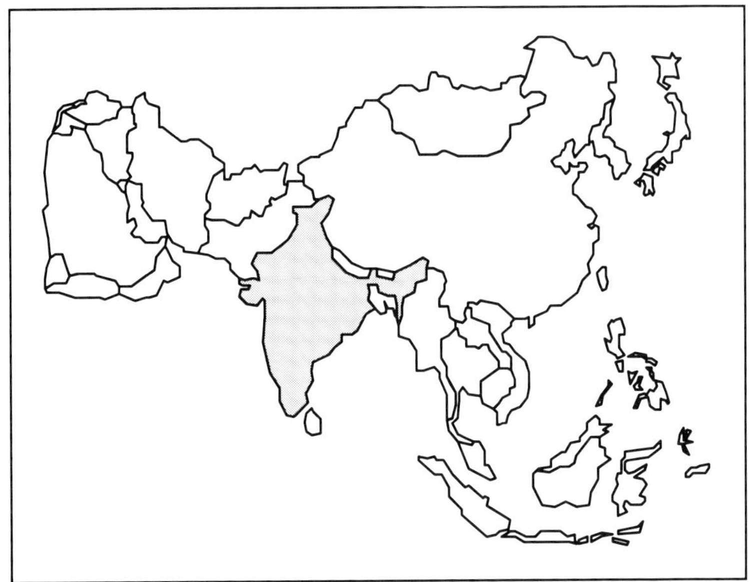

Figure 1. India and the Asia Pacific region.

Table 1. Indian food industry profile.

Value of raw materials produced	Rs 800 Billion (25.4 Billion USD)
Materials after primary processing	Rs 1080 Billion (34.3 Billion USD)
Materials undergoing secondary/ tertiary processing	Rs 220 Billion (7 Billion USD)
Net value addition	Rs 500 Billion (15.9 Billion USD)
Value of finished products	Rs 1300 Billion (41.3 Billion USD)

Table 2. Growth of India's poultry industry.

	1993	1995	2000
Broiler population, millions (approximately 20% annual growth)	300	430	950
Layer population, millions* (annual growth approximately 10%)	140	170	290
Broiler parent population, millions*	4	5.5	12.5
Layer parent population, millions*	1.75	2.1	3.5
Egg production, millions	27 000	34 000	52 000
Per capita egg consumption	30	34	48
Per capita poultry meat consumption, g	600	800	1 200

*Includes replacement stock.

GROWTH TRENDS

India has a robust and fast-growing animal agricultural industry. The layer population is currently 170 million birds; and there over 430 million broilers (Table 2, Figure 2). Milk production exceeds 66 million tonnes (Figure 3). Such figures create exciting market opportunities.

More interesting is the fact that each of these markets are growing at 12 to 15% per annum and rates of growth are expected to accelerate. Considering that red meat, table egg and pork consumption in several markets is leveling off, even in the face of a market contraction the robust growth in the Indian market spells opportunity for several food animal production-related businesses.

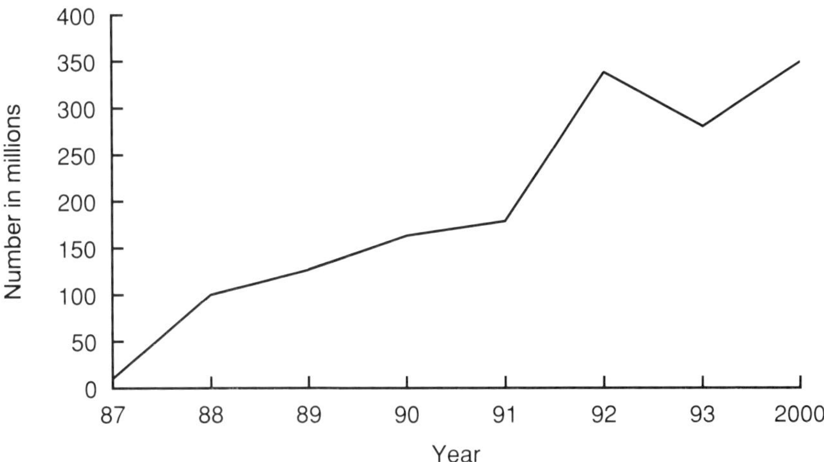

Figure 2. Growth of the broiler population in India.

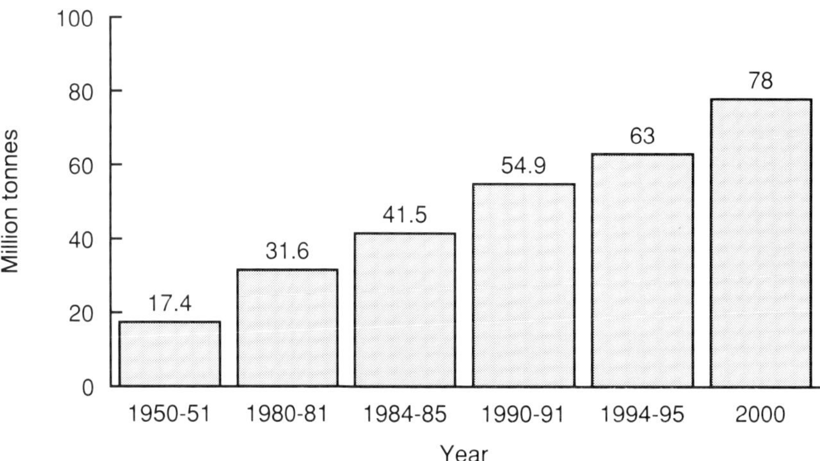

Figure 3. Milk production and availability: AD 1950–2000.

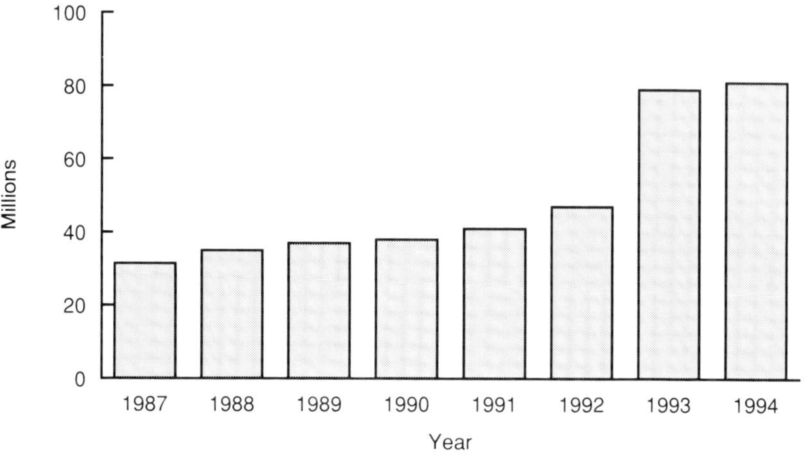

Figure 4. Number of lactating buffalo in India.

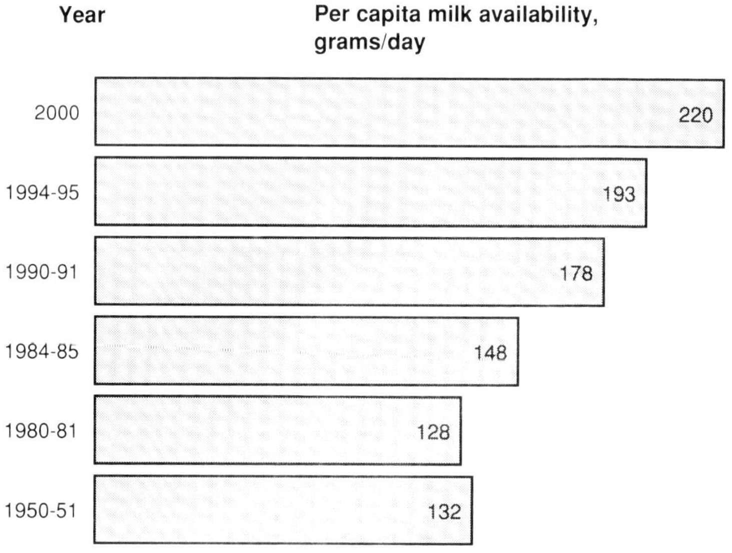

Figure 5. Per capita milk availability in India, grams per day.

FUTURE POTENTIAL

In spite of large absolute numbers in Indian animal production (Figure 4) and their rates of growth, the real future of the market is seen when the very low per capita consumption and availability of milk, meat and eggs are considered (Figures 5, 6 and 7).

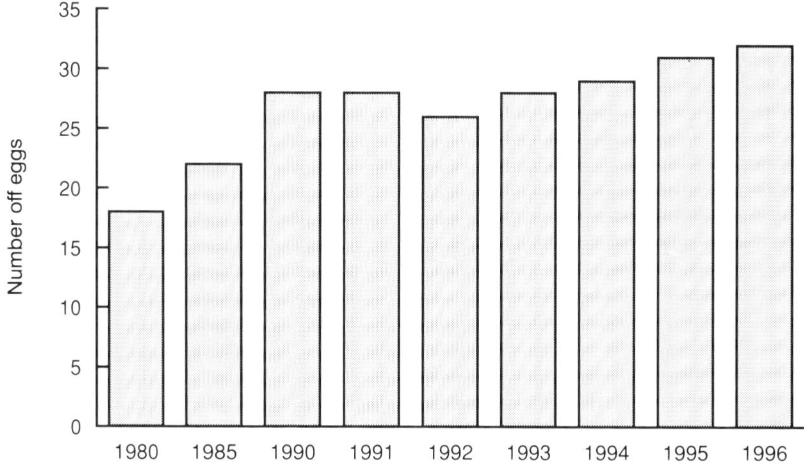

Figure 6. Per capita availability of eggs in India: 1980–1996.

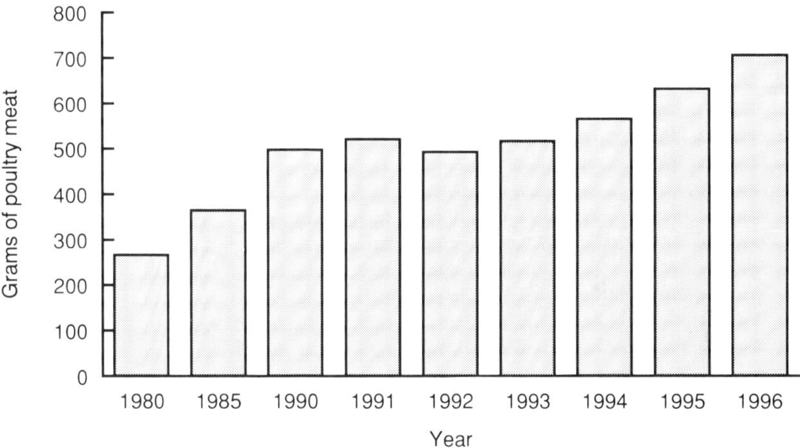

Figure 7. Per capita consumption of poultry meat in India: 1980–1996.

The reasons for growth in consumption of animal products are not hard to determine. A rapidly expanding middle class, currently estimated at 100 million (larger than entire population of several European countries), would be expected to have increasing education and nutritional standards, larger numbers of women in the work-force and more urbanization. All these factors contribute to the rapid change in food consumption patterns in India.

Notwithstanding religious pressures against meat and egg consumption, a recent survey conducted by the Livestock Feed Industry Association (compound feed manufacturers) dramatically revealed that over 66% of Indians would consume broiler meat if they could afford to!

Thus the notion that Indians are by and large vegetarians for religious reasons is mistaken. However it is true that religion substantially inhibits growth of the pork and beef industry in India due to the large number of people who practice Hinduism and Islam. Religion does not appear to be a barrier to poultry meat and egg consumption. Imagine then a small change in per capita consumption caused by any of the several qualitative factors. The multiplier effects on the total population and requirements of the animal production industry would be tremendous (Figure 8).

Barriers to growth

Having studied the vast potential, growth rates and major indicators of India's economy, what then prevents India from becoming a major global player in animal production?

COMPETITION FOR LAND

As land is scarce and under constant pressure from the human population, considerable pressure is placed on planners to choose between allocating land for animal production or for food grain crop production.

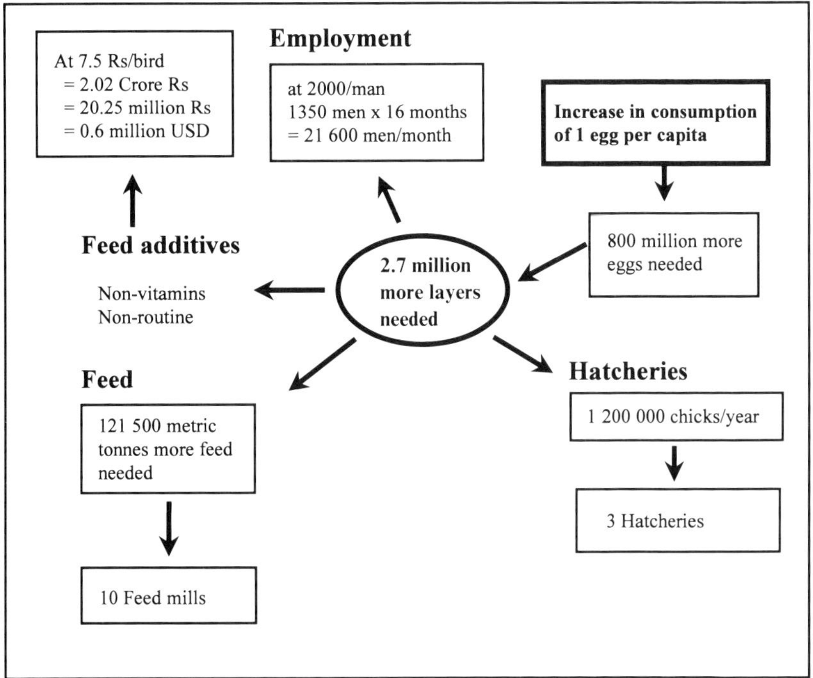

Figure 8. Multiplier effects on the feed industry resulting from an increase in egg consumption of one egg per capita.

There is virtually no grazing land developed in a planned fashion in India. All animals must be intensively managed.

COMPETITION FOR FOOD GRAINS

There is considerable pressure to allocate cereal grains for direct human consumption rather than for animal feeds. India is self-sufficient in food grain production, producing over 130 million tonnes. Additionally, food grains are exported to neighboring countries. As a result, feed manufacturers must rely entirely on minor grains and by-products along with crop residues. This often results in challenges to provide adequate metabolizable energy in feed. Low grade protein sources, high fiber ingredients and anti-nutritional aspects of various minor grains are encountered. This challenge provides opportunities for biotechnology tools such as microbial products, enzymes, etc., to increase production.

POOR POST-HARVEST PRACTICES

Due to small land holdings and low technological inputs, Indian agriculture tends to be somewhat medieval in practice. This results in valuable grain stores being spoiled by rodents, parasites and fungal growth. Mycotoxins are a major problem. Several efforts have been made to ameliorate effects of these toxins in poultry diets. Most recently, work at the University of Agricultural Sciences in Bangalore demonstrated that addition of viable yeast culture to broiler diets which were contaminated with aflatoxin reduced mortality, improved performance and resulted in higher titers to Newcastle Disease (Table 3). Additionally, it was observed that incubation of aflatoxin with viable yeast culture resulted in increased aflatoxin disappearance with time (Figure 9).

Table 3. Effect of Yea-Sacc on body weight, feed conversion, mortality and HI titer to Newcastle disease in broiler chickens fed diets containing aflatoxin.

Aflatoxin	Yea-Sacc	Weight (g)	F:G	Mortality (%)	Total protein (g/dl)	HI titer, $\log_2$
0	0	1099	2.21	3	2.75	2.41
500	0	975	2.25	13	2.38	1.3
500	0.1	1105	2	6	2.51	2.1
500	0.2	1106	2.03	0	2.78	2
1000	0	830	2.21	20	1.52	1
1000	0.1	854	2.13	16	2.1	2
1000	0.2	945	2.07	10	2.46	2.1

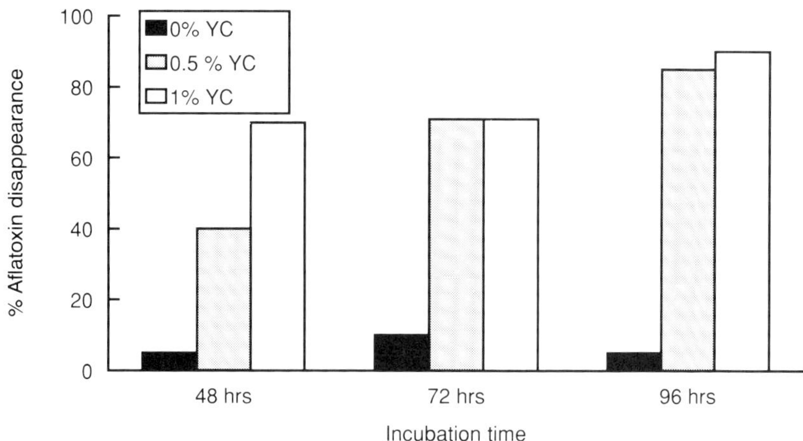

Figure 9. Percent aflatoxin disappearance after 48, 72 and 96 hours incubation with 0, 0.5 or 1% yeast culture (YC).

Table 4. Effect of Yea-Sacc 1026 fed lactating buffalo during weeks 1–10 on milk yield.

	Control	Yea-Sacc	Increase (*n*)	Increase (%)
Milk yield weeks 1–10, kg/day	7.94	9.01	1.07	13.46
Fat-corrected milk, kg/day	12.03	14.25	2.22	18.45
Milk yield weeks 2–7, kg/day	8.22	9.5	1.28	15.57
Fat-corrected milk, kg/day	12.14	14.55	2.31	19.03

LOW GRADE GENETIC STOCK

Dairy production capacity depends both on cattle numbers and the genetic potential of the animals involved. After 25 years of a centralized dairy cooperative movement (known as 'Operation Flood') that involved upgrading genetics, artificial insemination, improved nutrition with balanced diets and bypass protein, etc., there is still considerable dependence on the traditional buffalo for milk production. Over 50% of India's milk comes from the buffalo.

Biotechnology inputs have demonstrated clear benefits in this area, as well. In several studies addition of viable yeast culture has provided significant improvements in milk yield and composition owing to its ability to stimulate fermentation in the rumen (Tables 4, 5 and 6).

Table 5. Effect of Yea-Sacc 1026 on average milk composition during its period of supplementation (1 through 10 weeks) in diets fed buffalo*.

Components	Control	Yea-Sacc	Increase, *n*(%)
Fat, %	7.40	7.69	0.29 (3.92)
Protein, %	4.37	4.59	0.22 (5.08)
Lactose, %	5.06	5.21	0.15 (3.04)
Total solids, %	17.92	18.52	0.6 (3.36)
Fat yield, kg/day	0.59	0.71	0.12 (23.73)
Protein yield, kg/day	0.35	0.42	0.07 (20.00)
Lactose yield, kg/day	0.4	0.48	0.08 (20.00)
Total solids, kg/day	1.4	1.71	0.29 (20.42)

*Means differ, $P<0.01$

Table 6. Effect of Yea-Sacc 1026 on milk production and fertility in buffalo.

	Control	Treatment	Difference
Milk yield, kg/d	8.85	9.27	0.42
Fat-corrected milk, kg	9.58[a]	10.46[b]	0.88
Fat, %	6.67[a]	7.02[b]	0.35
Protein, %	3.91	3.94	0.03
Lactose, %	4.78	5.13	0.35
Solids non-fat %	9.48[a]	9.68	0.2
Total solids, %	16.14[a]	16.79[b]	0.65
Total yield, kg	2099.86	2337.2	237.62
Lactation length, days	274.4	314.2	–39.8
Days post-partum to oestrus	85.4	81	– 4.4

[ab] Means differ, $P<0.05$

Challenges for the future

To summarize, the market opportunities emerging in India place major challenges on the scientific community. Many difficulties, including *Escherichia coli* and salmonella on the farm, aflatoxin in the feed and anti-nutritional factors in available feedstuffs must be overcome if Indian agriculture is to achieve major growth.

To close on a personal note, my own current focus involves bioremediation – in this instance growth of huge quantities of water hyacinth on city sewage in order to form a biofilter. This would result in not only clean water downstream but would make available low cost biomass which could be fermented into valuable animal feed. I realize that many communities are experimenting with these concepts, however the real challenge is to overcome ground-level difficulties with engineering solutions and biotechnology applications to put into place an eco-friendly system in India.

MINERAL METABOLISM: A DECADE LATER

METAL IONS, CHELATES AND PROTEINATES

MICHAEL J. HYNES[1] and MICHAEL P. KELLY[2]

[1]*Chemistry Department and*
[2]*European Biosciences Centre, University College Galway, Ireland*

Introduction

Of the 93 naturally occurring elements, most mammalian species must procure approximately 50 from dietary sources in order to maintain a normal healthy state. In addition to the six core elements, carbon, hydrogen, nitrogen, oxygen, sulphur and phosphorus, of which carbohydrates, fats, proteins and nucleic acids are comprised, many other elements are essential to the nutritional requirements of higher animals. Some of these elements are required in relatively large quantities, in excess of 100 mg/day. These are termed 'macro' elements. Others are required in much smaller quantities, of the order of a few mg per day. These are termed 'micro' or 'trace' elements.

Over the past 50 years, continuous efforts have been made to meet the nutritional requirements of animals through the scientific formulation of feed rations. Traditionally, trace mineral supplementation of feed rations was achieved by the addition of simple inorganic salts such as copper(II) sulphate. Research showed however that the bioavailability and hence the performance enhancement achieved by trace metal supplementation was significantly improved if the metal was added in the form of a metal complex or chelate (Ashmead, 1993; Ashmead, *et al.*, 1985). The important transition elements in biological process are the redox catalysts iron, copper, cobalt and molybdenum. Manganese and chromium are also important; and zinc, although strictly not a transition metal, is also usually included and is an important element in a large number of enzymes. From the foregoing it is apparent that an understanding of the efficacy of chelated or complexed minerals requires an understanding of co-ordination chemistry.

Complexes and chelates

'Complex' is the generic term used to describe the species formed when a metal ion reacts with a ligand. A ligand is a molecule or ion that contains an atom which has a lone pair of electrons. The metal ion in a complex is bonded to the ligand through donor atoms such as oxygen, nitrogen or sulphur. Ligands that contain only one donor atom are termed 'monodentate' ligands. Ligands that contain two or more donor atoms capable of bonding to a metal ion are termed bi-, tri- or tetradentate ligands. These may also be called polydentate ligands. When such ligands bond to a metal ion via two or more donor atoms, the complex formed contains one or more heterocyclic rings which contain the metal atom. Such species are called 'chelates' (Greek: χηλη, *chele*, a crab's claw). Amino acids are examples of bidentate ligands which bond to metal ions via an oxygen of the carboxylic acid group and the nitrogen of the amino group. (Figure 1).

Ethylenediaminetetraacetic acid (EDTA) is a example of a hexadentate ligand containing six donor atoms. It forms stable complexes with most metal ions. For example, the complex formed with Cu^{2+} has the following formula, $[Cu(EDTA)]^{2-}$. This demonstrates two aspects of complexes, the use of square brackets to indicate the complex species and the fact that chelates need not necessarily be electrically neutral.

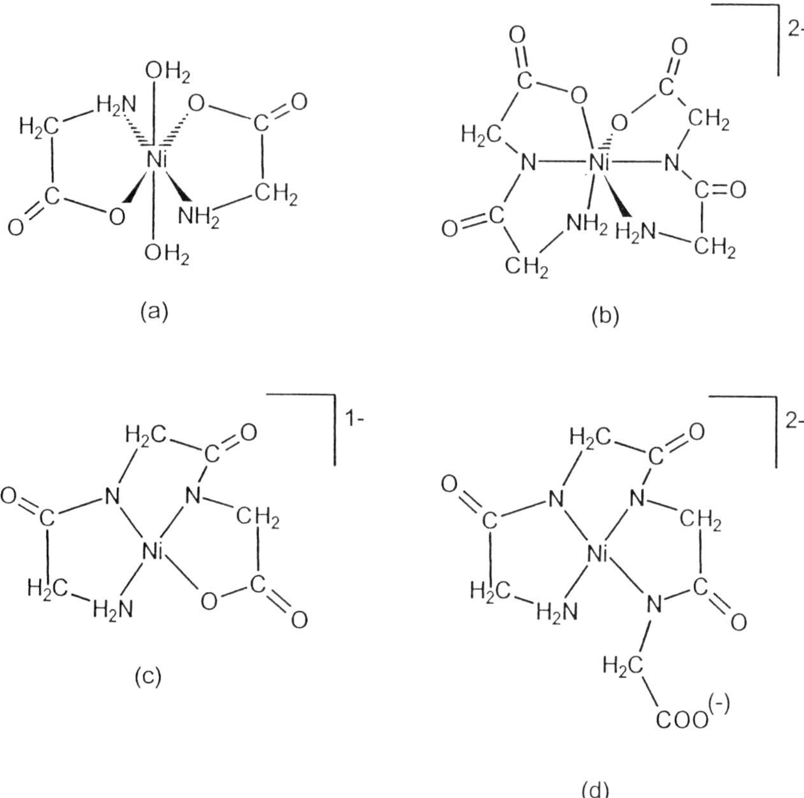

Figure 1. Structures of a nickel(II) complex of glycine and glycine-based peptides (Freeman *et al.*, 1968), (**a**) diaquobiglycinatonickel(II); (**b**) bisglycylglycinatonickelate(II); (**c**) diglycylglycinatonickelate(II) and (**d**) triglycylglycinatonickelate(II).

Although chelates having four, five- six- and seven-membered rings may be formed, it has been shown that chelates having five-membered rings have the greatest stability (Graddon, 1968).

Amino acids and peptides as ligands

Each of the naturally occurring amino acids may form a stable five-membered ring with a metal ion. In general, the donor groups will be the amino and the carboxylate groups as shown in Figure 1. At lower pH however, the amino acid may co-ordinate as a neutral ligand and in certain circumstances four-membered rings may be formed. The enhanced stability of metal complexes of histidine and cysteine may be ascribed to the involvement of the side chains (imidazole and thiol) in complex formation. It has often been asserted that metal complexes of peptides have enhanced stabilities compared with the comparable complexes formed by simple amino acids. However as is evidenced by stability constant data, peptides combine less strongly with metal ions than do amino acids (Sillén and Martell, 1964; Smith and Martell, 1989). If only the terminal groups of peptides ($-NH_2$ and $-COO^-$) were involved in complex formation, very large rings (11-membered in the case of a simple tripeptide) of reduced stability would be formed. Thus, it is unlikely that the terminal groups alone are involved in complex formation and this is supported by X-ray structural data of peptides (Margerum and Dukes, 1974). However, it does appear that the non-deprotonated N atoms of the peptide linkage are not used in bonding to the metal ion. At higher pH values, some of the peptide N atoms are deprotonated (Smith and Martell, 1989) and in the case of the copper(II) complex of tetraglycine, the first, second and third peptide hydrogens are lost in the region of pH 5.4, 6.8 and 9.1, respectively. The structure of the comparable complex with nickel(II) is shown in Figure 1(d).

A number of aspects of the chemistry of metal complexes are important in understanding the modes of action of mineral supplements. These include:

1. A description of the solution equilibria involving the metal ion, proton and the ligand.
2. The kinetics of the substitution reactions of the hydrated metal ion and the complexes present.
3. The redox behaviour of the metal ion and its complexes.
4. Reactions involving the co-ordinated ligands.

Only the equilibrium and kinetic aspects are relevant to the present discussion.

Proteinates

Commercial mineral supplements are frequently described as proteinates. These are normally produced by first partially hydrolysing a protein source using either acid or enzymatic procedures. This results in the formation of a hydrolysate containing a mixture of amino acids and peptides of varying chain length. The average chain length will depend on a number of factors including the degree of hydrolysis. Reaction of a metal salt with the hydrolysate under the appropriate conditions results in the formation of complexes containing chelated metal ions.

Characterization of complex species in solution

EQUILIBRIA IN SOLUTION

When a metal salt such as copper(II) sulphate is dissolved in water, the copper(II) ion exists in solution as $[Cu(H_2O)_2]^{2+}$. Addition of a bidentate ligand such as an amino acid results in the formation of a series of complexes as shown in equations (1) – (4) where β_1 and β_2 are the overall stability constants of the metal complexes as normally defined.

$$Cu(H_2O)_6]^{2+} + O\text{-}L^- = [Cu(H_2O)_4(O\text{-}L)^+] + 2H_2O \quad (1)$$

$$\beta_1 = [Cu(H_2O)_4(O\text{-}L)^+]/[Cu(H_2O)_6{}^{2+}][O\text{-}L^-] \quad (2)$$

$$[Cu(H_2O)_6]^{2+} + 2\ O\text{-}L^- = [Cu(H_2O)_2(O\text{-}L)_2] + 4H_2O \quad (3)$$

$$\beta_2 = [Cu(H_2O)_2(O\text{-}L)_2{}^+]/[Cu(H_2O)_6{}^{2+}][O\text{-}L^{-2}] \quad (4)$$

In addition, the protonation equilibrium of the amino acid must be taken into account, equation (5). Complex formation involves competition between the proton and the metal ion for the free ligand, the latter being the form of the ligand present in the complex. It may be shown that in a given solution containing Cu^{2+} and a ligand, the fraction of the metal present as the cth complex α_c depends only on the stability constant of the complex, β_c and the free ligand concentration (L), (equation (6)) when n is the maximum number of ligands bound to the metal ion. Various compilations of stability constants are available (Martell and Smith, 1982; Sillén and Martell, 1964; Sillén and Martell, 1971)

$$H^+ + L^- = HL \quad (5)$$

$$\alpha_c = \frac{\beta_c[L]^c}{\sum_{i=0}^{i=n} \beta_i[L]^i} \quad (6)$$

Equation (6) is an extremely important equation in that it allows the calculation of the concentrations of all species present in a solution at equilibrium once the total concentrations of the reactants and the stability constants of the complexed species present are known. In a solution containing a mixture of Cu^{2+} and an amino acid the distribution of the copper species will depend on the concentrations of metal ion and ligand present as well as the pH of the solution. Figure 2 shows such a distribution for a mixture containing 0.001 M Cu^{2+} and 0.002 M glycine while Figure 3 shows the distribution for the zinc(II)–glycine system. Figure 4 shows the species distribution for the complexes formed by Cu^{2+} with tetraglycine as a function of pH. Figures 2, 3 and 4 highlight a number of significant features:

1. The distribution of metal species present at given concentrations of metal and amino acid depends on the pH of the solution.
2. Complexed forms (chelates) of the dipositive metal ions are not necessarily neutral.
3. Different metal ions have different stability constants and thus the percentage of a metal present as a particular species will depend not only on the pH of the solution, but also on the stability constant of the complex (compare Figures 2 and 3).

The lower stability of the copper(II)–tetraglycine complexes compared with the corresponding glycine chelate is clearly evident (compare Figures 2 and 4). This is further illustrated in Figure 5 where the fraction of copper(II) complexed by various amino acids, peptides and EDTA is plotted as a function of pH. From this it is evident that increasing the chain length of peptides results in decreased stability of the complexes formed. The fraction of metal complexed by EDTA is included in Figure 5 for comparison purposes. It is evident that in the pH range investigated, virtually all the copper(II) is complexed. However, it has been demonstrated that addition of mineral supplements as the EDTA chelates does not result in increased bioavailability. The complexes formed are both too stable and are also kinetically inert

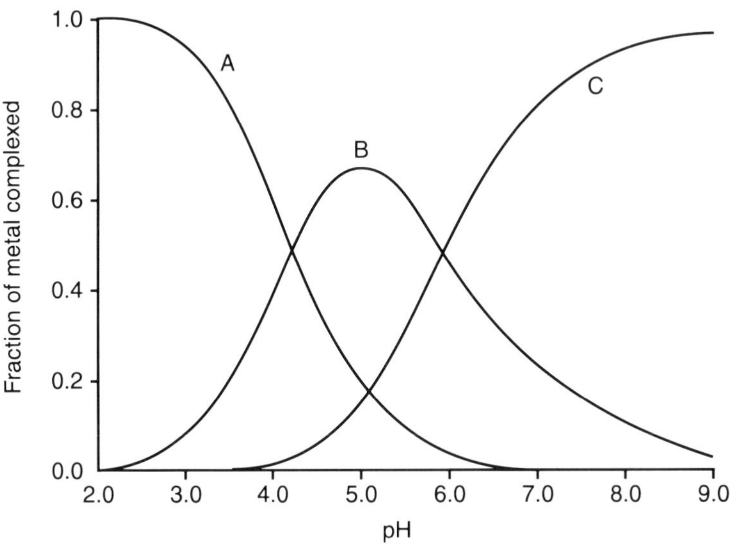

Figure 2. Species distribution curves for solutions containing copper(II) (0.001 M) and glycine (0.002 M). A, $[Cu(H_2O)_6]^{2+}$; B, $[Cu(H_2O)_4L]^+$; C, $[Cu(H_2O)_2L_2]$ where L = $H_2N\text{-}CH_2\text{-}COO^-$.

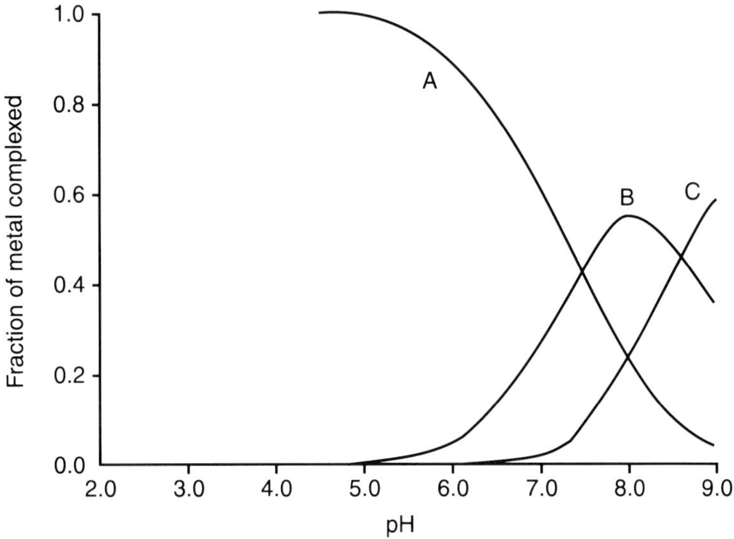

Figure 3. Species distribution curves for solutions containing zinc(II) (0.001 M) and glycine (0.002 M). A, $[Zn(H_2O)_6]^{2+}$; B, $[Zn(H_2O)_4L]^+$; C, $[Zn(H_2O)_2L_2]$ where L = $H_2N\text{-}CH_2\text{-}COO^-$.

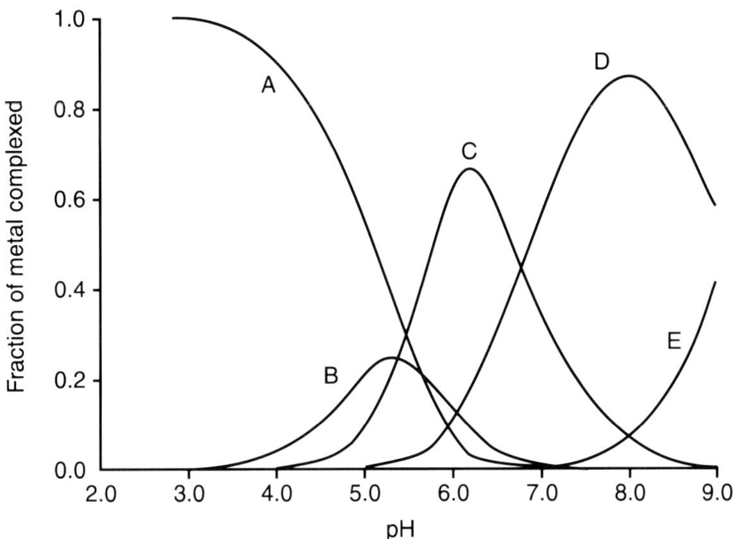

Figure 4. Species distribution curves for solutions containing copper(II) (0.001 M) and tetraglycine (0.002 M). A, $[Cu(H_2O)_6]^{2+}$; B, $[Cu(H_2O)_4L]^+$; C, $[Cu(H_2O)_4L(H_{-1})]$, D = $[Cu(H_2O)_3L(H_{-2})]^{1-}$, E = $[Cu(H_2O)_2L(H_{-3})]^{2-}$ where L = $H_2N\text{-}(CH_2\text{-}C(O)NH)_3\text{-}CH_2\text{–}COO^-$.

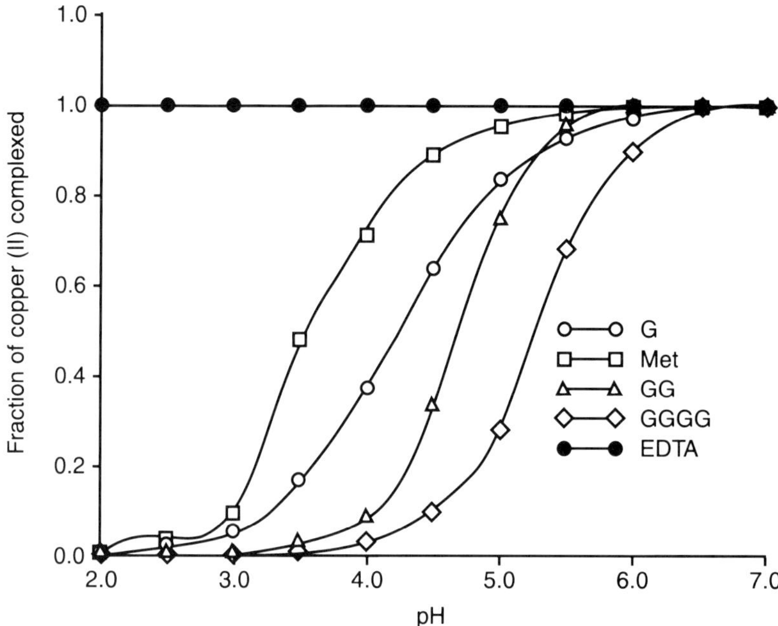

Figure 5. Comparison of the total fraction of copper(II) complexed by glycine, methionine, diglycine tetraglycine and EDTA as a function of pH (L_{total} = 0.002 M, M_{total} = 0.001 M).

with the result that even though they are initially absorbed, they are subsequently quantitatively excreted intact (Ashmead, 1993; Miller, 1967). It appears that chelates of amino acids and peptides have stability constants of such magnitude as to allow the metal ions to be transferred to the host's biological system. The amino acids and peptides may also have an additional role in the possible provision of a 'biological vector' which enhances the absorption process.

KINETIC FACTORS RELEVANT TO METAL COMPLEX EQUILIBRIA

Another factor that must be considered in equilibrium studies of metal chelates is the lability or inertness of a complex. The original classification proposed by Henry Taube (Taube, 1952) is still widely used. Complexes are labile when they exchange ligands at rates for which the $t_{1/2}$ is less than 1 min while inert complexes exchange ligands at rates for which $t_{1/2}$ is greater than 1 min. An appreciation of the concept of solvent exchange on metal ions is fundamental to an understanding of ligand exchange reactions in metal complexes (Wilkins and Eigen, 1965). Solvent exchange is described by equation (7)

$$[M(H_2O)_6]^{n+} + \underset{\text{bulk solvent}}{H_2O^*} = [M(H_2O)_5(H_2O^*)]^{n+} + H_2O \qquad (7)$$

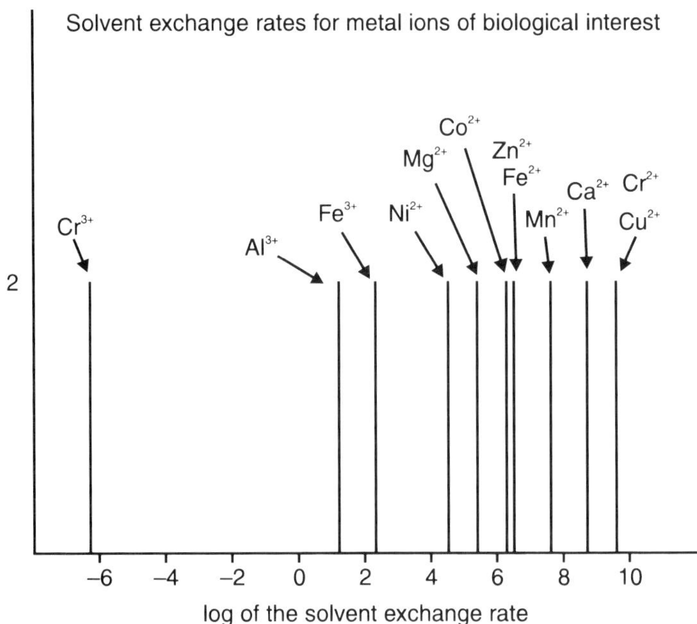

Figure 6. Rate constants (s^{-1}) for solvent exchange on a range of metal ions of biological interest.

Figure 6 shows the water exchange rates for a selection of metal ions of biological interest. It is apparent that there is approximately a factor of 10^{16} difference in the rate at which $[Cu(H_2O)_6]^{2+}$ and $Cr(H_2O)_6]^{3+}$ exchange water molecules with the bulk solvent. Consequently, a labile solvated metal ion such as $[Cu(H_2O)6]^{2+}$ exchanges its water molecules thousands of times per second. For many metal ions including Mn^{2+}, Fe^{2+}, Co^{2+}, Ni^{2+}, Cu^{2+} and Zn^{2+}, the rate at which a metal complex is formed parallels the rate of solvent exchange and is mainly independent of the nature of the ligand. Arising from this, it can be readily demonstrated that simple metal complexes of these metal ions with monodentate ligands such as NH_3 and bidentate ligands such as amino acids will also be dissociatively labile, i.e. they will also lose ligands very rapidly. On the other hand, complexes of metals such as chromium(III) will tend to dissociate very slowly on time scales of minutes or hours at room temperature.

However, in the case of multidentate ligands a somewhat different situation pertains. Metal complexes with such ligands have very high stability constants. These have often been ascribed to an entropy effect. However, in many instances it is due to the relatively slow rate at which these complexes dissociate. At neutral pH values, complexes of multidentate ligands can have half-lives of many hours. However, in acid solution, the acid catalysed dissociative process (k_2 pathway) is greatly accelerated (Scheme 1).

$$[M(H_2O)_6]^{n+} + HL \underset{k_2}{\overset{k_1}{\rightleftharpoons}} [M(H_2O)_5L]^{(n-1)+} + H^+$$

$$+H^+ \updownarrow -H^+$$

$$[M(H_2O)_6]^{n+} + L^- \underset{k_4}{\overset{k_3}{\rightleftharpoons}} [M(H_2O)_5L]^{n-1)+}$$

Scheme 1

Factors such as the above must be taken into account when attempting to characterize metal complexes in solution using chromatographic techniques. A recent study (Brown and Zeringue, 1994) describes a method for the 'laboratory evaluation of solubility and structural integrity of complexed and chelate trace mineral supplements' using gel filtration chromatography. In this study it is stated that $[Cu(EDTA)]^{2-}$ is eluted intact in a gel filtration experiment. Unfortunately, the authors do not state clearly at what pH the gel filtration experiments were carried out. However, Figure 5 shows the distribution of $[Cu(EDTA)]^{2-}$ as a function of pH while Table 1 shows the approximate half lives for dissociation of a $[Cu(EDTA)]^{2-}$ at different pH values. While these data have been derived from studies of CyDTA, the results for EDTA will not differ greatly. It is quite clear that the results obtained by Brown and Zeringue for $[Cu(EDTA)]^{2-}$ are consistent with those predicted from a consideration of stability and kinetic factors.

In the case of the copper(II) and zinc(II) amino acid chelates the situation is quite different. First, the metal species present in solution will depend greatly on the pH of the solution as shown in Figures 2 and 3. Second, amino acid chelates of both zinc(II) and copper(II) are labile from the point of view of both complex formation **and** dissociation. Thus when some of the free metal ion is retained by the beads of the gel, the equilibrium will adjust according to equations (1) and (2) so that over

Table 1. Half-life for hydrolysis of $[Cu(CyDTA)]^{2-}$ at various pH values.

pH	Half-life/s
1	1.77
2	1.77×10^1
3	1.77×10^2
4	1.77×10^3
5	1.77×10^4
6	1.77×10^5 (49.3 hours)
7	1.77×10^6 (493 hours)

a relatively short column length, complete separation of the hydrated metal ion and the amino acid (or peptide) will be effected. Thus the experiments as carried out give no information regarding the structural integrity of the metal chelates.

Other attempts to assay metal proteinates in solution (Parks and Harmston, 1994) contain similar shortcomings and provide no information regarding the structural integrity of the complex in solution. This is strictly a function of the concentrations of the reacting species, the medium, the stability constants of the complexes formed and the pH of the medium.

Characterization of metal chelates in the solid state

While infrared spectroscopy can be used to demonstrate the presence of chelate formation in solids containing amino acids and transition metal cations (Chow and McAuliffe, 1975; Nakamoto, 1963), quantitation can be difficult. As part of our continuing attempt to improve methods for the characterization of metal chelates of amino acids and peptides we have investigated alternative techniques such as solid state nuclear magnetic resonance (NMR), X-ray powder diffractometry and near infrared spectroscopy. While solid state NMR has certain attractions, the relative unavailability of instrumentation and the cost of both the instrumentation itself and consequently the cost of obtaining spectra from service laboratories renders the technique unsuitable for in-process quality control.

Figure 7 shows powder diffraction patterns of methionine, $CuSO_4 \cdot 5H_2O$ and the neutral bis-chelate formed between copper(II) and methionine. It is immediately clear that this technique provides a ready method for evaluating the structural integrity of such chelates in the solid state. The diffraction patterns of all three compounds are readily distinguishable even by the non-expert. Furthermore, the presence of even small quantities of un-chelated $CuSO_4 \cdot 5H_2O$ in the solid sample will be immediately obvious. This approach can thus be used to demonstrate in an unambiguous fashion the presence or absence of uncomplexed copper sulphate in metal proteinates. This is an important consideration in view of the allegation that crystals of copper sulphate could be observed in samples of metal proteinates (Parks and Harmston, 1994). The results for zinc methionine (Figure 8) are equally conclusive. Powder diffractometry has the advantages of widespread availability of instrumentation, rapid analysis and ready interpretation of the results.

One of the most rapidly growing analytical techniques in a wide range of industries is near infrared spectroscopy (McClure, 1994; Wetzel, 1983). The near infrared (NIR) spectral region lies between 700 and 2500 nm (14 287–4000 cm^{-1}) where samples have lower absorbtivity than in the mid-infrared region usually found on traditional infrared spectrophotometers, 400–4000 cm^{-1} (25 000–2500 nm). The NIR region contains chemical information in the form of overtone and combination bands from the mid-infrared region. Absorptions in the NIR region are

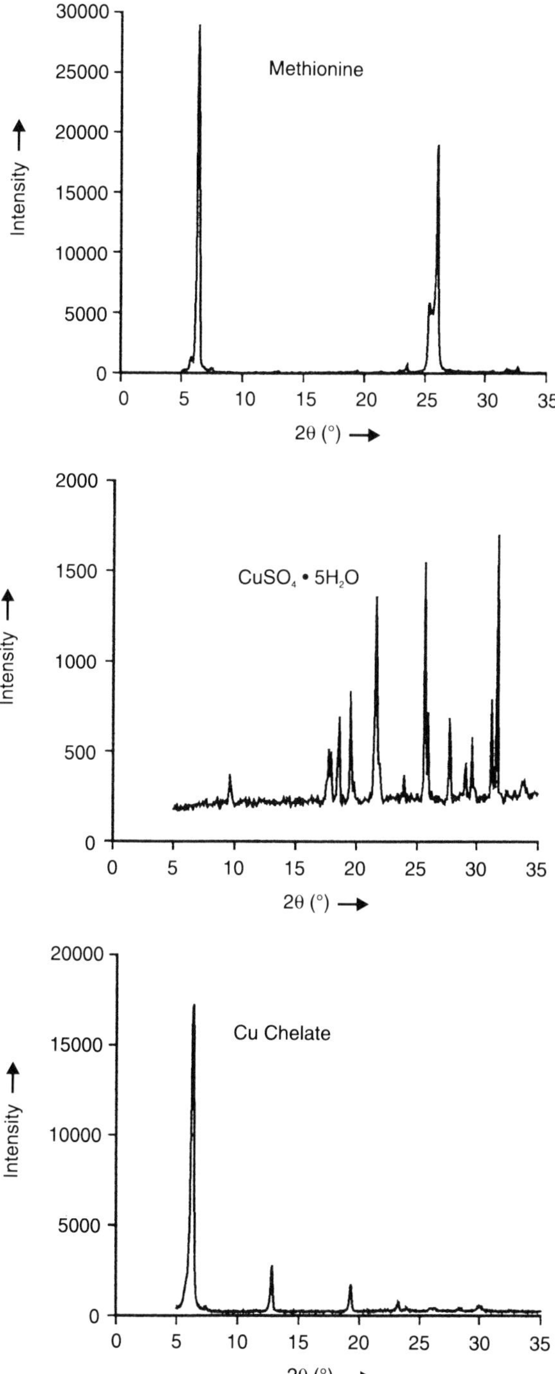

Figure 7. X-ray powder diffraction pattern for methionine (L), $CuSO_4.5H_2O$ and CuL_2.

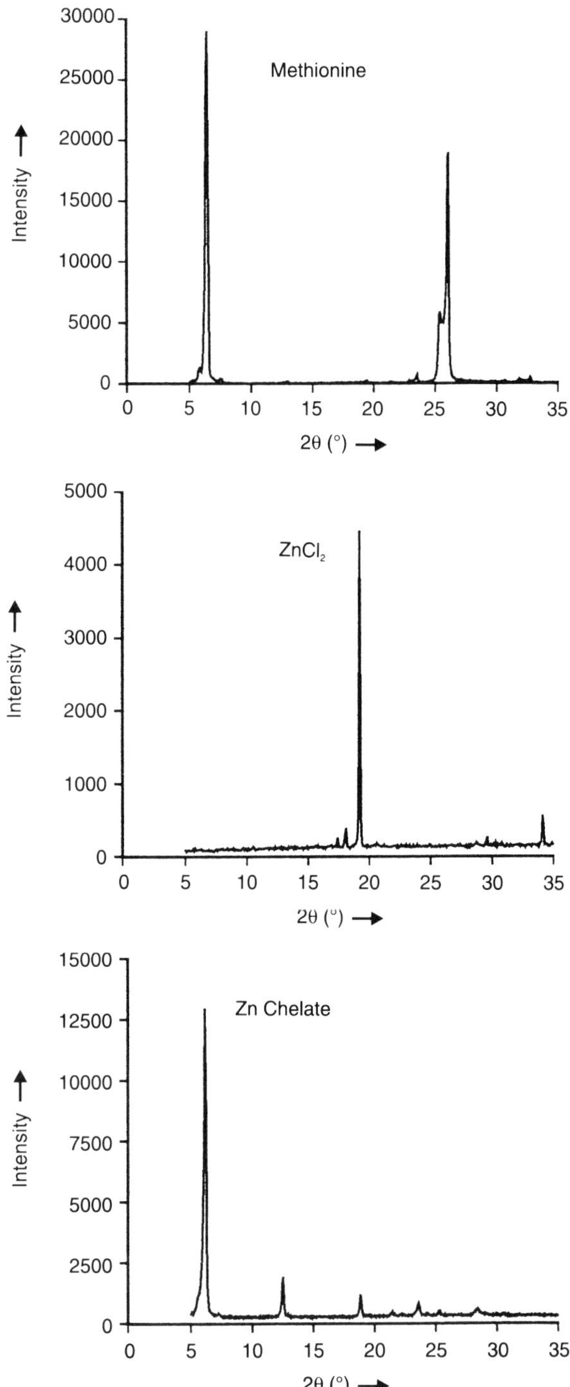

Figure 8. X-ray powder diffraction pattern for methionine (L), $ZnCl_2$ and ZnL_2.

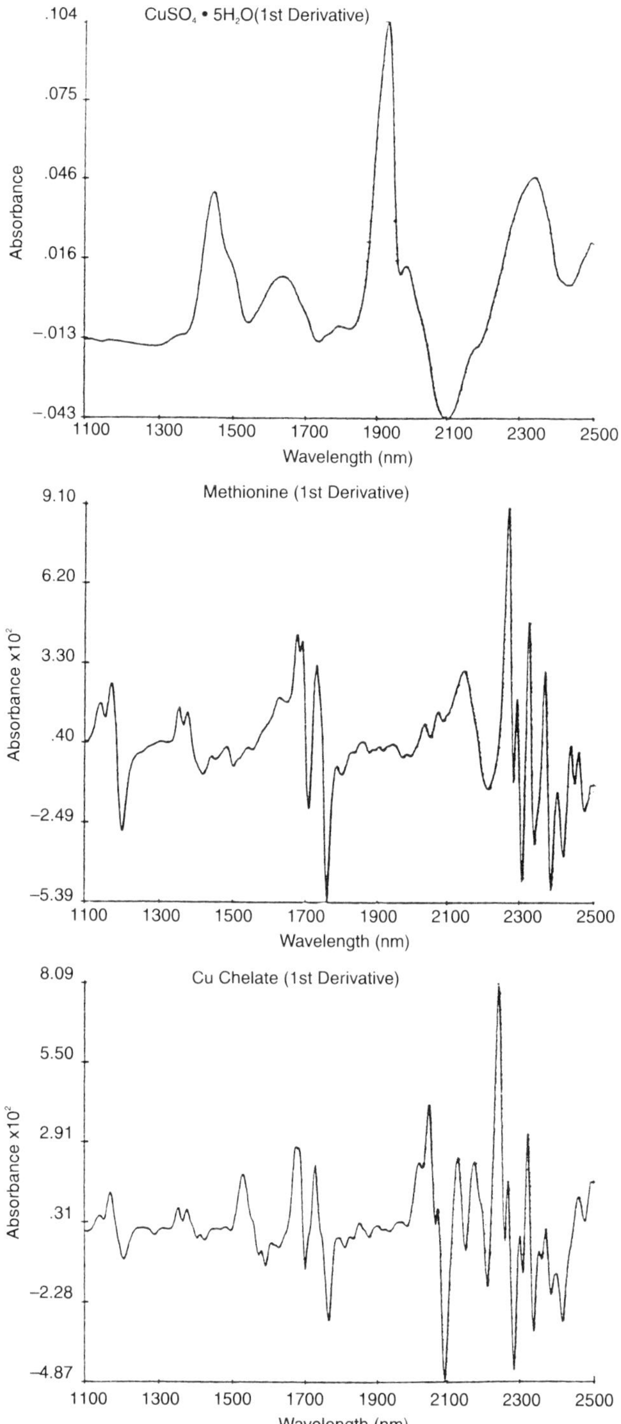

Figure 9. First derivative NIR spectra of methionine (L), $CuSO_4{\cdot}5H_2O$ and CuL_2

usually 100 to 1000 times weaker than the fundamental absorptions found in the mid-infrared. Low absorbtivity allows radiation to penetrate opaque samples and this allows the use of flexible sampling techniques. In particular, reflectance measurements can be made directly without sample preparation. First and second-order derivative spectra are commonly used and advanced chemometric software packages are used to extract vast amounts of both qualitative and quantitative information from such spectra.

We have recently carried out preliminary analysis of metal chelates of amino acids in order to ascertain the applicability of the technique for the characterization and evaluation of such compounds. Figure 9 shows the first-derivatives of the NIR spectra of methionine, $CuSO_4{\cdot}5H_2O$ and [Cu(methionine)$_2$]. Unlike the coventional infra-red spectra, there are clearly discernable differences between the NIR spectra of the three species. Similar results are found in the corresponding zinc(II) and manganese(II) chelates. Additional experiments are required in order to demonstrate that these bands can be used for quantitative analysis of these compounds.

Conclusions

In ruminant nutrition, recent research has shown that amino acid absorption from the rumen in the form of peptides is more important than absorption of free amino acids (Webb *et al.*, 1992; Webb *et al.*, 1993). Ruminal microorganisms may also play an important role in determining the compostion of the peptides available for absorption. Furthermore, the authors state that most peptides are transported as zwitterions having no net overall charge. This is hardly surprising in view of the pH of the rumen and the pK values of the acidic groups on the peptide. Protons and in certain instances, metal ions may also be required to assist in the absorption process. These observations have necessitated considerable re-evaluation of previous concepts and theories concerning the mode of absorption of ‘amino acids’ and accordingly mineral supplementation in ruminant nutrition.

The preferential absorption of peptides invites the question as to what is the nutritionally optimal metal chelate. It would appear that there is no single answer. While amino acid and peptide complexes of dipositive transition metal ions have been demonstrated to be superior to inorganic salts, it appears that the optimum form of chromium(III) supplementaion is in the form of the *tris*-piccolinato complex (Passwater, 1992). Based on the findings of Webb *et al.* it might be concluded that a neutral peptide complex would have maximum efficiacy, for example complex (C) in Figure 4. Even though this would be the predominant form of a copper(II) tetraglycine complex present in the rumen, in the case of labile metal ions, it is not necessary for such a complex to predominate in order to provide an efficient absorption mechanism. Indeed it would appear that metal ions are not absorbed by any single mechanism. They could be absorbed intact as either peptide or amino acid complexes.

They could also be absorbed by a mechanism in which the chelated ligands are replaced by other ligands (other proteins and peptides) at the absorption interface. Such is the situation in the case of the *tris*-piccolinato complex of chromium(III) (Hynes and Clarke, unpublished work). They could also be incorporated into the microbial biomass in the rumen for later utilization.

In summary, metal ion uptake involves a complex series of reactions and it is clear that considerable additional research is required in order to provide answers to the questions raised both here and in other recent work (Webb *et al.*, 1992; Webb *et al.*, 1993). Metal ion equilibria, kinetic factors, pH gradients and in the case of redox active ions such as copper(II), redox equilibria must be considered.

Acknowledgements

The assistance of the X-ray crystallography unit at University College Galway in carrying out the X-ray analyses is gratefully acknowledged. We are indepted to Mr P. Roche and Ms K. McKenzie of Mason Technology for the provision of NIR facilities. We gratefully acknowledge useful discussions with Dr P. Dauvillier of UFAC.

References

Ashmead, H.D. 1993. The Roles of Amino Acid Chelates in Animal Nutrition. Noyes Publications, Park Ridge, Il.

Ashmead, H.D., D.J. Graff and H.H. Ashmead. 1985. Intestinal Absorption of Metal Ions and Chelates. Charles C. Thomas, Springfield, Il.

Brown, T.F. and L.K. Zeringue. 1994. Laboratory evaluations of solubility and structural integrity of complexes and chelated trace mineral supplements. J. Dairy. Sc. 77:181.

Chow, S.T. and C.A. McAuliffe. 1975. Transition metal complexes containing tridentate amino acids. In: Progress in Inorganic Chemistry, 19. S.J. Lippard (Ed). J. Wiley & Sons, New York, p. 51.

Freeman, H.C., J.M. Guss and R.L. Sinclair. 1968. Crystal structures of four nickel complexes of glycine and glycine peptides. J. Chem. Soc. Chem. Commun., 485.

Graddon, D.P. 1968. An Introduction to Co-ordination Chemistry (2nd edn.). Pergamon Press, Oxford.

Margerum, D.W. and G.R. Dukes. 1974. Kinetics and mechanisms of metal-ion and proton-transfer reactions of oligopeptide complexes. In: Metal Ions in Biological Systems. H. Sigel (Ed.). Marcel Dekker Inc. New York, p. 157.

Martell, A.E. and R.M. Smith. 1982. Critical Stability Constants. Plenum Press, New York.

McClure, W.F. 1994. Near-infrared spectroscopy – The giant is still running. Anal. Chem., 66(1):43.

Miller, R. 1967. Chelating agents in poultry nutrition. In: Del Marva Nutrition Short Course.

Nakamoto, K. 1963. Infrared Spectra of Inorganic and Coordination Compounds. John Wiley & Sons Inc., New York.

Parks, F.P. and K.J. Harmston. 1994. Judging organic trace minerals. Feed Management 45(10):35.

Passwater, R.A. 1992. Chromium Picolinate. Keats Publishing Inc., New Canaan, Ct.

Sillén, L.G. and A.E. Martell. 1964. Stability Constants of Metal-ion Complexes. The Chemical Society, London.

Sillén, L.G. and A.E. Martell. 1971. Stability Constants of Metal-ion Complexes. The Chemical Society, London.

Smith, R.M. and A.E. Martell. 1989. Critical Stability Constants. Plenum Press, New York.

Taube, H. 1952. Rates and mechanisms of substitution in inorganic complexes in solution. Chem. Rev., 50:69.

Webb, Jr., K.E., J.C. Matthews and D.B. RiRienzo. 1992. Peptide Absorption: A review of current concepts and future perspectives. J. Anim. Sci. 70:3248.

Webb, Jr, K.E., D.B. DiRienzo and J.C. Matthews. 1993. Symposium: Nitrogen metabolism and amino acid nutrition in dairy cattle. J. Dairy Sci. 76:351.

Wetzel, D.L. 1983. Near-infrared reflectance analysis. Anal. Chem. 55(12):1165.

Wilkins, R.G. and M. Eigen (Ed.). (1965). The kinetics and mechanisms of formation of metal complexes. In: Mechanisms of Inorganic Reactions. Americal Chemical Society., Washington, D.C.

THE EFFECT OF CHROMIUM SUPPLEMENTATION ON METABOLIC RESPONSE TO EXERCISE IN THOROUGHBRED HORSES

J.D. PAGAN, S.G. JACKSON and S.E. DUREN

Kentucky Equine Research, Inc., Versailles, KY 40383, USA

Introduction

Chromium has been recognized as an essential nutrient in humans for many years (Mertz, 1992). Though chromium may well be involved in other physiological processes, at present its only known role is as a component of glucose tolerance factor (GTF) which potentiates the action of insulin. Therefore, it is involved in carbohydrate metabolism and other insulin dependent processes such as protein and lipid metabolism. A number of recent experiments have evaluated the effect of supplemental chromium in a wide range of species including humans, cattle, pigs, and turkeys. In these studies, chromium supplementation has increased lean muscle deposition (humans and swine), improved immune response (cattle), and reduced cortisol production caused by heat and transport stress (cattle and swine). No research, however, has been conducted with chromium in horses. Therefore, the following study was conducted to determine the effect that chromium supplementation has on metabolic response to exercise in trained Thoroughbred horses.

Material and methods

Six trained Thoroughbred horses (four gelded males, two females) were used in this two period switch-back design experiment. The horses were divided into two groups which received identical diets with one group also receiving 5 mg of chromium from a chromium yeast product (Co-Factor III, Alltech, Inc.). The diet consisted of a textured grain mix, forage cubes and orchard grass hay. Amounts and compositions are listed in Tables 1 and 2. At the conclusion of the first period, the diets were switched and the horses repeated the same exercise and testing regime.

During each 14-day period, the horses were exercised on a high speed treadmill (Beltalong, Eurora, Australia) inclined to 3° according to the schedule outlined in Table 3. Heart rate was monitored on a daily basis (Hippocard PEH 200). At the end of each period, the horses performed

Table 1. Nutrient concentrations of experimental feeds.

Nutrient	Sweet feed	Orchard grass hay	Forage cubes	Chromium yeast
Dry matter (%)	85.5	91.0	91.6	92.3
Crude protein (%)*	14.2	15.8	16.7	48.6
Acid detergent fiber (%)*	9.8	34.9	37.3	3.7
Neutral detergent fiber (%)*	19.3	54.5	49.9	17.9
Lignin %*	2.4	5.9	8.0	–
Ether extract (%)*	4.7	3.2	2.2	0.9
Soluble CHO (%)*†	55.85	17.08	21.47	26.33
Calcium (%)*	0.71	0.76	1.40	0.15
Phosphorus (%)*	0.62	0.40	0.27	1.22
Magnesium (%)*	0.22	0.16	0.24	0.21
Potassium (%)*	0.86	3.20	1.96	1.84
Sodium (%)*	0.241	0.012	0.03	0.084
Iron (ppm)*	363	135	460	157
Copper (ppm)*	30	7	7.9	51
Zinc (ppm)*	98	19	22	192
Chromium (ppm)*	2.6	0.30	0.74	1008

*dry matter basis
†soluble CHO = 100–CP-EE-NDF-Ash

Table 2. Daily feeding schedule.

	Sweet feed	Orchard grass hay	Forage cubes*	Chromium yeast
7 am	1.81 kg		1.36 kg	
12 pm		1.13 kg		
5 pm	1.81 kg		1.36 kg	5 grams
10 pm		1.13 kg		

*AlfaOats, Canadian Agra Bio-cube, Kincardine, Ontario

Table 3. Exercise schedule.

	Period 1	Period 2
3 times per week	1600 m @ 7–8 m/s	1600 m @ 8–9 m/s
4 days before SET	1600 m @ 10 m/s	1600 m @ 10 m/s
3 days before SET	no exercise	no exercise
2 days before SET	1600 m @ 8 m/s	1600 m @ 8 m/s
1 day before SET	no exercise	no exercise

a standardized exercise test (SET) on the treadmill. On the test day, each horse received 1.36 kg of hay cubes 5 hours before the SET and 1.81 kg grain 3 hours before the test. Blood samples were collected from an indwelling jugular catheter after an 8 hour overnight fast, before grain was fed and hourly after feeding for 3 hours. The SET began at 3 hours after feeding and consisted of a 2 min warm-up walk followed by an 800 m trot (~4 m/s), then 800 m gallops of 8 m/s, 9 m/s, 10 m/s and

11 m/s. These gallops were followed by an 800 m warm-down trot and 2 min walk. During the SET, a blood sample was taken after the warm up trot, during the last 15 sec of each gallop and after the warm-down walk. Blood samples were also collected at 15 and 30 min post exercise. Heart rate was recorded at each speed.

Each blood sample was placed in a sterile tube containing EDTA and centrifuged immediately. The plasma was pipetted into glass tubes, measured for lactate and glucose and then frozen. Plasma glucose was measured using an automated glucose analyzer (YSI, 2300 STAT). Lactate levels were measured using an automated L-lactate analyzer (YSI,1500 Sport). The following day, the frozen plasma was analyzed for triglycerides and cholesterol (Eppendorf 5060 Automated Analyzer, Gibbstown, NJ 08027). At the conclusion of the study, all of the frozen plasma samples were analyzed for insulin and cortisol using commercially available radioimmunoassay (RIA) kits which had been validated for specificity and accuracy in equine plasma (BET Labs, Lexington, KY). Variables were analyzed by analysis of variance using animal, period and treatment as main effects (NCSS, Kaysville, Utah 84037).

Results

Plasma glucose (Figure 1) in both groups peaked 1 hour after the grain meal the morning of the SET. Glucose tended to be higher in the control horses at this time (121.8 vs. 116.8 mg/dl, P=0.18). At the onset of exercise, glucose was still elevated above fasting levels in both

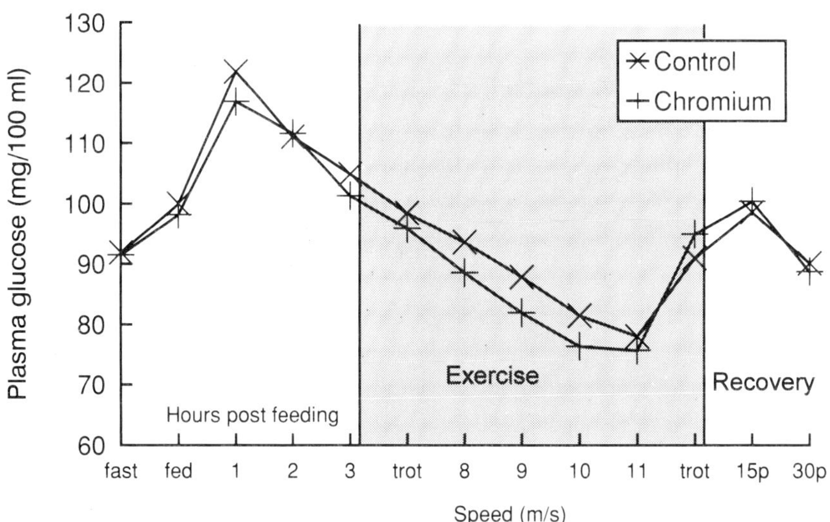

Figure 1. Effect of chromium supplementation on plasma glucose in exercised Thoroughbreds.

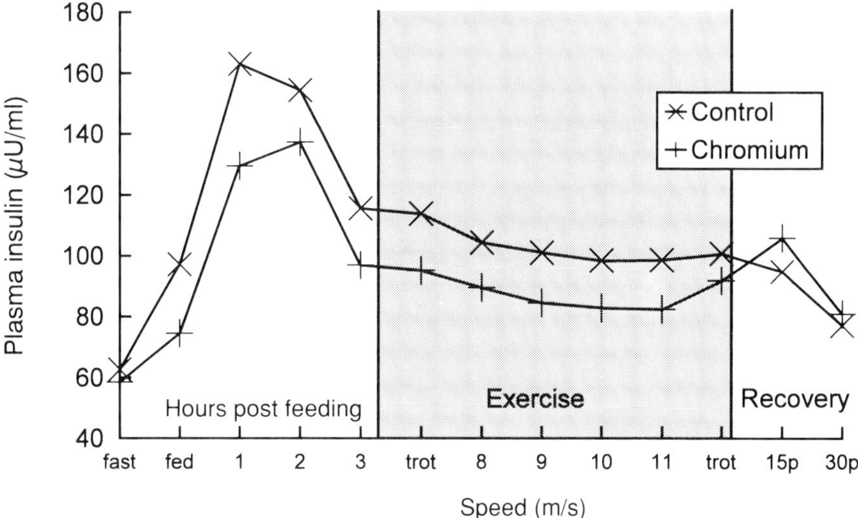

Figure 2. Effect of chromium supplementation on plasma insulin of exercised Thoroughbreds.

groups. During exercise, glucose decreased with each step. Glucose was significantly lower in the chromium group after the 8 m/s and 9 m/s steps ($P<0.05$). During the warm-down phase of the SET, glucose rebounded to fasting levels in both groups and glucose was not different ($P>0.10$) 15 or 30 min after exercise.

Plasma insulin (Figure 2) increased following feeding in both groups. Insulin peaked in the control group 1 hour after grain, while the peak in the chromium group did not occur until 2 hours after grain was fed. Blood insulin was lower in the chromium group (163 vs. 129 μU/ml, $P<0.10$) 1 hour after grain feeding.

Plasma lactate (Figure 3) increased in both groups throughout the SET, peaking in both groups following the final gallop step. Lactate was lower in the chromium group (8.65 vs. 7.55 mmol/l, P=0.08) after the 11 m/s step.

Cortisol (Figure 4) was significantly lower in the chromium group 3 hours after feeding ($P<0.01$), after the warm-up trot ($P<0.05$) and after the 8 and 11 m/s steps ($P<0.10$).

Triglycerides (Figure 5) tended to be higher in the chromium group following the 10 and 11 m/s steps and were significantly higher 15 and 30 min after exercise ($P<0.05$). Heart rate (Figure 6) and cholesterol were unaffected by diet.

Discussion

Supplementation with an organic form of chromium affected insulin response to a grain meal in these exercised horses. This response is

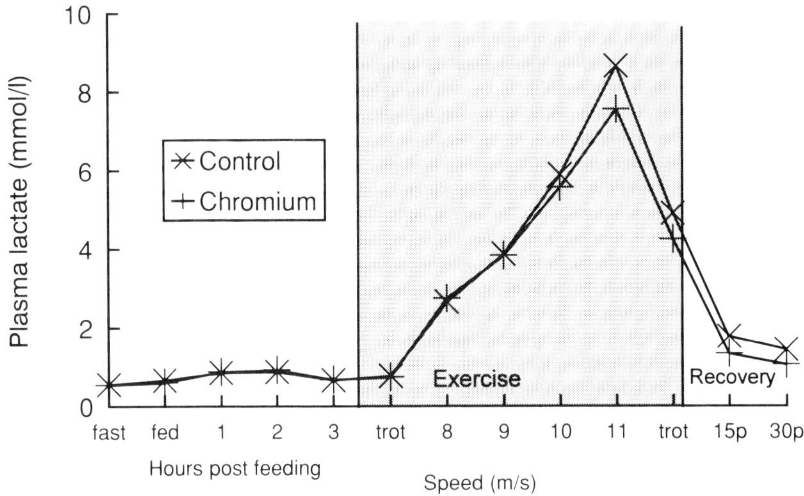

Figure 3. Effect of chromium supplementation on plasma lactate in exercised Thoroughbreds.

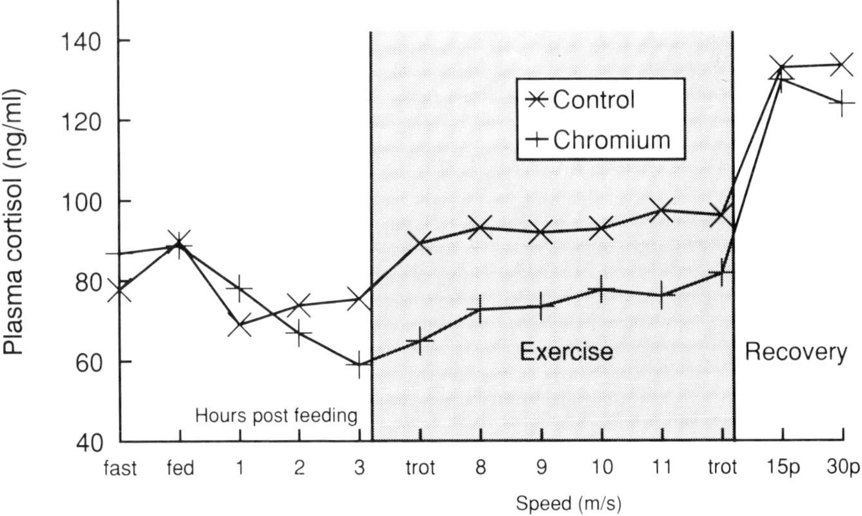

Figure 4. Effect of chromium supplementation on plasma cortisol of exercising horses.

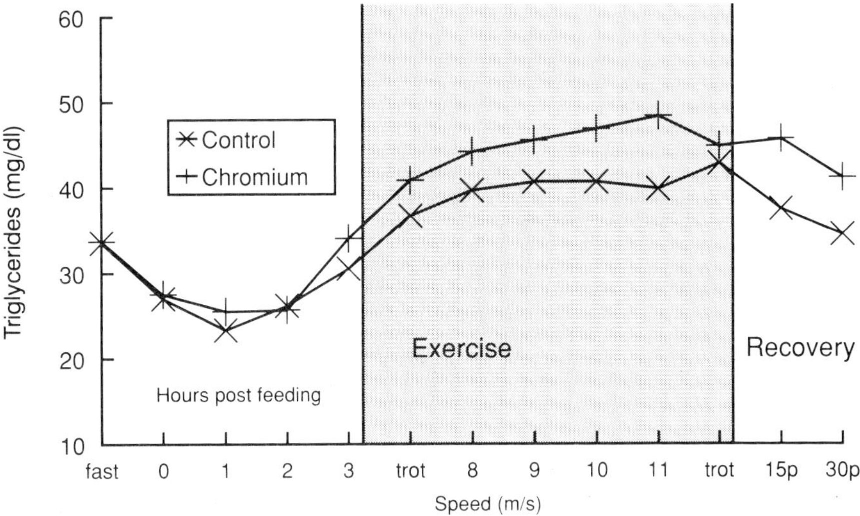

Figure 5. Effect of chromium supplementation on plasma triglyceride concentrations of exercising horses.

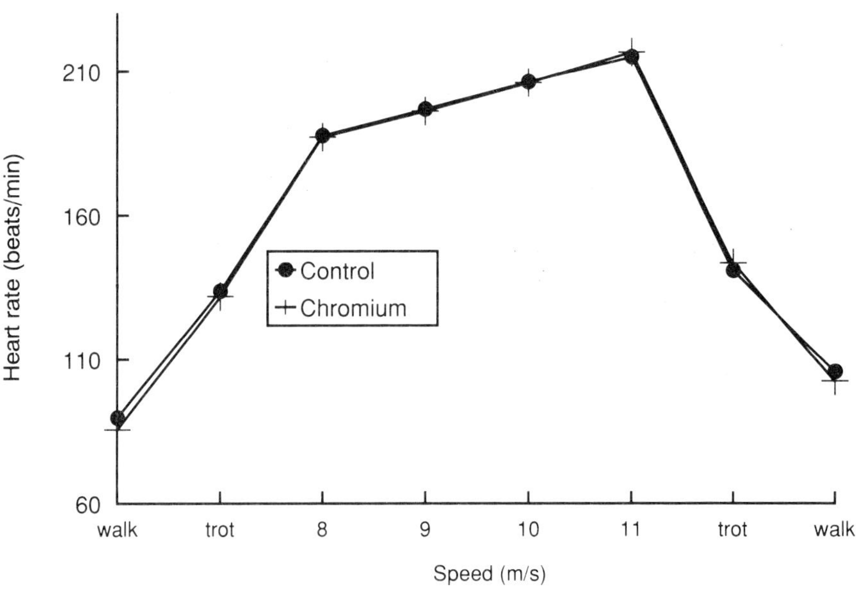

Figure 6. Effect of chromium supplementation on heart rate in exercised Thoroughbreds.

consistent with findings in humans (Anderson *et al.*, 1991a) and pigs (Evock-Clover *et al.*, 1993) where chromium supplementation increased glucose tolerance and insulin sensitivity. Trivalent chromium, as a constituent of glucose tolerance factor (GTF), is thought to facilitate interactions between insulin and insulin receptors on target tissues such as muscle and fat.

Plasma cortisol was also affected by chromium supplementation. Cortisol concentrations were lower in the chromium group 3 hours after feeding and during exercise and this effect may have also been related to differences in insulin status. Cortisol acts antagonistically to insulin and this may be the underlying metabolic role of cortisol in stress (Munck *et al.*, 1984). The function of cortisol may be to prevent insulin from causing dangerous hypoglycemia. In a previous study with trained Thoroughbreds (Pagan *et al.*, 1994), high insulin levels after feeding were associated with high cortisol production.

Elevated triglycerides in the chromium group during and after exercise may have also been related to insulin production. Insulin inhibits lipid mobilization from adipose. The increase in triglycerides in the chromium group may have been the result of high lipid mobilization during exercise. Free fatty acids mobilized during exercise that were not taken up and oxidized by the working muscle may have been used to resynthesize triglycerides in the liver.

Plasma lactate was lower after the fastest step of the SET in the chromium supplemented horses. The reason for this decrease is unknown, but may be related to a change in carbohydrate and lipid metabolism brought about by a reduction in insulin production or increase in tissue insulin sensitivity. Since lactate accumulation has been implicated as a contributing factor to fatigue during strenuous exercise, this reduction in lactate accumulation can be interpreted as beneficial for the performance horse.

The horses used in this experiment had been in training for several months before the start of this experiment and they continued to exercise throughout the study. Research in humans has shown that chromium excretion is related to exercise intensity (Anderson *et al.*, 1991b). This may explain why these horses responded to chromium supplementation while untrained, sedentary horses in a subsequent study (Pagan, unpublished data) did not. Chromium is a nutrient, not a drug. Therefore, it is reasonable to only expect a metabolic response in horses that are deficient in chromium. More research is needed to determine which factors affect chromium status in horses and what levels of supplementation are effective.

References

Anderson, R.A., M.M. Polansky, N.A. Bryden and J.J. Canary. 1991a. Supplemental-chromium effects on glucose, insulin, glucagon, and urinary chromium losses in subjects consuming controlled low-chromium diets. Am. J. Clin. Nutr. 54:909–916.

Anderson, R.A., N.A. Bryden, M.M. Polansky and J.W. Thorp. 1991b. Effect of carbohydrate loading and underwater exercise on circulating cortisol, insulin and urinary losses of chromium and zinc. Eur. J. Appl. Physiol. 63:146.

Evock-Clover, C.M., M.M. Polansky, R.A. Anderson and N.C. Steele. 1993. Dietary chromium supplementation with or without somatotropin treatment alters serum hormones and metabolites in growing pigs without affecting growth performance. J. Nutr. 123:1504–1512.

Mertz, W. 1992. Chromium, history and nutritional importance. Biol. Trace Elem. Res. 32:3.

Munck, A., P.M. Guyre and N.J. Holbrook. 1984. Physiological functions of glucocorticoids in stress and their relation to pharmacological actions. Endocr. Rev. 5:25–44.

Pagan, J.D., I. Burger and S.G. Jackson. 1994. The long-term effects of feeding fat to 2 year old Thoroughbreds in training. 4th ICEEP, Queensland, Australia (Abstr.).

SELENIUM METABOLISM IN ANIMALS: WHAT ROLE DOES SELENIUM YEAST HAVE?

D.C. MAHAN

Animal Sciences Department, Ohio State University, Columbus, Ohio, USA

Selenium metabolism

Since the discovery of selenium (Se) as a component of glutathione peroxidase (GSH-Px) by Rotruck *et al*. (1973), there have been a number of seleno-compounds identified in animal tissue. Arthur (1993) has listed 10 Se-containing proteins that have been fully or partially characterized. The biological functions of these different compounds have not been clearly established, but it is becoming clearer that the role of Se in the animal's body exceeds the cytosolic antioxidant function initially assigned to this element.

Glutathione peroxidase is generally recognized for its antioxidant protective function. There are, however, different forms of this enzyme which function at different locations (cytosolic, plasma, phospholipid hydroperoxide, intestine, lung), each perhaps with specificity to the antioxidant systems needed for that tissue. Phospholipid hydroperoxide GSH-Px appears to be involved with antioxidant activity at the cellular membrane level; whereas cytosolic GSH-Px is associated with antioxidant activity within the cytosol of the cell. The distribution of the GSH-Px types differ by tissue and by species. Consequently the clinical symptoms demonstrated from the observed tissue deficiencies of various species may reflect different distributions of the GSH-Px antioxidant systems in these species.

A selenoprotein (deiodinase), which is involved in the conversion of T_4 to the active T_3 form of thyroxine, has been identified (Arthur, 1990). In Se deficiency, plasma thyroid stimulating hormone (TSH) and T_4 levels were found to increase while plasma T_3 concentrations decreased. When Se and iodine deficiencies occur simultaneously, the results are exacerbated, resulting in even higher plasma TSH and T_4 concentrations than when either element was singly deficient. The importance of this is that under many field conditions one or both elements may be inadequate (Mee *et al*., 1994).

Sperm from Se deficient animals have poor motility with abnormal development characteristics in the sperm tail. Results from Marin-Guzman and Mahan (1989a) have demonstrated that Se deficiency not only precipitated this problem in boars, but that the mis-shaped sperm were less effective in their subsequent fertilization of the ovulated oocyte (Marin-Guzman and Mahan, 1989b).

There are other seleno-proteins identified in animal tissue but their biological function is largely unknown. Seleno-protein P synthesized in the liver is transferred into the plasma and may act as a Se-transport protein (Arthur, 1993). Other identified seleno-proteins are found in various tissues, but their function is even less clear.

Determining the selenium requirement

Fisher *et al.* (1994) stated that messenger RNA for plasma GSH-Px is regulated by the available supply of Se, whereas other tissue messenger RNAs were unaffected by Se supply. This suggests a priority usage of Se by body tissue and that GSH-Px activity in different tissues may have a later priority in their formation. Tissue storage, excretion and transfer of Se through the placenta and mammary tissue may therefore be achieved after the animal's more critical requirements for the element are attained.

The various body seleno-protein containing compounds generally occur after the activation of Se to the reduced form (selenide) in the liver where it undergoes enzymatic incorporation with cysteine forming selenocysteine. The biologically important seleno-proteins arising from selenocysteine and many of the biologically active compounds may contain a few or several residues of this seleno-amino acid (Arthur, 1993). The utilization of Se for the formation of the active form of thyroxine (T_3) and for sperm development appears to have a higher priority for Se than the formation of GSH-Px. Consequently, when the supply of Se is limited Se may be diverted from those seleno-compounds of lower biological need to more biologically important

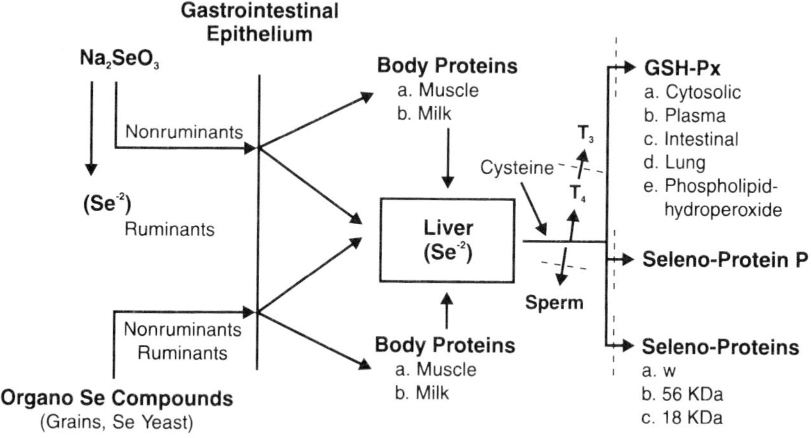

Figure 1. Utilization of absorbed selenium. Dotted lines refer to where inadequacies will affect formation of the seleno-proteins.

seleno-proteins. These results therefore imply that GSH-Px may be the best measurement criteria for determining the animal's Se requirement. The general utilization for absorbed Se to the various seleno-proteins is presented in Figure 1.

Species differences: largely related to absorption

The delivery of dietary Se from the digestive system to body tissue, however, varies greatly by species, namely between the ruminant and the non-ruminant. The absorption and utilization of organic Se from grain or an enriched Se yeast product (Sel-Plex 50, Alltech, Inc.) appears to be high for both groups, whereas there is a great difference in how inorganic Se is absorbed between ruminants and non-ruminants. In general, the rumen reduces some of the inorganic selenite to selenide by the hydrogenation process that occurs in the rumen. The resulting selenide is not absorbed in the rumen nor later in the intestinal tract. The non-ruminant does not have the extensive microbial reducing capacity in the fore-part of the digestive tract, and therefore most of the selenite remains in the oxidized form as it enters the small or large intestine. The intestinal excrement from both species would be expected to be in a non-available (selenide) form because of microbial hydration of oxidized Se that is in the hindgut.

Because organo-seleno compounds (amino acids or their analogs) are bound within the protein molecule, they are either utilized by the microorganisms within the rumen or 'by-passed' into the small intestine where they are hydrolyzed by intestinal enzymes. The organic seleno-compounds are actively transported through the enterocyte of the small intestine. The non-ruminant absorbs selenite passively while organo-seleno compounds are actively absorbed in the small intestine. In the non-ruminant the inorganic Se form appears to be better absorbed than the organic form. The general process is depicted in Figure 1.

Organic selenium: grain vs enriched grain

The production of enriched Se yeast is presented in a general schematic in Figure 2. The process involves feeding a yeast strain that has a high sulfur (S) requirement with sources of inorganic and organic Se. Because S and Se are chemically similar, the Se is incorporated into yeast cell protein structures, essentially replacing the S. The resulting yeast is killed but has a high Se content. The yeast is dried, analyzed for Se content and subsequently standardized to 1000 ppm Se for incorporation into diet formulas. The Se yeast product contains little or no inorganic Se with approximately 50% of the Se identified as selenomethionine (Powers, personal communication). The remainder of the organic fraction is in one of several seleno-amino acids or their analogs (selenocysteine, selenocystathione, methylselenocysteine, selenocystine). The organic form of Se in grains also contains 50% of its Se as selenomethionine, similar to that present in Se yeast.

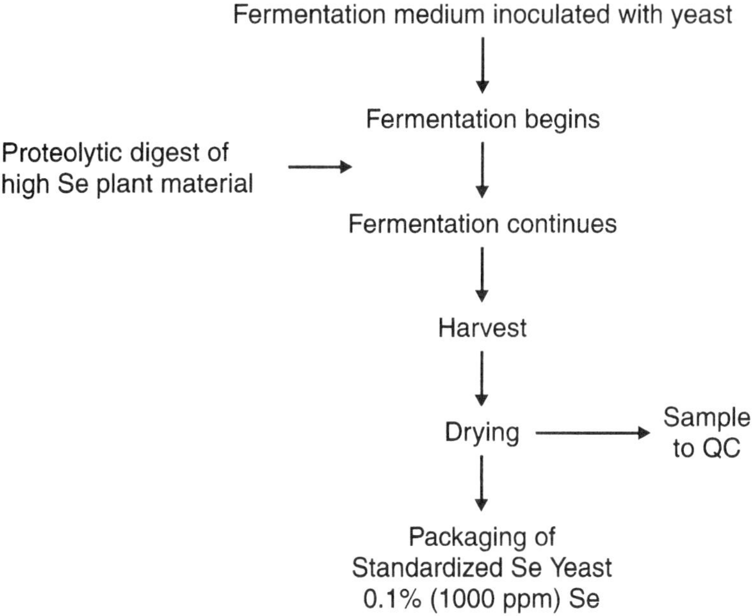

Figure 2. Production of Se-enriched yeast (Sel-Plex 50).

The ability of an animal to use any particular Se source will therefore be reflected in the production of the biologically important seleno-compounds in animal tissue. On a clinical basis, if we could measure those biologically important seleno-compounds that best reflect the overall Se requirement of the animal, then relative merit could be awarded to the various Se sources with at least some degree of assurance. We must, however, also be able to evaluate the potential role for the seleno-compounds stored in body tissue for later use, or transferred to non-functional tissue (body muscle, placental and mammary transfer) and the amount of Se which would be excreted and placed into the environment.

Based on the scientific knowledge to date, it appears that GSH-Px activity may be the best reflection of the animal's dietary requirement for Se. Tissue and serum Se concentrations appear to increase beyond the point where blood or serum GSH-Px plateau and therefore do not best reflect the Se requirement. Under marginal or deficient Se conditions there is a high correlation between blood Se and GSH-Px; but when the diet contains higher than required levels the correlation between these two variables declines (Meyer *et al.*, 1981). Blood Se values can, however, reflect the current Se status of a herd on a short-term basis since these values reflect the difference between absorption and retention by body tissue or excretion. This, however, does not negate the value of stored Se depots for later use during periods of higher Se demand which may not be apparent by blood Se content. These Se depots may have a somewhat delayed impact but can have an important function in Se homeostasis.

Glutathione peroxidase levels are most easily measured from blood samples. Within this circulatory fluid compartment the relative distribution of GSH-Px between plasma (serum) and RBC differs with various species. For example, most of the GSH-Px activity in cattle is found in the erthrocyte, whereas in the pig a large amount of GSH-Px activity is present in the serum or plasma fraction.

A recent study with pigs demonstrated that when either inorganic selenite or Se yeast (Selo-Plex 50) were fed to grower pigs, both inorganic and organic Se sources were equally effective in supporting serum GSH-Px activity (Figure 3). The data also demonstrated that when both Se sources were above 0.1 ppm in the diet there was no increase in serum GSH-Px activity. This suggests that a dietary requirement of 0.1 ppm Se meets the Se requirement of the grower-finisher pig when serum GSH-Px is used as the measurement criterion.

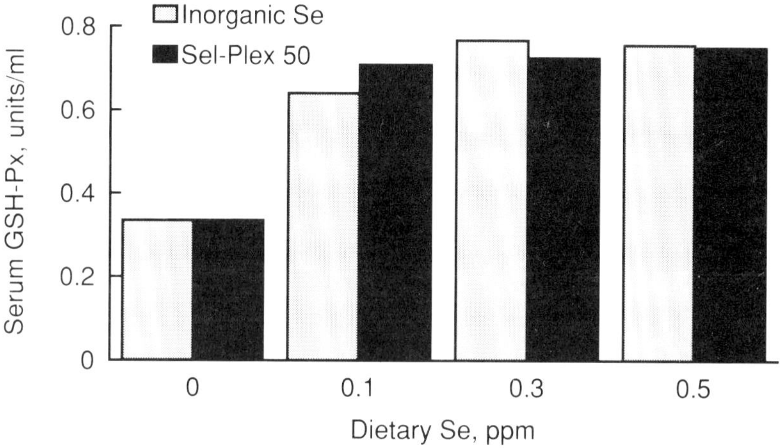

Figure 3. Effect of dietary Se source (inorganic Se or organic Se (Sel-Plex 50)) and level on serum GSH-Px activity in growing pigs.

As indicated above, the ruminant differs in the absorption and utilization of Se yeast compared with sodium selenite and consequently in resulting blood GSH-Px activity. Erythrocyte GSH-Px activity was approximately doubled when selenomethionine or Se yeast was fed to dairy heifers as compared to when selenite was provided (Table 1).

A study by Fisher *et al.* (1994) feeding either inorganic or organic Se sources at 0.1 ppm to lactating dairy cows demonstrated a higher increase in both whole blood and serum Se concentrations when the Se yeast product was fed. This implies that the organic Se source is more highly absorbed (Table 3). Although blood Se level is probably not as accurate in determining the Se status of the animal, it is probably the most widely used measurement tool used in the field to detect

Table 1. Effect of Se source on change in erythrocyte GSH-Px activity in dairy heifers.*

Se Source	Daily dosage (mg Se)	GSH-Px activity
Sodium selenite	560	253
Cobalt selenite	640	280
Selenomethionine	470	495
Se Yeast	570	747

*Pehrson *et al.*, 1989.

the Se status of a herd. This is because of the lack of uniformity in measuring GSH-Px activity among laboratories and the lowered stability of the enzyme when collected under differing field situations. Therefore the GSH-Px measurement has not been the tool of choice under field conditions. When Fisher *et al.* (1994) evaluated whole blood and serum Se concentrations in the milking herd it was clear that a higher percentage of cows that were classified as being in an adequate Se status were fed Se yeast compared with selenite, and that serum Se was a more sensitive measure of the herd's Se status than whole blood Se concentration (Table 2).

Table 2. Selenium status evaluation using blood Se content in lactating dairy cows fed sodium selenite or Se yeast.*

Blood measurement	Selenite (% of cows)	Se Yeast (% of cows)
Marginal status		
Whole blood Se (50–90 ng/ml)	0	0
Serum Se (40–70 ng/ml)	73.7	37.5
Adequate status		
Whole blood Se (>100 ng/ml)	100	100
Serum Se (70–100 ng/ml)	26.3	62.5

*Fisher *et al.*, 1994

Non-functional selenium

Non-functional Se is described in this report as that Se which lends itself to the formation of Se compounds that are not used immediately by body tissues for biologically functional seleno-proteins. Selenium stored in tissue may be a source of non-functional Se even though it may eventually be diverted to biologically important compounds. This non-functional Se will initially be bound within the tissue until it is loosened or is mobilized by tissue turnover. The released Se can subsequently be converted into a metabolic useful form.

The pig stores most of the absorbed inorganic Se in the liver and muscle tissue. The chemical form(s) in which it is stored in these tissues are unknown; but muscle apparently does not have the capacity to store large quantities of Se when the primary dietary source is inorganic Se.

Table 3. Effect of inorganic selenium on tissue selenium in market pigs.*

	Feed origin:	Michigan	South Dakota	
	Natural Se, ppm:	0.04	0.40	0.40
	+ Selenite, ppm:	0.40	0	0.1
No. pigs		4	4	4
Tissue Se, ppm				
Longissimus		0.12	0.48	0.45
Liver		0.61	0.84	0.92
Kidney		2.14	2.17	2.33

*Adapted from the data of Ku *et al.* (1973)

Ku *et al.* (1973) demonstrated in grower-finisher pigs that loin Se content was not increased when sodium selenite was fed in combination with organic Se contributed from grains (Table 3).

In contrast, loin Se content was increased approximately four-fold when pigs were fed organic Se from grain compared with supplemental selenite. A recent study (Mahan, unpublished data) evaluated the efficacy of inorganic Se and organic Se from an enriched Se yeast product (Sel-Plex 50) with grower pigs. The results demonstrated that muscle Se was increased when both Se forms were fed (compared with a non-fortified basal diet) but was increased by approximately 60% when the organic Se source was provided (Figure 4). The combined results of these studies suggest that absorbed inorganic Se is retained at lower concentrations by muscle tissue, and there is no further increase when the dietary level of Se exceeds 0.1 ppm. On the other hand, when organic Se was provided in the diet, larger quantities of organic Se were incorporated within the muscle (and/or liver) tissue. Although these latter reservoirs will not be mobilized until tissue turnover occurs, they

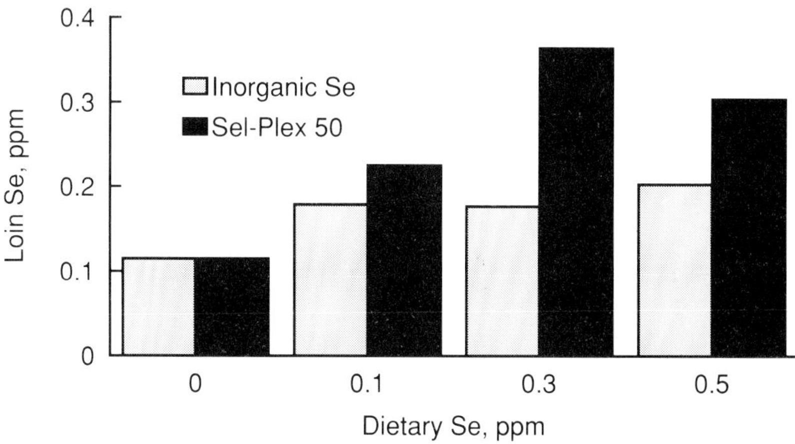

Figure 4. Effect of organic Se (Sel-Plex 50) and inorganic Se (sodium selenite) level on Se concentration in loin muscle of growing pigs.

may have a biologically important role later, and certainly the stored Se has merit for humans consuming pork.

Although milk is a production output of the lactating female, it can be considered as a pool of Se that serves in a non-functioning role for the animal secreting it, but yet can serve as an important dietary Se source for the animal (or person) consuming it. The efficacy of transferring Se through mammary tissue differs with dietary form (organic or organic Se) and species. Several independent research studies (Suoranta *et al.*, 1993; Fisher *et al.*, 1994) have demonstrated that organic Se fed to lactating dairy cows increased milk Se content substantially more than when inorganic Se was provided. Pehrson (1989) also demonstrated an approximate five-fold increase in milk Se content when an enriched Se yeast source was compared with selenite (cited by Lyons, 1994). Both Pehrson (Lyons, 1994) and Suoranta *et al.* (1993) demonstrated that the increased milk Se concentration occurred very shortly after being fed the enriched Se yeast source and that milk Se plateaued within 3 weeks of feeding. Inorganic Se is not absorbed as effectively as organic Se by the ruminant and this is undoubtedly one of the major reasons for the different milk Se concentrations for these two dietary sources. Weiss *et al.* (1990) demonstrated that when dietary Se was provided in combination with vitamin E the somatic cell count (SCC) and number of clinical mastitis cases declined. The form of Se in cows' milk has not been determined, but the organic Se yeast would be expected to contribute a major portion of milk Se as selenomethionine in the milk protein fraction, whereas such may not be the case when sodium selenite is fed.

Several countries of the world have Se deficiencies affecting both animal and humans. Because the growing animal and/or infant would be more susceptible to Se deficiencies than the mature organism, milk could serve as a valuable dietary Se source for the young. Consequently, the Se requirement of the lactating dairy cow should include an assessment for the Se contribution in the milk. It would appear that a major proportion of dietary Se must be in an organic form for the lactating ruminant if higher milk Se content is desired.

Although it was indicated earlier that both inorganic and organic Se were effectively absorbed and equally effective in GSH-Px activity of the pig, the pig has a lower tissue Se storage capacity when fed inorganic Se (Figure 4). The limitation of this restricted storage capacity is evident in older reproducing animals. Mature milk (21 day) samples were collected ($N = \sim 75$ observations/mean) from adult lactating sows of different parities when fed dietary Se at 0.3 ppm throughout their reproductive history (Figure 5). The data suggested that from parity 3 to 5, milk Se content was lower when compared with earlier milk Se concentrations. Consequently the progeny from older sows would be more prone to encounter the deficiency onset that the progeny of younger sows. Although all sows had their diets supplemented with sodium selenite at 0.3 ppm, this level was apparently inadequate to maintain milk Se concentrations of older sows compared with when these sows were younger. It is also possible that retained tissue Se

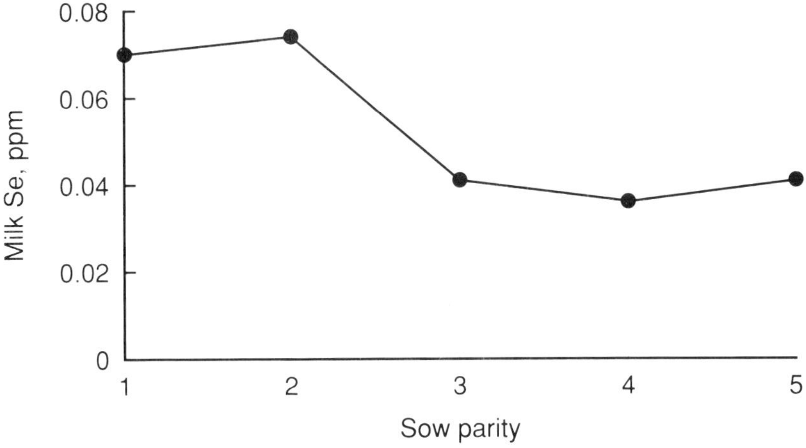

Figure 5. Selenium content of milk from mature sows 21 days post-partum.

was depleted as the sows matured. Studies are underway to evaluate the efficacy of organic Se yeast fed to lactating sows.

Selenium source and the environment

Perhaps an area of increasing concern by the public is the effect excreted Se has on the environment. A recent report by CAST (1994) addressed the dietary level that was approved by FDA (1987) for livestock diets and the concern about the toxicity reports in the Kesterson Reservoir in California. The CAST report was not questioning the animals' need for

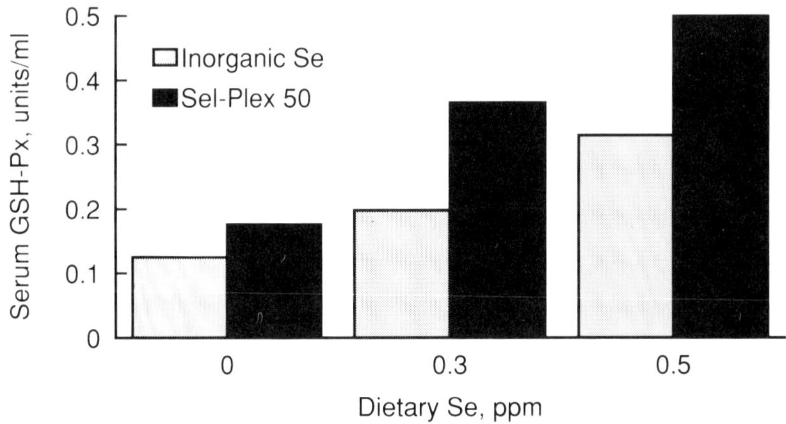

Figure 6. Effect of inorganic and organic selenium (Sel-Plex 50) on selenium retention in growing pigs.

the element but for subsequent environmental safety levels that would be acceptable. Although many factors can influence the 0.3 ppm Se requirement (i.e., phase of production, feed intake, etc.) the 0.3 ppm dietary level is clearly near that which would support optimum health and performance responses in domestic livestock. A recent digestibility study with pigs demonstrated, however, that total body Se retention is approximately doubled at the various dietary Se levels evaluated (0.1 to 0.5 ppm) and therefore Se excretion was less when the enriched Se yeast was fed compared to sodium selenite (Figure 6). Consequently less Se would be excreted when the organic Se source is fed to livestock. Although the form of Se in animal excrement may not have a high availability for plant tissue from either Se source, the dietary inorganic Se source would contribute more Se to the soil.

References

Arthur, J.R., F. Nicol and G.J. Rechette. 1990. Hepolic iodothyronine 5′–deiodenase: The role of selenium: Biochem. J. 272:537.

Arthur, J.R. 1993. The biochemical functions of selenium relationships to thyroid metabolism and antioxidant systems. Rowett Research Institute Annual Report. Bucksburn, Aberdeen, UK.

CAST. 1993. Risks and benefits of selenium in agriculture. Issue paper No. 3, June pp. 6.

Fisher, D.D., S.W. Saxton, R.D. Elliott and J.M. Beatty. 1994. Effects of selenium source on lactating cow selenium status. (in press).

Ku, P.K., E.R. Miller, R.C. Wallstram, A.W. Grace, J.P. Hitchcock and D.E. Ullrey. 1973. Selenium supplementation of naturally high selenium diets for swine. J.Animal. Sci. 37:501.

Lyons, T.P. 1994. Biotechnology in the feed industry: 1994 and beyond. In: Biotechnology in the Feed Industry. Proceedings 10th Annual Symosium. T.P. Lyons and K.A. Jacques (Eds). Nottingham University Press, UK, p.23.

Marin-Guzman, J. and D.C. Mahan. 1989a. Effect of dietary vitamin E and selenium on semen quality of boars. Ohio Swine Research and Industry Report. Animal Sci. Series 89–1, 20–23.

Marin-Guzman, J. and D.C. Mahan. 1989b. Sow fertilization rates as affected by dietary selenium and vitamin E levels in boar diets. Ohio Swine Research and Industry Report. Animal Sci. Series 89–1, 24–25.

Mee, J.F., K.J. O'Farrell and P.A.M. Rogers. 1994. Bare-line survey of blood trace element status of 50 dairy herds in the south of Ireland in the spring and autumn of 1991. Irish Vet. J. 47:115.

Meyer, W.R., D.C. Mahan and A.L. Moxon. 1981. Value of dietary selenium and vitamin E for weanling swine as measured by performance and tissue selenium and glutathione peroxidase activities. J. Anim. Sci. 52:302–311.

Pehrson, B., M. Knutsson and M. Gyllenswoad. 1989. Glutathione peroxidase activity in heifers fed diets supplemented with organic and inorganic selenium compounds. Swedish J. Agr. Res. 19:53.

Rotruck, J.A., A. Pope, H. Ganther, A. Swanson, D. Hafeman and W. Hoekstron, 1973. Selenium: biochemical role as a component of glutathione peroxidase. Science 179:588.

Suoranta, K., E. Sinda and R. Pihlah. 1993. Selenium of the selenium yeast enters the cow's milk. Nor. J. Agr. Sci. Suppl. No. 11:215.

Weiss, W.P., J.S. Morgan, K.L. Smith and Hoblet. 1990. Relationship among selenium, vitamin E and mammary gland health in commercial dairy herds. J. Dairy Sci. 73:381.

ROUNDTABLE INTERACTIVE PROGRAM

COMPARATIVE EFFECTS OF INORGANIC AND ORGANIC SELENIUM SOURCES (SELENIUM YEAST) ON SELENIUM STATUS OF LACTATING COWS

D.D. FISHER

Co-operative Feed Dealers, Inc., Chenango Bridge, New York, USA

Summary

Two selenium (Se) sources, sodium selenite and Se yeast, were fed as topdress feeds at a rate of 2.2 mg Se per head per day to two groups of lactating Holstein cows on a commercial northeastern Pennsylvania dairy farm for 90 days. Forages and commercial feed mixtures were low in Se. Within the treatment, whole blood and serum Se concentrations increased during the trial (final vs. initial). Cattle supplemented with Se yeast had higher serum Se concentrations at the end of the trial. Individual serum and milk Se concentrations tended to be uniformly higher in animals given the Se yeast. Individual serum Se concentrations were negatively related to log transformed somatic cell counts at completion of the trial. Organic Se appeared to be more efficiently absorbed than inorganic Se in lactating dairy cattle.

Introduction

Low intake of dietary Se is associated with nutritional muscular dystrophy (white muscle disease) in calves, unthriftiness, impaired killing ability of phagocytic neutrophils, and impaired fertility in cattle (NRC, 1989; Hogan *et al.*, 1990; Minson, 1990). Inadequate dietary Se intake is also related to increased incidence of clinical mastitis, retained placenta and possibly cystic ovaries in cattle (Harrison *et al.*, 1984; Smith *et al.*, 1984). Dairy herds with inadequate Se intakes also tend to have elevated milk somatic cell counts (SCC, Erskine *et al.*, 1987).

Forages and grains cultivated on soils in the northeastern (NE) United States are known to be quite low in Se content (<0.10 mg/kg DM). When FDA decided in 1993 to return to the pre-1987 allowable level of Se supplementation, only 0.1 ppm Se could be added to the diet. Since the typical NE dairy ration is composed of forages and cereal grains that are cultivated locally, the total ration would ordinarily contain less than 0.20 mg Se/kg DM after it had been supplemented with Se at the legal 0.1 mg/kg limit. As this would in many cases mean feeding rations known to be insufficient in Se to support health or production, evaluating selenium sources which might potentially be more bioavailable or bioactive became an important goal.

Inorganic vs organic Se sources

Evidence exists to suggest that organic and inorganic sources of Se are differently absorbed. In experiments with growing lambs and steers, Ullrey and others (1977) reported that serum and tissue Se concentrations were higher when animals were fed diets consisting of natural Se sources compared with concentrations resulting from diets fortified with sodium selenite (selenite). Anderson and Scarf (1983) reported that most of the Se in plants occurs as selenoether amino acids. Selenium in alfalfa, wheat, and probably the other grains is considered to be primarily in the form of selenomethionine (Levander, 1986). Selenocystine, selenocysteine, and Se-methylselenomethionine appear to be the other major organic Se compounds in seeds and forages (Anonymous, 1983).

Pehrson (1993) reported that Se yeast was approximately twice as active as selenite for increasing glutathione peroxidase (GSH-Px) activity in red blood cells (RBCs) of Se-deficient young cattle. Nearly half of the Se in Se yeast is composed of selenomethionine with a mixture of other unidentified Se-containing organic compounds and selenocysteine constituting the remainder (K. Jacques, personal communication).

Objectives

The primary objective of this trial was to compare the effects of organic and inorganic Se sources on Se-containing components of blood from lactating cows, milk Se, and Dairy Herd Improvement (DHI) SCC linear score (DHI-SCC LS). A secondary objective consisted of studying the relationships between Se-containing blood component levels and DHI-SCC LS.

Materials and methods

ANIMALS AND DIETS

The 90-day trial was conducted on a commercial dairy farm located in northeast Pennsylvania. This high-producing Holstein herd (>10 000 kg rolling herd average) is housed in a two-row tie stall barn with a narrow side-to-side aisle that conveniently partitioned the herd into four subgroups such that one subgroup of each treatment was on either side of the barn.

Forages fed daily to the herd were corn silage, mixed mainly grass hay, and mixed mainly legume hay (Table 1). A protein supplement and an energy mixture were fed together via an 'around-the-barn style'

Table 1. Feeding frequencies and Se content of diet ingredients.

Ingredient	Feeding frequency	Se content, ppm
Corn silage	Twice daily	0.02
Mixed (mainly grass) hay	Once daily	0.03
Mixed (mainly legume) hay	Once daily	0.06
Protein supplement (SBM 44% CP, flaked soybeans, distillers grains with solubles)	Eight times daily	0.09
Energy mixture (dry rolled corn, hominy, soy hulls, molasses, minerals)	Eight times daily	0.11

Table 2. Formula of the topdress supplement.

Ingredient	Percent
Distillers grains with solubles	97.6
Se premix (0.02% Se)*	2.4
Total	100.0

*One topdress feed was manufactured with the Se yeast premix (0.02% Se) whereas the other was manufactured with the sodium selenite premix (0.02% Se). (Both premixes were routinely assayed for Se content as part of the manufacturing quality control program that was required under previous existing FDA regulations.)

electronic rail feeder. Selenium content of the feed ingredients was typical of those reported for local feedstuffs. When deemed necessary due to changes in forage composition, representative composite samples were sent to Northeast DHI Forage Lab (Ithaca, NY) for analysis. The dairy ration was then reformulated to meet the nutrient requirements and to provide updated guidelines for feeding the forages and commercial feeds. Additional representative composite samples of the forages and feeds were sent to Michigan State University Animal Diagnostic Lab for Se analysis.

One treatment group received once daily 1 lb of a topdress feed containing selenite whereas the other group received the topdress feed containing Se yeast (Sel-Plex 50, Alltech, Inc; Table 2). Each topdress feed contained 2.2 mg Se such that total dietary Se was 0.10 mg Se/kg DM for the average animal.

MEASUREMENTS

Two blood samples were collected from the coccygeal vein of each animal during the late morning on both the initial and final days of the trial. Immediately upon collection, these samples were placed on dry ice and later sent to the Michigan State University Animal Diagnostic Lab for whole blood and serum Se analysis. Red blood cell Se was estimated by difference (whole blood Se minus serum Se). [It is acknowledged that a small quantity of Se in this RBC fraction actually would be found in the platelets or thrombocytes (Johansson and Lindu, 1985), and possibly in the other clotting factors.] At the end of the trial, 10 individual milk samples per treatment were collected for Se analysis. The procedures for sample preparation and Se analysis were as described by Reamer and Veillon (1983) and Whetter and Ullreg (1978).

The trial began and ended on the Pennsylvania DHI test days when milk yield weights and samples were collected from individual cows. Thus, monthly DHI reports were used to obtain data regarding the mean body weight, days in milk (DIM), age at calving, parity, milk yield, milk fat and protein contents, and DHI-SCC linear score (SCC LS) of the individual animals. Fat-corrected milk (3.5% FCM), fat and protein yields of animals were also calculated from these data.

Student's *t*-test procedure for unpaired data was employed to test for treatment effects on parity; age at calving; DIM; milk, 3.5% FCM, fat and protein yields; fat and protein percentages; and whole blood, serum, and RBC Se concentrations. Within treatment, the paired *t*-test was used to detect differences between initial and final measurements for milk, 3.5% FCM, fat and protein yields; fat and protein percentages; and whole blood, serum, and RBC Se concentrations (Steele and Torrie, 1960).

The Kruskal–Wallis non-parametric test was used to test for treatment differences in whole blood Se:serum Se ratios (Steele and Torrie, 1960). Linear regression and correlation were utilized to determine the degree of association between Se concentrations of whole blood or serum and DHI-SCC LS. DHI-SCC linear score is the log transformed SCC. DHI-SCC must be transformed in order to meet certain criteria for hypothesis testing (normal distribution of errors and homogenous subclass variances; Ali and Shook, 1980).

Results and discussion

PERFORMANCE EFFECTS

Effects on milk yield and composition

Initial parity (2.6), days in milk (Selenite, 160 days; Se Yeast, 125 days), and age at calving (46 months) did not differ between treatments. Mean estimated body weight (using a tape) of the lactating cows was 612 kgs. Initial and final milk yields, 3.5% FCM yields, milk component yields, and fat and protein contents were unaffected by treatment. Milk, 3.5% FCM and protein yields of animals on both treatments declined significantly during the trial. [The initial and final 3.5% FCM mean daily yields were respectively 34.8 kg and 31.9 kg (selenite), and 37.3 kg and 32.1 kg (Se Yeast).] The fat yield of the Se Yeast supplemented animals decreased ($P<0.05$) during the trial. Milk protein content increased in both treatment groups whereas milk fat content increased only in the selenite group at the completion of the trial.

EFFECTS OF SE SOURCE ON BLOOD AND MILK SE

Whole blood and serum Se

During the three-month trial, whole blood and serum Se concentrations increased significantly for both treatments (Table 3). In a two-lactation-gestation cycle experiment, Stowe and others (1988) reported that lactating animals peaked in mean serum Se concentration at seven months postpartum in cycle I and at four months postpartum in cycle II. The whole blood Se: serum Se ratio tended to be reduced during the trial on the Se Yeast treatment.

Final serum Se concentration of the Se Yeast-supplemented animals was significantly higher ($P<0.05$) than that of the selenite-supplemented group (Table 3). The whole blood Se:serum Se ratio was also significantly lower ($P<0.05$) for the Se Yeast supplemented animals at the termination of the trial. Stowe and Herdt (1992) reported that this ratio would tend to narrow after an increase in oral Se intake. In this case, the quantity of Se fed daily to the animals remained constant throughout the trial. However, the ratio would also be expected to narrow if these animals were fed Se from a source that would be more efficiently absorbed into the animal's system from the digestive tract.

Approximately 40% of the supplemental Se as selenite passes across the intestinal wall by passive diffusion (CAST, 1994; NRC, 1989). A portion of the selenite Se may have been reduced to insoluble forms by the microorganisms in the rumen (Anonymous, 1983). In contrast, selenomethionine probably passes across the intestinal wall by an active transport mechanism (Pehrson, 1993). It is not known how selenocysteine and other organic Se compounds are absorbed into the body. [After absorption, selenocysteine would be catabolized by selenocysteine B-lyase, thus releasing Se (Levander and Burk, 1990).]

Table 3. Effects of selenium source and lactational stage on lactating animal blood and milk selenium concentrations.

Item	Whole blood Se (ng/ml)	Serum Se (ng/ml)	RBC Se[a] (ng/ml)	Whole blood Se: serum Se	Milk Se (ng/ml)
Selenite:					
Initial	139.4[c] ± 18.2	59.6[i] ± 9.5	79.8[g] ± 16.6	2.4	
Final	155.2[f] ± 23.2	65.7[jk] ± 8.7	89.5[h] ± 19.6	2.4[k]	17.9 ± 4.3
Se Yeast:					
Initial	134.5[c] ± 23.2	59.4[e] ± 13.5	75.1 ± 24.9	2.4[i]	
Final	153.8[d] ± 20.3	71.9[fl] ± 9.0	81.9 ± 19.4	2.2[jl]	20.5 ± 5.1
Increment:[b]					
Selenite	+ 15.8	+ 6.1	+ 9.7	–	–
Se Yeast	+ 19.3	+ 12.5	+ 6.8	–	–

[a]RBC Se is the quantity of Se that is contained in the red blood cells (and in the thrombocytes and other blood clotting factors; whole blood Se – serum Se).
[b]Mean increment in blood component value within treatment during the trial (final value – initial value.)
[cd, ef, gh, ij]Paired initial and final means (± sd) of each blood component within treatment differ ([cd]$P<0.001$, [ef]$P<0.005$, [gh]$P<0.05$, and [ij]$P<0.10$; respectively).
[kl]Final treatment means (± sd) differ ($P<0.05$) for serum Se, and 'whole blood Se: serum Se' ratio.

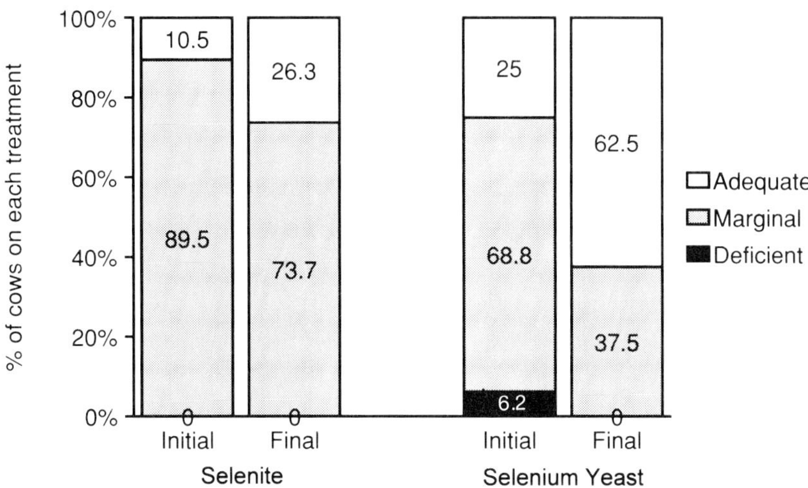

Figure 1. Effect of Se source on Se status at the beginning and end of the three month trial as evaluated using serum Se.

RBC Se

RBC Se concentration of the animals on the selenite treatment increased significantly ($P<0.05$); it also tended to increase on the Se Yeast treatment during the trial (Table 3). In ruminants, most of the Se in RBC appears to be present in the form of the selenoenzyme, GSH-Px (Scholz and Hutchinson, 1979; Anonymous, 1983; Levander, 1986).

There were no significant treatment effects on final RBC Se. Final RBC Se tended to be slightly lower in cows given Se Yeast. Since Se from absorbed dietary selenomethionine is only available for RBC GSH-Px synthesis after is has been catabolized (in the liver, Levander and Burk, 1990), RBC Se could be expected to increase more gradually in the lactating cow on a marginal methionine intake. (Methionine is often cited as being the most limiting amino acid for lactating dairy cows, NRC, 1989.)

According to Levander (1986), after being absorbed into the body, Se in the inorganic selenite form is reduced into the selenide form. Furthermore, Se in the selenide form can be transformed into selenocysteine which is the main form that is utilized to synthesize GSH-Px.

CAST (1994) reports that it has been recently proposed that GSH-Px messenger RNA is regulated by Se supply. 'During Se deficiency, GSH-Px messenger RNA levels are low, and free Se is diverted to other selenoproteins whose messenger RNA levels are not reduced significantly. Such selenoproteins include phospholipid hydroperoxide glutathione peroxidase, selenoprotein P, and Type I 5'- iodothyronine deiodinase. Once these critical needs for Se are met, intracellular Se concentrations can increase and lead to the incorporation of Se into GSH-Px. These effects would limit the presence of free and possibly toxic concentrations of selenide within the cell.'

Milk Se

Milk Se values obtained were similar to those observed in other studies at similar Se intakes (Maas *et al.*, 1980; NRC, 1989; Pehrson, 1993). There was no significant treatment effect on milk Se, however it is noteworthy that 90% of the milk samples from Se-Yeast-supplemented cows equaled or exceeded a level of 18 ng/ml (Table 3).

In the Nordic countries, a desirable Se concentration in milk for human consumption is 'about 20 ng/ml'. (Dairy products supply approximately 25% of the dietary Se in this region of Europe; Pehrson, 1993.) Since selenomethionine may substitute for methionine during milk protein synthesis in the mammary gland, it is reasonable to expect a higher Se concentration in milk of cows supplemented with an organic Se source containing selenomethionine.

USING WHOLE BLOOD AND SERUM SE TO DETERMINE SE STATUS

Individual whole blood and serum Se values were categorized as deficient, marginal, or adequate (Van Saun, 1989) in order to compare Se status of cattle in this study to that of cattle in other studies. All cattle in this study had acceptable whole blood Se concentrations (>100 ng/ml, Table 4). Whole blood Se levels are expected to rise and fall moderately over a 3 to 4 month period for the following reasons (Stowe and Herdt, 1992).

1. A significant portion of the whole blood Se is within RBCs (mostly as GSH-Px).
2. Se apparently only enters RBCs during erythropoiesis.
3. Life span of the bovine RBC is 90 to 120 days.

In contrast, serum Se is known to respond rapidly to changes in the available Se supply (Van Saun, 1989; Stowe and Herdt, 1992). Approximately 1/10 and 1/4 of the animals in the selenite and Se Yeast groups, respectively, had adequate initial serum Se values (>70 ng/ml; Van Saun, 1989; Stowe and Herdt, 1992). At the end of the

Table 4. Se status of the two treatment groups of lactating animals during the trial.

Treatment	Deficient		Marginal		Adequate	
	Whole blood (10–40)	Serum (<)	Whole blood (50–90)	Serum (40–70)	Whole blood (>100)	Serum (70–100)
Selenite:						
Initial	0%	0%	0%	89.5%	100%	10.5%
Final	0%	0%	0%	73.7%	100%	26.3%
Se Yeast:						
Initial	0%	6.2%	0%	68.8%	100%	25.0%
Final	0%	0%	0%	37.5%	100%	62.5%

*These ranges of whole blood and Se values (ng/ml) have been proposed for evaluating the Se status of cattle (Van Saun, 1989).

trial, about 1/4 and 2/3 of the animals in the selenite and Se Yeast groups, respectively, had adequate serum Se values (Table 4, Figure 1). All animals were estimated to have ingested between 250 and 550 IU of supplemental vitamin E during this study. Weiss and others (1990) have suggested that a higher minimum 'reference' plasma (or serum) Se level may be required as an indicator or adequate Se status in cattle if no supplemental vitamin E is fed.

SOMATIC CELL COUNT LINEAR SCORE

No treatment effects were observed on DHI-SCC LS ($P>0.10$). Overall final and mean final (of last two test days) DHI-SCC LS were 2.9 ± 1.8 and 3.0 ± 1.8, respectively ($n=35$). Final individual cow serum Se concentrations in the range of 55 to 91 ng/ml were inversely correlated to the last DHI-SCC LS $r=-0.371$, $P<0.03$, $n=35$. Final serum Se concentrations were also negatively correlated to the mean DHI-SCC LS of the last two test days ($r=-0.365$, $P<0.035$, $n=35$). Final whole blood Se concentrations were not correlated to DHI-SCC LS ($P>0.10$).

In a study that involved two groups of commercial dairy herds (16 herds per group) which had annual mean SCC of <150 000 cells/ml and >700 000 cells/ml (high SCC) respectively, Erskine and others (1987) observed significantly lower whole blood Se levels and GSH-Px activities in the high-SCC group of herds. The authors suggested that Se may be involved in the resistance of the mammary gland to infections with *Streptococcus agalactiae* and *Staphylococcus aureus*.

In nine commercial herds, Weiss and others (1990) reported that mean herd serum Se concentrations and GSH-Px activities were both negatively correlated to bulk tank SCC. When comparing healthy and mastitic animals, Atroshi and others (1986) reported that the mastitic animals had significantly lower whole blood GSH-Px activities than healthy animals; whole blood Se levels were similar between the two groups.

Massive influx of neutrophils into the mammary gland from the circulatory system is considered to be a primary defense mechanism against mastitis (Guyton, 1961; Smith *et al.*, 1988). Microbial activity of neutrophils is decreased by Se deficiency (Boyne and Arthur, 1985; Hogan *et al.*, 1990). Neutrophils of cows receiving additional Se either from the diet or direct injection had improved bactericidal activity (Weiss *et al.*, 1990).

Conclusions

Lactating cows supplemented with SEL-PLEX 50 (Se Yeast) had significantly higher serum Se concentrations than those supplemented with selenite at the completion of this trial. These data support previous work (Pehrson, 1993; Ullrey *et al.*, 1977) in which seleno amino acids and other organic Se compounds were demonstrated to be apparently more efficiently absorbed by ruminants than was selenite.

Milk Se concentrations of cows supplemented with Se-yeast were more uniform and slightly higher than those of selenite-supplemented animals. (Selenomethionine is the predominant form of organic Se in Se Yeast; thus, more of the absorbed selenomethionine would be available as a methionine substitute in the synthesis of milk proteins.)

Final serum Se concentration of individual animals was negatively correlated to the DHI-SCC LS. These results support those of previous studies (Erskine *et al.*, 1987, Weiss *et al.*, 1990) demonstrating that Se can play a key role in reducing the incidence of mastitis in commercial dairy herds.

Under current conditions in the US and other countries, the total quantity of Se is closely regulated for use in animal feeds. The potential adverse impact of animal excreta Se on the environment appears to be the main justification for current rigid regulation of this nutrient in animal rations. Since organic Se (the natural form in plants) is more efficiently absorbed than inorganic Se, less Se in this form would be excreted from the animal into the environment. However, the high cost of Se Yeast (organic Se) appears to preclude its widespread use as the sole supplemental Se source for animal diets. In situations where legitimate concern exists regarding suboptimal Se status of cattle, partial supplementation with organic Se (at 0.10 mg/kg dietary DM) would definitely have a positive impact on body Se reserves. Furthermore, no long-term detrimental health effects on animals would be anticipated at this level of supplementation since the natural or organic Se levels of ruminant diets in the central US would routinely exceed this level (Ullrey *et al.*, 1977; Maas *et al.*, 1980; Anonymous, 1983).

References

Ali, A.K.A. and G.E. Shook. 1980. An optimum transformation for somatic cell concentration in milk. J. Dairy Sci. 63:487.

Anderson, J.W. and A.R. Scarf. 1983. Selenium and plant metabolism. In: Metals and Micronutrients: Uptake and Utilization by Plants. D.A. Robb and W.S. Pierpoint (Eds). Academic Press, New York, NY, 241.

Anonymous. 1983. Distribution. In: Selenium in Nutrition. National Academy Press, Washington, DC, 10.

Atroshi, F., J. Tyopponen, S. Sankari, T. Kanganiemi and J. Parantainen. 1986. Possible roles of vitamin E and glutathione metabolism in bovine mastitis. Internat. J. Vit. Nutr. Res. 57:37.

Boyne, R., and J.R. Arthur. 1985. The effects of selenium deficiency on the function of neutrophils in cattle and peritoneal macrophages in rats. In: Proc. 5th International Symposium on Trace Elements in Man and Animals. C.F. Mills, I. Bremner, and J.K. Chesters (Eds.). CAB, Farnham Royal, UK, 123.

CAST. 1994. Risks and Benefits of Selenium in Agriculture, Issue Paper Suppl. No. 3. Page 8. CAST, Ames, IA.

Erskine, R.J., R.J. Eberhart, L.J. Hutchinson and R.W. Scholz. 1987. Blood selenium concentrations and glutathione peroxidase activities in dairy herds with high and low somatic cell counts. JAVMA 190 (11):1417.

Fisher, D.D. 1980. Use of dietary composition, body fluids and tissues to prognosticate and diagnose production diseases in ruminants. Ph.D., Diss., The Pennsylvania State University, University Pk, PA.

Guyton, A.C. 1961. Medical Physiology, 2nd edition. W.B. Saunders Co., Philadelphia, PA, 160.

Harrison, J.H., D.D. Hancock and H.R. Conrad. 1984. Vitamin E and selenium for reproduction of the dairy cow. J. Dairy Sci. 67:123.

Hogan, J.S., K.L. Smith, W.P. Weiss, D.A. Todhunter and W.L. Shockey. 1990. Relationships among vitamin E, selenium and bovine blood neutrophils. J. Dairy Sci. 73:2372.

Johansson, E. and U. Lindu. 1985. Comparison for models of selenium status expressed as plasma selenium, erythrocyte glutathione peroxidase or blood cell selenium. In Proc. 5th International Symposium on Trace Elements in Man and Animals. C.F. Mills, I. Bremner and J.K. Chesters (Eds). CAB, Farnham Royal, UK, 598.

Levander, O.A. 1986. Selenium. In: Trace Elements in Human and Animal Nutrition-2, 5th ed. W. Mertz (Ed.). Academic Press, Inc., New York, NY, 209.

Levander, O.A., and R.F. Burk. 1990. Selenium. In: Present Knowledge in Nutrition, 6th ed. II, SI, Washington, DC, 268.

Maas, R.W., F.A. Martz, R.L. Belyea and M.F. Weiss. 1980. Relationship of dietary selenium in plasma and milk from dairy cows. J. Dairy Sci. 63:532.

Minson, D.J. 1990. Forage in Ruminant Nutrition. Academic Press, Inc., New York, NY, 369.

NRC. 1989. Nutrient Requirements of Dairy Cattle, 6th revised ed. (Update). National Academy Press. Washington, DC.

Pehrson, B.G. 1993. Selenium in nutrition with special reference to the biopotency of organic and inorganic selenium compounds. In: Proc. Alltech's 9th Annual Symposium. T.P. Lyons (Ed.) Alltech Technical Publications, Nicholasville, KY, 171.

Reamer, D.C. and C. Veillon. 1983. Elimination of perchloric acid in digestion of biological fluids for fluorometric determination of selenium. Anal. Chem. 55:1605.

Scholz, R.W. and L.J. Hutchinson. 1979. Distribution of glutathione peroxidase activity and selenium in the blood of dairy cows. Am. J. Vet. Res. 40(2):245.

Smith, K.L., J.H. Harrison, D.D. Hancock, D.A. Todhunter and H.R. Conrad. 1984. Effect of vitamin E and selenium supplementation on incidence of clinical mastitis and duration of clinical symptoms. J. Dairy Sci. 67:1293.

Smith, K.L., J.S. Hogan, W.P. Weiss, W.L. Shockey and H.R. Conrad. 1988. Vitamin E for prevention of mastitis in dairy cattle. In: Proc. 3rd Animal Nutrition Forum on BASF, Lufwingshofen, 2.

Steele, R.G.D. and J.H. Torrie. 1960. Principles and Procedures of Statistics. McGraw-Hill Book Co., New York, NY.

Stowe, H.D. and T.H. Herdt. 1992. Clinical assessment of selenium status of livestock. J. Anim. Sci. 70:3928.

Stowe, H.D., J.W. Thomas, T. Johnson, J.V. Marteniuk, D.A. Morrow and D.E. Ullrey. 1988. Responses of dairy cattle to long-term and short-term supplementation with oral selenium and vitamin E. J. Dairy Sci. 71:1830.

Ullrey, D.E., P.S. Brody, P.A. Whetter, P.K. Ku and W.T. Magee. 1977. Selenium supplementation of diets for sheep and beef cattle. J. Anim. Sci. 46:559.

Van Saun, R.J. 1989. A rational approach to selenium supplementation in cattle. In: Proc. Cornell Nutr. Conf., Syracuse, NY, 113.

Weiss, W.P., J.S. Hogan, K.L. Smith and K.H. Hoblet. 1990. Relationships among selenium, vitamin E, and mammary gland health in commercial dairy herds J. Dairy Sci. 73:381.

Weiss, W.P., D.A. Todhunter, J.S. Hogan and K.L. Smith. 1993. Relationships between dietary vitamin E and selenium in dairy cows. In: OARDC-OSUE Dairy Sci. Res. Highlights, Ohio State University, 37.

Whetter, P.A. and D.E. Ullrey. 1978. Improved fluorometric method for determining selenium. J. Assoc. Off. Anal. Chem. 61:927.

A COMPARATIVE STUDY OF SELENITE AND SELENIUM YEAST (SEL-PLEX 50) AS FEED SUPPLEMENTS FOR MULTIPAROUS DAIRY COWS

BO PEHRSON and KERSTIN ORTMAN

Experimental Station, Veterinary Institute, Swedish University of Agricultural Sciences, Skara, Sweden

Introduction

Plants in many countries of the world are selenium deficient. As a consequence, selenium is often added to commercial feeds to prevent clinical and subclinical diseases in animals. Most often sodium selenite is used for this supplementation, but there is a growing interest in using organic selenium compounds as alternatives because of some disadvantages of selenite. The most obvious is its pro-oxidative effect, but also the fact that selenite is not naturally a feed ingredient favours organic selenium compounds (Pehrson, 1993).

Organic and inorganic selenium have in many respects different bioavailability and biological activity. Therefore, recommendations aimed at inorganic compounds must not be used uncritically for organic compounds. There is an urgent need for research work aimed at finding optimal supplemental levels of organic compounds for different animal species. The intention of the present investigation was to compare the metabolic pathways of selenium from selenite with those from an organic selenium yeast product in dairy cows.

Materials and methods

Thirty-one Swedish Red and White cows were divided into three groups as equal as possible with respect to age and lactation stage. The present article is based on results from 18 of these cows – all them multiparous. (More detailed results, including data also from primiparous cows, concentrations of selenium in blood and possible relations between selenium concentrations and lactation stage, will be published later.)

The cows were fed according to the official Swedish standards for metabolizable energy and protein in relation to their daily milk production. Before the experiment they also received 200 g daily of a commercial mineral feed which contained vitamins and trace elements, including 30 mg selenium/kg as sodium selenite. The level of selenium supplementation was thus 6.0 mg per cow daily. During the experimental period selenium was excluded from this mineral feed.

The experimental period was from November 1993 to September 1994. Group 1 (n=10) received a daily supplement of 3.0 mg of selenium per cow as sodium selenite; group 2 (n= 11) the same amount of selenium as selenium yeast (Sel-Plex 50, Alltech, Nicholasville, Kentucky), and group 3 (n=10) 0.75 mg selenium as Sel-Plex 50. All selenium supplements were mixed in glucose such that 15 ml contained a daily dose which was fed on the concentrate. The selenium contents of the other feedstuffs used were 0.04, 0.04, 0.02, 0.06, 0.06 and 0.52 mg/kg dry matter (DM) for hay, silage, grain, dried beet pulp, soya meal and commercial protein feed, respectively.

The mean selenium content of total feed DM was calculated to be 0.51 mg/kg during the pre-experimental period. During the trial the corresponding figures were 0.35 mg/kg for cows in groups 1 and 2 and 0.20 mg/kg for cows in group 3.

Jugular blood and milk samples were taken at the start of the trial and again after 50, 119, 163 and 273 days. Glutathione perosidase activity (GSH-Px) in erythrocytes was measured *ad modum* Paglia and Valentine (1967) and selenium in milk according to Sari *et al.* (1975). To avoid influence of a high selenium concentration in colostrum (Weiss *et al.*, 1984; Pehrson *et al.*, 1990) no milk samples from the first week after calving were included.

Results and discussion

The GSH-Px activity in erythrocytes was lower during the pre-experimental period than during the trial (Figure 1), indicating a reduced selenium bioavailability when selenite is mixed with other minerals. At the high selenite supplementation during the pre-experimental period – corresponding to 0.4 mg selenium/kg feed DM at a daily feed intake of 15 kg DM – the mean selenium concentration in milk was 15.3 μg/kg. Similar values (14.9–17.1μg/kg) were reached when selenite at a lower dosage level was added separately (Group 1).

The results preserited in Figure 2 show that – irrespective of the method of supplementation – selenium from Sel-Plex 50 was considerably more available than selenium from selenite for increasing the selenium content of milk. This is accordance with earlier reports (Conrad and Moxon, 1979; Aspila, 1991; Suoranta *et al.*, 1993).

Aspila (1991) proposed that a recommendable selenium concentration of milk for human consumption should be at least 20 μg/kg, a concentration which can easily be reached if an organic selenium is used to supplement a deficient diet. On the contrary, it does not seem possible to reach such a level without unacceptably high supplemental levels of selenite.

The basic diet for lactating cows in the present investigation contained about 0.15 mg selenium per kg DM. The supplemental levels were about 0.20 mg/kg for cows in groups 1 and 2 and about 0.05 mg/kg for cows in group 3. A comparison between Figures 1 and 2 indicates

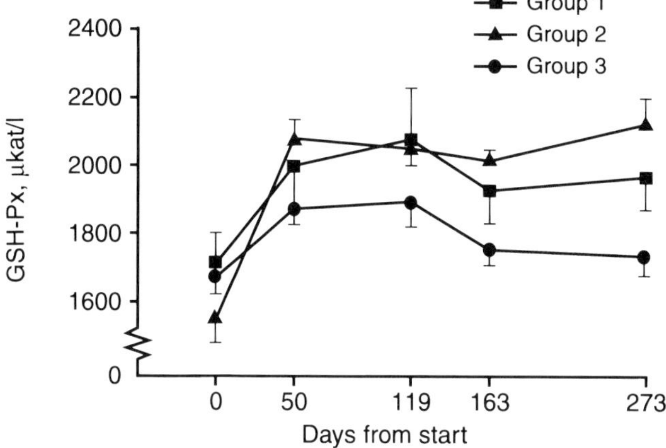

Figure 1. GSH-Px activity in erythrocytes ($\bar{X} \pm$ se) of lactating dairy cows after supplementation of the diet with 0.20 ppm Se from selenite (group 1), 0.20 ppm Se from Sel-Plex 50 (group 2) and 0.05 ppm Se from Sel-Plex 50 (group 3). The basic diet contained 0.15 ppm Se. During the pre-experimental period the basic diet was supplemented with 0.4 ppm Se from selenite as an ingredient of a commercial mineral feed.

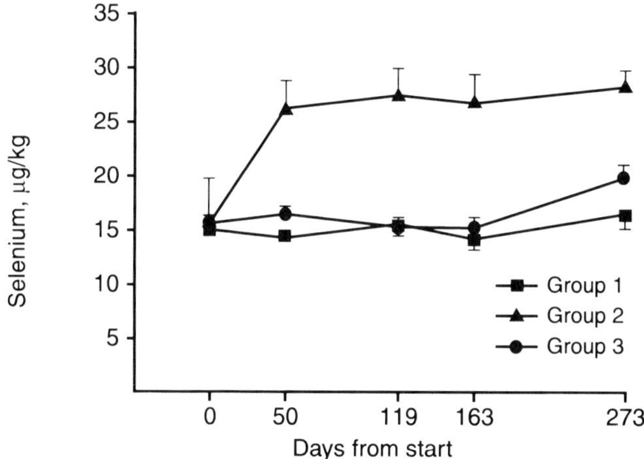

Figure 2. Selenium concentration in milk before and at different times during the experiment. Pre-experimental group = basic diet + 0.40 ppm from selenite in a commercial mineral feed. Group 1 = basic diet + 0.20 ppm Se from selenite given separately. Group 2 = basic diet + 0.20 ppm Se from Sel-Plex 50. Group 3 = basic diet + 0.05 ppm Se from Sel-Plex 50.

differences in the metabolism of selenium compounds depending on whether they will be used for GSH-Px synthesis or for increasing the selenium concentration in milk. Thus, the very low supplemental level of 0.05 mg/kg from selenium yeast in group 3 resulted in a slightly higher selenium concentration in milk but in a slightly lower GSH-Px activity in erythrocytes than the four times higher supplemental level from sodium selenite. At the higher supplemental level selenium yeast resulted both in a higher GSH-Px activity and in a higher selenium content in milk than sodium selenite.

Conclusions

The experiment verifies earlier results that organic selenium is a much more efficient means to increase the selenium content of milk than sodium selenite. Also the conclusion made earlier (Pehrson, 1993) that organic compounds might be preferable as feed supplements and can be used at lower dosage levels than inorganic compounds is verified, even if an additional level of just 0.05 mg selenium/kg DM from selenium yeast might be a bit too low for dairy cows.

References

Aspila, P. 1991. Metabolism of selenite, selenomethionine and feed-incorporated selenium in lactating goats and dairy cows. J. Agric. Sci. Finland. 63:1–74.

Conrad, H.R. and A.L. Moxon. 1979. Transfer of dietary selenium to milk. J. Dairy Sci. 62:411.

Paglia, D.E. and W.N. Valentine. 1967. Studies on the quantitative and qualitative characterization of erythrocyte glutathione peroxidase. J. Lab. Clin. Med. 70:158–169.

Pehrson, B. 1993. Selenium in nutrition with special reference to the biopotency of organic and inorganic selenium compounds. Proceedings of Alltech's Ninth Annual Symposium. Alltech Technical Publications, Nicholasville, Kentucky, USA. pp. 71–89.

Pehrson, B. J. Hajjarainen and L. Blomgren. 1990. Vitamin E status in newborn lambs with special reference to the effect of dl-a-tocopheryl acetate supplementation in late gestation. Acta Vet. Scand. 31:359–367.

Sari, E., D. Siemmer and L. Hageman. 1975. Modified atomic absorbtion spectrophotometric determination of selenium in foodstuffs by hydride generation. Anal. Letters 8:323–337.

Suoranta, K., E. Sinda and R. Pihlak. 1993. Selenium of the selenium yeast enters the cow's milk. Norw. J. Agric. Sci. Supplement No 11:215–216.

Weiss, E.P., W.P. Colenbrander and M.D. Cunningham. 1984. Maternal transfer and retention of supplementation selenium in neonatal calves. J. Dairy Sci. 67:416–420.

TRACE MINERALS AND FERTILITY IN DAIRY CATTLE

PAUL E. JOHNSON

Veterinary Consultant, Enterprise, Alabama, USA

Introduction

The interaction of nutrition on reproductive efficiency is a complex and ill-defined relationship that has received much attention in the literature in recent years. The current economic trend in the dairy industry will require the dairy manager to become more efficient in all aspects of milk production. Paramount in this process will be the need to maintain and increase the longevity of the cattle in the herd. Reproductive performance affects milk production per cow per day of herd life, total number of replacements available to the farm, and the ability to invoke voluntary vs involuntary culling procedures.

Communicable diseases which have limited reproductive performance in past years have recently been less of a problem than nutritional imbalances which may have occurred in response to increase milk production and its interaction with environmental influences.

Nutritional implications concerning reproductive health are often hard to diagnose and many times involve complex nutrient interactions.

Actual nutrient intake is hard to calculate and can vary due to the source of feedstuffs being fed. Digestion and absorption of nutrients can vary according to environmental influences, mineral interactions, and nutrient composition. Even though there is some ambiguity to certain nutrient relationships with to reproductive performance, those with documented references in the literature will be discussed.

The microminerals most used in organic complexes include zinc, iodine, manganese, cobalt, and copper. It should be noted that macrominerals such as calcium, phosphorus, magnesium, and sulfur are important in reproductive physiology. The interactions of macrominerals with certain inorganic trace minerals can lead to complexes that are not bioavailable to the animal. The advent of organic sources of trace minerals has averted these imposed deficiencies. Organic trace mineral supplements have been shown to improve serum and tissue levels of these micronutrients as compared to the inorganic sources when complex interactions occur. The use of organic trace minerals in cattle diets is indicated to maintain adequate levels of copper, iodine, zinc, cobalt, and manganese.

The implication of organic trace minerals for the improvement of reproduction in dairy cattle has been discussed in the literature. The discussion of organic trace mineral supplementation for enhancing the immune system of the bovine has been reviewed extensively. Improvement of the immune system has a secondary implication in reproductive performance. Maintaining adequate immunocompetence improves vaccination response and disease prevention.

Copper

Copper has been implicated in health related problems that have been associated with immunosupression in cattle. The impact of copper deficiency is on enzyme systems and it thus affects the killing capacity of phagocytes. The cytochrome oxidase and copper-zinc superoxide dismutase enzymes are most affected by copper deficiency (Nockels, 1992).

Copper is required for hemoglobin formation and is important in many enzymatic reactions in metabolic function. Low hemoglobin can lead to anemia and this will result in poor conception rates and silent heats (Nockels, 1992). An increase in inactive ovaries and a greater incidence of retained placenta has been associated with reduced daily intakes of copper. Copper metabolism can be disrupted by excess intake of dietary iron and/or sulfur, as well.

Copper requirements can vary with age, stage of lactation, and the molybdenum content of the ration. A ratio of 4:1 of to copper molybdenum is advised. A ratio of cobalt, copper, and iron should be 1:10:100 respectively. Excess copper intake can lead to liver dysfunction and subsequent reduced fertility due to ill health and weight loss.

Serum copper is the traditional means by which copper analysis has been determined in the bovine. Serum copper only reflects severe deficiencies while copper content of liver tissue is a more accurate means to ascertain copper status of the animal. Superoxide dismutase and ceruloplasmin are copper containing proteins which can offer an alternative means to serum copper measurements (Nockels, 1992).

Copper supplementation must be closely monitored to avoid toxicity. Younger animals are more susceptible and high intakes of copper can lead to death.

Iodine

Iodine is required in the synthesis of thyroid hormones. The impact of iodine on fertility relates to thyroid and hormonal sequences. Manifestation of iodine deficiency seen in the reproduction system is delayed puberty, cessation of estrus and heat activity, and periods of anestrus (McDowell, 1985).

Adult cattle fed iodine deficient diets may have weak or stillborn calves at birth. Some cattle may abort and have an increased incidence of retained placenta. Cattle on diets such as kale or rape should receive extra iodine supplementation due to the goitrogenic effects of these forages. Excessive iodine intake has been reported as a cause of abortion, especially in the first three months of pregnancy. Excess iodine intake can also lead to immunosuppression (McDowell, 1985).

Zinc

The impact of low zinc intake has a profound effect on the male bovine. Zinc deficiency can lead to delayed testicular development in young bulls while adults may experience testicular atrophy. The seminiferous tubules atrophy and spermatogenesis ceases.

Adult females may have reduced fertility with zinc deficient diets. The exact cause is unknown, but reproductive failure precedes such signs as reduced growth rates and parakeratosis. Zinc is an important trace mineral in immune function and a deficiency of zinc can contribute to communicable disease outbreaks in the herd. Zinc metabolism is affected by iron and sulfur intake. Zinc is an important mineral in certain enzyme reactions which affect immunoglobulin formation. Clinical zinc deficiencies can affect vaccination response and lead to increased reproductive disease situations.

Serum zinc is used to diagnose zinc deficiencies; but as with copper, only prolonged deficiency can result in low serum zinc values. Alkaline phosphatase is a serum enzyme with zinc incorporated in its structure. It can also be used to measure zinc status in affected animals. When collecting serum samples for zinc status do not use red top serum tubes because there are soaps containing zinc which are used to clean the tubes. Only royal blue top vacuum blood tubes can be used.

Cobalt

The cobalt requirement of ruminants is actually a cobalt requirement of the rumen organisms. The rumen microorganisms incorporate cobalt into vitamin B_{12}, which is utilized by both microorganisms and animal tissues. The main sources of energy for ruminants are acetic and propionic acids. Vitamin B_{12} is a primary agent in the utilization of propionic acid.

The impact of cobalt deficiency is manifested as a disorder of propionate metabolism which can be indistinguishable from energy and protein deficiency. The impact of cobalt on reproductive efficiency will be reflected as anestrus, increased retained placenta, or any of a multitude of conditions that can be related to weight loss in cattle (McDowell, 1985).

Cobalt in the organic form is often substituted for a portion of the daily requirement of lactating cattle during periods of heat stress. The proposed effect of this supplementation is to increase appetite (vitamin B_{12} effect) and improve animal performance from the increased conversion of propionic acid.

Cobalt toxicity is rare and a safety margin up to 5 ppm is recognized.

The need for cobalt in cattle diets is dependent on cobalt content of the soils where the cattle graze or on which their forages were grown. Improving reproductive performance with cobalt supplementation is strictly through improved body condition with subsequent improvement in hormonal synthesis (McDowell, 1985).

Manganese

Manganese is needed for many enzymatic functions in the bovine. Manganese is required for luteal metabolism. Deficiencies of manganese can lead to anestrus, delayed onset of estrus post-calving, poor conception rates, and ovulation failure. Reduced birth weights, abortions, and stillborn calves have been associated with reduced manganese intake. Birth defects of the limbs including twisted legs and enlarged joints can be manifested with manganese deficient diets. Manganese deficiency in the male results in sterility (Gerloff and Morrow, 1986).

Manganese metabolism is dependent on iron intake and a deficiency could result even though adequate levels are present. Manganese supplementation should be utilized in cases where corn silage diets are fed. Corn silage is often low in manganese content (Gerloff and Morrow, 1986).

Conclusion

As milk production increases in today's dairies, there comes a need to ensure that ration formulation meets the trace mineral requirements of the dairy cow. Feeding strategies of the past will not function effectively in the modern dairy and attempts to resist change in feeding practices will lead to nutritional stresses on cattle which will affect reproductive health of the herd.

Nutrition is only one cause of reproductive failure, however, and one should not neglect poor estrus detection which is probably more important as a cause of reduced breeding performance than nutritional imbalance. Other causes of reproductive failure include environmental stress, sanitation of the calving area, vaccination status, and improper insemination technique. It should be noted that high milk production will not affect breeding performance if the nutrient requirements of the animal are met. If and when nutritional imbalances are implicated as a source of breeding inefficiency, often it can be recognized that management has failed to monitor and meet the nutritional requirement in question.

The use of organic sources of trace minerals is indicated when complex interactions among inorganic minerals and other nutrients occur. Improving reproductive performance with organic trace minerals will be accomplished by improving the immune status of the animal. Direct influence of organic sources of trace minerals on reproduction will be accomplished through improved ovarian activity and post-calving uterine involution and tone.

Many feeding practices, although convenient to the dairyman, can compromise nutrient status of the cow and it should be the dairyman's goal to meet the animal's requirement through good feeding management.

References

Gerloff, B. and D. Morrow. 1986. Effect of Nutrition on Reproduction in Dairy Cattle. Current Therapy in Theriogenology, pp. 317–319.

McDowell, L.R. 1985. Nutrition of Grazing Ruminants in Warm Climates.

Nockels, C.F. 1992. Mineral alterations associated with stress, trauma, and the effect on immunity. The Compendium 100:1133–1139.

THE EFFECT OF PROTEINATED MINERALS ADDED TO THE DIET ON THE PERFORMANCE OF POST-PARTUM DAIRY COWS

D. O'DONOGHUE, P.O. BROPHY, M. RATH and M.P. BOLAND

Faculty of Agriculture, University of College Dublin, Lyons Research Farm, Newcastle, Co. Dublin, Ireland

Summary

An experiment was carried out using spring-calving dairy cows fed proteinated minerals from 2 weeks pre-partum until about 80 days post-partum. There was no effect of treatment on milk production or composition. Treated cows had a slightly earlier development post-partum of the first dominant follicle and ovulated earlier (20 vs 25 days) and had a higher conception rate. Mean somatic cells counts were significantly lower in treated than in control cows.

Introduction

Mastitis is a significant management challenge to dairy producers. Nutritional factors related to immune status include protein/energy nourishment, trace mineral and vitamin nutrition (Spain, 1993). Selenium supplementation increased somatic cell response to intramammary infusion of *Escherichia coli*. Erskine *et al.* (1987) demonstrated lower blood selenium concentrations in cows with high SCC compared to cows with low SCC. Kellogg (1990) reported that chelated zinc decreased SCC and increased milk production. Perhaps epithelial cell integrity may have been involved. Epithelial cell integrity of the teat canal has been shown to be linked with mastitis prevention. Bitman *et al.* (1991) describe the keratin lining of the teat canal as a physical and chemical barrier for protection of the mammary gland. Keratin lining may physically trap bacteria and prevent migration into the mammary gland.

Objectives

Previous research from Teagasc (Fallon *et al.*, 1993) indicated higher fertility in heifers supplemented with a combination of trace mineral proteinates and yeast culture (Bio-Boost Plus, Alltech Ireland Ltd.)

before superovulation. In addition, a preliminary trial with a small number of autumn-calving lactating cows demonstrated that Bio-Boost reduced somatic cell counts; however fertility was unaffected. It is possible that addition of proteinated trace mineral supplements may have beneficial effects on reproductive parameters either by acting to enhance oocyte maturation via paracrine effects or via the endocrine route. Therefore, this experiment was designed to examine the role of proteinated minerals in post-partum dairy cows.

Materials and methods

ANIMALS AND DIETS

Fifty six spring calving Friesian cows were selected 2 weeks pre-partum and were paired on the basis of calving date, parity and milk yield (previous lactation). Each pair was randomly allocated to either control or Bioplex treatment groups. The supplement, called Bioplex Dairy (10 g/head/day, Table 1), was fed to 28 cows and a further 28 cows acted as controls. Two control animals and five treated animals did not complete the trial, leaving 49 (26 control and 23 treatment cows). Bioplex Dairy was pre-mixed with 100 grams of rolled barley and fed in the parlor at milking time. Control cows were offered a similar quantity of rolled barley daily. The feeding period commenced 14 days before estimated calving date and continued until 12 weeks post-partum or 30 days after first insemination, whichever was later. The calving period extended from the 15th of January to the 30th of April. Cows were housed indoors in straw bedded cubicles with a slatted passageway. Grass silage of approximately 70% dry matter digestibility (DMD) was offered *ad libitum* in an easy feeding system. Cows were milked twice daily at 06.00 hours and at 15.30 hours and were offered 3.2 kg of UCD Dairy 18 (Table 2) at each milking. Cows were turned out to pasture on the 26th of April and 2 weeks later were reduced to 0.9 kg of concentrate (Pasture Milkmore) at each milking.

Table 1. Bioplex trace mineral supplement.

Cu, mg/day	100 mg
Zn, mg/day	250 mg
Se, mg/day	2 mg

MEASUREMENTS

Milk yield was measured weekly for the first 12 weeks of lactation on 2 consecutive days. Milk composition (fat % and protein %) was determined on weeks 2, 4, 6, 8, 10 and 12 of lactation. Somatic cell counts were determined at fortnightly intervals. Fertility, as measured

Table 2. Chemical composition of the concentrate mixtures.*

U.C.D. Dairy 18		Pasture Milkmore	
Protein	18.2%	Protein	14.1%
Oil	3.0%	Oil	5.0%
Fiber	7.2%	Fiber	10.5%
Ash	8.8%	Ash	11.5%
Moisture	12.1%	Moisture	14.1%
Magnesium	0.91%	Magnesium	1.50%
Vitamin A	7.50 kiu/kg	Vitamin A	22.50 kiu/kg
Vitamin D_3	1.50 kiu/kg	Vitamin D_3	4.50 kiu/kg
Vitamin E	15.00 kiu/kg	Vitamin E	15.00 kiu/kg
Copper	73 mg/kg	Copper	205 mg/kg
Selenium	0.83 mg/kg	Selenium	2.43 mg/kg
Avoparcin	12.5 mg/kg	Avoparcin	37.5 mg/kg

*Analysis of rations as quoted by Waterford Foods

by conception to first service, was determined by ultrasound scanning at 25–30 days after first service. Days to first ovulation and follicular dynamics were determined by ultrasound scanning in 10 control and 10 treated cows which were scanned daily from 7 days post-partum until first ovulation. Time to first progesterone (P4) rise was determined by taking blood samples once per week and assaying the serum for progesterone concentrations. A rise above 0.5 ng/ml was indicative of an active corpus luteum and hence cyclicity. Blood copper, zinc and selenium levels were measured at day 0 (14 days pre-estimated calving date), day 35, and day 70.

Results and discussion

MILK PRODUCTION

There was no effect of treatment on milk production or composition in the spring calving herd, however there was a significant reduction in somatic cell count (by 40%) which is similar to that observed in the previous experiment (Table 3, Figure 1). Bioplex Dairy was fed for 4 weeks before the first somatic cell count measurements were taken and thus data must be treated with care in that treatment groups were not necessarily balanced for somatic cell counts at initiation of the trial.

Table 3. Milk production data for herd treated with Bioplex Dairy.

	Control	Bioplex Dairy
Number of animals	26	23
Average daily yield, kg	24.5	24.75
Fat, %	3.18	3.11
Protein, %	3.04	3.00

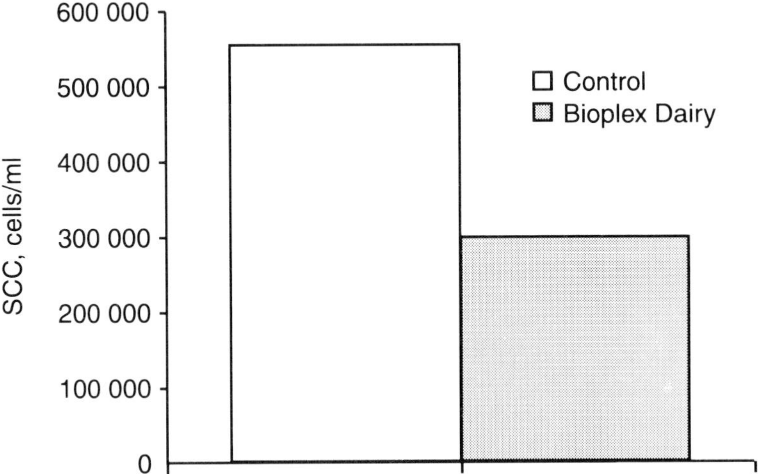

Figure 1. Effect of Bioplex Dairy on somatic cell counts of lactating cows.

REPRODUCTIVE PARAMETERS

Cows treated with Bioplex Dairy had a non-significant reduction in days to emergence of the first dominant follicle and to first ovulation (Table 4). Conception rate of first service, while satisfactory in both groups, was higher in the treated group. Results from this component indicate that Bioplex minerals may have a beneficial role in dairy cow reproduction in that the cows ovulated sooner after calving and conception rates to first service may be increased in cows which ovulate soon after calving.

BLOOD MINERAL CONTENT

Mineral status, determined in blood collected from cows at days 0, 35 and 70 of feeding, indicated normal mineral status. This suggested that Bioplex minerals have a beneficial effect in cows that have normal blood mineral concentrations (Table 5).

Table 4. Effect of organic mineral supplementation on reproduction in dairy cows.

	Control	Bioplex
Days to 1st dominant follicle (DF)	9.3	7.8
% which ovulated 1st DF	60	60
% which ovulated 2nd DF	10	30
% which ovulated 3rd DF	20	10
% which ovulated 4th DF	10	–
Conception rate, %	57.7	65.2
Days post-partum to ovulation	25.3 ± 3.1	20.4 ± 1.4
Days to first service	75.4 ± 6.1	68.8 ± 3.8

Table 5. Effect of Bioplex supplementation on blood mineral status of Cu, Zn and Se.

	Copper, µmol/l			Selenium (GPx), IU/g Hb			Zinc, µmol/l		
Day	0	35	70	0	35	70	0	35	70
Control	12.05	12.13	13.54	109.15	130.89	129.1	15.09	12.84	14.27
Bioplex	12.44	13.95	13.21	105	112.15	132.49	15.27	14.35	14.3

References

Bitman, J., D.L. Wood, S.A. Bright, R.H. Miller, A.V., Capuco, A. Roche and J.W. Pankey. 1991. Lipid composition of teat canal keratin collected before and after milking from Holstein and Jersey cows. J. Dairy Sci. 74:414.

Erskine, R.J., R.J. Eberhart, I.J. Hutchinson and R.W. Scholz. 1987. Blood selenium concentrations and glutathione peroxidase activities in dairy herds with high and low somatic cell counts. J. Amer. Vet. Med. Assoc. 178:704.

Fallon, R.J., R. Matovani, J.F. Roche and M.P. Boland. 1993. Effect of proteinated minerals and yeast culture on fertilization in superovulated heifers. Ir. J. Agric. Food Res. 32:111.

Kellogg, D.W. 1990. Zinc methionine affects performance of lactating cows. Feedstuffs 62:35.

Spain, J. 1993. Tissue integrity: A key defence against mastitis infection: The role of zinc proteinates and a theory for mode of action. In: Biotechnology in the Feed Industry. Proc. Alltech's 9th Sym. Alltech Technical Publications, Nicholasville, Kentucky, 53–60.

THE EFFECT OF FEEDING ZINC PROTEINATE TO LACTATING DAIRY COWS

BARNEY HARRIS, Jr.

Professor Emeritus, University of Florida, Gainesville, USA

Introduction and objectives

Supplementation with zinc (Zn) proteinate has been associated with reduced somatic cell counts (SCC) in milk and improved hoof condition in confined dairy and beef cattle. The changes in SCC have been particularly marked in herds with initially high somatic cell counts associated with contagious mastitis problems; however reductions due to organic Zn supplementation have been noted in herds with low SCC ($\leq$ 200 000 cells/ml) as well. The objectives of this field trial were to examine effects of zinc proteinate on SCC and mobility score in a commercial herd previously on inorganic zinc.

Procedures

The field trial was conducted at a 320-cow (Holstein) dairy in North Florida in late fall of 1994 and early winter of 1995. This herd had recently been moved to a new free-stall facility which would mean additional time spent on a cement surface. Prior to entering the new free-stall barn, the herd received zinc sulfate ($ZnSO_4$) in a balanced diet for a period of 6 months. The cows were divided into two groups of about 70 cows per group. The two groups averaged 120 days in milk at the beginning of the 90-day study. The zinc proteinate (Bio-Plex Zn, Alltech Inc.) was weighed and added daily to the total mixed ration (TMR) being fed to one group at a rate of 400 mg Zn per cow per day. The TMR contained corn silage, alfalfa hay, soybean meal, cottonseed, hominy feed, corn distillers grains, and a mineral-vitamin premix. Each group had access to round bales of fair quality bermuda grass hay in an outside corral. Groups were fed twice daily in the barn and milked three times daily. Milk production, composition and somatic cell counts were recorded at 0, 30, 60 and 90 days as part of the testing procedure of the Dairy Herd Improvement Association.

Table 1. Effects of zinc supplementation upon milk production, fat percentage and somatic cell count.

	Zinc proteinate				Control			
Time (days)	0	30	60	90	0	30	60	90
3.5% FCM (kg/d)	29.1	29.7	32.7	31.0	27.0	24.7	24.5	25.4
Fat, %	3.12	3.02	3.33	3.24	3.49	3.13	2.75	3.0
SCC (×1000 cells/ml)	169.0	102.0	128.0	129.0	220.0	329.0	285.0	300.0

Results

During the 90-day study, fat percentage changed from 3.12 to 3.24 in the zinc proteinate group and from 3.49 to 2.99 in the control group (Table 1). The somatic cell count declined from 169 000 to 129 000 cells/ml (40 000 less) in the zinc proteinate group whereas in the control group there was an increase from 220 000 to 300 000 (80 000 more). While milk production changes were similar for both groups, the changes in fat percentage resulted in an increase of 1.86 kg/day fat-corrected milk (FCM) for the zinc proteinate group and a decrease of 1.64 kg FCM/day for the control group. Milk protein percentage values were similar for both groups. In addition, the dairyman commented that there were fewer cows with feet or leg problems in the group fed Bioplex Zn.

Conclusions

Antioxidant nutrients such as vitamin E and the trace minerals selenium, copper, zinc and manganese in enzymes are very important in protecting tissues from oxidative destruction. Zinc is an essential component of enzyme systems of the body and is an integral component of keratin, the fibrous protein in hoof, hair and horn. In this study the addition of zinc proteinate in the diet of lactating dairy cows reduced somatic cell counts and improved production performance.

ORGANIC CHROMIUM IN GROWING PIGS: OBSERVATIONS FOLLOWING A YEAR OF USE AND RESEARCH IN SWITZERLAND

CASPAR WENK

Institute for Animal Sciences, Zürich, Switzerland

Introduction

Chromium (Cr) is an essential trace element needed in only very small amounts to meet the requirements of man and animals. It is present in small quantities in almost all feedstuffs; and in general nutrient recommendations it is not taken into consideration. Due to the interaction with carbohydrate and therefore insulin metabolism, Cr is increasingly discussed in both human and animal nutrition.

The essentiality of Cr was discovered by Schwarz and Mertz in 1959. They discovered the function of the Cr-bearing 'glucose tolerance factor' (GTF). The essentiality of Cr is well documented. The following physiological observations have been made with several animal species and man under Cr deficiency (Anderson, 1987 and Gokel *et al.*, 1986):

- reduced glucose tolerance, hyperglycemia, glucosuria;
- reduced amounts of circulating insulin;
- increased plasma lipid concentrations;
- reduced growth, reduced life span.

Chromium exists in various forms. Only Cr (III) is biologically active and absolutely non-toxic. Cr (IV) itself is toxic, but can be transformed to Cr (III) in the digestive tract of an animal. In nutritive doses metallic Cr is not toxic, but is not absorbed at all. Compared with Cr chloride ($\approx$2–3% absorbed), organic Cr sources such as Cr yeast are absorbed 5- to 10-fold better (Anderson and Kizlovsky, 1985). The availability of Cr in the digestive tract can be increased with the supplementation of vitamin C (Wang *et al.*, 1985), niacin (Urberg *et al.*, 1986) or some animo acids (Mertz and Roginski, 1971). This is an indication that Cr is absorbed in complexes like chelates.

Mertz *et al.* (1974) demonstrated the positive effect of Cr on insulin and therefore on carbohydrate metabolism. They concluded that it could also have a positive effect on protein synthesis. Seerley (1993) found that protein deposition could be increased and fat deposition reduced when Cr is supplemented. The faster the absorption of glucose and amino acids from the blood into muscle and other organs, the more efficient is liver metabolism. Evock-Clover *et al.* (1993) found furthermore a clear interaction between Cr supplementation and the use of somatotropin in growing pigs, indicating a contribution of Cr to the partitioning between protein and fat deposition. Finally Mowatt (1993) demonstrated that growth parameters of stressed animals supplemented with organic Cr were beneficially affected.

The supplementation of pig diets with Cr in the form of Cr yeast has been permitted since January 1994 in Switzerland. The fact that Cr is essential, in addition to the positive influence of Cr on protein and fat deposition of growing pigs, have proven to be persuasive arguements for its use and have already led to wide use of Cr yeast in the first months. One major argument for rapid introduction was the positive outcome of the following experiment carried out in 1993 at our institute.

The influence of Cr in different forms on growth performance and some physiological parameters was investigated with growing pigs. The experimental design is shown in Table 1 (Wenk *et al.*, 1995).

Table 1. Experimental treatments.

Treatment	Cr supplement
1	Control (no Cr)
2	0.5 ppm Cr chloride
3	0.5 ppm Cr yeast†
4	0.5 ppm Cr picolinate*

† Dr K. Guatschi, University Hospital of Zurich, CH

* Alltech Inc. Kentucky, Nicholasville, USA 'Sel-Plex 50'

Materials and methods

Chromium in the form of Cr chloride, Cr yeast or Cr picolinate was added to a basal diet (Table 2) at 0.5 ppm. All other nutrients including the trace elements were added in amounts to meet requirements. The nutrient content of the four experimental diets was very consistent with the exception of crude protein. Digestible energy content was 14.7 MJ/kg DM and crude protein content about 190 g/kg DM. The Cr content of the four experimental diets is given in Table 3.

The unpelleted basal diet contained an average of 1.39 ppm Cr. Only dried yeast, which was included at 2.5%, might have contained some organic Cr. Further processing (mixing and pelleting) caused an increase of about 0.55 ppm Cr. It can be assumed that this increase did not influence the amount of available Cr in the basal diet. Supplementation with the different Cr sources caused a further increase of between 0.3 ppm (Cr chloride) and 0.8 ppm (Cr picolinate). Analytical differences are assumed to be the result of the small quantities of Cr added and the difficulty of the sample collection and analysis.

Table 2. Composition and nutrient content of the experimental diets.

	Basal	Cr source		
		Cr chloride	Cr yeast	Cr picolinate
Composition of the experimental diets (%)				
Barley	40.0			
Wheat	21.5			
Middlings	5.0			
Dried potatoes	7.5			
Soybean meal	10.5			
Sunflower meal with hulls	2.5			
Dried yeast	2.5			
Meat and bone meal	2.5			
Molasses	3.0			
Bone fat	2.0			
Celite 545	1.0			
Minerals, amino acids, premix *	2.0			
Nutrient content of the experimental diets (g per kg dry matter)				
Dry matter (%)	88.8	88.6	88.4	88.1
Organic matter	802	808	803	803
Crude protein	216	194	189	190
Crude fiber	53	52	55	53
Gross energy, MJ	18.3	18.2	18.2	18.0
Digestible energy†	14.7	14.7	14.7	14.5

*Per kg feed: lysine-HCl 3 g, DL-methionine 0.6 g, Cu 10 mg, Fe 30 mg, Zn 60 mg, Mn 40 mg, I 1 mg, Se 0.3 mg, Vit.A. 10 000 IE, Vit.D3 1000 IE, Vit.E. 40 mg, Vit.K_3 1.5 mg, Vit.B_2 4 mg, Vit.B_6 4 mg, Vit.B_{12} 0.015 mg, biotin 0.1 mg, niacin 20 mg, Ca-pantothenate 15 mg, folic acid 0.3 mg.
†Measured in the experiments.

Table 3. Cr content of the basal diet and the pelleted experimental diets.

Cr supplement	Basal diet		Cr chloride	Cr yeast	Cr picolinate
Physical form of the feed	(mash)	(pelleted)	(pelleted)		
Cr, ppm	1.39	1.87	2.17	2.33	2.64

Forty female pigs of the Swiss Landrace breed housed in individual pens were distributed over the four treatments. Initial weight was 27 kg. Slaughter weight was 106 kg bodyweight (BW). Feed was offered according to a diet plan (about 105 g feed per kg $BW^{.75}$) based on weekly weighing of the pigs. Additionally, digestibilities of energy and organic matter were measured using the indicator method (HCl-unsoluble ash, celite 545 as indicator carrier) at body weights of 41 and 73 kg.

Statistical analyses were carried out with Statgraphics (version 5). The results are given as mean values with the corresponding standard deviations (if not differently indicated then with 10 single values). The statistically significant differences between treatments ($P \leq 0.005$) were calculated with the range-test method of Bonferroni.

Results and discussion

PERFORMANCE

In the grower period body weight gain in all treatments was very uniform with a mean of 666 g per day and a feed conversion ratio of 2.37 kg feed per kg gain (Table 4). No effect of treatment on growth rate and feed conversion was noted. In the finishing period increased growth rates were observed in the treatments with Cr supplementation. Owing to restricted feeding, feed conversion ratio was also positively affected by the Cr supplements.

Lindemann *et al.* (1993) did not observe a positive effect of Cr picolinate on growth performance during the whole experimental period (15–102 kg BW). On the other hand Page *et al.* (1992) and Page *et al.* (1993) found an increased weight gain with 0.2 ppm Cr picolinate but not with Cr chloride.

Table 4. Growth performance of the pigs with different Cr supplements.

		Cr supplement		
	Basal	Cr chloride	Cr yeast	Cr picolinate
Body weight, kg				
Start	27.5±1.3	27.6±0.7	27.3±1.4	27.2±1.1
Middle	61.1±1.7	60.4±2.0	60.6±1.8	60.9±1.8
End	106.5±2.3	107.3±2.8	107.1±1.6	105.0±2.5
Feed intake, kg/day	2.07±0.03	2.04±0.03	2.05±0.04	2.02±0.03
Body weight, kg/day				
Grower period	667±23	671±38	673±42	653±43
Finishing period	829±56[a]	916±52[b]	875±47[ab]	878±86[ab]
Relative, %	100	111	106	106
Whole fattening period	752±31	795±35	775±25	763±60
Relative, %	100	106	103	102
Feed conversion ratio, kg/kg				
Grower period	2.37±0.10	2.35±0.11	2.36±0.12	2.38±0.16
Finishing period	3.05±0.22[b]	2.73±0.17[a]	2.86±0.15[ab]	2.88±0.28[ab]
Relative, %	100	90	94	95
Whole fattening period	2.75±0.12	2.58±0.12	2.65±0.09	2.66±0.21
Relative, %	100	94	96	97

CARCASS COMPOSITION

Carcass composition and back fat thickness were not influenced significantly by the Cr supplements (Table 5). The subjective classification of the carcass gave generally good results. Higher scores were achieved with the Cr supplements (especially Cr yeast and Cr picolinate) compared with the control treatment. Because of the wide variation these differences were not significant. However in our inquiries at

Table 5. Effect of Cr source on carcass composition and intramuscular fat content of the longissimus dorsi muscle.

		Cr supplement		
	Basal	Cr chloride	Cr yeast	Cr picolinate
Dressing percentage (%)				
Warm	75.0±1.87	75.4±1.6	76.2±1.5	74.8±1.7
Subjective classification*	2.1±0.6	2.2±0.4	2.3±0.5	2.4±0.5
Relative	100	105	110	114
Percentage of cold carcass (%)				
Lean cuts	50.7±1.1	51.1±1.9	51.8±1.9	50.9±1.4
Back	19.4±0.4	19.5±0.9	20.0±1.2	19.5±0.7
Shoulder	9.3±0.5	9.4±0.4	9.1±0.5	9.2±0.3
Ham	22.0±0.8	22.2±0.8	22.8±0.6	22.2±1.0
Belly	19.3±0.7	19.2±1.1	18.8±0.9	19.3±0.8
Fat layer above back	6.0±0.5	6.2±0.6	6.1±0.9	6.3±0.5
Back fat thickness (cm)				
Sacral region†	1.6±0.3	1.5±0.2	1.7±0.3	1.6±0.3
Back‡	1.8±0.3	1.9±0.3	2.0±0.3	1.9±0.3
Longissimus dorsi				
Area, cm†	49.5±4.5	51.2±6.7	53.8±4.2	48.6±5.1
Relative §	100	102	106	99
Intramuscular fat (%)	1.45±0.3	1.61±0.44	1.58±0.60	1.87±0.58

*Subjective classification at slaughter: 1= low meat content, 2= normal, 3= high meat content
†Thinnest part in the sacral region
‡Thinnest part in the middle of the back between two whirls
§To cold slaughter weight, %

Swiss pig units this observation was always the argument to add Cr to the diets.

Intramuscular fat content should not be reduced as a result of increased growth performance if meat quality is to be maintained (Bejerholm and Barton-Gade, 1986; Wenk, 1991). Therefore the fat content of the longissimus dorsi at the 10th rib was measured. Mean fat content was found to be 1.6%. Page *et al.* (1992 and 1993) found, in agreement with Lindemann *et al.* (1993), a reduced back fat thickness and increased surface area of the longissimus dorsi muscle at the 10th rib in response to organic Cr. We did not observe reduced fat deposition, but did find a trend toward increased muscle surface particularly in the Cr yeast treatment. In additional studies (Wenk *et al.*, 1995) we found no effect of Cr supplements on fatty acid composition of the complex and neutral lipids of intramuscular and adipose lipids in the longissimus dorsi. We therefore concluded that effects of supplemental Cr would not be expected to have a negative effect on meat quality.

ENERGY DIGESTIBILITY

Energy utilization as measured by calculating digestible energy was unaffected by Cr supplementation though the increase in growth rate

Table 6. Digestibility of feed energy and amount of digestible energy required per kg body weight gain.

		Cr supplement		
	Basal	Cr chloride	Cr yeast	Cr picolinate
Digestibility of energy				
At 41 kg body weight*	0.801±0.009	0.810±0.013	0.807±0.017	0.802±0.012
At 73 kg body weight*	0.812±0.012	0.809±0.009	0.810±0.014	0.808±0.011
Total	0.806±0.0011	0.810±0.009	0.809±0.009	0.805±0.009
Digestible energy (DE) needed per kg body weight gain, MJ				
DE per kg dry matter	14.7	14.7	14.7	14.5
DE per kg body weight	39.0	36.2	36.6	37.3
Relative, %	100	93	94	96

*5 animals per treatment

noted during the finishing period numerically lowered the energy required for gain (Table 6).

Summary

An experiment with 40 growing pigs from 27.4 to 106.5 kg body weight in individual cages was conducted to evaluate the effect of different Cr supplements (Cr chloride, Cr yeast and Cr picolinate) at 0.5 ppm Cr in the diet compared with a control diet without any additional Cr. The influence on growth performance and carcass as well as meat composition was studied.

Organic chromium supplementation tended to improve weight gain in the finishing period. In contrast to previous work, addition of Cr chloride was also effective. Carcass composition and the fatty acid profile of neutral and complex lipids in the longissimus dorsi muscle at the 10th rib were unaffected. Furthermore, energy utilization as evaluated by digestibility was not affected by Cr supplementation. The Cr supplements tended also to increase longissimus dorsi area while maintaining the intramuscular fat content needed to preserve meat quality.

Acknowledgments

Cr yeast was received from Alltech, Inc., Kentucky USA, and all other Cr sources by Dr K. Gautchi, University Zürich. Dr R. Anderson, Human Nutrition Research Center, Beltsville, MA did the Cr analysis in the diets. The experiments were conducted at the experimental farm 'Bühl' at Hendschken of UFA. The evaluation of the experiments was done by Stefan Gebert in his graduate thesis. All the contributions are highly appreciated.

References

Anderson, R.A. and A.S. Kozlovsky. 1985. Cr intake, absorption and excretion of subjects consuming self-selected diets. Am. J. Clin. Nutr. 41:1177.

Anderson, R.A. 1987: Cr. In: Trace Elements in Human and Animal Nutrition Vol. 1. Mertz (Ed.), Academic Press, NY, p.225.

Bejerholm, C. and P.A. Barton-Gade. 1986. Effect of intramuscular fat level on eating quality of pig meat. 32nd Europ. Congr. of Meat Research Workers; Gent.

Evock-Clover, C.M., M. Polansky, R.A. Anderson and N.C. Steele. 1993. Dietary Cr supplementation with or without somatotropin treatment alters serum hormones and metabolites in growing pigs without affecting growth performance. J. Nutr. 123:1504–1512.

Gokel, E.M., M. Kirchgessner and H.P. Roth. 1986. Alimentär induzierter Chommangel bei wachsenden Ratten. J. Anim. Physiol. Nutr. 56:251.

Lindemann, M.D., C.M. Wood, A.F. Harper and E.T. Kornegay. 1993. Cr picolinate additions to diets for growing finishing pigs. J. Anim. Sci. 71:14.

Mertz, W. and E.E. Roginski. 1971. Newer trace elements in nutrition. Dekker, NY.

Mertz, W., E.W. Töpfer, E.E. Roginski and M.M. Polansky. 1974. Present knowledge of the role of Cr. Fed. Pro. 33:2275.

Mowat, D.N. 1993. Organic Cr: New nutrient for stressed animals. The Feed Compounder 9.93.3pp.

Page, T.G., C.C. Southern, T.C. Ward, J.E. Pontif, T.C. Sinder and D.C. Thompson. 1992. Effect of Cr picolinate on growth, serum and carcass traits and organ weights of growing-finishing pigs from different ancestral sources. J. Anim. Sci. 70:235.

Page, T.G., C.C. Southern, T.C. Ward and D.C. Thompson. 1993. Effect of Cr picolinate on growth and serum and carcass traits of growing finishing pigs. J. Anim. Sci. 71:656.

Schwarz, K. and W. Mertz. 1959. Cr (III) and the glucose tolerance factor. Arch. Biochem. Biophys. 85:292.

Seerley, R.W. 1993. Organic Cr and manganese in human nutrition: important possibilities for manipulating lean meat deposition in animals. In: Biotechnology in Feed Industry. Proceedings of the 9th Symposium T.P. Lyons (Ed.) Alltech Technical Publications, Nicholasville, Kentucky, 41–51.

Urberg, M., M. Parent, D. Mill and M. Zemel. 1986. Evidence for synergism between Cr and nicotinic acid in normalizing glucose tolerance. Diabetes 35:37a.

Wang, M.N., Y.C. Ci, K. Odalut and B.J. Stoecker. 1985. Cr and ascorbate deficiency effects on serum cholesterol, triglycerides and glucose of guinea pigs. Fed. Proc. 44:751.

Wenk, C. 1991. Fütterung und Schweinefleischqualität. In: Schriftenereihe aus dem Institut für Nutztierwissenschafter, Gruppe Ernährung, ETH Zürich (Eds). Heft 5:23.

Wenk, C., S. Gebert and H.P. Pfirter. (1995). Chromzulagen zum Schweinemastfutter: Eingluss auf Wachstum und Fleischqualität (Cr supplements in the feed for growing pigs: Influence on growth and meat (quality). Archive for Anim. Nutr. (in press).

CHROMIUM DEFICIENCY IN FIRST PARITY COWS

D.N. MOWAT, A. SUBIYATNO and W.Z. YANG

University of Guelph, Canada

It took until the 1990s for animal nutritionists to recognize the need for supplementation with organic chromium (Cr), particularly during stress periods (Chang and Mowat, 1992; Page *et al.*, 1993). However, a few animal producers were aware of some practical benefits of supplemental Cr in the early 1970s, capitalizing on clues in the human literature.

There are both physiological and dietary circumstances that indicate a need for supplemental Cr in the diet as a result of either primary deficiency or depletion (Table 1). Stress depletes body stores of Cr. In humans, pregnancy has been shown to predispose Cr deficiency due to placental transport and increased urinary losses (Anderson, 1994). Additionally, high carbohydrate diets or diets which contain interfering substances increase the need to supplement Cr.

No test is presently available to specifically diagnose Cr status. A method that has been commonly used to evaluate deficiency of Cr

Table 1. Circumstances which dictate a need for supplemental chromium in the diet.

Chromium-deficient diet

- Low level of feeds rich in bioavailable chromium.
- High level of chromium-deficient feeds.
- High level of feeds grown on low chromium soils.
- High level of interfering minerals, iron and/or zinc.
- Low level or low rumen synthesis of dietary precursors of bioavailable chromium (e.g. amino acids, niacin).

Chromium depletion

- High intake of chromium-robbing diets (e.g. high simple sugars or lactose, high propionate or insulin-producing, supplemental fat).
- Aging
- Heat stress
- Pregnancy
- Lactation
- Acute exercise
- Physical trauma
- Obesity
- Other stresses

in human nutrition and glucose metabolism in animals is the glucose tolerance test.

First parity cows deficient in Cr

In our glucose tolerance tests conducted two weeks prepartum, first parity cows appeared to be particularly deficient in Cr as shown by markedly decreased plasma insulin, insulin to glucose ratio, triglyceride and Cr with Cr supplementation (Figure 1; Subiyatno *et al.*, 1993). Little, if any, effect on these parameters occurred during late pregnancy with multiparous cows. This confirmed previous studies which suggested that first parity cows were more insulin resistant than multiparous cows during late pregnancy.

In tests conducted two weeks postpartum during negative energy balance, supplemental Cr appeared to decrease insulin sensitivity with first parity cows (Figure 1). This increase in insulin resistance, if correct, is opposite to the effects of supplemental Cr during prepartum.

Supplemental organic Cr increased milk yield by 11% during the first 14 weeks of lactation with first parity cows (Figure 2) (Mowat *et al.*, 1995). No effect on milk production or composition with multiparous cows was obtained. However, a reduction in metabolic disorders may occur in aged cows fed supplemental Cr. Despite the increased milk production observed in first parity cows, no adverse effects on reproduction were clearly established.

First parity cows may be more stressed than mature cows, nutritionally, psychologically and physiologically. When group housed, younger animals have to compete for available bunk space with older larger animals. First parity animals are attempting to adapt to new phenomena of pregnancy and lactation, new social interaction and often new surroundings or environment. In addition, first parity animals need to direct some nutrients for growth in order to achieve mature size. Finally, first parity cows appeared to be beginning lactation already depleted in Cr stores. However, supplemental Cr may only be required for first parity cows during the heavy stress period of late gestation and early lactation. Moreover, as dry matter intake increases to maximum at 10–12 weeks postpartum, Cr intake accordingly increases. Also, adaptive mechanisms may eventually lead in first parity cows to increased concentration of Cr in-body stores or redistribution of Cr in specific tissues as shown with trained athletes (Anderson, 1994).

Cr and insulin resistance during early lactation

Increased milk production in first parity cows fed supplemental Cr could be explained by the proposed increased insulin resistance during the negative energy balance of early lactation, probably due to hormonal manipulations. Recently, a marked increase occurred in maternal as well as fetal plasma concentrations of IGF-1 in pregnant gilts periodically

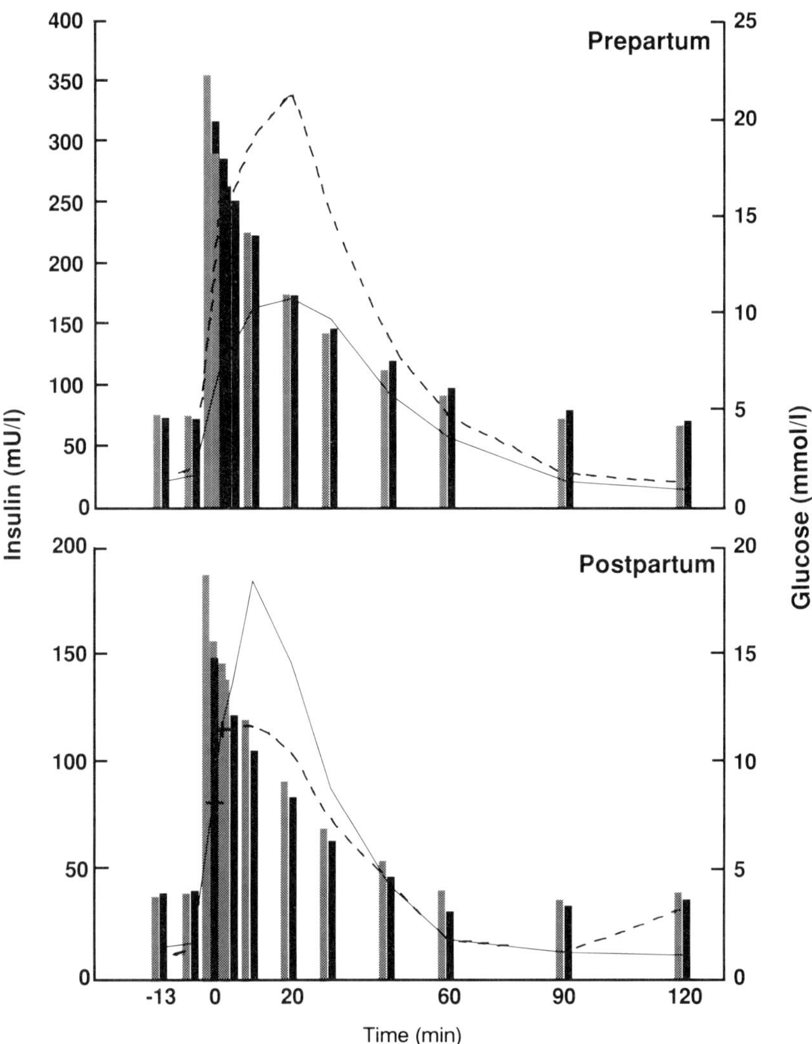

Figure 1. Plasma insulin (lines) and glucose (bars) after i.v. glucose challenge in primiparous cows fed either control (broken lines and light bars) or chromium-supplemented diets (solid lines and dark bars) prepartum and postpartum (Subiyatno *et al.*, 1993).

injected with chelated Cr (Okere, 1994). Furthermore, recent propionate-loading tests in early lactating first parity cows showed a tendency with supplemental Cr towards increased serum IGF-1 levels (Yang and Mowat, 1994). If supplemental Cr increases plasma IGF-1, it may cause the suggested increased resistance noted during early lactation. IGF-1 can bind to insulin receptors in peripheral tissues without necessarily producing its biological effects. This latter possibility would be similar to mechanisms proposed for bST action (NRC, 1994). Although we

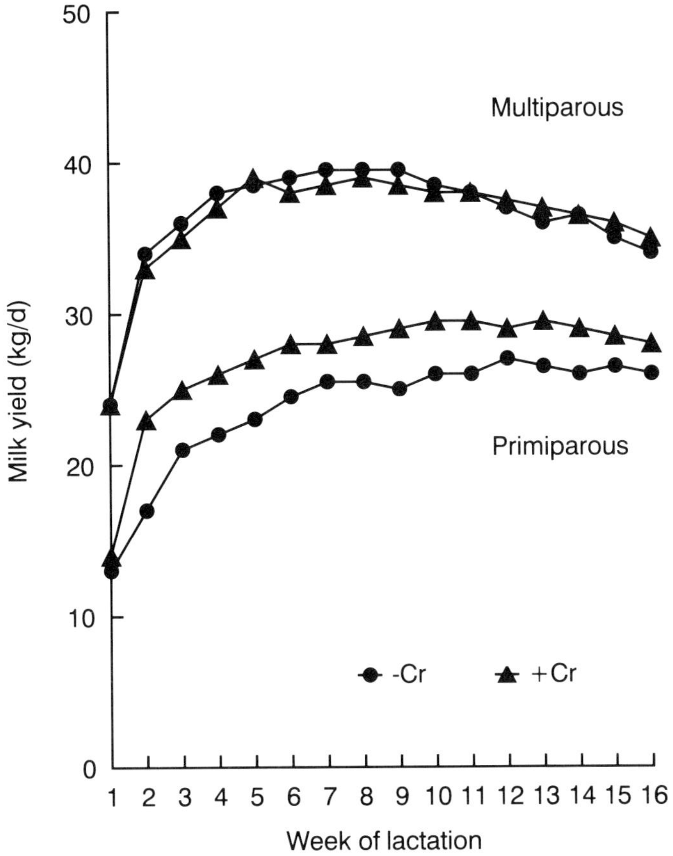

Figure 2. Effect of chromium supplementation on milk yield during early lactation over two experiments (Mowat *et al.*, 1994).

do not have a clear understanding of how the IGF complex is able to mediate mammary function, it is apparent that the changes in circulating concentrations of IGF-1 are closely tracking the biological events and magnitude of milk responses that occur with bST treatment.

Recently we showed that Cr supplementation dramatically increased serum glucose peak by 73% for week 2 and 34% for week 6 postpartum in first parity cows following propionate infusion (Figure 3). It was speculated that this may be due to increased gluconeogenesis caused by insulin resistance. However, more refined or authoritative metabolism techniques are needed to verify the proposed insulin resistance (i.e., hyperinsulinemic/euglycemic clamps) and hormonal manipulations with supplemental Cr during early lactation.

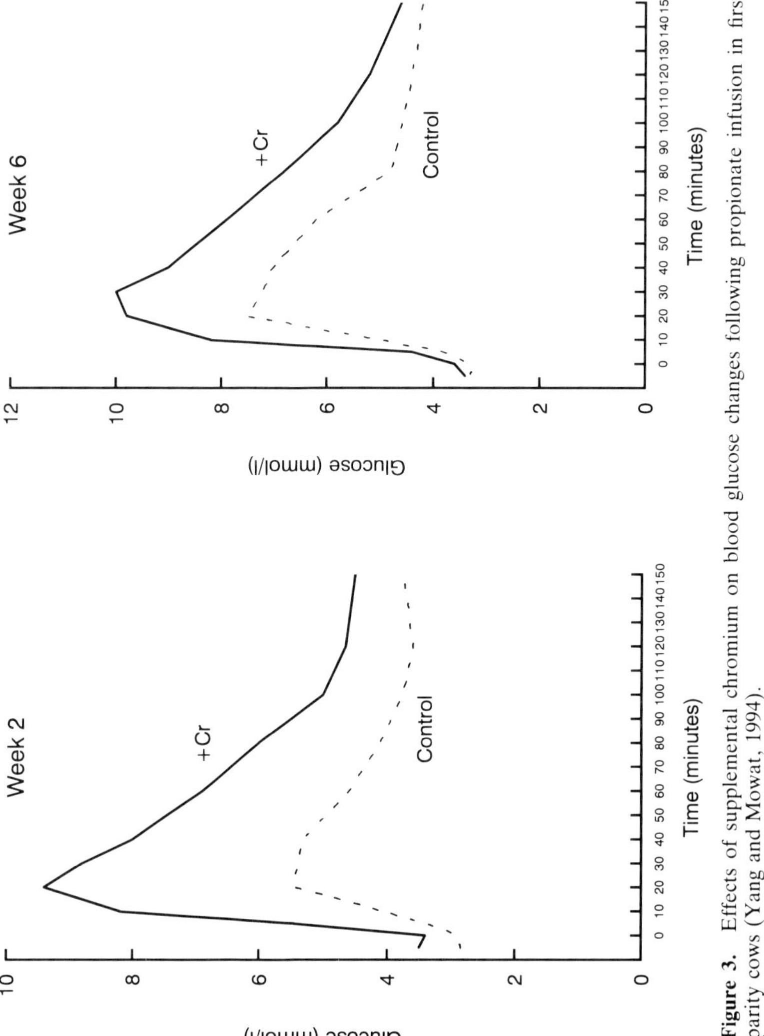

Figure 3. Effects of supplemental chromium on blood glucose changes following propionate infusion in first parity cows (Yang and Mowat, 1994).

Conclusions

First parity cows appear to be particularly vulnerable to suboptimal Cr nutrition during the critical and stressful period of late pregnancy and early lactation. Supplemental organic Cr increased milk yield by 11% during the first 14 weeks of lactation with first parity cows. Verification of proposed increased insulin resistance during early lactation with supplemental Cr is needed along with clarification of hormonal manipulations.

References

Anderson, R.A. 1994. Stress effects on chromium nutrition of humans and farm animals. In: Proc. Alltech's Tenth Ann. Symp. T.P. Lyons and K.A. Jacques (Eds). Nottingham University Press, Loughborough, 267–274.

Chang, X. and D.N. Mowat. 1992. Supplemental chromium for stressed and growing feeder calves. J. Anim. Sci. 70:559.

Mowat, D.N., W.Z. Yang, A. Subiyatno and J.J. van der Broek. 1995. Effects of chromium supplementation on early lactation performance of Holstein cows. J. Dairy Sci. (Submitted).

NRC. 1994. Metabolic Modifiers. Effects on the Nutrient Requirements of Food-Producing Animals. National Academy Press, Washington, D.C.

Okere, C. 1994. Reproductive and endocrine responses of gestating gilts to hormone [PST and insulin] or micronutrient supplementation. Ph.D. Thesis. University of Guelph, Ontario, Canada.

Page, T.G., L.L. Southern, T.L. Ward and D.L. Thompson. 1993. Effect of chromium picolinate on growth and serum and carcass traits of growing-finishing pigs. J. Anim. Sci. 71:656.

Subiyatno, A., W.Z. Yang, D.N. Mowat and G.A. Spiers. 1993. Chelated chromium alters plasma metabolite responses to glucose infusion in dairy cows. Proc. ASDA Ann. Mtg.

Yang, W.Z. and D.N. Mowat. 1994. Supplemental chromium on gluconeogenesis in lactating cows following propionate infusion. Proc. ADSA/ASAS Ann. Mtg.

COPPER PROTEINATE MAY BE ABSORBED IN CHELATED FORM BY LACTATING HOLSTEIN COWS

Z. DU, R.W. HEMKEN and T.W. CLARK

Department of Animal Sciences, University of Kentucky, Lexington, Kentucky, USA

Introduction

Copper (Cu) availability can be significantly reduced by dietary antagonistic elements such as iron (Fe), molybdenum (Mo) and zinc (Zn) (Bremner *et al.*, 1987). High levels of these Cu-antagonistic elements can be ingested by farm animals when soils are consumed during grazing or as contaminants in silage. As a result, Cu deficiency can develop. To prevent this adverse effect, Cu supplementation is often needed. However, increasing the amount of inorganic Cu added to a ration to compensate for a deficiency could lead to Cu toxicity (Smart *et al.*, 1992). Since different forms of copper can react differently with copper antagonists, they can be absorbed and(or) utilized with varying efficiency. Providing copper in a readily available form which is not markedly influenced by other elements would be a logical approach to overcome these problems. Previous studies have shown that absorption rate of copper in the form of copper amino acid chelate was much higher than that in the inorganic forms for rats (Kirchgessner and Grassmann, 1970; Du *et al.*, 1993; 1994). However, the results of studies on availability of copper chelates for ruminants are not consistent. Some studies have reported that copper availability of Cu chelates was higher than that of inorganic copper, especially under stressful situations (Kincaid *et al.*, 1986; DeBonis and Nockels, 1992). Others obtained contradictory results (Wittenberg *et al.*, 1990; Ward *et al.*, 1993; Kegley and Spears, 1994). Very little information is available comparing the bioavailability, absorption and metabolism of copper chelates with copper sulfate ($CuSO_4$) in the presence of high levels of dietary iron. The objective of this experiment was to determine whether copper proteinate and cupric sulfate were differently metabolized.

Materials and methods

ANIMALS AND EXPERIMENTAL TREATMENTS

Sixteen lactating Holstein cows (35 days post-calving) were randomly assigned to one of four treatments in a 2 × 2 factorial design with

5 mg Cu/kg feed DM supplementation from either $CuSO_4$ or from Cu proteinate (Bioplex Cu, Alltech, Inc.). Iron was supplemented at either 0 or 1000 mg Fe per kg feed DM as $FeSO_4$. The cows were individually fed a total mixed ration of corn silage, alfalfa silage and concentrate at the ratio of 7 : 6 : 5.6 (as fed basis) *ad libitum*. The basal diet contained 6.9 mg Cu and 630 mg Fe per kg DM of feed and met 1989 NRC recommendations for other nutrients. The iron supplement was mixed with the concentrate and copper was supplemented individually after the first liver biopsy. The experimental period was 90 days.

SAMPLING PROCEDURES AND ANALYTICAL METHODS

Liver samples were taken via biopsy and blood samples were collected via jugular vein on days 0 to 90 of the experiment. Liver and plasma samples were analyzed for Cu and Fe concentration.

Plasma ceruloplasmin (Cp) oxidase activity was measured with o-dianisidine dihydrochloride as substrate described by Schosinky *et al.* (1974). The procedure was modified by stopping the reaction at 10 and 20 min rather than 5 and 15 min.

Results and discussion

Because of technical difficulties, liver samples were not obtained from some experimental animals. The number of liver samples for each

Table 1. Effects of Cu sources and Fe levels on liver Cu and Fe, plasma Cu, Fe and ceruloplasmin

Fe supplementation	0 ppm Fe		1000 ppm Fe		C.V.
Cu source	$CuSO_4$	Cu-P†	$CuSO_4$	Cu-P†	%
Day 0					
Number of cows*	3	3	4	3	
Hepatic Cu (μg/g DM)	462	318	314	365	37.3
Hepatic Fe (μg/g DM)	273	274	262	255	24.6
Plasma Cu (μg/ml)	1.16	1.36	1.25	1.12	20.8
Ceruloplasmin (U/l)	19.4	20.8	24.8	19.8	31.0
Day 90					
Number of cows†	2	1	3	3	
Hepatic Cu (μg/g DM)	356	265	228	235	35.5
Hepatic Fe (μg/g DM)‡§	217	448	509	584	11.5
Plasma Cu (μg/ml)	0.59	0.59	0.63	0.54	11.5
Ceruloplasmin (U/l)¶	18.9	13.1	19.8	14.9	27.1

*Number of cows for hepatic Cu and Fe samples. For other blood parameters, there were four animals sampled per treatment.
†Cu-P is abbreviation of Cu proteinate.
‡$P<0.01$ for Fe levels.
§$P<0.05$ for Cu sources.
¶$P=0.055$ for copper sources.

treatment is presented in Table 1. Hepatic and plasma Cu concentrations were unaffected by either Cu source or Fe supplementation level during the experiment (Figures 1 and 2, $P<0.05$). Supplementation with Cu proteinate increased hepatic iron content (516 vs 363 mg/kg dry weight) on day 90 compared with $CuSO_4$ (Figure 1, $P<0.05$), suggesting that Cu proteinate did not interfere with absorption of Fe when compared with $CuSO_4$.

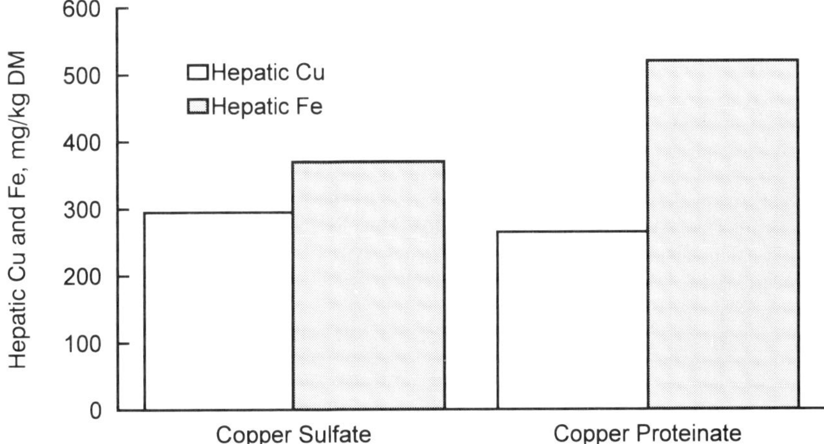

Figure 1. Hepatic Cu and Fe content on day 90 in cows fed copper proteinate and $CuSO_4$.

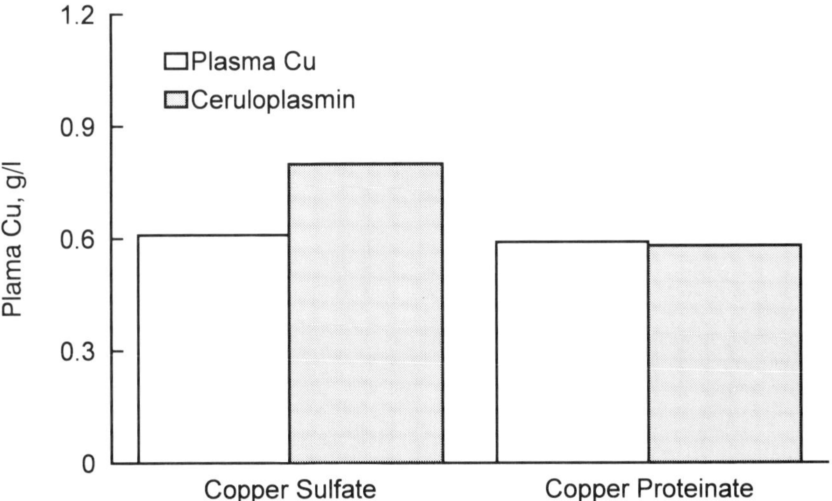

Figure 2. Plasma Cu concentration and ceruloplasmin on day 90 in cows fed Cu proteinate and $CuSO_4$.

On day 90, the cows fed Cu proteinate had lower plasma ceruloplasmin activity (Figure 2, 14.0 vs 19.4 U/l) than cows fed $CuSO_4$ (P=0.055), while plasma Cu was the same for Cu proteinate and $CuSO_4$ treatments (P>0.05). This indicated that the ratio of the ceruloplasmin form of plasma Cu to other forms of plasma Cu was lower for cows fed Cu proteinate than for cows fed $CuSO_4$ and suggested that Cu proteinate was absorbed in the organic form and transported in blood without binding to ceruloplasmin.

Implications

The results of this study suggest that Cu proteinate may be absorbed in the organic form and be transported in blood without binding to ceruloplasmin. This is based on firstly that cows fed Cu proteinate had higher liver Fe at day 90 than cows fed $CuSO_4$; and secondly that plasma ceruloplasmin activity was lower for cows fed Cu proteinate than cows fed $CuSO_4$, while no difference appeared in plasma Cu concentration at day 90. This needs to be further verified as no differences were observed in liver and plasma Cu content in this study. Additionally, other supportive data such as fate of the chelates in the rumen and the various forms of Cu in blood and possibly other tissues were not known.

References

Bremner, I., W. R. Humphries, M. Phillippo, M. J. Walker and P. C. Morrice. 1987. Iron-induced copper deficiency in calves: dose-response relationships and interaction with molybdenum and sulphur. Br. Anim. Prod. 45:403–414.

DeBonis, J. and C. F. Nockels. 1992. Stress induction affects copper and zinc balance in calves fed organic and inorganic copper and zinc sources. J. Anim. Sci. 70(Suppl. 1):314.

Du, Z., R. W. Hemken and T. W. Clark. 1993. Effects of copper chelates on growth and copper status of rats. J. Dairy Sci. (Suppl.) 76:306.

Du, Z., R. W. Hemken and S. Trammell. 1994. Comparison of bioavailabilities of copper in copper proteinate, copper lysine and cupric sulfate and their interaction with iron. J. Anim. Sci. (Suppl.) 72:273.

Kegley. E. B. and J. W. Spears. 1994. Bioavailability of feed-grade copper sources (oxide, sulfate or lysine) in growing cattle. J. Anim. Sci. 72:2728–2734.

Kincaid, R. L., R. M. Blauwiekel and J. D. Cronrath. 1986. Supplementation of copper as copper sulfate or copper proteinate for growing calves fed forages containing molybdenum. J. Dairy Sci. 69:160–163.

Kirchgessner, M. and E. Grassmann. 1970. The dynamics of copper absorption. In: C. F. Mills (Ed.) Trace Element Metabolism in Animals, Proc. WAAP/IBP Int. Symp. pp 277–287. Aberdeen, Scotland.

SAS. 1990. SAS User's Guide: Statistics. SAS Inst. Inc., Cary, NC.

Schosinsky, K. H., H. P. Lehmann and M. F. Beeler. 1974. Measurement of ceruloplasmin from its oxidase activity in serum by use of o-dianisidine dihydrochloride. Clin. Chem. 20:1556–1563.

Smart, M.E., N.F. Cymbaluk, D. A. Chistensen. 1992. A review of copper status of cattle in Canada and recommendation for supplementation. Can. Vet. J. 33:163–170.

Ward J. D., J. W. Spears and E. B. Kegley. 1993. Effect of copper level and source (copper lysine vs copper sulfate) on copper status, performance, and immune response in growing steers fed diets with or without supplemental molybdenum and sulfur. J. Anim. Sci. 71:2748–2755.

Wittenberg, K. M., R. J. Boila and M. A. Shariff. 1990. Comparison of copper sulfate and copper proteinate as copper sources for copper-depleted steers fed high molybdenum diets. Can. J. Anim. Sci. 70:895–904.

ALLZYME PHYTASE: THE SWISS EXPERIENCES

STEFAN GEBERT and CASPAR WENK

Institute for Animal Sciences, Zürich, Switzerland

Introduction

Phosphorus (P) is an essential element with more known functions in the body than most other mineral elements. In addition to its vital participation in the development and maintenance of skeletal tissue, it also controls many metabolic processes (NRC, 1980).

Feeds for pigs and poultry principally contain ingredients of plant origin. A traditional Swiss ration for growing and fattening pigs includes 60–80% cereals (mainly barley) and cereal by-products such as wheat bran and middlings. Available P in these ingredients is not sufficient to obtain good performance and may negatively affect certain aspects of animal husbandry (Jongbloed, 1987). Therefore, additional inorganic phosphorus is supplied.

It has been recognized that excretion of P can lead to environmental problems. This is predominantly in areas with a high density of pigs and poultry or in areas under delicate environmental conditions like in some regions of Switzerland. Application of manure in large quantities in certain areas leads to accumulation of P in the soil together with leaching and run-off. The result is eutrophication of ground water and surface water resources. To minimize environmental pollution, excretion of P in faeces and urine of livestock should be reduced as much as possible and must be a central aim of feed formulation.

First of all, the supply of P in the feed should be in agreement with the animal's requirement. Therefore adequate knowledge of the digestibility of P in the feed ingredients is needed, as well as the requirement known for P at any stage and type of production. As a second step, P excretion can be reduced by enhancement of P digestibility.

Digestible phosphorus in feedstuffs

Until recently the P availability in feedstuffs of plant origin was assumed to be about 30–35%. Current evidence indicates, however, a wide range in P digestibility among plant-derived feeds. Phosphorus availability in

Table 1. Apparent digestibility of P in feedstuffs for pigs.

Feedstuff	Digestibility of P (% of total) (mean ± sd)
Barley	39 ± 4
Wheat	47 ± 2
Wheat bran	41 ± 11
Maize	17 ± 5
Maize gluten feed	20 ± 6
Soybean meal extract	37 ± 1
Sunflower meal extract	16 ± 1
Meat meal	74 to 85
Meat and bone meal	80
Monocalcium phosphate	83 ± 4
Dicalcium phosphate (dihydrate)	69 ± 3

ingredients of animal origin (DLG, 1987) and in feed phosphates (NRC, 1988) is often assumed to be 95–100%. However, even these feedstuffs have recently been shown to have a wider range of P digestibility than previously thought. In Table 1 the apparent P digestibilities of some typical ingredients for pig diets are listed (Jongbloed, 1987; Jongbloed and Kemme, 1990).

These data suggest that the digestibility of P in feedstuffs of plant origin varies substantially. Some of the factors which cause P digestibility to vary among feedstuffs include origin (plant or animal), the concentrations of phytate and total P, and the presence of phytase (activity of plant phytase). The digestibility of P in maize and sunflower seed meal was found to be less than 20%, therefore lower than the generally accepted value. In the case of wheat (phytase-rich) and some legumes these values are largely underestimated. The apparent P digestibility in feedstuffs of animal origin and of inorganic phosphates is high. A high availability and a substantial contribution to the supply of P for pigs can be expected in feeds containing such ingredients. Meat and bone meal or other by-products from the slaughter and meat industry as well as calcium phosphates are cheap and popular ingredients for pigs. However since bovine spongiform encephalopathy (BSE) has been also manifested in our country, feedstuffs of animal origin are no longer allowed in feed formulation for ruminants in Switzerland.

Effect of microbial phytase

The low digestibility of P in cereal grains is due to the fact that about two-thirds of it is found in the form of poorly soluble salts of phytic acid (myoinositol hexaphosphates, phytases), which have very limited digestibility/availability for pigs. In contrast to ruminants, which take advantage of phytase produced by ruminal microorganisms, non-ruminants lack phytase, the enzyme that cleaves phosphate groups from the phytate molecule.

In several experiments it has been shown that the addition of microbial phytase to diets fed growing pigs enabled up to 50% reduction in P excretion (Jongbloed, 1989; Kessler and Egli, 1991; Näsi, 1991). Further studies have shown that microbial phytase also improved growth rate and feed conversion ratio (Larimier *et al.*, 1994). Additionally, the increased degradation of phytate by added exogenous phytases caused not only a better utilization of P but also of other minerals such as calcium (Ca), magnesium (Mg), iron (Fe) and zinc (Zn) (Pallauf *et al.*, 1992).

Since most commercial phytases contain not only phytase activity but also fairly high amounts of carbohydrase and protease, a higher digestibility of energy and other macronutrients can be expected with phytase supplementation (Wenk *et al.*, 1993a). On the other hand, effects of exogenous enzymes on organic matter and energy digestibility of diets rich in dietary fiber are uncertain (Wenk *et al.*, 1993b).

In the last few years we have investigated supplementation of microbial phytase to pig diets at our institute. The influence on growth performance and blood parameters as well as the apparent digestibility of energy, nitrogen and some minerals has been investigated. The latest project is not yet finished and will last about two more years.

EXPERIMENTS 1 AND 2: *AD LIBITUM* VS RESTRICTED FEEDING OF PHYTASE AND CARBOHYDRASE-SUPPLEMENTED DIETS

Materials and methods

The treatments employed and the basal diets were similar in Trials 1 and 2 (Table 2). The diet contained 30% barley, 20% wheat, 34% wheat bran, 6% soybean meal, 3.5% potato protein, 2% animal fat, minerals and vitamins. Wheat bran was included as a source of dietary fiber and phytic P. Feed phosphates were not added to the diets.

Table 2. Enzyme treatments.

Treatment	Basal	Carbohydrase	Phytase	Phytase + carbohydrase
Carbohydrase	–	+	–	+
Phytase	–	–	+	+

Eight male castrates (Large White) with an initial body weight (BW) of 45 kg were used in the 1st experiment. The pigs were kept in individual pens and fed restricted diets (105 g per kg $BW^{0.75}$). Water was offered *ad libitum*. Two animals were assigned to each treatment and rotated over four periods in a Latin square design with two replications. The digestibilities of organic matter, energy, nitrogen, P and Zn were measured with the indicator method (HCl-insoluble ash as indicator, celite 545 as indicator carrier). The second experiment used 24 individually penned growing pigs (male castrates, Swiss Landrace) from 25 to 100 kg BW. These animals were fed *ad libitum*. In addition to performance parameters, nutrient digestibility was estimated with the indicator method at 30, 60 and 90 kg BW for each animal individually. Both experiments were concluded without any disturbances in normal performance.

The nutrient content of the diets (per kg dry matter; organic matter 916 g, crude protein 180 g, crude fiber 68 g, digestible energy 13.4 MJ and P 8.3 g) was identical in both experiments, except for the phytase activity (e.g. Table 3).

Table 3. Phytase activity (Units (U) per kg dry matter).

Treatment	Basal	Carbohydrase	Phytase	Carbohydrase/phytase
Carbohydrase	–	+	–	+
Phytase	–	–	+	+
1st experiment, units	2050	1770	2910	1900
2nd experiment, units	2270	1230	3380	3020

The phytase activity was unexpectedly high in all diets. Nevertheless, a substantial increase could be seen in both the phytase and carbohydrase/phytase (only in Trial 2) treatments with phytase supplementation. The basal diet contained a quite considerable phytase activity obviously due to the high amount of wheat bran. In any case addition of the carbohydrase reduced the phytase activity markedly, and there existed a negative interaction between the two enzymes. It is not yet clear whether this effect occured during the determination of the phytic activity or in the diets during storage.

Table 4. Apparent digestibility of nutrients (mean ± se).

Treatment	Basal	Carbohydrase	Phytase	Carbohydrase/ phytase
Carbohydrase	–	+	–	+
Phytase	–	–	+	+
1st experiment*				
Number of animals	8	8	8	8
Organic matter	0.732 ± .011	0.747 ± .018	0.747 ± .015	0.735 ± .020
Energy	0.711 ± .016	0.726 ± .017	0.725 ± .017	0.712 ± .022
Nitrogen	0.766 ± .024	0.796 ± .034	0.787 ± .035	0.776 ± .035
Phosphorus	0.287 ± .112	0.321 ± .086	0.382 ± .102	0.298 ± .087
Zinc	0.124 ± .085^{a}	0.211 ± .117^{a}	0.405 ± .060^{b}	0.144 ± .086^{a}
2nd experiment†				
Number of animals	6	6	6	6
Organic matter	0.741 ± .014^{b}	0.738 ± .012^{ab}	0.737 ± .013^{ab}	0.728 ± .010^{a}
Energy	0.724 ± .016^{b}	0.719 ± .015^{b}	0.716 ± .014^{b}	0.704 ± .012^{a}
Nitrogen	0.709 ± .029	0.718 ± .020	0.723 ± .034	0.709 ± .027
Phosphorus	0.317 ± .047^{ab}	0.280 ± .057^{a}	0.357 ± .067^{b}	0.334 ± .051^{b}

*Means with different superscripts differ, $P \leq 0.05$
†Means with different superscripts differ, $P \leq 0.1$

Results and discussion

The digestibility of organic matter, energy and nitrogen was generally low as a result of the high fiber content of the diets (Table 4). Despite this high fiber content, addition of the carbohydrase had no significant effect on digestibility. Supplemental phytase increased digestibility of organic matter, energy and nitrogen along with dietary minerals. In both experiments the combination of the enzymes had a detrimental effect on nutrient digestibility. A clear explanation for this antagonistic effect could not be found.

Growth performance was only measured in the second experiment. Body weight gain was 859, 854, 869, and 842 g per day for the control, carbohydrase, phytase and combination treatments, respectively. Despite the low content of digestible nutrients in the diet, a mean feed conversion ratio of 2.75 kg feed or 26.8 MJ digestible energy per kg weight gain occurred. Although treatment effects were not statistically significant, lowest values were obtained when adding both the the phytase and carbohydrase enzymes.

EXPERIMENT 3: EFFECTS OF PHYTASE ON BLOOD PARAMETERS, GROWTH AND DIGESTIBILITY OF GROWING/FINISHING PIGS

Material and methods

In this experiment 40 individually penned growing pigs from 25 to 100 kg BW were used. Growth performance and apparent digestibility of energy, nitrogen, P, Ca, Fe and Zn as well as some blood parameters were investigated. The animals were fed a grower ration (25–65 kg BW) *ad libitum* and later fed restricted amounts (2.5 kg feed/day) of a finisher ration. Both diets were based on barley, maize, soybean meal and fish meal. Treatments A and B were formulated using inorganic P and other minerals. In treatments C and D calcium phosphate as well as Fe and Zn were replaced by Opalite, a limestone mineral. Diets B and D were

Table 5. Nutrient content of the experimental diets (per kg dry matter).

Treatment	A	B	C	D
Phytase	–	+	–	+
Grower period				
Digestible, energy, MJ	13.6	14.0	13.7	14.0
Crude protein, g	178	177	180	179
Calcium, g	8.2	7.4	6.7	7.1
Phosphorus, g	5.7	5.7	5.0	5.0
Iron, mg	245	215	377	371
Finisher period				
Digestible energy, MJ	13.5	13.9	13.7	13.9
Crude protein,	153	152	151	156
Calcium, g	8.9	8.8	7.0	7.5
Phosphorus, g	5.5	5.5	4.6	4.6
Iron, mg	175	172	395	381

supplemented with phytase (1000 U/kg). Table 5 presents the nutrient content of the diets. Energy and nutrient digestibilities were estimated at 42, 63 and 92 kg BW with the indicator method. The trial was concluded without disturbances related to the experimental design. The data were analyzed using analysis of variance.

Results and discussion

Body weight gain was rapid due to *ad libitum* feeding during the growing period (>950 g/day). In the finishing period (restricted feeding) only slightly lower values were observed. The addition of phytase had no statistically significant effect on growth rate (Table 6). Pigs in treatment C (low minerals, no phytase supplementation) grew more slowly, especially in the finishing period. The negative influence of low mineral content could be fully compensated by the phytase addition.

A marked increase in apparent digestibility of energy and almost all nutrients could be observed owing to supplementation with phytase (Table 6). In treatment D (without supplemental P, Fe and Zn) the same effect was found. The increase in nutrient digestibility with addition of phytase explains the better growth performance of the pigs.

Table 6. Growth performance and apparent digestibility of energy, nitrogen and minerals.

Treatment	A	B	C	D
Phytase	–	+	–	+
Number of animals	10	9	9	9
Body weight gain, g/day				
Grower period	959 ± 95	967 ± 78	971 ± 46	1017 ± 105
Finishing period	923 ± 54	917 ± 67	854 ± 101	918 ± 52
Whole fattening period	935 ± 46	940 ± 50	907 ± 57	961 ± 57
Feed conversion, kg/kg				
Grower period	2.1 ± 0.1	2.1 ± 0.1	2.1 ± 0.2	2.0 ± 0.1
Finishing period	2.7 ± 0.2	2.7 ± 0.2	2.9 ± 0.3	2.7 ± 0.2
Whole fattening period	2.4 ± 0.1	2.4 ± 0.1	2.5 ± 0.2	2.4 ± 0.1
Apparent digestibility (mean of 3 observations per animal)				
Energy	.824 ± .009	.846 ± .011	.834 ± .017	.844 ± .013
Nitrogen	.780 ± .025	.816 ± .025	.795 ± .035	.821 ± .024
Phosphorus	.522 ± .024	.583 ± .038	.561 ± .029	.567 ± .012
Calcium	.468 ± .059	.524 ± .058	.531 ± .089	.551 ± .025
Iron	.160 ± .038	.185 ± .038	.203 ± .037	.234 ± .014
Zinc	.168 ± .038	.236 ± .037	.258 ± .034	.246 ± .030

Phosphorus status in blood plasma did not change significantly in response to treatment, although the phytase supplementation numerically increased alkaline phosphatase (Table 7).

Table 7. Effect of phytase supplementation on blood parameters at 92 kg BW.

Treatment	A	B	C	D
Phytase	–	+	–	+
Number of animals	10	9	9	9
Alkaline phosphatase, μl	141	189	162	199
Iron, mmol/l	18.4	20.5	19.5	20.2
Zinc, mmol/l	9.0	12.2	10.5	14.1
Phosphorus, mmol/l	2.7	2.9	2.8	2.6

EXPERIMENT 4: EFFECTS OF PHYTASE ON PERFORMANCE AND P AVAILABILITY

Materials and methods

Preliminary results of a fourth trial investigating effects of phytase supplementation revealed a similar pattern of response. Twenty-four growing pigs (castrates, Large White) from 26 to 106 kg BW were kept in individual pens. Effects of phytase both on growth performance and apparent digestibility of phosphorus were investigated.

Composition and nutrient content of the experimental diets are given in Table 8. The premix contained only the essential vitamins without any trace elements.

Table 8. Composition and analyzed nutrient content of the experimental diets.

Ingredient	%	Treatment / Phytase	Basal –	Phytase +
Barley	35	Crude ash, g/kg	70	70
Maize	30	Crude fiber, g/kg	54	55
Soybean meal	10	Crude protein, g/kg	196	197
Sunflower meal	10	Heat of combustion, MJ	19.3	19.3
Dried potatoes	5	Phosphorus, g/kg	4.0	3.9
Lysine-HCl	0.2	Phytase activity, U/kg	<15	990
Ca carbonate	1.5			
Salt	0.5			
Free fatty acids	2.5			
Bone fat	2.5			
Celite 545	1.5			
Premix*	1.31			

*per kg feed: Vitamins B_2, 4 mg; B_6, 4 mg; AD_3, 4 mg; Ca-pantothenate, 15 mg; Niacin, 20 mg; phytase treatment, phytase: 240 mg.

The animals were fed *ad libitum*. Body weight gain and feed intake were measured weekly. The digestibility of P was estimated at 40 and 75 kg BW using the indicator method (Celite 545 as indicator carrier). The experiment was concluded without problem.

Statistical analyses were carried out by Statgraphics (version 5.0). In Table 9 the results are given as mean values with the corresponding standard deviations. When preceded by a significant *F*-test, treatment means were tested for the significance using Bonferroni *t*-test ($P \leq 0.5$).

Results and discussion

As a result of *ad libitum* feeding, the pigs in treatment P grew very quickly (Table 9). Phytase supplementation significantly affected all investigated performance parameters. Intake of dietary P was generally higher than that in the basal diet and explains both the faster body weight gain and the better feed conversion ratio. The apparent digestibility of P was significantly improved by phytase supplementation. Phytase supplementation resulted in improved P utilization and a marked reduction in P excretion along with improved growth performance. Carcass composition was unaffected.

Table 9. Growth performance and apparent digestibility of phosphorus*.

Treatment Number of animals	Basal 12	+ Phytase 12
Body weight, kg		
Start	25.6 ± 1.8	26.0 ± 2.9
Middle	58.6 ± 4.4	67.1 ± 3.3
End	105.4 ± 2.8	105.7 ± 3.4
Feed intake, kg/day	1.98 ± 0.11[b]	2.16 ± 0.13[a]
Body weight gain, g/day		
Grower period	670 ± 60[b]	840 ± 30[a]
Finishing period	830 ± 90[b]	950 ± 70[a]
Whole fattening period	760 ± 40[b]	890 ± 30[a]
Feed conversion ratio, kg/kg		
Grower period	2.36 ± 0.29[b]	2.11 ± 0.17[a]
Finishing period	2.82 ± 0.17	2.80 ± 0.23
Whole fattening period	2.62 ± 0.12[b]	2.44 ± 0.13[a]
Apparent digestibility (mean of 2 observations per animal)		
Phosphorus	0.331 ± 0.064[b]	0.559 ± 0.058[a]

*Values are means ± se

[ab]Means with different superscripts differ *P<0.05*.

Implications

For nutritional, physiological and ecological reasons phosphorus content in the feed clearly in excess of the animal's requirement must be rejected. This will prevent unnecessary physiological problems, reduce environmental pollution, and avoid negative interactions among diet minerals.

The results clearly show that microbially-derived phytases are an efficient means of increasing availability of P from cereals and other feedstuffs of plant origin in which P is bound as phytate. Diets based on 60–80% cereals and their by-products as used in Switzerland do not contain enough available P and therefore additional inorganic P must be supplied. In areas with a high number of pigs per hectare excretion of P leads to environmental problems. A beneficial effect of enzyme supplementation on total P excretion can be achieved by a substantial reduction in the P content of the diets without risking P deficiency of the pigs.

Summary

Several investigations with growing pigs from 25 to 106 kg body weight (BW) were conducted to evaluate effect of phytase supplementation to diets mainly containing ingredients of plant origin. Effects on apparent digestibility of energy, nitrogen, P and other minerals as well as growth performance were studied.

Phytase supplementation increased availability of P along with digestion of other minerals such as Ca, Fe and Zn. Furthermore, phytase addition resulted in a higher growth rate and a better feed conversion ratio.

References

Gesellschaft Für Ernährungsphysiologie. 1987. Energie- und Nährstoffbedark landw. Nutztiere Nr. 4, Schweine. DLG Verlag, Frankfurt/Main.

Jongbloed, A.W. 1987. Phosphorus in the feeding for pigs. Thesis IVVO no. 179, Lelystad.

Jongbloed, A.W. 1989. Phytase can increase P digestibility. Pigs – Misset 21.

Jongbloed, A.W. and P.A. Kemme. 1990: Apparent digestible phosphorus in the feeding of pigs in relation to availability, requirement and environment. 1. Digestible phosphorus in feedstuffs from plant and animal origin. Netherlands J. Agric. Sci. 38: 567.

Kessler, J. and K. Egli. 1991. Phosphor sparen dank Phytase: Erste Ergenbnisse beim Mastschwein. Landwirtschaft Schweiz 5: 5.

Larimier, P., A. Pointillart, A. Coriouer and C. Lacroix. 1994. Influence de l'incorporation de phytase microbienne dans les aliments, sur les performances, la résistance osseuse et les rejects phoshorés chez le porc charcutier. J. Rech. Porcine en France 26: 107.

Näsi, M. 1991. Plant phosphorus responses to supplemental microbial phytase in the diet of growing pigs. Proc. 5th Symp., 'Digestive Physiology in Pigs', Wageningen, 114.

NRC, 1988. Mineral tolerance of domestic animals. National Academy of Science, Washington DC, 577 pp.

Pallauf, J., D. Höhler and G. Rimbach. 1992. Effekt einer Zulage mikrobieller Phytase zu einer Mais-Soja-Diät aus die scheinbare Absorption von Mg, Fe, Cu, Mn und Zn sowie auf Parameter des Zinkstatus beim Ferkel. J. Anim. Physiol. Anim. Nutr. 68: 1.

Wenk, C., E. Weiss, G. Bee and R. Messikommer. 1993a. Interactions between a phytase and a carbohydrase in a pig diet. In: Enzymes in Animal Nutrition. Proceedings of the 15th Symposium Kartause Ittingen, Switzerland, 160.

Wenk, C., E. Weiss, G. Bee and R. Messikommer. 1993b. Phytase in rations for growing pigs: Influence on the nutrient utilization. In: Enzymes in Animal Nutrition. Proceedings of the 15th Symposium Kartause Ittingen, Switzerland, 226.

IN-FEED ASSAY OF ENZYMES BY RADIAL ENZYME DIFFUSION – RECENT DEVELOPMENTS AND APPLICATION TO ANALYSIS IN PELLETED FEED

GARY WALSH

Department of Industrial Biochemistry, University of Limerick, Ireland

Introduction

A large body of evidence exists to support the theory that addition of selected hydrolytic enzyme activities to monogastric animal feed improves animal performance. Assay of enzyme activity after addition to animal feed has proven difficult, and to date no widely acceptable method has been reported. The major difficulty regarding assay development relates to the presence naturally in feed of high levels of enzyme products (Power and Walsh, 1994). The extremely high catalytic efficiency of enzymes also ensures product efficacy even when incorporated in feedstuffs at very low levels. While this is fortuitous from an economic standpoint, it dictates that any in-feed assay developed be extremely sensitive.

While the practice of enzyme addition to feed has grown steadily in popularity over the last decade or so, to date the technical difficulties outlined above have thwarted attempts to develop a universally acceptable, validated assay system for in-feed analysis. None the less, several promising approaches to assay development have been reported, including the use of chromogenic substrates or assay by radial enzyme diffusion (Power and Walsh, 1994).

Development of assay systems for in-feed enzyme analysis has become a priority for many industrial and academic scientists whose interests impinge on this area. The requirement for sensitive, accurate and ideally inexpensive and technically straightforward assays is self-evident from both a quality control and regulatory standpoint. Failure to develop satisfactory assay systems could ultimately result in the enforced withdrawal of enzyme products from animal feed, as European, North American and many other regulatory bodies demand inclusion of suitable assay systems in product registration dossiers. The availability of such assay systems would also provide conclusive data regarding topics such as enzyme stability during pelleting, etc.

As previously described, enzyme assay by radial diffusion techniques provides one method by which enzyme activity in feed might be detected and quantified (Power and Walsh, 1994; Walsh *et al.*, 1995). This technique is based on the radial diffusion of an enzyme-containing feed extract through an agar gel in which a suitable substrate has been dissolved. A linear relationship between the diameter of the zone of substrate hydrolyzed (which may be visualized by staining with a suitable dye), and the log of the enzyme activity present in the extract is observed.

This method is sufficiently sensitive to detect several enzymes typically added to feed, when present at normal inclusion levels.

The major advantages of this system at a practical industrial level are its simplicity and ease of application. It also requires no sophisticated equipment and all reagents needed are easily obtainable. As the zone of hydrolysis observed is proportional to the logarithm of the enzyme activity present, accurate measurement of zone diameter is crucial as relatively large differences in enzyme activities may yield relatively small differences in zone diameter of substrate hydrolyzed. This has led some to suggest that the technique is best regarded as semi-quantitative. However, various diverse publications over many years have demonstrated successful use of this technique in the quantitation of various enzymatic activities.

Table 1 summarizes various medically-significant enzymes for which quantitative assays based on radial diffusion have been developed (Schumacher and Schill, 1972). The authors tested the experimental error and reproducibility of this method by assaying 25 identical samples of appropriate enzyme present in a biological sample, with good results. For example, in the case of elastase, they reported levels of 65.8 μg/ml standard elastase equivalent, with a standard deviation of ± 4.5 μg/ml. A coefficient of variation of the reproducibility of 6.9% was quoted. Similar values of standard deviation and variation in reproducibility was recorded in the case of the other enzymes assayed.

Table 1. Some enzymes of medical significance which have been quantified in various biological materials using radial diffusion techniques.

• Muramidase	• α-amylase	• Acid phosphatase
• Alkaline phosphatase	• DNase I	• RNase A
• Plasminogen activator	• Elastase	• Non-specific protease

Wilkstrom and co-workers (Wilkstrom *et al.*, 1981) reported the sensitive quantitation of the proteolytic enzymes trypsin and pronase with good reproducibility using the radial diffusion method, permitting quantitative determination of protease activity in crude samples.

This method has also been successful in the quantitative determination of several other enzymes, including cellulase (Carder, 1986) and endo-beta-D-mannanase activity (Downie *et al.*, 1994). The technique of enzyme assay by methods of radial diffusion through a substrate-containing gel therefore is a well established one. In previous papers the assay of β-glucanase after its addition to animal feed by this method was reported (Power and Walsh, 1994; Walsh *et al.*, 1995). In this paper the applicability of the method to other enzymes often added to feed is discussed along with some pelleting stability data obtained in other laboratories using this technique.

Summary of methods

Gel plates were generally prepared by dissolving relevant substrate and agar in appropriate buffer yielding a final substrate concentration of 0.1% and a final agar concentration of 1.5%. Substrates used included carboxymethyl cellulose (cellulase) and gelatine (protease). Molten gel was poured into Petri dishes (140 mm diameter) and allowed to solidify. Circular wells (10 mm diameter) were subsequently punched in and removed from the agar. In these studies the enzyme preparations used were; Allzyme C (cellulase) and Allzyme PB (bacterial protease). Recovery of enzyme from feed was achieved by sequential extraction of a 5 g sample with two 50 ml lots of buffer. Control feed (i.e., a feed sample taken immediately before enzyme addition) was extracted in the same manner. Standards for the plate assay were prepared by direct dilution of the same enzyme concentrate as was added to the feed.

Enzyme assays were carried out as follows: In each case, aliquots (200 μl) of feed extract (control and supplemented), in addition to suitable standard dilutions, were applied to the wells in the gel. The gel plates were then incubated at suitable temperatures for an appropriate time period (typically overnight). Zones of protease hydrolysis were evident by direct inspection while zones of cellulase hydrolysis were visualized by flooding the gel surface with 0.3% w/v Congo red for 15 min. After rinsing with water, the diameter of the zones of hydrolysis was measured at several orientations using a Vernier calliper. Standard curves were constructed by plotting log enzyme inclusion level vs diameter of zone of hydrolysis.

The sensitivity of the radial diffusion assay relating to each enzyme was determined by assaying a range of dilutions made directly from the enzyme concentrate. The dilution factors take into account: (a) the direct dilutive effect of feed on enzymes, and (b) the further dilutive effect incurred when 5 g feed is extracted in a total final volume of 100 ml.

Results and discussion

ASSAY SENSITIVITY

As examples of the assay sensitivity for the radial enzyme diffusion assay, plots relating zone diameter to the logarithm of the enzyme inclusion (kg/t equivalent) for cellulase and protease are shown in Figure 1. From this figure, it is evident that the assay method is sufficiently sensitive to detect enzyme preparations at levels below normal inclusion levels in feed. In the case of all preparations, assay of enzyme dilutions

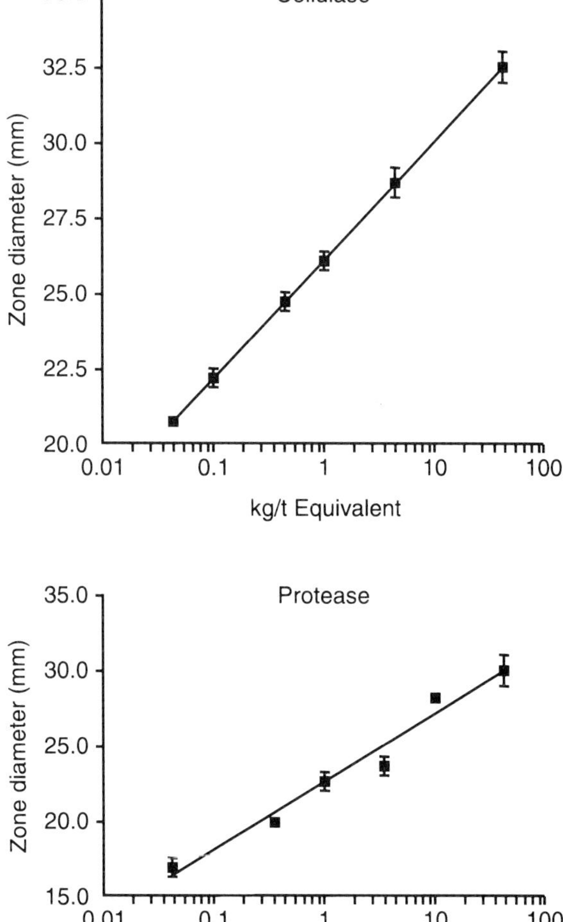

Figure 1. Plot of the relationship between enzyme levels (expressed here as kg/t equivalent) vs diameter of the zone of substrate hydrolyzed.

significantly greater than those reported in Figure 1 resulted in a non-linear relationship between enzyme activity and diameter of the zone of substrate hydrolyzed. Furthermore, at very low enzyme levels, determination of the exact boundary of the zone became more difficult, making its accurate measurement difficult.

IN-FEED ASSAYS

The performance of the radial diffusion enzyme assay was assessed by spiking an unsupplemented (commercially obtained) animal feed with the relevant enzyme preparation to an inclusion level of 1 kg/t. Subsequent extraction and assay was carried out as described in the summary of methods. The % recovery of activity added was calculated to be as follows: cellulase: 110 ± 8; protease: 91.8 ± 10.2, as averaged from triplicate measurements assayed on two separate occasions. The results clearly illustrate that enzyme assay by radial diffusion techniques can be successfully applied to assay of these enzymes in feed.

PELLETING STABILITY OF ENZYMES

Recent studies have underlined the significance of developing suitable in-feed enzyme assay systems. The effect of various pelleting temperatures on the activity of different enzymes has been recently assessed, with interesting results (P. Spring, C. Wenk, K. Newman, personal communications). In these studies, samples of a wheat-barley-soybean diet containing either no enzyme (control) or cellulase, pentosanase, bacterial amylase, and fungal amylase were pelleted at temperatures varying from 60 to 100°C. Amylase and cellulase activity was assessed using the radial enzyme diffusion technique. Pentosanase activity, in this case, was quantified using the chromogenic substrate 4–0–methyl-D-glucorono-D-xylan coupled to Remazol brilliant blue. The results obtained are outlined in Table 2. These data suggest that cellulase, fungal amylase and pentosanase can be pelleted at temperatures up to at least 80°C, and bacterial amylase up to 90°C without considerable loss in activity.

The improved stability of bacterial amylase over the other tested enzymes was in agreement with the stability data provided by the manufacturer, who determined their data after temperature exposure of the enzymes in different buffered solutions. Inactivation in buffered solutions, not surprisingly, occurs at lower temperatures than inactivation during the pelleting process. The difference is most likely explained on the basis of different water activity and likely protective effects of the feed matrix on enzyme activity.

Table 2. Enzymatic activity after pelleting (% of original activity).

Temperature, °C	Cellulase	Pentosanase	Bacterial amylase	Fungal amylase
60	100	100	100	100
70	100	100	100	100
80	100	90	100	100
90	10	10	80	10
100	0	0	65	0

Conclusions

The development of suitable assay systems to facilitate detection and quantification of enzyme activity present in finished feed is of obvious importance. Although some thought it likely that the obvious technical challenges involved would render development of such assay systems almost impossible, to date a number of potential assay systems have been reported. Although no one assay method is likely to prove universally popular, enzyme detection and quantification by methods of radial enzyme diffusion exhibits many advantages. Additional methods will no doubt be reported in the literature in time. While the efficacy of enzyme addition to feed to achieve certain defined goals is no longer seriously questioned, widespread acceptance of this technology will only come about when questions relating to topics such as stability of enzymes during pelleting or in the animals' digestive tract are satisfactorily addressed. Assay techniques such as radial enzyme diffusion should allow many such questions to be answered.

References

Carder, J. 1986. Detection and quantitation of cellulase by congo red staining of substrates in a cup-plate diffusion assay. Anal. Biochem. 153: 75–79.

Downie, B., H. Hilhorst and J. Bewley. 1994. New assay for quantifying endo-beta-D-mannanase activity using congo red dye. Phytochemistry 36 (4): 829–835.

Power, R.P. and G. Walsh. 1994. Registration of enzymes and biological products around the world: Current analytical methods for feed enzymes and future developments. In: Biotechnology in the Feed Industry. T.P. Lyons and K.A. Jacques (Eds) Nottingham University Press, 117–127.

Schumacher, G. and W.B. Schill. 1972. Radial diffusion in gel for microdetermination of enzymes. Anal. Biochem. 48: 9–26.

Walsh, G., R. Murphy, G. Killeen, D. Headon and R. Power. 1995. Technical note: Detection and quantification of supplemental fungal β-glucanase activity in animal feed. J. Anim. Sci. (In press.)

Wikstrom, M., H. Elwing and A. Linde. 1981. Determination of proteolytic activity: A sensitive and simple assay utilizing substrate adsorbed to a plastic surface and radial diffusion in gel. Anal. Biochem. 118: 240–246.

THE PRODUCTION OF, AND RATIONALE FOR, DIGESTS IN PET FOOD DESIGN

JOHN A. LOWE[1] and STEVEN WOODGATE[2]

[1]*Gilbertson & Page, P.O. Box 321, Welwyn City, Hertfordshire, UK*
[2]*Beacon Research Ltd., Clipston, Market Harborough, UK*

Introduction: 'Palatability'

The enormity of the pet food industry today, both in terms of animal numbers and business value, has resulted in a plethora of commercially prepared, nutritionally complete and balanced foods presented in many forms and varieties. However it is self-evident, though still worth stating, that the nutritional quality of a pet food is of no consequence if the dog or cat will not eat it. Making pet foods that are consistently consumed over extended periods requires a great deal of experience and a basic understanding of the factors affecting appetite and food intake.

The domestic dog (*Canis familiaris*) and cat (*Felis catus*) occupy an interesting and unusual position to those interested in diet selection. On the one hand domestic pets have food readily available; yet on the other hand they have had little to do with its initial selection. The majority of the food is preselected by the pet owner. This often results in a wide variety of products chosen at random for the animal, particularly in the case of the cat whose owners perceive it to be a finicky eater with a need for variety.

The extent to which pets thus choose to consume their food is thought to respond largely to 'palatability', a term loosely used to describe the interrelated properties of taste, smell and texture. Clearly, the wild ancestors of the domestic dog and cat were more concerned with locating and catching prey and selecting food that was safe and met their immediate nutritional needs. Palatability, if involved at all, would have been well down the agenda. In the modern pet food situation, however, palatability is all. For if the animal will not eat the product the pet owner will not value the product for the future and is unlikely to repurchase. The dog or cat will, nevertheless, survive as it will most likely be fed an alternative product.

Taste preferences of dogs and cats

Simple statements about the kinds of foods preferred by either dogs or cats are unlikely to have universal application in palatability enhancement, though some generalizations for diet preferences can be made. Taste preferences/differences in carnivores have been extensively reviewed by Mugford (1977). One may conclude that the dog and cat are almost opposite in their preferences since the dog generally prefers lean while the cat prefers fat.

Palatability for the pet food company is pivotal in the development of diets; and large sums of research and development money are spent in this area. However the science of palatability is vague, and comparisons or evaluations of improvements in palatability of a diet are prone to errors in interpretation. The results of palatability tests can also be confounded by such influences as stress, neophilia, neophobia, previous experiences and food temperature.

Dogs and cats use both taste and smell in the detection and selection of food. The vomeronasal organ appears only to be involved in the perception of social odors (Hart and Leedy, 1987). While smell may be an important sense, there is a much greater wealth of literature on the taste systems of dogs and cats (Bradshaw, 1991). In addition to these chemosensory systems, there are also behavioral factors such as food availability and novelty involved in food preference. The sense of smell appears important in distinguishing among food sources and varieties (Houpt *et al.*, 1978); however taste appears to be the overriding factor in long-term acceptance. This is based on the information of Bradshaw (1986) where only whole meat flavor, and not the odors alone, would overcome a fear of new foods (neophobia) in cats. Similarly, meaty odors in a bland food would not sustain dogs' initial interest in that food (Houpt *et al.*, 1978). Thus any means used to alter or enhance the 'palability' of a food must cater to both factors.

Whilst the food preference environment of the wild ancestors of the modern domesticated canids and felids bears little resemblance to the presentation of modern prepared pet food, some indication of basic preferences may remain. Thus a short examination of those preferences is helpful here. Wolves, according to a study by Ewer (1973), consume large ungulates and small mammals. Plant materials such as grass and berries are deliberately eaten, but form an insignificant part of the diet. In calorie terms this is minimal, but may account for up to 18% of the diet of other canids. The dentition is not specialized for eating meat indicating an adaptation to, if not flexibility in, the adoption of a varied diet. Wild canids also cache food for times of shortage, as seen in the domestic dog when it buries a bone. This indicates tolerance to the products of decay and degeneration of foodstuffs.

In the case of cats it is preferable to look at feral colonies as opposed to big cats for clues. The diet is composed mainly of small mammals, followed by birds, reptiles and insects. Very little if any plant material is eaten; though occasionally grass is consumed. Feral cats prefer young lagomorphs to murids, the latter being often caught but not eaten. Shrews are the least palatable of all prey caught. All material caught which is intended for consumption is eaten almost immediately. This indicates a preference for fresh food.

To take what may be considered an overly simplistic view (but one that can at least highlight the differences in flavor design for the dog and cat), cats prefer 'fresh' material. They respond positively to amino acids such as L-proline and L-cysteine while rejecting the monophosphate nucleotides which accumulate post-slaughter and give a bitter taste. This would also account for the cat's dislike of carrion. There are also a number of other amino acids with hydrophobic side chains, for example L-tryptophan, that are a negative influence on palatability to the cat. The belief commonly held that cats prefer fish to meat was supported by Houpt and Wolski (1982), but other studies have shown that this is not universally true (Bradshaw, 1991). This may be explained by the 'freshness' of the fish in the study, or may be due to behavioral factors.

In the dog the most abundant taste buds are those which respond to sugars, accounting for the 'sweet tooth'. Many sugars and artificial sweetness will trigger these taste buds. They also respond to the fruity sweet (furaneol) and to the sweet amino acids. It is worth remembering that the cat, while responding to the sweet amino acids, shows a complete lack of response to sugar (Carpenter, 1956). This may be an extreme adaptation to meat eating.

Dogs tend to prefer meat to cereal. The meat will be affected by freshness and animal status at slaughter. Meats for dogs in order of preference are tripe, beef, pork, mutton, chicken, horse, rabbit (Houpt *et al.*, 1978). Unfortunately, this palatability hierarchy does not necessarily correlate to digestibility. The same meat is preferred canned over cooked and either are preferred to raw (Lohse, 1974). A study by Walker (1971) indicated that short chain fats are unpalatable to dogs. This is not so for cats, which reject medium chain fatty acids (8:0) (MacDonald *et al.*, 1985). In general, the amino acids that are inhibitory in the cat are either neutral or stimulatory in the dog. Carnivores in general lack a taste system sensitive to salt.

The second most abundant group of taste receptors are the acid units. Cats and dogs respond similarly to phosphoric acid, carboxylic acid, and other Brönsted acids (nucleotide triphosphates, histidine). Also the amino acids L-taurine and L-cysteine have substantial positive effects on palatability response. The pH of raw fresh meat is around 5.5 to 7.0.

These palatability responses do not entirely explain the food selection preferences of the dog or cat. One other important factor, interpretation by the cat or dog's brain of the initial taste or aroma sensation, is poorly understood. These response patterns do, however, indicate a basis for flavor addition or enhancement in pet foods. It is apparent that merely 'topdressing' an aroma will not resolve our problems. A more full flavor throughout the product, which matches the animal's preferences is required. An aroma, if present, is perhaps better addressed to the pet owner and then not at a level which would offend either owner or animal, the latter having a much greater sense of smell (up to 10^6 times more sensitive than humans to certain substances). Such preferences as do exist can be built into the design of a flavor enhancing agent, such as a digest, so that it complements the pet food flavor profile and ensures both initial and long term animal acceptance.

Digests

Overall flavor addition or enhancement can be achieved by the use of a 'digest' applied either onto or throughout the pet food. A digest is the result of an enzymatic digestion of an animal substrate. It can be formulated to include those chemicals to which the target animal responds favorably. Furthermore, it can be based on the meat of choice for the target animal or for a flavor/species label claim and should be regarded as natural.

MANUFACTURE OF DIGESTS

A digest can be manufactured by one of two different procedures, but the principle is the same for both. In the majority of digests enzymes found in the raw material itself (endogenous enzymes in the gut mucosa) will effectively hydrolyze the structured protein into soluble protein when activated. Some digests, however, are manufactured from dried or rendered products in which any endogenous enzyme has been deactivated at the high processing temperatures. To produce digests from these materials, addition of exogenous enzyme (probably a protease mixture from either plant, bacterial or fungal origin) is required.

The choice of starting material has consequences not only for flavor and aroma, but commercially in terms of soluble protein yield and consequently final cost of the product. Poultry digest, made from the viscera of poultry species such as chicken, turkey or duck, is a particularly effective product with reference to all of these parameters. It can be processed from the raw state as it has the correct amount of water for the autolysis to proceed. If it is manufactured as a liquid, with a similar level of moisture added to the starting material, the yield can be almost 100%. Liquid digest can be manufactured from other 'raw materials' or mixtures of raw materials such as livers, hearts, fish (various species), rabbit, game birds. Commonly, raw materials are co-manufactured with poultry viscera to produce products such as 'poultry and liver digest' or 'poultry and rabbit digest', for example. The variations are wide and reinforce the palability benefits and ingredient claims.

The principle features of the digest manufacturing process are shown in Figure 1.

The key factors in each production stage are as follows:

- **Size reduction**. This is important for both initially fresh or dry raw materials. As small a size as possible is needed to provide a large surface area for enzyme action. At this stage, water and enzyme may be added if necessary. An antioxidant is added to protect the fat from oxidation and to give a long shelf life.
- **Process and sterilize**. Agitation is required to ensure homogeneity during the process, while temperature control between 40 and 60°C maximizes enzyme function. After the digestion is complete, sterilization follows to ensure product stability and safety, vital features of the process. A sterilization temperature of between 105 and 120°C for a minimum of 20 minutes is typical.

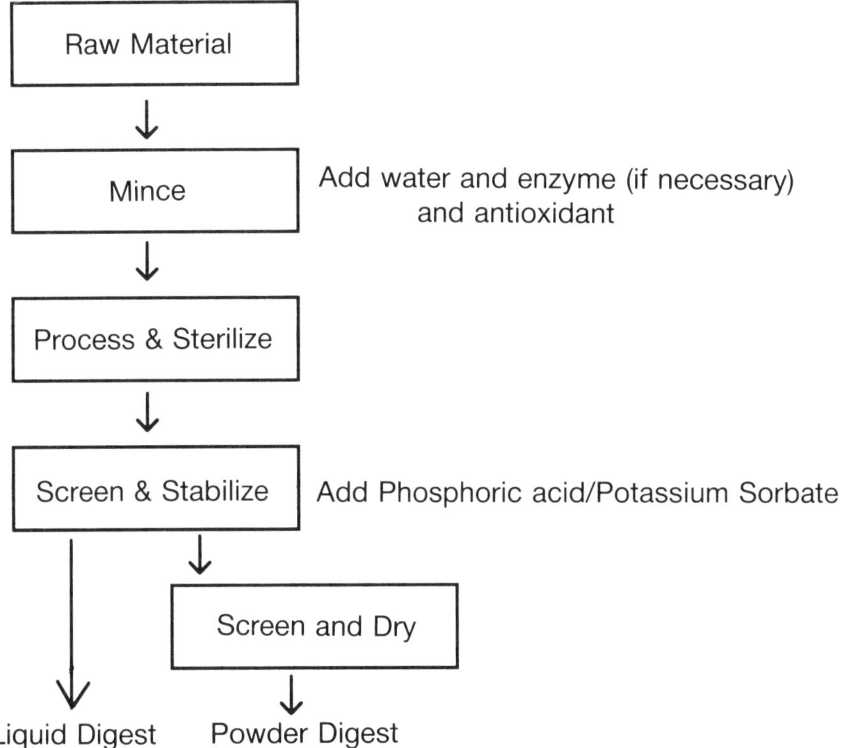

Figure 1. Manufacture of liquid or powder digest.

- **Screening and stabilizing**. Screening removes all of the undigested particles and any contaminants. Stabilizing agents are added to digests that are used in liquid form. Typically phosphoric acid is used to reduce pH to below 3.2; although some products are manufactured with a pH of 1.8–2.0. Potassium sorbate may be included to prevent mold colonizing the liquid digest. The liquid digest is then cooled and drained or pumped into bulk for delivery. Digest for immediate drying may not require the addition of stabilizing agents to the same level as for liquid digest.
- **Screening and Drying**. This second screening reduces the product to a suitable particle size (for example less than 70 microns) so that the product can be sprayed using atomizing spray nozzles. The small particle size allows for better coating of the finished feed. Digests are normally dried in a spray dryer without loss of flavor or aroma. Spray drying is expensive in terms of capital and operational costs. As a result, spray dried digest is a high cost product which must achieve cost effective results in terms of palatability.

Apart from the process control, the prime assurance of quality to be considered in digest manufacture is the starting raw material. If this is substandard then the digest process will not convert a 'sow's ear into a silk purse'. This is particularly relevant to fresh raw material such as poultry viscera, which must be used as fresh as possible without suffering any uncontrolled decomposition. Thereafter process controls ensure the preparation of the digest to standards agreed between the supplier and pet food manufacturer. Such standards will include, for example, targets for solids/moisture content, pH and analytical values. A typical analysis of liquid poultry digest is shown in Table 1.

Table 1. Liquid poultry digest analysis.

Solids	35%
Moisture	65%
Protein	15%
Fat	15%
Ash	5%
pH	3.0

For liquid digests pH is the primary means of controlling stability. Because of the buffering capacity of the initial raw material, quality assurance should rely on monitoring the actual pH rather than the amount of acid added per batch. The shelf life in cool, dry conditions should be at least 3 months for dried digest. Liquid digest would be expected to have a shelf life of over 6 months.

METHODS OF APPLYING DIGEST TO PET FOODS

The purpose of an added digest is to affect the entire meal, consequently it is important that the pet food is completely and evenly coated. This will allow the digest to be in direct contact with the animal's taste buds when the food is consumed. A low level applied evenly will be detected by the dog or cat as their sensitivity to flavour is about 200 000 times greater than our own. The methods of application for liquid and dry digest are of course quite different; so it is relevant to consider them separately.

Dry digests are normally sold as very fine powder (as a result of spray drying). Application should be completed very carefully so as to avoid waste and ensure uniformity. Normally this would be via a powder applicator set to dose between 0.5 and 2.0% (w/w) relative to the finished dry pet food, while the food travels on the belt or in a screw conveyor or in a drum coater. Most effective adherence of the digest can take place if the recipient product is slightly sticky, perhaps directly after fat coating.

Liquid digests are usually applied directly with an atomizing nozzle spray. In order to effect pumping and spraying, the digest should be heated to between 50 and 60°C before application. Levels of up to 6% (w/w) can be added with ease to dry or partially dried pet food made using extrusion techniques. Typically this application is achieved using a digest/fat coater composed of a spray bar onto the tumbled product whereby the digest is absorbed into or absorbed onto the expanded product. Digest can also be included as an ingredient if there is a demand for water in the recipe so as to effect a taste throughout the product. The main area for caution with liquid digests is the level of moisture of the final product; so the point of application is of great importance. A typical application rate would be 1.0–2.5% (w/w) post-drying, pre-cooling, so that the final product is able to comply with the desired moisture specification of typically less than 10%.

In conclusion, digests offer the pet food manufacturer a simple and natural means of cost effective palatability enhancement based on animal preference and the type of pet food being produced.

References

Carpenter, J.A. 1956. Species differences in taste preferences. J. Comp. Physiol. Psychol. 49: 139–144.

Bradshaw, J.W.S. 1986. Mere exposure reduces cats' neophobia to unfamiliar food. Anim. Behav. 34:613–614.

Bradshaw, J.W.S. 1991. Sensory and experimental factors in the design of foods for domestic dogs and cats. Proc. Nutr. Soc. 50:99–106.

Ewer, R.F. 1973. The Carnivores. London, Weiderfield & Nicolson.

Hart, B.L. and M.G. Leedy. 1987. Stimulus and hormonal determinants of flehmen behavior in cats. Horm. Behav. 21:44–52.

Houpt, K.A., H.F. Hintz and P. Shepherd. 1978. The role of olfaction in canine food preferences. Chem. Senses Flavor 3:281–290.

MacDonald, M.L., Q.R. Rogers and J.G. Morris. 1985. Aversion of the cat to dietary medium chain triglycerides and caprylic acid. Physiol. Behav. 35:371–375.

Mugford, R.A. 1977. External influences on the feeding of carnivores. The Chemical Senses in Nutrition. Academic Press, London.

Walker, A.D. 1971. Nutritional studies in the domestic dog and cat. Ph.D. Thesis, University of London, England.

EFFECTS OF β-GLUCANASE (ALLZYME BG) SUPPLEMENT TO A BARLEY BASED DIET ON BROILER CHICK PERFORMANCE

J.B. SCHUTTE

TNO-Institute of Animal Nutrition and Physiology (ILOB), Wageningen, The Netherlands

Introduction

It is well established that the β-glucans in barley have an anti-nutritive activity in broiler chicks. It is thought that these anti-nutritional effects, which are manifested by depression in performance and wet droppings, are connected with the high viscosity of these polysaccharides.

The improvements in performance together with improved litter quality following β-glucanase supplementation of barley-based broiler diets are well known. Alltech, Inc. manufactures an β-glucanase preparation for feed application (Allzyme BG). The efficacy of this enzyme was tested in a trial with broiler chicks on a diet containing 50% barley.

Materials and methods

Day-old sexed broiler chicks ('Ross') housed in litter floor pens (4.3 m^2) in an insulated broiler house with concrete flooring were used. The nutritionally complete basal diet was based on barley (50%) and soybean oilmeal, and calculated to contain 20.5% crude protein (CP) and 2980 kcal metabolizable energy (ME)/kg. The basal diet was supplemented with an antibiotic (20 ppm virginiamycin). The batch of barley used was analyzed to contain 4.3% β-glucans.

With the basal diet four treatment groups were formed containing 0, 500, 1000 and 2000 ppm Allzyme BG, respectively. Each experimental diet was fed to three pens of 50 male chicks each and three pens of 50 female chicks each for 39 days (1–39 days of age). Diets (pelleted) and water were available *ad libitum* to the birds. The pelleting temperature was approximately 70°C. At the end of the trial, chicks were weighed individually, and feed consumption of each pen was recorded.

In addition, quality of the bedding was judged visually at 24, 30 and 38 days of age. After termination of the trial, carcass quality of chicks of the control group and the group with 2000 ppm Allzyme BG was examined. The criteria studied were slaughter weight and yield of oven-ready, griller, breast meat, abdominal fat, thighs, drumsticks and edible organs.

Results

Performance data are summarized in Table 1. The effect of Allzyme BG on feed conversion efficiency was more pronounced than on weight gain. Weight gain was only improved significantly at a dietary level of 1000 ppm Allzyme BG. For feed conversion efficiency this was true at all three dietary levels of Allzyme BG. The improvement in feed conversion through Allzyme BG was mainly a result of a decrease in daily feed intake. The latter may be the result of an increase of the energy digestibility by including β-glucanase in barley-based diets as reported by Leong *et al.* (1962), Potter *et al.* (1965) and Friesen *et al.* (1992). Bedding quality was improved significantly by including Allzyme BG in the diet. This was true for all dose levels of the enzyme preparation. The differences in carcass yield characteristics between the control group and the group with 2000 ppm Allzyme BG were small and of no significant importance.

Table 1. The effect of dietary inclusion levels of Allzyme BG on broiler performance (1–39 days of age).

Allzyme BG addition (ppm)	Body weight (g)	Daily feed intake (g/bird)	Feed:gain (g:g)
0	2163[a]	96.5[a]	1.742[a]
500	2193[ab]	95.0[ab]	1.692[b]
1000	2215[b]	95.3[ab]	1.678[b]
2000	2160[a]	93.2[b]	1.685[b]
LSD ($P \leq 0.05$)	50	2.4	0.023

[a,b] Mean values with no common superscript within a column differ significantly ($P \leq 0.05$)

Conclusions

The results of the trial demonstrated that performance of broiler chicks fed on a diet containing 50% barley and supplemented with an antibiotic can be improved substantially by including Allzyme BG in the diet. Best performance in this trial was achieved with a dietary inclusion level of 1000 ppm Allzyme BG.

References

Friesen, O.D., W. Guenter, R.R. Marquardt and B.A. Potter. (1992). The effect of enzyme supplementation on the apparent metabolizable energy and nutrient digestibilities of wheat, barley, oats and rye for young broiler chicks. Poultry Sci. 71:1710–1721.

Leong, K.C., L.S. Jensen, and J. McGinnis. (1962). Effect of water treatment and enzyme supplementation on the metabolizable energy of barley. Poultry Sci. 41:36–39.

Potter, L.M., M.W. Stutz, and L.D. Matterson. (1965). Metabolizable energy and digestibility coefficients of barley for chicks as influenced by water treatment or by presence of fungal enzyme. Poultry Sci. 44:565–573.

USING ENZYMES TO INCREASE PHOSPHORUS AVAILABILITY IN POULTRY DIETS

AUSTIN H. CANTOR

Department of Animal Sciences, University of Kentucky, Lexington, Kentucky, USA

Introduction

There has been considerable interest in recent years in examining ways to decrease the amount of phosphorus excreted by animals in order to minimize water pollution. Much of the phosphorus in feed ingredients obtained from plants is in the form of phytic acid, a compound not degraded by the endogenous enzymes found in animals. Consequently, approximately two-thirds of this phosphorus is considered unavailable for monogastric animals. Thus, it is often necessary to add sources of inorganic phosphorus to diets for poultry and swine in order to meet the animals' requirements for this element.

One method of reducing the amount of phosphorus added to diets is to increase the availability of the phosphorus in the feed ingredients by the addition of microbially produced enzymes to the diet. This idea was introduced over 25 years ago, but was not applied in the feed industry for a variety of practical reasons. However, in the past several years, because of increasing environmental concerns and regulations, this concept has received considerable attention among researchers and application in the field.

We recently conducted a study to compare the efficacy of acid phosphatases from yeast and *Aspergillus niger* with that of a commercial phytase (from *A. niger*) for improving availability of phosphorus in a corn/soybean meal broiler starter diet. The organisms used for production of acid phosphatase were genetically modified to increase their yield of the enzyme.

Methods

The study involved 468 day-old male broiler chicks housed in cages (61 × 51 cm). Four replicate groups of nine chicks were assigned to each of 13 dietary treatments. A basal broiler starter diet was formulated (Table 1) to meet all of the broilers' nutrient requirements except for calcium and available phosphorus (calcium = 0.65%, available phosphorus =

Table 1. Composition of the basal diet.

Ingredient	Per cent of diet
Corn	57.00
Soybean meal	35.00
Limestone	1.00
Dicalcium phosphate	0.70
Vitamin-mineral mix	0.25
Cellulose	1.35
Vegetable oil	4.00
Calculated nutrient composition	
ME, mcal/kg	3.12
Protein	21.90
Met + Cys	0.91
P, total (analyzed)	0.54
P, available	0.27
Ca	0.65

0.27%). This diet was fed alone or with graded levels of calcium and phosphorus (providing 0.74, 0.83 and 0.92% calcium and 0.33. 0.39 and 0.45% available phosphorus) to establish standard dose response curves for the variables measured in the trial. The levels of calcium and phosphorus were adjusted by varying the amounts of limestone, dicalcium phosphate and cellulose in the diet. In addition to these four standard diets, there were nine experimental diets which consisted of the basal diet supplemented with 400, 800 or 1200 phytase units per kg provided by yeast acid phosphatase, *A. niger* acid phosphatase or Allzyme Phytase. Chicks were fed the diets on a free-choice basis for 14 days.

Results

Variables measured in this study included weight gain, feed intake, feed efficiency (gain:feed ratio), plasma inorganic phosphorus, % toe ash, % tibia ash and tibia breaking strength. Compared with the basal diet (no added phosphorus or enzyme), weight gain to 14 days was significantly ($P < 0.05$) increased by 1200 units per kg of all three enzymes (Table 2). Feed intake was also increased by supplementing the diet with the three enzymes. There were no differences among these treatments in feed conversion.

Adding inorganic phosphorus to the basal diet did not result in linear responses in plasma phosphorus and toe ash. Therefore, these parameters were not useful in evaluating the efficacy of the three enzymes. However, enzyme supplementation at 1200 units per kg resulted in plasma phosphorus levels equivalent to that obtained with the highest level of dietary phosphorus (Table 3).

The addition of enzymes to the basal diet led to significant improvements in both breaking strength and per cent ash of the tibia (Table 4).

Table 2. Effect of dietary phosphorus and enzymes on broiler growth performance.

Dietary supplement	Weight gain, g	Feed intake, g	Gain/feed
(none)*	320[b]	391[b]	0.82
0.18% P†	336[ab]	424[a]	0.79
YAP‡	349[a]	428[a]	0.82
ANAP§	348[a]	422[a]	0.82
PHY¶	339[a]	418[a]	0.80
SEM	6	7	0.01

*Basal diet: 0.27% available P, 0.54% total P, 0.65% Ca
†Provided 0.45% available P, 0.70% total P, 0.92% Ca
‡Yeast acid phosphatase, 1200 units/kg
§*A. niger* phosphatase, 1200 units/kg
¶*A. niger* phytase, 1200 units/kg
[a,b]$P < 0.05$

Table 3. Effect of dietary phosphorus and enzymes on plasma phosphorus and toe ash.

Dietary supplement	Plasma inorganic P, mg/dl	Toe ash, %
(none)*	5.61	10.7
0.18% P†	6.61	11.9
YAP‡	7.01	11.0
ANAP§	6.48	10.9
PHY¶	7.49	12.2
SEM	0.45	0.5

*Basal diet: 0.27% available P, 0.54% total P, 0.65% Ca
†Provided 0.45% available P, 0.70% total P, 0.92% Ca
‡Yeast acid phosphatase, 1200 units/kg
§*A. niger* phosphatase, 1200 units/kg
¶*A. niger* phytase, 1200 units/kg

Table 4. Effect of dietary phosphorus and enzymes on tibia breaking strength and tibia ash.

Dietary supplement	Breaking strength, kg	Tibia ash, %
(none)*	5.53[c]	38.3[c]
0.18% P†	10.06[a]	46.1[a]
YAP‡	7.84[b]	42.7[b]
ANAP§	7.98[b]	42.7[b]
PHY¶	8.47[b]	43.6[b]
SEM	0.34	0.5

*Basal diet: 0.27% available P, 0.54% total P, 0.65% Ca
†Provided 0.45% available P, 0.70% total P, 0.92% Ca
‡Yeast acid phosphatase, 1200 units/kg
§*A. niger* phosphatase, 1200 units/kg
¶*A. niger* phytase, 1200 units/kg
[a,b,c]$P < 0.05$.

However, the improvements noted with the highest levels of enzyme supplementation were not as great as those obtained with the addition of 0.18% inorganic phosphorus.

We were interested in calculating the increase in available phosphorus due to using the enzymes. Per cent tibia ash was the parameter that had the most linear response to graded levels of supplemental calcium and phosphorus (Figure 1). Based on a linear regression equation for tibia ash vs. % available phosphorus in the diet ($Y = 4.16X + 27.5$, $r^2 = 0.82$), the equivalent available phosphorus was calculated for the various enzyme supplements. Supplementing the basal diet with 1200 units of yeast acid phosphatase, *A. niger* acid phosphatase and phytase increased the available phosphorus in the diet by 0.10, 0.10 and 0.12%,

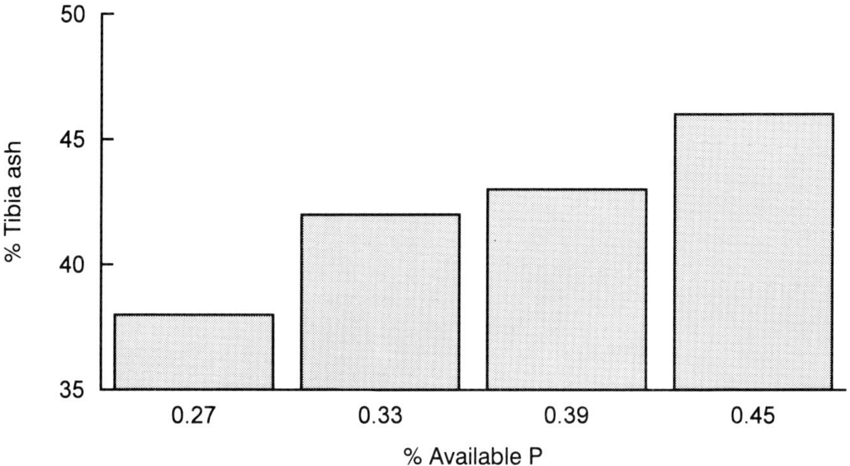

Figure 1. Tibia ash as a function of dietary available phosphorus.

Table 5. Calculated increased available phosphorus based on tibia variables.

	Increased available P(%) based on:	
Enzyme	Breaking strength*	Per cent ash†
YAP‡	0.09	0.10
ANAP§	0.10	0.10
PHY¶	0.11	0.12

*Strength = 28.4 X (available P) − 2.4; $P < 0.001$, $r = 0.89$

†% ash = 41.6 X (available P) + 27.5; $P < 0.001$, $r = 0.90$

‡Yeast acid phosphatase, 1200 units/kg

§*A. niger* phosphatase, 1200 units/kg

¶*A. niger* phytase, 1200 units/kg

respectively (Table 5). A similar calculation was made using the results obtained with tibia breaking strength. Increases of 0.09, 0.10 and 0.11% available phosphorus were obtained from using 1200 units of yeast acid phosphatase, *A. niger* acid phosphatase and phytase, which is in close agreement with the values calculated from the tibia ash data.

The basal diet was calculated to contain 0.27% available phosphorus and 0.52% total phosphorus. Therefore, the unavailable phosphorus, which was presumably mostly in the form of phytic acid, was equal to 0.25%. Based on the assumption that the increases in available phosphorus were due to hydrolysis of phytate, we then calculated the percentage of phytate phosphorus liberated by the three enzymes using the values from Table 5. This value ranged from 36 to 44% and from 40 to 48% using breaking strength and ash data, respectively (Table 6).

Table 6. Percentage of phytate phosphorus hydrolyzed based on tibia variables.

	% Phytate phosphorus hydrolyzed based on:	
Enzyme	Breaking strength	Per cent ash
YAP†	36	40
ANAP‡	40	40
PHY§	44	48

*Calculated phosphorus values: total = 0.52%, available = 0.27%, phytate = 0.25%
†Yeast acid phosphatase, 1200 units/kg
‡*A. niger* phosphatase, 1200 units/kg
§*A. niger* phytase, 1200 units/kg

Conclusions

The commercial phytase and the two experimental acid phosphatases were all effective in increasing phosphorus availability in the present study. The increases in available phosphorus due to enzyme supplementation were approximately 0.10%, which corresponds to roughly 40% of the phytate phosphorus. Thus, it appears that enzyme supplementation can be an effective method of replacing some of the supplemental phosphorus in poultry diets.

THE IMPACT OF PHYTASE ON PIG PERFORMANCE

NAHEEDA KHAN and D.J.A. COLE

Faculty of Agricultural and Food Sciences,
University of Nottingham, Sutton Bonington, UK

Introduction

Phytase has received much attention in recent years due to environmental constraints being imposed on the livestock production sector. Around two-thirds of the total phosphorus in pig diets exists in the form of insoluble phytates which must be degraded before the phosphorus can be absorbed in the intestine. Until recently, dietary supplementation with inorganic phosphates has been an unavoidable practice. But while these phosphates are mainly absorbed, much of the phytate phosphorus passes undigested into the faeces. Application of slurry to the land, and eventual leaching and run-off leads to eutrophication of water sources, and in many countries has prompted legislation aimed at controlling phosphate output from the farming sector. Maximum levels set for soil phosphate levels are rigorously imposed and fines levied on any offending producers. The outcome has been a limitation on the number of pigs that may be kept in an area, and has fuelled the recent interest into commercial application of the phytase enzyme.

Pioneering phytase use

Under experimental conditions, the use of microbial phytase in pig diets has demonstrated consistent improvements in phosphorus digestibility. The enzyme hydrolyses phytate by stepwise removal of orthophosphates, releasing them for absorption in the gut. Initial work carried out in the Netherlands showed an increase of 24% in phosphorus digestibility (Simons *et al.*, 1990). Since then, phytase from *Aspergillus niger* has been tested in pigs by several groups of workers (e.g. Ketaren *et al.*, 1993; Jongbloed *et al.*, 1992; Cromwell *et al.*, 1993). Responses range from an increase in apparent digestibility of phosphorus and other nutrients to an improved growth rate, feed conversion efficiency and protein deposition rate. Perhaps the most meaningful aspect of work to date is the indication that performance can be maintained even in the absence of phosphate supplements. One could anticipate that if used correctly, phytase could largely replace inorganic phosphate, particularly in regions where soil phosphate levels sources are of concern. This laid the foundation of a research programme at the University of Nottingham. Throughout a three-year period, a series of trials was undertaken to assess the potential of phytase; its effect on digestibility of phosphorus and other nutrients, optimum inclusion level, and effect on growth and bone development of growing pigs.

ESTABLISHING A PHYTASE EFFECT

Initial digestibility work used high-phytate diets to examine the effects of phytase on dietary availability of phosphorus and other nutrients associated with phytate. Determination of both ileal and total tract digestibility of nutrients was made possible by using surgically modified animals fitted with a simple T-piece cannula. Phytase produced by *Aspergillus niger* (var. *ficuum*) was added to the feed to provide a resultant activity of 1000 units/ kg feed.

Addition of phytase resulted in an increased ileal and total tract digestibility of phosphorus. Pigs receiving phytase had a significantly lower phosphorus concentration in the faeces (1.12% vs 1.63% P = 0.05), which was reflected in the total amount of phosphorus excreted in the faeces. There were no discernable effects of the enzyme on calcium digestibility. However, an increased protein digestibility was observed.

THE OPTIMUM INCLUSION RATE

Up until this point, around 800–1000 phytase units/kg feed were considered to be the optimum (one unit being defined as the amount of enzyme necessary to liberate 1 μmol of inorganic phosphate per minute under standard assay conditions of 37°C and pH 5.5). But variation in phytase source and experimental conditions had been the cause of some 'grey areas' in work from different research groups; and it was therefore considered necessary to look more closely at the relationship between phosphorus digestibility and level of phytase in the diet.

Phosphorus digestibility at four inclusion levels of phytase was tested (0, 500, 1000 and 1500 units/kg). The quadratic response obtained is illustrated in Figure 1. To explain the curve in biological terms, it is known that phosphorus uptake, unlike, for example calcium, is not finely regulated at the gut level in pigs. The changes in apparent digestibility reflect changes in availability of phosphorus in the digestive tract rather than uptake *per se*, and in this case, could be attributed to enzyme activity in the gut. There may be a cumulative limiting effect of the enzyme either directly, for example by end-product inhibition, or indirectly by altering the gut environment which could become unfavourable to phytase, so that as the enzyme builds up the effect becomes limiting. While maximum phosphorus digestibility was achieved with 1000 units of enzyme per kg feed, a level of 400–500 units gave a notable increase in digestibility

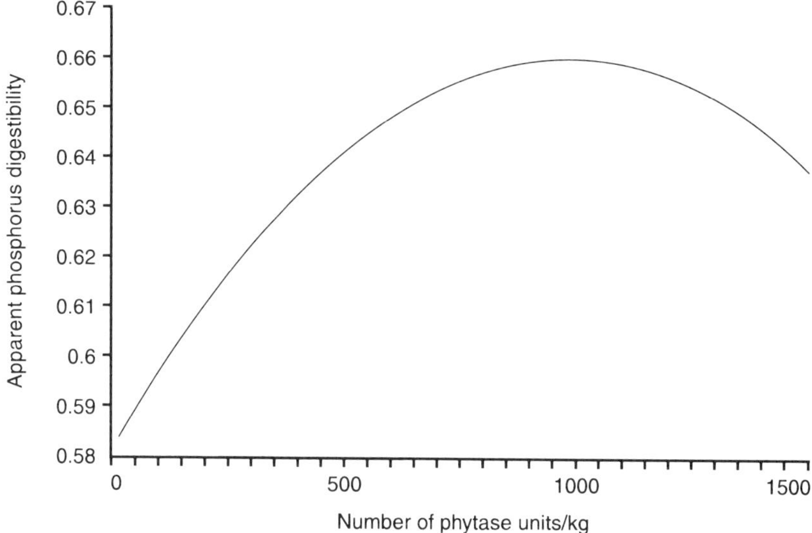

Figure 1. Effect of phytase level on phosphorus digestibility.
$y = 0.5832 + 0.000162x - 0.84E^{-7}x^2$

and from an economic viewpoint would seem the sensible dosage rate to use.

It was interesting to note that while urinary phosphorus excretion was minimal in animals receiving no phytase, it increased dramatically with increasing levels of the enzyme. In pigs, phosphorus surplus to requirement is excreted via the urine, hence urinary phosphate concentration can be used as a reliable indicator of phosphorus status. While it may be argued correctly that phosphorus excreted via the urine is no less harmful than that excreted in the faeces, it should be pointed out that in this experiment the basal level of phosphorus was near to the animal's requirement. Had a lower level been used one could expect that the newly available phosphorus would be retained rather than shunted into the urine as occurred in this experiment. However, these results were encouraging as they provided further evidence that adding phytase to the diet increased the dietary availability of phosphorus.

Evaluating the phosphorus requirement

While the role of phosphorus in many metabolic functions is well documented, perhaps the most fundamental requirement is for development and maintenance of the skeleton, in which the mineral is co-precipitated with calcium in the hydroxyapatite bone complex. In order to satisfy requirements of the growing pig, relatively large amounts of dietary phosphorus must be provided. However, discrepancies over variation in estimated phosphorus requirements of growing pigs continue, and have received particular attention during the 'matching feed to requirement' crusade of recent years.

Having established a repeatable increase in phosphorus digestibility with the phytase enzyme, observations were extended to overall performance. Our aims were threefold; to determine the phosphorus requirement for growth, to look at the effects of phytase on growth, and as quantification of phytase/phosphorus substitution was as yet undetermined, to calculate the phosphorus equivalence of phytase. The question 'how much phosphorus could be replaced by phytase?' had yet to be answered. Experiments were designed which used graded dietary phosphorus levels. Response criteria included growth performance, bone breaking strength and phosphorus content of the body. The response to increasing dietary phosphorus could be compared with the response to phytase at each level of phosphorus, thus providing a reliable estimate of the phosphorus/ phytase equivalence.

MONITORING GROWTH OF PIGLETS

The first experiment of this nature looked at pigs growing from 10 kg to 25 kg. For pigs at this stage of growth the phosphorus requirement is not well established. Although it is recognized that maximum bone mineralization is not necessary for optimum growth, it is important that skeletal development in later life should not be impaired by an early deficiency of phosphorus. Nine levels of available (non-phytate) phosphorus were achieved by incremental addition of monoammonium phosphate to a basal maize/wheat/soya diet containing 2 g non-phytate phosphorus/kg. Levels of non-phytate phosphorus were 2, 2.5, 3, 3.5, 4, 4.5, 5, 5.5 and 6 g/kg. Seventy-two piglets were penned individually and assigned to one of the diets, fed *ad libitum* either with or without phytase at 1000 units/kg feed.

THE PHOSPHORUS/PHYTASE EQUIVALENCE

Growth of the piglets is represented in Figure 2. At the lowest level of phosphorus, a reduced feed intake of pigs was apparent. However, addition of phytase counteracted this reduction ($P=0.085$). Indeed, feed intake of pigs receiving the lowest phosphorus level with supplementary phytase approached that of pigs on the 4 g/kg diet (Figure 2). A linear increase in total body phosphorus was obtained with increasing dietary phosphorus ($P<0.001$). Addition of phytase increased the phosphorus content of the body at all except the highest level of dietary phosphorus. It appeared that an intake of 3.5–4 g non-phytate phosphorus/day maximized the bone strength of young pigs; any further intake interfering with mineralization to decrease bone strength.

Use of the enzyme increased femur breaking strength at all levels of dietary phosphorus ($P<0.001$), and femur breaking strength of pigs receiving the basal phosphorus level was equivalent to that of pigs receiving 3 g non-phytate phosphorus. Our calculations showed that

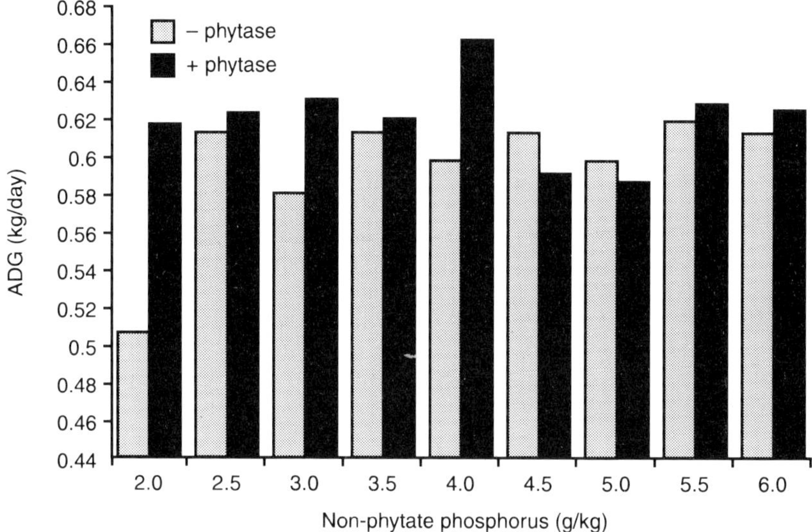

Figure 2. Effect of phosphorus level and phytase on growth of piglets.

adding phytase to a phytate-rich diet at low levels of available phosphorus made available approximately 70% of the phytate phosphorus for bone accretion.

Performance of growing pigs

Work was continued in a similar trial using growing pigs from 25 to 60 kg liveweight to look at the effects of phosphorus level and phytase. Growth performance, carcass measurements and bone development were used as response criteria. By comparing the performance with and without phytase, further assessment of the amount of the phosphorus/phytase equivalence was possible.

Seventy-two male pigs were penned individually and assigned to one of nine levels of non-phytate phosphorus; 0.85 (basal diet), 1.25, 1.65, 2.05, 2.45, 2.85, 3.25, 3.65 or 4.05 g/kg. Diets were based on maize/oat/soya and contained 13.4 MJ digestible energy, 164 g crude protein and 8 g calcium per kg freshweight. As in the previous experiment, diets were fed *ad libitum* either with or without phytase added at 1000 units/kg feed. Pigs were slaughtered at 60 kg, and the left third and fourth metatarsal removed for breaking force determination and thereafter for phosphorus determination.

Mean daily gains are presented in Figure 3. With a striking resemblance to data of the previous trial, results showed a reduced daily gain of pigs receiving the lowest phosphorus diet which was overcome by phytase addition (501 vs 669 g/day; P=0.015), approaching that of pigs on the 1.65 g/kg diet.

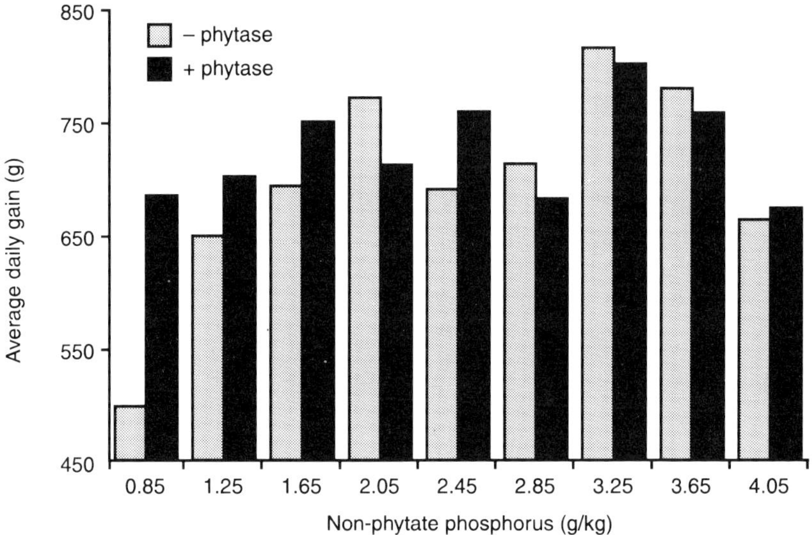

Figure 3. Effect of phosphorus level and phytase on growth of pigs.

RELEASE OF PHYTATE PHOSPHORUS FOR BONE ACCRETION

Quadratic responses of metatarsal strength and phosphorus content to daily intake of non-phytate phosphorus were used to assess the dietary phosphorus intake at which optimum bone integrity was achieved. While a daily intake of 5.5 g non-phytate phosphorus was sufficient for maximum strength of the third metatarsal, 6.5 g/day was required for maximum for strength of the fourth metatarsal. Addition of phytase to the diet resulted in an increased phosphorus content and breaking strength of the metatarsals at levels of up to 2.45 g non-phytate phosphorus/kg. The third metatarsal of pigs receiving the basal phosphorus level was as strong as pigs receiving 2.05 g/kg non-phytate phosphorus, and the fourth metatarsal strength was equivalent to that of pigs receiving 1.65 g non-phytate phosphorus. From this trial it was concluded that adding phytase to a phytate-rich diet at low levels of available phosphorus enabled 50% of the phytate phosphorus to be utilized for bone accretion.

Of notable interest in both piglets and growers was that differences in performance (growth, bone strength, bone mineral content) achieved with the enzyme diminished as the dietary phosphorus level increased towards requirement. This may have been due to a homeostatic effect. In other words, the same amount of phytate phosphorus was always made digestible by the phytase, regardless of the inorganic phosphorus content of the diet. However as the animal reached its required level of phosphorus, more was excreted via the urine. On the other hand, it could be that mineral phosphates exert a negative effect on phytase, decreasing the resultant activity of the enzyme. This area certainly merits further investigation as it has clear implications as to the way in which phytase can be used.

The overall picture?

Since the initiation of phytase work at Nottingham University, a profusion of experimental data have become available from other working groups. Increases in phosphorus digestibility are nearly always obtained, but valid information on other aspects of performance has so far been scarce. Our work goes a step further in showing that the enzyme can liberate up to 70% of the phosphorus which up until now has been excreted in the faeces. In light of increasing environmental pressure, therefore, phytase will fast become an essential component of pig production.

Acknowledgements

The authors wish to thank Alltech, Inc. for funding the work on phytase in pig diets.

References

Cromwell, G.L., T.S. Stahly, R.D. Coffey, H.J. Monegue and J.H. Randolph. 1993. Efficacy of phytase in improving the bioavailability of phosphorus in soybean meal and corn-soybean meal diets for pigs. J. Anim. Sci. 71:1831.

Jongbloed, A.W., Z. Mroz and P.A. Kemme. 1992. The effect of supplementary *Aspergillus niger* phytase in diets for pigs on concentration and apparent digestiblility of dry matter, total phosphorus and phytic acid in different sections of the alimentary tract. J. Anim. Sci 70:1159.

Ketaren, P.P., E.S. Batterham, E.B. Dettmann and D.J. Farrell. 1993. Effect of phytase supplementation on the digestibility and availability of phosphorus in soya-bean meal for grower pigs. Br. J. Nutr. 70:289–312.

Simons, P.C.M., H.A.J. Versteegh, A.W. Jongbloed, P.A. Kemme, P. Slump, K.D. Bos, M.G.E. Wolters, R.F. Buedeker and G.J. Verschoor. 1990. Improvement of phosphorus availability by microbial phytase in broilers and pigs. Br. J. Nutr. 64:525.

NEW INSIGHTS ON THE MODE OF ACTION OF YEAST CULTURES

IVAN D. GIRARD and KARL A. DAWSON

Animal Science Department, University of Kentucky
Lexington, Kentucky, USA

Introduction

Live yeast culture supplements have been shown to increase milk production (Hoyos *et al.*, 1987; Williams *et al.*, 1991), and to improve weight gain and the efficiency of feed utilization (Fallon and Harte, 1987; Hughes, 1988; McLeod *et al.*, 1990; Edwards, 1991) in ruminants. Many of the beneficial production responses associated with the use of live yeast culture supplements in ruminants may be related to their stimulatory effects on specific groups of microorganisms in the rumen. Previous work (Dawson, 1990) established a model which explains the action of yeast culture supplements in the rumen based on both *in vivo* and *in vitro* observations (Figure 1). This model suggests that stimulation of ruminal bacteria by the yeast culture is important in the processes that lead to improved digestive function. Although several studies have suggested that this stimulation is mediated by either live yeast cells or a heat labile component found in yeast culture preparations (Dawson *et al.*, 1990), the specific stimulatory mechanisms are still not

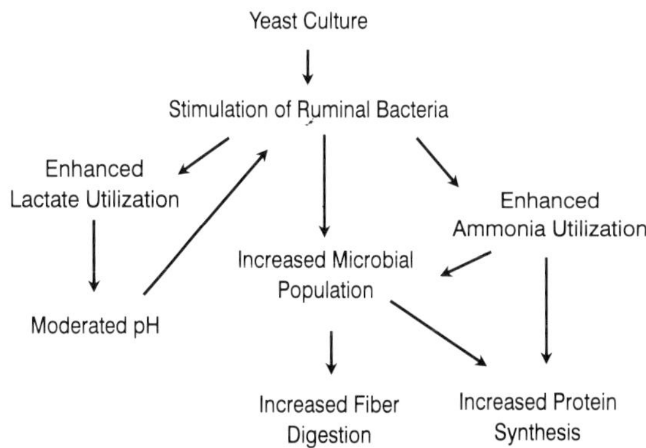

Figure 1. Model for mode of action of yeast culture in the rumen.

completely understood. Recent studies have provided new insights into these stimulatory mechanisms. The aim of this paper is to present new research data that support the proposed model for the mode of action of yeast cultures and help describe the stimulatory effects of live yeast cells on ruminal bacteria.

Increased microbial population

Many studies have shown that yeast culture supplements can increase the concentrations of strictly anaerobic bacteria in the rumen (Dawson, 1990; Dawson *et al.*, 1990). Recent studies have shown that different yeast strains are able to differentially stimulate specific bacterial populations in rumen-simulating continuous cultures fed various rations (Table 1). Yea-Sacc[1026] was able to increase the concentrations of total, cellulolytic and lactate-utilizing bacteria in rumen-simulating cultures fed a high concentrate diet. In contrast, Yea-Sacc[8417] showed a superior ability to stimulate lactate-utilizing bacteria when the supplements were provided with high concentrate diets (Girard *et al.*, 1993). It is clear that responses to yeast are both yeast strain and diet dependent. This work also suggests that a variety of stimulatory actions may be involved and that one mechanism may not completely explain all of the stimulatory activities observed in the rumen. More importantly, these differential effects of different yeast strains may provide nutritionists with new tools for improving animal performance by matching a specific yeast culture to a diet.

Table 1. Responses of ruminal bacterial populations to yeast culture supplementation in rumen-simulating continuous cultures.

Diet	Yeast culture	Bacterial group	Increase in concentrations
High concentrate ration	Yea-Sacc[1026]	Total anaerobes	79%*
		Cellulolytics	79%*
		Lactate-utilizers	23%*
High concentrate ration	Yea-Sacc[8417]	Total anaerobes	17%†
		Lactate-utilizers	44%†
High roughage ration	Yea-Sacc[8417]	Total anaerobes	27%†
		Cellulolytics	56%*
		Lactate-utilizers	21%

*Indicates statistically significant increases ($P<0.10$) with yeast culture supplements.
†Indicates statistically significant increases ($P<0.05$) with yeast culture supplements.

Enhanced lactate utilization

Yea-Sacc[1026] supplementation has also been shown to increase the activities of lactate-utilizing bacteria in rumen-simulating continuous cultures fed a 60% concentrate ration. When batch cultures were estab-

Table 2. Effect of Yea-Sacc[1026] on the rate of DL-lactic acid disappearance (mmol x l^{-1} x h^{-1}) between 2 and 4 h following a 10 mM DL-lactic acid pulse in batch cultures established from rumen-simulating cultures fed a 60% concentrate diet*.

Period	Control	+ Yeast	Difference
Day 4	1.89	2.23†	18%
Day 9	2.44	3.32†	36%

*Girard *et al.*, 1993.
†Indicates statistically significant increases ($P<0.05$) with the yeast culture supplement.

lished from rumen-simulating continuous cultures and supplemented with a solution providing 10 mM of DL-lactic acid, rates of lactate disappearance within the 4 h of incubation were greater in the cultures containing yeast cells (Table 2). These results suggest that yeast cells are able to stimulate both the growth and the ability of lactate-utilizing bacteria to metabolize lactic acid to propionate. The pH of rumen-simulating cultures receiving Yea-Sacc[1026] was also slightly greater than that in the unsupplemented control cultures. In other studies where Yea-Sacc[8417] supplementation increased the concentration of lactate-utilizing bacteria in continuous cultures fed an alfalfa-based dairy diet, the mean concentrations of propionate were greater ($P<0.05$) in cultures receiving the yeast supplement than in the controls (24.5 vs 22.8 mM). This may have been the result of enhanced conversion of lactic acid to propionic acid. The use of specific yeast cultures to stimulate lactic acid utilization and propionate production in the rumen may provide a tool for controlling ruminal fermentations in fast growing animals fed high concentrate diets.

Stimulation of ruminal bacteria

Stimulation of ruminal bacteria is important to the overall beneficial effects of yeast cultures in the diets of ruminants. The effects of any specific yeast culture are a function of its stimulatory action on the individual groups of bacteria in the rumen. Therefore, it was necessary to develop systems which allow for the study of the effects of yeast culture on specific strains of representative ruminal bacteria. The yeast *Saccharomyces cerevisiae* strain 1026 was examined in detail since previous studies have shown beneficial increases in the concentrations of total anaerobic, cellulolytic and lactate-utilizing bacteria. The time required to initiate growth (lag time) was decreased when live yeast cells were added to several cultures of ruminal bacteria including cellulolytic and lactate-utilizing bacteria (Table 3). An example of the stimulatory effects of strain 1026 on the growth of the cellulolytic organism *Ruminococcus albus* is shown in Figure 2. The addition of yeast cells

Table 3. Decrease in the lag time of representative strains of ruminal bacteria by a yeast preparation (strain 1026) in batch cultures.

Strains of bacteria	Lag time (min)		Change (%)
	Control	+ Yeast*	
Prevotella ruminicola, GA-33	92†	59	–36
Butyrivibrio fibrisolvens, D-1	113	89	–21
Ruminococcus flavefaciens, FD-1	193	135	–30
Ruminococcus albus, 7	168	115	–31
Megasphaera elsdenii, T-8	154	99	–36

*The yeast preparation consisted of yeast cells grown on Tryptic Soy Broth (TSB) and provided 2.8 x 10^4 cells/ml. The bacterial basal medium did not support the growth of the yeast.
†Values are the mean of three observations. Lag time changes of 15 min or less were not detectable.

to the cultures of lactic acid-producing bacteria such as *Streptococcus bovis* and *Lactobacillus plantarum* did not affect lag time nor cell yield (Girard and Dawson, 1994). These studies have clearly established the differential stimulatory effects of yeast on specific beneficial groups of ruminal bacteria.

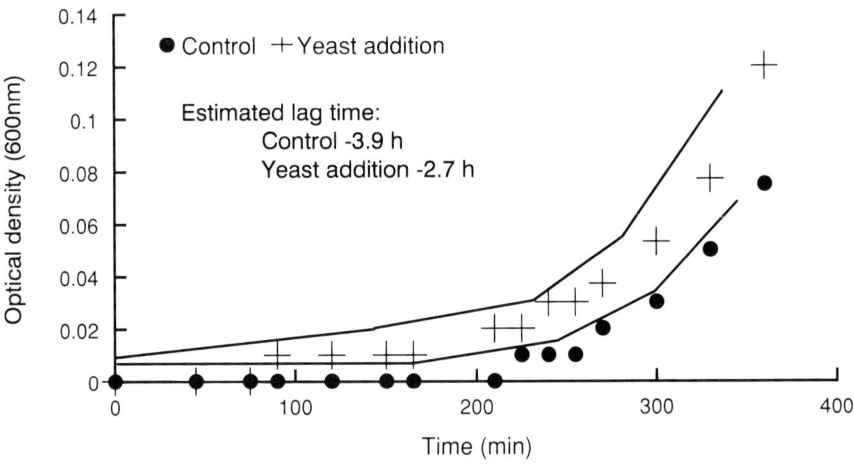

Figure 2. Stimulatory effects of live yeast on the growth of *Ruminococcus albus*.

Increased fiber digestion

Recent studies (Dawson and Hopkins, 1991) have confirmed stimulatory effects of yeast on the activities of isolated cellulolytic bacteria and have shown that the time required to initiate fiber digestion was decreased by 30% and that the rate of fiber digestion during the initial incubation periods was increased in the presence of live yeast cells (Figure 3). These

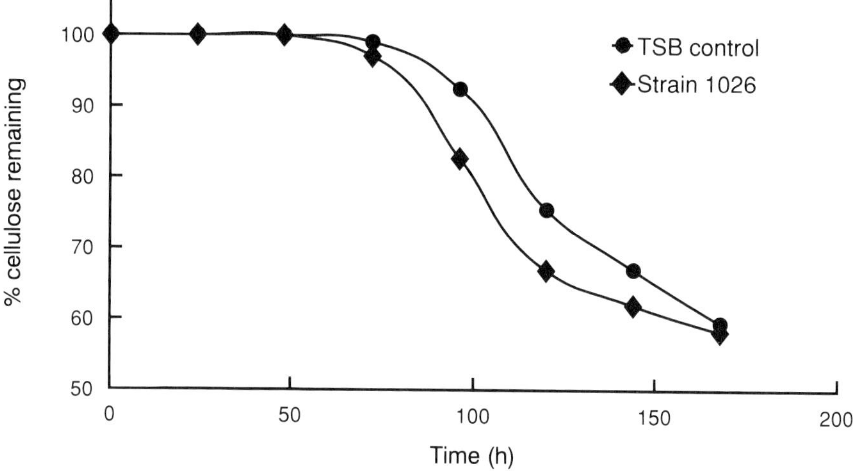

Figure 3. Effect of coculture with *S. cerevisiae* 1026 on lag time to cellulose digestion by *Fibrobacter succinogines* stain S85.

data confirm a direct stimulatory effect of yeast on cellulolytic bacteria in the rumen and support the hypothesis that yeast culture supplementation may have a significant impact on fiber digestion in the rumen.

Cofactor isolation

Studies in coculture systems have been used to further characterize stimulatory interactions between yeast culture and bacteria from the rumen. These systems are currently being used to identify the components in yeast culture preparations that are responsible for the stimulatory activities. Recent experiments have examined a number of different yeast cell fractions and have shown that yeast cells are able to stimulate specific ruminal bacteria by providing both a heat-stable cofactor and a heat-labile cofactor. These cofactors are currently being characterized (Table 4). However, presence of the live yeast cells is required to obtain the maximum stimulatory effect.

Table 4. Characteristics of the two cofactors responsible for stimulation of bacterial growth from *Saccharomyces cerevisiae* strain 1026*.

Characteristics	Cofactor A	Co-factor B
Heat (121°C for 20 min)	Stable	Unstable
Isolation fractions	Supernatant	Intracellular
Size	< 10 000 MW	< 10 000 MW
Activity in mitochondria	No	No
Activity in respiratory deficient 1026 mutants	Yes	Yes

*Unpublished data.

Conclusion and implications

Yeast cultures are able to stimulate specific bacterial populations in the rumen leading to increased fiber digestion and lactate utilization. Experiments using *Saccharomyces cerevisiae* strain 1026 and cultures of ruminal bacteria support the proposed model of stimulatory activities in the rumen and suggest that yeast cells are able to provide two cofactors that are responsible for some of these stimulatory actions. It does not appear that all yeast strains produce the cofactors needed to stimulate all of the important microbial populations in the rumen. Further research will allow for the isolation of the cofactors produced by different yeast cells and may provide the information needed to optimize the stimulation of specific microbial populations in the rumen.

References

Dawson, K.A. 1990. Designing the yeast culture of tomorrow – Mode of action of yeast culture for ruminants and non-ruminants. In: Biotechnology in the Feed Industry, Vol. V. Alltech Technical Publications, Nicholasville, KY.

Dawson, K.A., K.E. Newman and J.A. Boling. 1990. Effects of microbial supplements containing yeast and lactobacilli on roughage fed ruminal microbial activities. J. Anim. Sci. 68:3392.

Dawson, K.A. and D.M. Hopkins. 1991. Differential effects of live yeast on the cellulolytic activities of anaerobic ruminal bacteria. J. Anim. Sci. 69(Suppl. 1):531.

Edwards, I.E. 1991. Practical uses of yeast culture in beef production: Insight into its mode of action. In: Biotechnology in the Feeds Industry, Vol.VI. Alltech Technical Publications, Nicholasville, KY.

Fallon, R.J. and F.J. Harte. 1987. The effects of yeast culture inclusion in the concentrate diet on calf performance. J. Dairy Sci. 70(Suppl. 1):143.

Girard, I.D. and K.A. Dawson. 1994. Effects of a yeast culture on the growth characteristics of representative ruminal bacteria. J. Anim. Sci. 72(Suppl. 1):300.

Girard, I.D., C.R. Jones and K.A. Dawson. 1993. Lactic acid utilization in rumen-simulating cultures receiving a yeast culture supplement. J. Anim. Sci. 71(Suppl. 1):288.

Hoyos, G., L. Garcia and F. Medina. 1987. Effects of feeding viable microbial feed additives on performance of lactating cows in a large dairy herd. J. Dairy Sci. 70(Suppl. 1):217.

Hughes, J. 1988. Effect of high strength yeast culture in diets of early-weaned calves. Anim. Prod. 46:526.

McLeod, K.R., K.J. Karr, K.A. Dawson, R.E. Tucker and G.E. Mitchell, Jr. 1990. Rumen fermentation and nitrogen flow in lambs receiving yeast culture and(or) monensin. J. Dairy Sci. 73(Suppl. 1):266.

Williams, P. E. V., C. A. G. Tait, G. M. Innes and C. J. Newbold. 1991. Effects of the inclusion of yeast culture (*Saccharomyces cerevisiae* plus growth medium) in the diet of dairy cows on milk yield and forage degradation and fermentation patterns in the rumen of steers. J. Anim. Sci. 69:3016.

USING YEAST CULTURE AND LACTIC ACID BACTERIA IN BROILER BREEDER DIETS

REYNALDO GUERRERO MARTIN

Apligén, S.A. de C.V., México City, México

The Mexican poultry industry, as well as other livestock industries, is facing very important challenges this decade. The world economy is difficult and complicated and is affecting even the most powerful countries. The recent opening of Mexico to free markets and the commercial partnership with USA and Canada through the North American Free Trade Agreement (NAFTA) have caused national producers to search for new ways to increase productivity and competitive ability. Changes in the economic situation are challenging producers to improve through the application of modern techniques in areas like management, nutrition, preventive medicine and genetics. New feed additives are playing a role in bringing about increased productivity in the Mexican poultry industry. One type of additive, viable microorganisms, is the subject of this chapter.

Yeast culture and productivity

Yeast are unicellular fungi and have been historically recognized for their fermentative ability in a number of industries. *Saccharomyces cerevisiae* is the yeast used for brewing and baking purposes; and to date is the species of most use in the animal feed industry. When using a yeast culture as an additive for animal feeds, it is very important to choose a strain that has demonstrated positive effects within the bird. *S. cerevisiae* strain 1026 has been proven to have such qualities in laboratory and field trials (Guerrero and Hoyos, 1991; McDaniel and Sefton, 1991). Other strains of *S. cerevisiae* have not proven useful in this regard (Brake, 1991).

One major concern of broiler breeder producers is to maximize the number of chicks per hen. Good performing day-old broiler chicks should:

- be free of *Mycoplasma gallisepticum* and *M. sinoviasis*
- be Salmonella free
- have high livability during the first week

- have high potential for growth
- have efficient feed utilization.

Total chick production depends on: 1) egg production of the female; 2) sperm cell production of the male; 3) mating efficiency of the male; 4) sperm survival in the oviduct; 5) fertility; 6) embryonic survival to hatching; and 7) hatchability (McDaniel,1991).

Shell quality and integrity are important factors affecting embryonic development through incubation. Specific gravity has been used as a practical method for estimating shell thickness. A research paper from the University of Saskatchewan in Canada demonstrated specific gravity variation with age in broiler breeders. Hunton (1992) found that the percentage of eggs with specific gravity less than 1.080 increased from 51% in young hens to 88% in hens between 55 and 56 weeks of age.

Fertility problems and hatchability in females normally start after week 45 of age. Obesity and poor egg shell quality are the main factors associated with hen infertility and early dead embryos (Robertson and McDaniel, 1992). Sefton (1991) reported that the use of yeast culture (Yea-Sacc 1026, Alltech) in Dekalb breeder diets improved egg specific gravity in hens between 60 and 65 weeks of age and improved hatch by 4%. In another study, McDaniel and Sefton (1991) observed in Indian River breeders supplemented with yeast culture (Yea-Sacc 1026) an increase of 3.5% in hatchability. Working with Arbor Acres females from 20 to 50 weeks old using artificial insemination techniques McDaniel reported that adding yeast culture to breeder diets significantly improved hatching percentages in eggs collected 1 to 5 and 6 to 9 days post-insemination (Figure 1). However there was no change in egg production, egg weight, egg shell quality or fertility.

As a general rule, fertility problems in flocks prior to 45 weeks of age are related to the male and are usually due to timidity, delayed sexual maturity or leg problems (Robertson and McDaniel, 1992). Mating

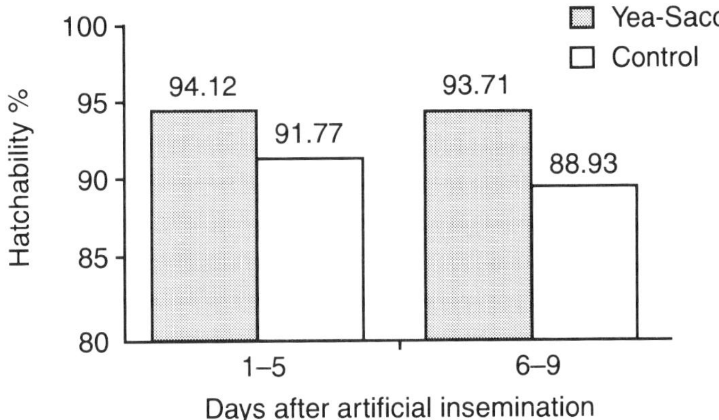

Figure 1. Effect of Yea-Sacc 1026 (1 kg/t) on hatch of fertile eggs days 1–4 and days 5–9 after artificial insemination.

efficiency in males is indispensible; however, sperm number present in the oviduct is a decisive factor in fertility (McDaniel,1991). Behtina (cited by Eslick and McDaniel, 1992) showed that while only one sperm cell penetrates to the ovule to form a zygote, many spermatozoa enter the vitelline membrane. This suggests that many spermatozoa may play a role in the early development of the embryo. Eslick and McDaniel (1992) reported that in insemination of Arbor Acres breeders with different sperm concentrations, the number of sperm had a significant effect on fertility and that early embryonic mortality (1–7 days) is greatly increased as the number of inseminated sperm decrease. Therefore, it is desirable to have an adequate number of active males producing high quality semen.

McDaniel (1991) demonstrated that addition of yeast culture to diets of Ross males at 20 weeks of age increased the percentage of birds in semen production. In addition, sperm concentration was increased in treated males (Figure 2).

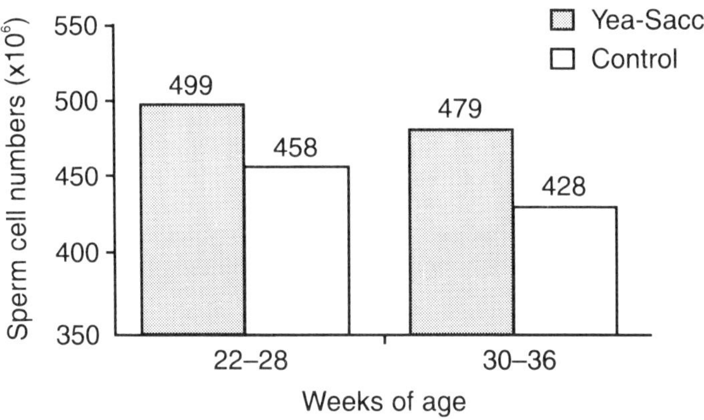

Figure 2. Effect of Yea-Sacc 1026 on sperm cell production.

Profitability of yeast culture and lactic acid bacteria in commercial operations

EFFECT OF LACTO-SACC IN BROILER BREEDERS

Effects of yeast culture in broiler breeders noted in research studies have been corroborated in field trials. Positive results have been achieved particularly in the production phase of broiler breeders; however, in a few cases responses were insufficient to prove cost effective. This is particularly true in flocks that perform very close to genetic potential (Guerrero and Hoyos, 1991). The following field trial was conducted using a commercial product called Lacto-Sacc, a combination of yeast culture (1026), *Lactobacillus acidophilus, Streptococcus faecium* and enzymes (Alltech Inc.).

PROCEDURES

Lacto-Sacc (1 kg/t) was added to diets fed to 4344 Shaver Starbro broiler breeders as part of a technical evaluation at a commercial facility in Córdoba, Veracruz. An additional 4344 broiler breeders served as controls. The test period was from 18 to 40 weeks of age; and the

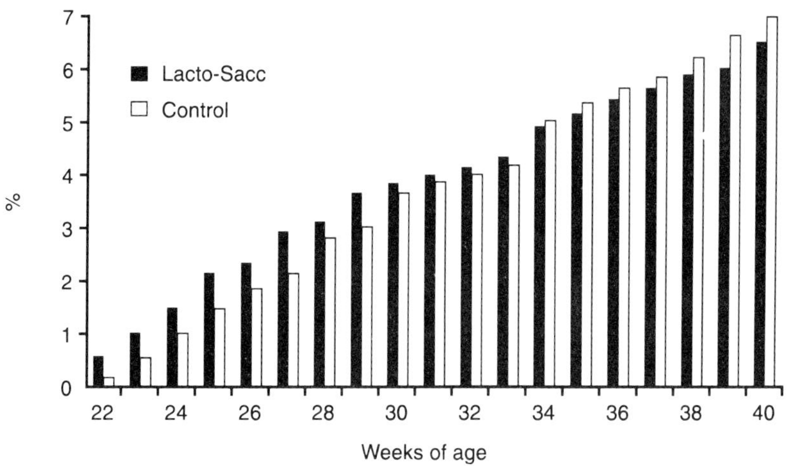

Figure 3. Effect of Lacto-Sacc cumulative mortality of female broiler breeders.

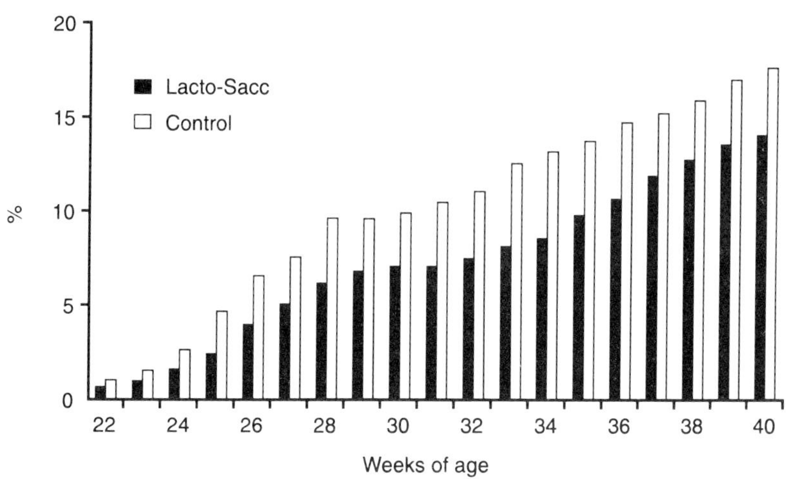

Figure 4. Effect of Lacto-Sacc on cumulative mortality of male broiler breeders.

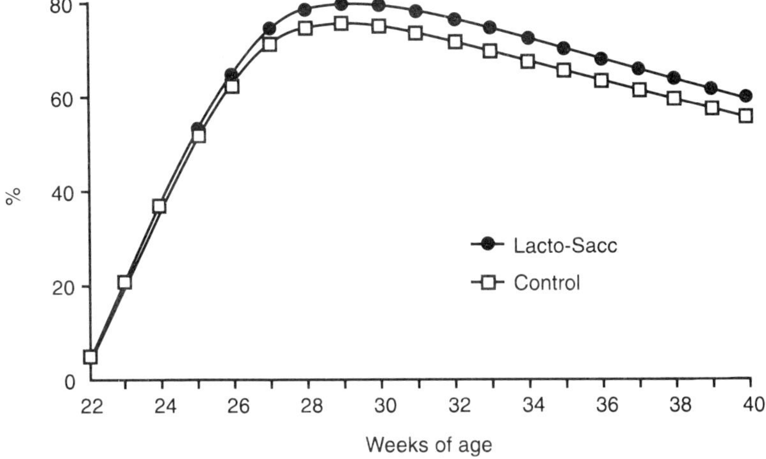

Figure 5. Effect of Lacto-Sacc on egg production (per hen day).

trial took place from March to July of 1991. Cumulative female mortality was reduced by 6.4%, while male mortality declined by 23.5% (Figures 3 and 4). Egg production (per hen day) in the period was improved from 59.9 to 63% (5.1% difference, Figure 5).

First quality chick (total hatch minus culls) production per hen housed was 61.92 in the treated group versus 58.78 in the control group (Figure 6).

At the end of the period, the amount and cost of Lacto-Sacc was calculated based on cumulative feed comsumption per bird. Weekly differences in first quality chicks produced per hen housed and value were also calculated. At the end of the period the relationship between cost and profit was 1 : 3.7 after paying for the product (Figure 7).

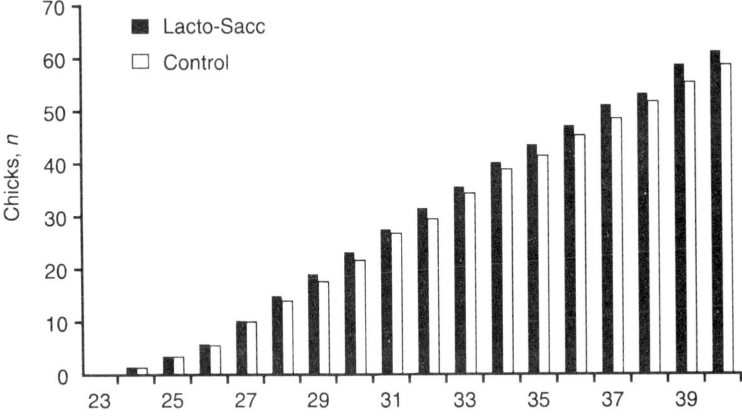

Figure 6. Effect of Lacto-Sacc on production of first quality chicks per hen.

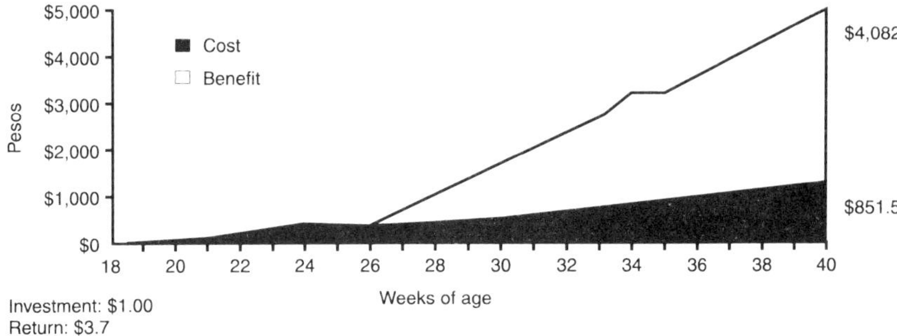

Figure 7. Economic impact of Lacto-Sacc.

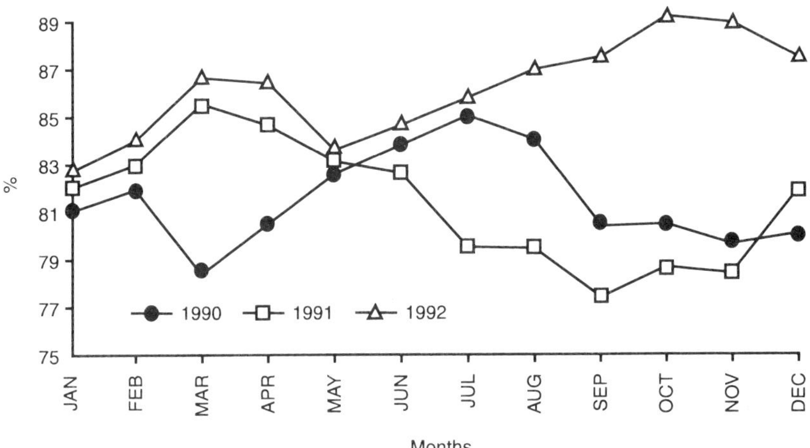

Figure 8. Historical comparison of monthly hatch (Lacto-Sacc started Nov 1991).

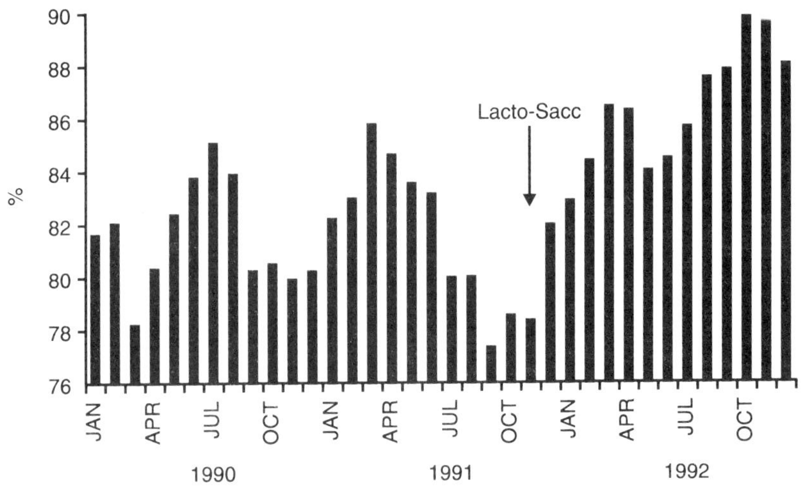

Figure 9. Historical comparison of monthly hatch.

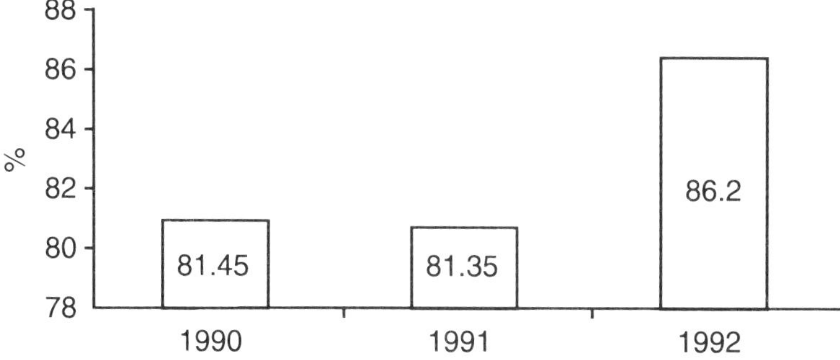

Figure 10. Historical comparison of annual hatch.

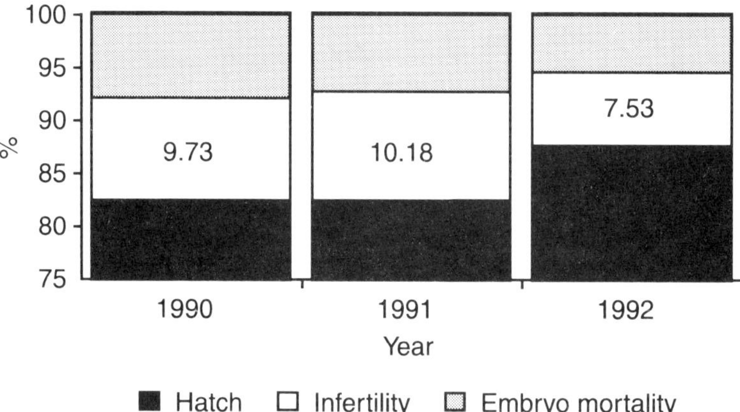

Figure 11. Annual incubation parameter averages.

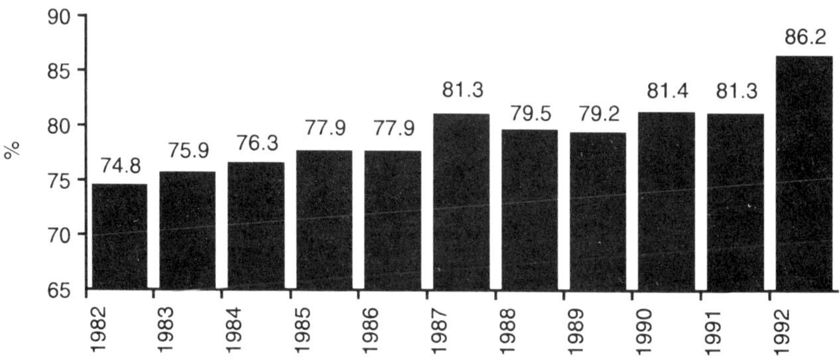

Figure 12. Average annual hatch over a 10-year period.

Based on these data the company began a program for all flocks in production. This program involved 40 000 birds using Lacto-Sacc at 1 kg/t of feed from 18 to 64 weeks of age beginning at the end of October, 1991. After one year of continuous use, monthly and annual results were compared with two previous years (1990 and 1991) with the following results (Figures 8 to 12).

As shown in Figure 11, while hatch was improved by 5.8%, infertility and embryo mortality were reduced by 22.6 and 26%, respectively. The most significant improvement was observed in 1992 (Figure 12).

References

Brake, J. 1991. Lack of effect of a live yeast culture on broiler breeders and progeny performance. Poultry Sci. 70:1037–1039.

Eslick, M.L. and G.R. McDaniel. 1992. Interrelationships between fertility and hatchability of eggs from broiler breeder hens. J. Appl. Poultry Res. 1:156–159.

Guerrero, R. and G. Hoyos. 1991. Biotecnología aplicada en la avicultura. Mecanismos de acción. Resultados en México. In: Biotecnología en la Industria de la Alimentación Animal, Vol. II Apligén, 109–129.

McDaniel, G.R. 1991. The importance of biological products in poultry operations, small improvements, major benefits. In: Biotechnology in the Feed Industry.Proceedings of Alltech's Seventh Annual Symposium. Alltech Technical Publications, Nicholasville, Kentucky, 293–300.

McDaniel, G.R. and T. Sefton. 1991. Effect of yeast culture (Yea-sacc[1026]) supplementation of broiler breeders. Poultry Sci. 70(Suppl. 1):172.

Robertson, R. and G.R. McDaniel. 1992. Solución de problemas de infertilidad en reproductoras. Correo Avícola, Noviembre.

Sefton, T. 1991. El concepto de los probióticos y la producción avícola, evaluación de datos de rendimiento. X ciclo de Conferencias Internacionales de Avicultura. AMENA.

SACCHAROMYCES BOULARDII: MORE APPLICATIONS FOR YEAST CULTURE IN MONOGASTRICS

K.E. NEWMAN

North American Biosciences Center, Alltech Inc., Nicholasville, Kentucky, USA

For years the advantages of supplementing animal diets with yeast culture have been noticed. These cultures have normally consistèd of different strains of *Saccharomyces cerevisiae*. Although numerous trials have proven the benefits of *S. cerevisiae* supplementation in monogastrics, the work of Dawson and others has convincingly shown that this strain of yeast can influence the response in livestock species (Dawson, 1994; Newman and Dawson, 1987; Newman and Spring, 1993). One of the yeast strains receiving a great deal of recognition in monogastrics is *Saccharomyces cerevisiae* var. *boulardii* (known in the industry as *S. boulardii*). *S. boulardii* is an non-pathogenic yeast with an optimum growth temperature of 37°C. It is used in some countries to prevent secondary gastrointestinal infections associated with antibiotic therapy in humans (Surawicz *et al.*, 1989). In animal models, administration of *S. boulardii* has demonstrated marked efficacy in protection from pseudomembranous colitis – a life-threatening inflammation of the colon associated with the overgrowth of pathogenic strains of *Clostridium difficile* (Thoothaker and Elmer, 1984; Corthier *et al.*, 1986). In trials with laboratory animals there were significantly lower concentrations of *C. difficile* in the cecum and a near abolition of toxin B in *S. boulardii* treated animals (Table 1).

Responses observed from *S. boulardii* supplementation are not limited to health and disease. In poultry, studies suggest performance improvements. In broilers receiving a commercial product (Yea-Sacc Mono-

Table 1. Effect of *S. boulardii* on *C. difficile* and toxin B concentrations in hamsters receiving *S. boulardii* (12 weeks following removal of vancomycin from animals).

Treatment	*C. difficile* [log CFU/g]	Toxin B (# positive/total)
Control	8.8[a]	10/10
S. boulardii	5.7[b]	0/10

[a,b] Values differ ($P<0.05$)
Adapted from Elmer and McFarland, 1987.

Table 2. Effect of Yea-Sacc Monogastric containing *S. boulardii* (treatment) on broiler performance compared to virginiamycin at 21 days.

Treatment	Gain (grams)	Feed conversion	Mortality (*n*)
Control	664	1.581	4
Treatment	698	1.505	2

Adapted from Compton, 1993.

Table 3. Effect of *S. boulardii* on performance and intestinal integrity in turkey poults.

Treatment	21 d weight (g)	Feed:gain ratio	Goblet cells (#/mm)	Crypt depth
Control	464^a	1.64^a	40.7^a	0.19^a
Treatment	527^b	1.58^b	31.9^b	0.14^b

^{a,b} Values in columns differ ($P<0.05$).
Adapted from Bradley *et al.*, 1994.

gastric) containing *S. boulardii*, improvements in performance were noted. Although not significant, trends were observed which indicated increased gain, improved feed conversion, and decreased mortality (Table 2).

Other investigators have found improvements in feed utilization with the inclusion of this strain of yeast in broiler rations. No improvements were noticed in body weight, mortality, incidence or severity of leg disorders, dressing percentage or abdominal fat content (Madrigal *et al.*, 1993).

In turkey poults, similar responses have been noted. Bradley and co-workers found improvements in weight gain, feed to gain ratio, goblet cell numbers and crypt depth (Table 3). The reduction in goblet cell numbers was attributed to an alteration in the microbial ecosystem due to the yeast and the decreased crypt depth could be indicative of a decreased turnover rate of these cells due to lower concentrations of toxins from certain bacterial populations. Decreased turnover rate of epithelial cells is thought to conserve energy that could in turn be used for gain (Bradley *et al.*, 1994).

Summary

Saccharomyces boulardii has been used successfully for the prevention of diarrhea and other disturbances caused by antibiotic administration. This strain of yeast, like many others, has been shown to remain viable in the gastrointestinal tract (Albert *et al.*, 1977) and inhibit the growth of a number of bacteria (e.g. *Staphylococcus aureus, Proteus vulgarus)* and yeasts (*Candida albicans*) *in vitro* and *in vivo* (Bizot, 1955; Massot *et al.*, 1977). Improvements in animal performance have been noted in broilers and poults.

References

Albert, O., J. Massot, and M.C. Courtois. 1977. Vie Med. 18:1604.
Bizot, M. 1955. La Press Med. 63:1251.
Bradley, G.L., T.F. Savage, K.I. Timm. 1994. Poultry Sci. 73:1766.
Compton, M.R. Masters Thesis. Stephen F. Austin University, Texas.
Corthier, G., F. Dubos and R. Ducluzeau. 1986. Can. J. Microbiol. 32:894.
Dawson, K.A. 1994. California Nutrition Conference.
Elmer, G.W. and L.V. McFarland. 1987. Antimicrob. Agents Chemother. 31:129.
Massot, J., M. Desconclois and F. Patte. 1977. Bull. Soc. Mycol. Med. 6:277.
Newman, K.E. and K.A. Dawson. 1987. Rumen Function Conference, Chicago.
Newman, K.E. and P. Spring. 1993. J. Anim. Sci. 71(Suppl. 1):289.
Surawicz, C.M., G.W. Elmer, P. Speelman, L.V. McFarland, J. Chinn and G. van Belle. 1989. Gastroenterology. 96:981.
Toothaker, R.D. and G.W. Elmer. 1984. Antimicrob. Agents Chemother. 26:552.

COMPETITIVE EXCLUSION OF SALMONELLA USING BACTERIAL CULTURES AND OLIGOSACCHARIDES

P. SPRING

North American Biosciences Center, Nicholasville, Kentucky, USA

Introduction

Each food borne disease outbreak clearly demonstrates that the responsibility for safe food handling cannot be left to food processing plants and consumers alone. Egg and poultry meat producers themselves must take all possible measures to improve the safety of their products entering the human foodchain. Although most producers are aware of their responsibility and extensive measures have been taken to control food borne pathogens, the problem is not under control. A recent survey conducted in different European countries on 1700 samples of poultry meat revealed that nearly 50% of all tested samples were infected with either campylobacter or salmonellae or both (International Testing, 1994). However, large differences were revealed among countries. No salmonella-infected birds and less than 10% campylobacter-infected birds have been reported from Sweden and Norway. This clearly demonstrates that the right combination of measures taken can reduce the occurrence of food-borne pathogens to an acceptable level. The concept of competitive exclusion is one link in the chain of measures that are taken to control food-borne pathogens in Scandinavian countries.

Competitive exclusion

The current intensive rearing conditions with no contact between the chick and the hen delays development of the intestinal flora in the young bird. The introduction of gastrointestinal microflora from a healthy adult bird into one-day-old chicks has been shown to accelerate maturation of the gut microflora in intensively-reared chicks and to increase the resistance of most treated chicks to salmonellae colonization. This concept, known as competitive exclusion (CE), is based on the fact that a mature intestinal microflora excludes certain enteric pathogens such as salmonellae from establishing themselves in the gastrointestinal tract. The regulatory mechanisms that exclude enteric pathogens from the gut are the result of complex bacterial interactions. Four general mechanisms have been proposed based on findings from many different studies:

- Creation of a restrictive physiological environment
- Competition for bacterial receptor sites
- Competition for, or depletion of, essential substrates
- Production and secretion of antibiotic-like substances

The early establishment of a mature intestinal microflora leads to increased volatile fatty acid (VFA) and lactate concentrations and to a lower pH in the gut of the young bird. The combination of a low pH and high VFA concentrations creates an unfavorable environment for many enteropathogens. A low pH enhances the antibacterial activity of VFAs by increasing the proportion present in the undissociated form. Undissociated VFAs penetrate bacterial cell walls more easily because they are uncharged and therefore exert a stronger antibacterial activity than their negatively charged dissociated forms (Barnes *et al.*, 1979; Hinton *et al.*, 1990). Nisbet *et al.* (1993) showed that birds containing higher cecal concentrations of undissociated propionic acid had lower numbers of salmonellae. Other restricting factors include the reduction in oxidation-reduction potential and increased hydrogen sulfide concentration.

The ability to adhere to mucosal surfaces is important in order to establish or maintain colonization in the gut, especially in areas of active peristalsis. Since the mucosal surface area is limited, the introduction of bacteria into the intestinal environment creates competition for mucosal attachment sites. Schneitz *et al.* (1993) convincingly demonstrated that the ability to inhibit enteropathogen colonization depended on the adhering ability of the CE culture. A wild strain of *Lactobacillus acidophilus* that adhered strongly to chicken epithelial cells was tested for its ability to displace *Salmonella infantis*. When the same strain was subcultured several times on a selective agar it lost its adhering ability. This change was accompanied by a marked loss of protective activity against colonization of *Salmonella* in newly hatched chicks.

The effect of nutrient competition is difficult to assess in relation to other inhibitory factors which may function in the environment (Rolfe, 1991). Investigation of the inhibitory activity of coliforms against *S. flexneri* in an *in vitro* culture showed that the inhibition could be reversed by the addition of glucose. This demonstrates that the inhibition of *S. flexneri* by coliforms was due to the competition for carbon and energy sources (Rolfe, 1991). Inhibition due to competition for substrates can exist for any limiting nutrient. In addition to nutrient competition, antibiotic-like substances have been identified that can inhibit enteropathogens including ammonia, hydrogen peroxide, bacterial enzymes, bacteriophages and bacteriocins.

BACTERIAL CULTURES

Nurmi and Rantala (1973) were the first to recognize the importance of early establishment of mature intestinal microflora in young chicks. Since their initial, work many different research groups have confirmed that the application of an undefined culture of intestinal microflora from a healthy adult bird to one-day-old chicks is effective in preventing salmonella colonization in young birds under laboratory conditions. Salmonella control programs that use CE to improve the hygienic conditions during production or processing of birds have been successful in controlling salmonella contamination of poultry meat.

Despite the fact that undesirable side effects have never been reported from the use of undefined CE cultures in the field, many regulatory agencies are still concerned about the safety of these products. Different research groups are therefore developing defined cultures to eliminate any safety concerns.

The application of pure cultures of streptococci, lactobacilli, clostridia, bacterioides or bifidobacteria have been shown to protect chicks against *Salmonella typhimurium* colonization in some cases. In other cases no protection was achieved or the treatment even led to a 10- to 100-fold increase in cecal concentrations of *S. typhimurium* (Impey *et al.*, 1982). Some pure cultures can disturb the ecological balance of the intestinal microflora and exclude beneficial bacteria. Exclusion of beneficial bacteria makes the intestinal microflora less complex and less competitive in the fight against enteropathogens. Many different mechanisms are involved in CE and it is therefore very unlikely that one single strain or a mixture of a limited number of strains could be as effective as an undefined culture. Nisbet *et al.* (1993) developed a defined culture combining 11 strains of bacteria and tested its ability to exclude *S. typhimurium* with or without addition of dietary lactose (Table 1). Protection against *S. typhimurium* was only satisfactory with the addition of lactose. Dietary lactose can be used to reduce the number of bacterial strains required in a defined CE culture to protect chicks against salmonellae. A reduction in the required numbers of bacterial strains facilitates production and quality control of such CE products.

Table 1. Effect of a defined CE culture and lactose on *S. typhimurium* colonization in chicks.

Treatment	Log_{10} CFU/g cecal content	% positive birds
Control	6.26	97
5% dietary lactose	4.57	83
CE culture	4.13	87
CE culture + lactose	2.21	60

OLIGOSACCHARIDES

The ability of simple sugars such as lactose, mannose and galactose to change the composition of the gut microflora has been known since the beginning of this century (Rettger, 1915) and has been thoroughly investigated. Lactose has been shown, when included at a rate of 5% in the diet or 2.5% in the drinking water, to decrease *Salmonella typhimurium* colonization in the ceca of chicks challenged shortly after hatching with different doses of this microorganism by two to four $\log_{10}$ units (Oyofo *et al.*, 1989; Hinton *et al.*, 1990; Hollister *et al.*, 1994a,b). Unfortunately the concentrations of simple sugars required to alter the gut microflora and to decrease colonization of pathogenic bacteria are relatively high and therefore not often used in commercial production units.

More recently it has been observed that certain complex carbohydrates such as mannan- (MOS), fructo- (FOS), galacto- (GOS) and isomalto-oligosaccharides (IMO) are effective in altering the gut microflora when added at lower inclusion rates. Oligosaccharides are not hydrolyzed during digestion because the gut enzyme complex does not contain the right enzymes to break their linkages. The absorption of other nutrients in the small intestine concentrates dietary oligosaccharides in the lower intestine, which may partly explain why these polymers are more effective in altering the composition of the gut microflora at lower inclusion rates than simple sugars (Küther, 1991). Oligosaccharides affect the GI microflora by:

- serving as substrate for beneficial intestinal bacteria,
- adsorbing to certain enteric bacteria,
- modulating the immune system.

As with simple sugars, oligosaccharides are selectively used by bacteria. FOS, GOS and IMO serve as substrates for many beneficial intestinal bacteria such as lactobacilli and bifidobacteria while being used less efficiently by coliforms. MOS, in contrast, is hardly broken down by intestinal bacteria.

Mannanoligosaccharides

A trial conducted on 100 birds at the Czech Research Institute (Sisak, 1994) showed promising effects of MOS on *Salmonella* colonization in chicks. One kilogram of dietary MOS has been shown to reduce the colonization of an invasive wild type of *Salmonella* in the cecum as well as in the liver and spleen. The number of birds displaying natural infections was reduced in the treatment group with 76% fewer cecal infections and 66% fewer organ infections being detected in the MOS group versus the control group. These results are the first reported that demonstrate that carbohydrates included as low as 1000 ppm in the diet significantly reduce colonization of enteric pathogens such as salmonella.

Table 2. Effect of mannanoligosaccharide* on Salmonella colonization in chicks.

Treatment	Cecum		Organs	
	Colonized birds	% colonization	Colonized birds	% colonization
Control	38/50	76	42/50	84
MOS, 1 kg/t	9/50	18	14/50	28

*BioMos, Alltech Inc, Nicholasville KY

Table 3. Fimbriae and adhesion of salmonellae.*

Serotype	No. of strains tested	Fimbriated	Mannose specificity	
			No. of strains	% of fimbriated strains
S. paratyphi A	78	2	2	2.6
S. paratyphi B	135	125	106	78.5
S. typhimurium	775	668	668	86.2
S. typhi	150	122	122	81.3
S. gallinarum	14	14	0	0
S. pullorum	30	11	0	0
S. enteritidis	21	21	21	100.0
Other salmonellae	250	221	220	88.0
Total	**1453**	**1184**	**1139**	**78.4**

*Adapted from Duguid *et al.*, 1966

While bacteria from CE cultures inhibit adherence of enteropathogens by competing for attachment sites on the mucosa, MOS inhibits adherence of bacteria by blocking their lectins. Bacteria with blocked lectins cannot adhere to the intestinal wall and can therefore not establish themselves in the gut. The use of MOS to inhibit salmonellae colonization is a promising approach since many salmonellae attach to the mucosal cells by mannose specific lectins (Table 3).

The use of MOS to prevent enteropathogen colonization is a very promising concept. Further research is required to evaluate the effect of MOS from different sources on the colonization of different enteropathogens. Specific treatment methods to alter the structure of MOS could improve its effect in the gastrointestinal tract and on the immune system. A better understanding of the exact oligomer structure that is required to bind bacteria or to enhance immune response would allow a better selection of the MOS source and would allow the use of purposeful modification procedures.

References

Barnes, E.M., C.S. Impey and B.J.H. Stevenson. 1979. Factors affecting the incidence and anti-salmonella activity of the anaerobic cecal flora of the young chick. J. Hyg. Camb. 82:263–283.

Duguid, J.P., E.S. Anderson and I. Campbell. 1966. Fimbriae and adhesive properties in salmonellae. J. Path. Bact. 92:107–138.

Hinton, A., D.E. Corrier, G.R. Spates, J.O. Norman, R.L. Ziprin, R.C. Ross and J.R. Deloach. 1990. Biological control of *Salmonella typhimurium* in young chickens. Avian Dis. 34:626–633.

Hollister, A.G., D.E. Corrier, D.J. Nisbet, R.C. Beier and J.R. DeLoach. 1994a. Comparison of effects of chicken cecal microorganisms maintained in continuous cultures and provision of dietary lactose on cecal colonization by *Salmonella typhimurium* in turkey poults and broiler chicks. Poult. Sci. 73:640–647.

Hollister, A.G., D.E. Corrier, D.J. Nisbet and J.R. DeLoach. 1994b. Effect of cecal culture encapsulated in alginate beads or lyophilized in skim milk and dietary lactose on *Salmonella* colonization in broiler chicks. Poult. Sci. 73:99–105.

Impey, C.S., G.C. Mead and S.M. George. 1982. Competitive exclusion of salmonellae from chick caecum using a defined mixture of bacterial isolates from the caecal microflora of an adult bird. J. Hyg., Camb. 89:479–490.

International Testing. 1994. Microbiological safety of chicken meat. I. T. Marylebone Road, London GB.

Küther, K. 1991. Wirkstoffalternativen in der Ferkelfuetterung. In: *Aktuelle Themen der Tierernaehrung*. Lohmann (Ed.) Lohmann Cuxhaven. Cuxhaven, Germany.

Nisbet, D.J., D.E. Corrier, C.M. Scanlan, A.G. Hollister, R.C. Beier and J.R. DeLoach. 1993. Effect of a defined continuous-flow derived bacterial culture and dietary lactose on *Salmonella typhimurium* colonization in broiler chicks. Avian Dis. 37:1017–1025.

Nurmi, E. and M.W. Rantala. 1973. New aspect of salmonella infection in broiler production. Nature 241:210–211.

Oyofo, A.O., J.R. DeLoach, D.E. Corrier, J.O. Norman, R.L. Ziprin and H.H. Mollenhauer. 1989. Effect of carbohydrates on *Salmonella typhimurium* colonization in broiler chickens. Avian Dis. 33:531–534.

Rettger, L.F. 1915. The influence of milk feeding on mortality and growth, and on the character of the intestinal flora. J. Exp. Med. 21:365–388.

Rolfe, R.D. 1991. Population dynamics of the intestinal tract. Colonization Control of Human Bacterial Enteropathogens in Poultry. In: L.C. Blankenship (Ed.) Academic Press, San Diego, CA.

Schneitz, C., L. Nuotio and K. Lounatma. 1993. Adhesion of *Lactobacillus acidophilus* to avian intestinal epithelial cells mediated by the crystalline bacterial cell surface layer (S-layer). J. Appl. Bact. 74:290–294.

Sisak, F. 1994. Stimulation of phagocytosis as assessed by luminol-enhanced chemiluminescence and response to salmonella challenge of poultry fed diets containing mannanoligosaccharides. Poster presented at the 10th Annual Symposium on Biotechnology in the Feed Industry. Alltech Inc., Lexington, KY.

MANNANOLIGOSACCHARIDES: EXPERIENCE IN COMMERCIAL TURKEY PRODUCTION

RANDY OLSEN

Poultry Specialist, Northwestern Supply Company, St. Cloud, Minnesota, USA

Introduction

The northern midwestern area of the United States served by Northwestern Supply Company produces about 52 million turkeys annually in flocks grown by both integrators and large independent producers. Over the past year we have been experimenting with addition of Bio-Mos (mannanoligosaccharide derived from yeast cell wall) to diets fed to turkeys under commercial production conditions; a project which appeared promising based on the results seen in broiler diets. Bio-Mos addition to broiler diets improved efficiency and reduced condemnations. These data, along with the immune stimulation properties of the cell wall mannan, prompted us to begin examining effects of this product versus other feed additives on performance parameters, livability, condemnation, litter quality and vaccine titers as well as economic impact.

Methods

Fifteen thousand poults housed in four similar buildings at a midwestern integrator were used in this trial. All birds received coccidiostat for the first 8 weeks. Two buildings also received Bio-Mos at 2 lbs/t for the first 3 weeks then 1 lb/t through 17 weeks. Two other buildings served as a control and received the standard program which included Stafac 20 (virginiamycin) from 8 weeks through 17 weeks.

Results

Turkeys supplemented with Bio-Mos were an average of 0.42 lbs heavier at the end of the growing period (Table 1). At the time this trial was completed (December, 1994) the turkey market price was around \$0.73 per lb. An increase of 0.42 lbs gain represented a \$0.3066 net gain per bird. For the whole flock, this represents a \$2,026.32 advantage (6609 birds × 0.3066).

Table 1. Effect of Bio-Mos on turkey performance and slaughter parameters.

	Control			Bio-Mos			% Change
	Bldg 5	Bldg 10	Total/mean	Bldg 6	Bldg 9	Total/mean	
Head started, *n*	3848	3848	7696	3848	3848	7696	
Head marketed, *n*	3297	3169	6466	3328	3479	6807	
Net live, %	85.68	82.35	84.02	86.49	90.41	88.45	+5.27
Average weight, lbs	32.30	33.88	33.09	33.49	33.49	33.49	+1.21
Daily gain, lbs	0.2504	0.2606	0.2555	0.2596	0.2576	0.2586	+1.21
Condemned, %	3.65	2.27	2.96	2.62	3.19	2.91	–0.05
Conversion for 30.5 lbs	2.56	2.51	2.54	2.49	2.49	2.49	–0.05

Feed conversion was improved from 2.54 in the Stafac group to 2.49 in the group given Bio-Mos. This change saved a total of 1.67 lbs of feed per bird at a cost savings of $0.1169 per bird. For the whole flock, this is a $899.66 advantage (0.1169 × 7696 birds started; cost of feed estimated at $140/ton).

Livability increased in response to Bio-Mos by 4.43% while condemnations were reduced from 2.96% to 2.91%. This resulted in an extra 334 birds or 11,185 lbs of turkey worth $816.55 (6807 × 0.0291 = 6609 birds vs 6466 × 0.0296 = 6275 birds).

Discussion and Conclusions

The improvements in performance, efficiency, livability and condemnations combined to produce $3,742.53 from a $625.72 investment in Bio-Mos, a 6:1 return. These results are representative of those seen in other field trials in this region. Livability is typically slightly improved in birds given Bio-Mos; and has been associated with fewer enteric problems due to *Salmonella*, *Escherichia coli*, *Clostridia* or non-specific enteritis. In another trial turkeys were sacrificed at 6 weeks of age to examine spleen weights in birds fed Bio-Mos versus those fed Stafac. Spleen weights were lower in the treated group which is indicative of an improved response to disease challenge. Turkeys challenged with *E. coli*, erysipelas, or HE typically show enlargement. In addition, vaccine titers to HE and Newcastle have shown slight improvments in birds given diets containing Bio-Mos.

Performance responses in this trial were similar to those seen in other field trials. Body weights are not markedly affected, however feed conversion efficiency is typically improved. On average there has been an improvement of 5 to 12 points for heavy toms and hens in field trials and a 2 to 6 point improvement in trials with consumer hens. This provides a feed saving that contributes significantly to the economic benefit of Bio-Mos (Table 2).

Improvements in condemnation also contribute to the cost advantage of Bio-Mos in production trials. The reduction in condemnation percentage is thought to be due to the drier litter conditions noted. Drier litter aids in reducing leg problems, pathogen challenge and breast

Table 2. Cost savings per turkey produced based on improvements in feed conversion (FCR) noted in field trials.

	Target weight, lbs	Feed saved	Cost savings per bird
Heavy toms*	36 lbs	1.8 to 4.3 lbs	$0.126 to $0.301
Heavy hens*	22 lbs	1.1 to 2.6	$0.078 to $0.182
Consumer toms*	23 lbs	0.7 to 1.4	$0.05 to $0.098
Consumer hens*	14 lbs	0.42 to 0.7	$0.029 to $0.049

*Based on 5–12 point improvement in FCR
†Based on 2–6 point improvement in FCR

blisters. In addition, drier litter means lower ammonia and reduced airsacculitis.

In conclusion, improvements in FCR, livability and condemnation percentage provide the biggest part of the economic advantage of Bio-Mos. Further research will be required to determine the way in which the mannanoligosaccharide brings about this performance response.

ENZYME APPLICATIONS FOR PLANT PROTEINS: TIME TO LOOK BEYOND CEREALS

RON PUGH[1] and PAT CHARLTON[2]

[1]*Agricultural Development and Advisory Service*
[2]*Alltech Inc., UK*

Introduction

The value of enzyme supplementation in barley, wheat and rye-based diets for poultry is well documented. In addition, data from a variety of cereal crops and target species have accumulated in the literature. This technical background has resulted in widespread use of these preparations in the field, with 90–95% of all broiler feeds in such countries as the UK and Ireland using cereal enzymes.

However, the time is right to start looking at the 25–30% of broiler feed that is not derived from cereal grains. In a paper presented at an earlier Alltech Symposium, it was suggested that further advantage could be gained by matching an enzyme formulation to a broader range of the raw materials in the diet. From this perspective we consider also the neutral detergent fibre (NDF), acid detergent fibre (ADF) and soluble non-starch polysaccharide (NSP) levels in other raw materials, and it is clear that the potential for improvement in metabolizable energy (ME) from such raw materials as oilseed rape (canola), peas and beans is great. In the UK development of new or modified vegetable protein sources that can be used in diets at up to 15% has been a major turning point in ration formulation (Table 1). These products generally include full-fat oilseed rape (double zero) combined with peas or beans and

Table 1. Typical UK broiler (finisher) diet.

Wheat (11% CP)	62%
Soya (48% CP)	15%
Oilseed rape/pea/bean compound	10%
Fishmeal	2%
Meat and bone meal	3%
Fat	5%
Premix*	3%

*Including amino acids, minerals, vitamins, growth promoters, etc.

heated using various techniques to break open the oilseed and also to improve digestibility.

While canola, field peas and beans may be less widely used, inclusion of soya is common to most broiler diets around the globe. The high protein and energy content of soya makes it a desirable ingredient; and its expense demands optimum utilization. Enzyme supplementation has not proven especially effective, however, and on occasion negative responses have been noted. The lack of response in some studies has underscored the importance of formulating enzyme supplements on the basis of the specific nature of a given raw material.

Using this approach, enzyme supplements formulated to aid digestion of dietary protein sources in poultry diets were developed. Investigators from the Agricultural Development and Advisory Service (ADAS) and Roslin Research Institute tested the resultant enzyme complex through TME(N) trials with broilers. The results of this 18 month study follow.

USE OF TME(N) TO EVALUATE ENZYME RESPONSE

Evaluation of the enzyme supplements consisted of measuring true metabolizable energy corrected for nitrogen (TME(N)), a method developed by McNab and co-workers at the Roslin Research Institute in Edinburgh as an improvement upon both the apparent metabolizable energy and true metabolizable energy systems. Supporters of the apparent metabolizable energy (AME) system have always noted that true metabolizable energy (TME) analysis is done with older birds (i.e., 20-week-old cockerels) which have a more developed digestive system. In response, detractors of the AME system suggest that the food wastage generated by free-choice feeding undervalues energy levels.

The TME(N) (nitrogen corrected) system developed at Roslin combines the better aspects of each of the AME and TME analyses. To derive TME(N), young birds (3 weeks) are force fed three times daily. This method yields more realistic energy values for broiler applications. Force-feeding avoids the error created by food wastage. To ensure the birds are not overly stressed, they are fed three times daily instead of once a day, as with the older TME system.

Trial 1. Effect of Allzyme Vegpro on oilseed rape (OSR)/legume compounds and soya (extracted)

METHODS

In the first study, an enzyme was developed to enhance the TME(N) for the major non-cereal raw material in a broiler diet. The enzyme product, Allzyme Vegpro, was formulated to complement the legume/oilseed rape (OSR) compound and was added at 1 kg/tonne. Three commercial raw materials were tested: one pea/oilseed rape compound, one bean/oilseed rape compound and Hi-Pro soya (48% protein).

RESULTS

The results obtained were close to those predicted from raw material evaluation. The pea product was expected to respond better to the enzyme than the bean product based on their respective non-starch polysaccharide levels. Enzyme addition increased TME(N) by 14.3, 9.2 and 4.1% for the pea/OSR, bean/OSR, and soy materials, respectively (Table 2). It was also noted that the enzyme reduced the variability among birds in TME(N).

Table 2. Effect of Allzyme Vegpro addition on true metabolizable energy (N-corrected) value of three vegetable protein sources fed broilers.

	TME(N), MJ/kg		
	Control	Allzyme Vegpro	% Increase over control
Pea/OSR Compound	15.26	17.44	14.3
Bean/OSR Compound	15.16	16.56	9.2
Soya (48% CP)	11.87	12.36	4.1

McNab and Pugh (unpublished)

Trial 2. Effect of Allzyme Vegpro (II) on extracted and full fat soya

METHODS

The response to Allzyme Vegpro addition with soya in the first trial encouraged the research group to look further at this universally used raw material. Changes were made in the enzyme supplement (Allzyme Vegpro II) to better reflect the chemical structure of the soya. The revised formula was tested with extracted soya and full-fat (whole) soya at addition rates of 1 and 2 kg/t.

RESULTS

Allzyme Vegpro II increased TME(N) of 48% soybean meal by 7.2% when added at 1 kg/t (Table 3). Doubling the inclusion level did not

Table 3. Effect of Allzyme Vegpro II on true metabolizable energy (N-corrected) of soybean meal (48% CP) and full-fat soybeans

	TME(N), MJ/kg		
	Control	Allzyme Vegpro (1 kg/t)	Allzyme Vegpro (2 kg/t)
48% soya	12.03	12.90 (+7.2%)	13.14 (+9.2%)
Full-fat soya	16.5	16.49	NC

NC – Not calculated
Values in parentheses are percentage increase in TME(N) over control
McNab and Pugh (unpublished)

double the response; however it was worth noting that no ill effects from over-inclusion occured. There was no effect of the enzyme on full-fat soy TME(N). As in the previous study it was noted that variability among birds was lower in the group given the enzyme supplement.

Conclusions

The research outlined in this report illustrates that addition of Allzyme Vegpro is an effective means of improving energy value for soya and other legume proteins.

This information can be used by the feed compounder in either of two ways. Commercial feed producers will be able to increase ME values for soya in formulation matrices, which in turn will show a formulation cost saving of 2 or 3 to 1 over cost of the enzyme. Alternatively, livestock producers and integrators will be able to add the enzyme on top of the current diet and measure improved level of performance whether it be FCR, daily gain or days to slaughter.

Allzyme Vegpro is already being used effectively in the UK market in addition to cereal enzymes for broiler diets. Although cereal enzymes may not be of world-wide interest – the research with soya previously outlined most certainly will.

BORON – REVISITED ONE YEAR LATER

DENIS R. HEADON

Cell and Molecular Biology Group, Department of Biochemistry
University College of Galway, Ireland

Introduction

Boron, next to carbon in the periodic table, is a rare element present at a concentration of 3 ppm in the earth's crust (Muertterties, 1967). It has been known for over 50 years that boron is essential for plant life, however, until recently, the requirement for boron by animals was unknown. Since the early 1980s, it has become evident that boron has major effects on the metabolism of minerals in higher animals including humans (Nielsen, 1988).

Boron and animal nutrition

In 1981 conclusive evidence was obtained that boron deprivation depressed growth and elevated plasma alkaline phosphatase activity in chicks with inadequate vitamin D_3 (Hunt, 1993). From studies carried out on both chicks and rats, it has been suggested that boron has an effect on a least three separate metabolic sites. These conclusions were reached because dietary boron supplementation compensated for perturbation in energy substrate utilization induced by vitamin D_3 deficiency, enhanced the mineral content in bone with vitamin D_3, and enhanced some indices of growth cartilage maturation independently of vitamin D_3.

Nielsen (1988) carried out the first nutritional study involving boron with humans. In the study, 12 post-menopausal women were fed a diet providing 0.25 mg B/2000 kcal for 199 days and then the same diet with boron supplementation at 3 mg/day for 48 days. With boron supplementation, the total plasma concentration of calcium (Ca) was reduced as was urinary excretion of Ca and magnesium (Mg). Serum concentrations of β-estradiol and testosterone were elevated.

Additional studies were carried out with post-menopausal women, some receiving estrogen and some not (Nielsen, 1993). The results of these studies indicated that boron can both enhance and mimic some effects of estrogen. Boron deprivation reduced the mean corpuscular hemoglobin content and blood hemoglobin concentration, indicating that boron appears to have an essential function affecting macromineral and cellular metabolism.

Boron and inorganic borate at 3 mg/kg per day in humans increased the rate of bone formation and decreased the rate of bone resorption thereby enhancing skeletal growth and development (Nielsen, 1993). Such emerging molecular evidence indicates a highly significant role for boron in relation to the molecular actions of estrogen, testosterone and vitamin D_3 in enhancing and mimicking their actions and influencing macromineral metabolism in relation to bone development.

References

Hunt, C.D., 1993. The biochemical effects of physiological amounts of boron in animals. In: Abstracts, International Symposium on Health Effects of Boron and Its Components.

Muertterties, E. (Ed.), 1967. The Chemistry of Boron and its Components. John Wiley and Sons, New York.

Nielsen, F.H., 1988. Boron – an overlooked element of potential nutrition importance. Nutrition Today, Jan/Feb p. 4.

Nielsen, F.H., 1993. Abstracts, International Symposium on Health Effects of Boron and Its Compounds. Environmental Health Perspectives, 1994 Spring Supplement.

ENZYMES IN FEATHER PROCESSING

TED SEFTON

Alltech Inc., Guelph, Ontario, Canada

Introduction

Environmental, social and economic factors have combined over the last decade to call for the greater utilization of by-products and thus the elimination of waste. This is reflected in the naming of the scientific journal, *Bio-resource Technology*. This journal started its life as *Agricultural Waste*. This title was found unsuitable and changed to *Biological Waste*. It is now *Bio-resource Technology*. Society mandates that industry reduce its negative impact on the environment. One way this can be accomplished is through conversion of waste streams into valuable products.

Feathers, currently either under-utilized or not utilized at all, have the potential to be a valued feed ingredient (Table 1). The limitation has been low digestibility. Traditionally, feathers have been steam hydrolyzed. This system has the potential to raise the digestibility of feathers to 70% for non-ruminants and to make a high by-pass protein for ruminants. The product, however, has been noted for its variability (Bielorai *et al.*, 1982). This variation has limited its use in commercial rations.

Enzymes have been long used by man to upgrade food substrates and in more recent times have been used in the feed industry. It is only natural that their use in feather meal upgrading would be considered. This area has been reported previously by both Harvey (1992) and Woodgate (1993, 1994).

For enzyme technology to be applied two technical areas need to be addressed. There must first be a simple, accurate laboratory test for quality control evaluation of feather meal digestibility. Secondly there must be a commercially applicable methodology for handling the feathers, i.e., specially designed equipment.

Quality control

The AOAC pepsin digestion method has been the traditional laboratory method used to evaluate protein digestion. This method serves well the quality control function within a given ingredient or process, but is less useful when comparing different ingredients or processes. Clunies and Leeson (1984) modified the procedure to increase the accuracy of predicting dry matter and crude protein digestibility for poultry. This modified procedure gives greater accuracy when comparing feather meals produced by different methods.

Commercial application of enzymes to feathers

The method proposed by Woodgate (1994) requires only a modification to procedures used in steam hydrolysis, while the extrusion/enzyme method of Harvey (1992) requires equipment other than that used in steam hydrolysis. Both systems begin with feathers as they exit the processing plant at approximately 70% moisture. The feathers are treated and moisture of the final product is reduced to 10–12%. This is accomplished by heat energy in the steam hydrolysis system and a combination of mechanical and heat energy in the extrusion enzyme system.

STEAM HYDROLYSIS/ENZYME TREATMENT

Woodgate (1993, 1994) reviewed the supplemental use of enzymes with traditional steam hydrolysis. Enzyme treatment allows cooking at lower temperatures. Energy input is reduced as a result, and in addition less protein denaturation occurs. This will allow production of a more consistent feather meal with digestibility at the higher (70%) end of the range. Costs can then be lowered and average quality improved while no capital cost is incurred. Similar modification can be made to poultry by-product meal (feather plus offal) hydrolysis.

Part of the equipment designed for the enzyme/extrusion system is a chopper/dewaterer which takes the feathers from the processing plant, chops them to 2 mm lengths and reduces the water levels to approximately 50%. For every 1000 2 kg broilers, approximately 270 kg of feathers are produced with 70% moisture. Reduction to 50% moisture requires, when done via heat energy, approximately 175 kW of power. Only a fraction of that energy is needed when done mechanically.

Thus, there can be a reduction in power input by initially reducing the moisture level by mechanical removal of water. There would be an initial equipment investment, but a reduction in operating costs. Final product quality when subsequently treated in a enzyme steam system would be comparable to product produced from 70% moisture feathers with enzyme steam hydrolysis.

ENZYME EXTRUSION

Equipment, including the dewaterer previously mentioned, has been designed and built specifically to process feathers by this method. The technical details of the process were previously reported (Harvey, 1992). Feathers from the processing plant are chopped and the moisture reduced mechanically to 50%. They are then reacted with enzymes to break the disulfide bonds. A co-extrudant is added to bring moisture content to 30–35% before extrusion. The end product is then dried to 10–12% moisture. The nutrient content of the final product depends on the co-extrudant. A typical analysis for a 45–55% feather/wheat shorts meal is shown in Table 1.

Table 1. Analytical comparisons.

		Raw feathers	45% Feathers, +55% Wheat shorts	
			Raw	Extruded
Moisture	%	10	10	10
Ash	%	2	3.4	3.4
Protein	%	82	46	46
Fat	%	6	5.1	5.1
Fibre	%	0	3.3	3.3
NFE	%	0	32	32
Calcium,	%	0.26	0.17	0.17
Phosphorus	%	0.67	0.75	0.75
Sodium	%	0.70	0.33	0.33
Potassium	%	0.29	0.64	0.64
Magnesium	%	0.20	0.23	0.23
Lysine	%	1.97	1.11	1.11
Methionine	%	0.57	0.38	0.38
Cystine	%	6.18	2.60	2.60
Threonine	%	4.19	1.96	1.96
Arginine	%	5.80	2.93	2.93
Leucine	%	6.67	3.69	3.69
Isoleucine	%	4.41	1.99	1.99
Histidine	%	0.54	0.50	0.50
Tyrosine	%	2.51	1.24	1.24
Tryptophan	%	0.72	0.40	0.40
Phenylalanine	%	3.92	2.28	2.28
Valine	%	6.64	3.44	3.44

While initial capital costs are not insignificant, a typical cash flow would show about a one year pay back when the equipment is operated 16 hours per day, 5 days per week.

Digestibility studies show that a consistent product is produced with the feather portion of the meal having a 90% digestibility.

Conclusion

In summary, enzymes allow current steam hydrolysis to be upgraded. Further improvement in the economics can be made through capital investments which substitute mechanical water removal for heat removal.

With further capital investment the extrusion system yields an improved product (higher digestibility) with greatly reduced energy input.

References

Bielorai, B., B. Losif, H. Neumark and E. Alumot. 1982. Low nutritional value of feathermeal protein for chicks. J. Nutr. 112(2):249.

Clunies, M. and S. Leeson. 1984. In vitro estimation of dry matter and crude protein digestibility. Poultry Sci. 63:98–96.

Harvey, J.D. 1992. Changing waste protein from a waste disposal problem to a valuable feed protein source: a role for enzymes in processing offal, feathers and dead birds. In: Biotechnology in the Feed Industry. Proceedings of Alltech's 8th Annual Symposium. T.P. Lyons (Ed.) Alltech Technical Publications, Nicholasville, Kentucky, 109.

Woodgate, S.L. 1993. Animal by-products: The case for recyling and possible nutritional upgrading. In: Biotechnology in the Feed Industry, Proceedings of Alltech's 9th Annual Symposium. T.P. Lyons (Ed.) Alltech Technical Publications, Nicholasville, Kentucky, 395.

Woodgate, S.L., 1994. The use of enzymes in designing a perfect protein source for all animals. In: Biotechnology in the Feed Industry, Proceedings of Alltech's 10th Annual Symposium. T.P. Lyons and K.A. Jacques (Eds). Nottingham University Press, Nottingham, UK, 67.

PUTTING TO REST THE UREASE INHIBITION THEORY FOR THE MODE OF ACTION OF *YUCCA SCHIDIGERA* EXTRACTS

GERRY F. KILLEEN

Alltech European Biosciences Research Centre of Excellence, National University of Ireland, Galway, Ireland

Summary

Extracts and preparations of the desert plant *Yucca schidigera* Roezl ex Ortgies (Mohave yucca), family Lillaceae, have a variety of beneficial effects when included in the diet of humans and domestic animals. Such effects include reduced gastrointestinal and faecal ammonia levels. *Y. schidigera* is the principle active ingredient of De-Odorase (Alltech). A proposed mode of action is inhibition of microfloral urease (urea amidohydrolase E.C. 3.5.1.5). To investigate this hypothesis we describe a rigorous method of *in vitro* urease assay, in the presence of potential effectors such as De-Odorase preparations, using the phenolindophenol reaction to measure the ammonia product. For comparison, the effects of De-Odorase preparations on the activity of β–galactosidase (β–D-galactoside galactohydrolase E.C. 3.2.1.23) from *Aspergillus oryzae*, an unrelated hydrolase, were also determined. Urease and β–galactosidase were both weakly and non-specifically inhibited, in a fashion linearly related to the concentration of De-Odorase preparation. Linear regression of the relationship between De-Odorase preparation and enzyme activity yielded inhibition ratios of 3.2 ± 0.4 and 5.4 ± 1.6 nkat ml preparation^{-1} for urease and β–galactosidase, respectively. By comparing with reported *in vivo* rates of urea degradation in mammals it was concluded that the observed inhibitory properties of preparations are much too low to account for their *in vivo* effects at feed inclusion levels of as little as 100 g per tonne.

Introduction

The indigenous populations of the arid deserts of the Americas have long used extracts and preparations of the flowers, seed pods and stalks of *Yucca schidigera* Roezl ex Ortgies (Mohave yucca), family Lillaceae, as a therapeutic agent. It has been approved by the United States Food and Drug Administration as an additive for human consumption (Oser, 1966) and is commonly used as a flavouring or foaming agent.

More recently, such preparations have been effectively applied to the treatment of arthritis, hypercholesterolemia, hypertension and elevated triglyceride levels (Bingham *et al.*, 1978). Improved performance and health in livestock, by inclusion of *Y. schidigera* preparations in feedstuffs at 100 to 250 g/tonne, have also been reported (Rowland *et al.*, 1976; Goodall and Matsushima, 1979; Goodall *et al.*, 1979; Johnston *et al.*, 1981; Foster 1983; Cromwell *et al.*, 1984; Preston *et al.*, 1984b; Mader and Brumm, 1987). Such preparations have also been reported to reduce gastrointestinal or faecal ammonia levels *in vivo* (Rowland *et al.*, 1976; Gibson *et al.*, 1984; Preston *et al.*, 1984a; Headon *et al.*, 1991) and *in vitro* (Ellenberger *et al.*, 1984). Proposed mechanisms of action include the inhibition of microfloral urease (Goodall and Matsushima, 1979; Ellenberger *et al.*, 1984; Preston *et al.*, 1984a), the direct binding of ammonia (Headon *et al.*, 1991) and enhanced microfloral nitrogen utilization (Peestock, 1979). Proponents of urease inhibition as a mode of action argue that improvements in animal performance are caused by amelioration of the considerable metabolic load associated with the detoxification of ammonia, the cellular and physiological effects of which have been reviewed by Visek (1984). The aim of this study was to determine the validity of the urease inhibition theory as a plausible mode of action for *Y. schidigera* preparations such as De-Odorase.

Although the urease of the jack bean (*Canavalia ensiformis*) has been studied in the most detail, microbial ureases have also been the subject of extensive investigation (see Mobley and Hausinger, 1989; Zerner, 1991 for reviews). The majority of plant (Faye *et al.*, 1986) and bacterial (Jeffries, 1964a and 1964b; Friedrich and Magasanik, 1977; Mobley *et al.*, 1986; Jones and Mobley, 1988; Myles *et al.*, 1991; McCoy *et al.*, 1992) ureases are cytosolic proteins. However much of the active urease found in the intestine is cell free and is stable long after the death of the microbial cells from which it originates (Visek, 1978). Intestinal ureases are thus exposed to all dietary and endogenously secreted effectors and are free from normal cellular regulation. The *in vitro* activity of cell free bacterial urease may therefore be taken as a reasonable model of gastrointestinal urease activity. As almost all gastrointestinal urease is of bacterial origin (Visek, 1978) the urease of *Bacillus pasteurii* was chosen for this study. Although *B. pasteurii* is not a typical gut bacterium, its urease is similar to those from other bacterial sources and is commercially available in partially purified form (Larson and Kallio, 1953; Christians and Kaltwasser, 1986).

An abundance of unpublished claims of urease inhibition by *Y. schidigera* preparations and protocols for its determination exist. However, most commercially used procedures, designed to demonstrate inhibition of urease by *Y. schidigera* preparations, ignore some of the fundamental biochemical characteristics of urease and *Y. schidigera* preparations and are thus potentially flawed.

Herein we describe a rigorous method, using the Berthelot reaction method of ammonia assay, for determining the effect of *Y. schidigera* preparations on urease activity. In order to determine whether the observed effects were specific to urease, the effects of such preparations on the β–galactosidase (β–D-galactoside galactohydrolase, E.C. 3.2.1.23) of *Aspergillus oryzae* were also investigated by a similar procedure.

Procedures

Ammonia concentration and hence urease activity was determined by the Berthelot reaction method using the reagents of Chaney and Marbach (1962). Incubations of urease with 5 mM urea were carried out at 30°C and made up to an ionic strength of 0.90 mol l^{-1} with KCl. Incubations were terminated by the addition of trichloroacetic acid (TCA). Ammonia concentration in De-Odorase-containing incubates was determined by comparison with NH_4Cl standards in the same buffer including the same concentration of De-Odorase. β–Galactosidase activity was determined under similarly controlled conditions by a modification of the method of Tanaka *et al.* (1975). Detailed description and discussion of the methods used can be found in Killeen *et al.* (1994).

Results

As previously reported, De-Odorase reduces the amount of ammonia measured by the Berthelot reaction assay (Headon *et al.*, 1991; Figure 1). The colour generated is dependent on the length of time that an ammonia-containing sample is allowed to stand in the PNP reagent before the addition of AH, the second colour developing reagent (Figure 1). PNP must therefore be regarded as an unsuitable reagent for the

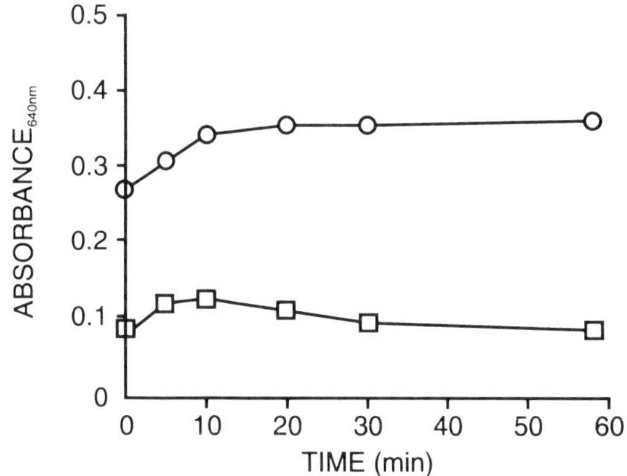

Figure 1. Dependence of colour development in the Berthelot reaction ammonia assay on presence of De-Odorase preparation in the reaction mixture (See also Headon *et al.*, 1991) and on incubation time in PNP reagent before the addition of the second colour developing reagent AH. (See Procedures)
○: 4.5 mM NH_4Cl; □:4.5 mM NH_4Cl plus 100 ml l^{-1} De-Odorase preparation).

termination of ammonia generating reactions, including the hydrolysis of urea by urease. As the order and timing of addition of NH_4Cl,TCA and extract have no detectable effect on the final absorbance, TCA was used as a stopping reagent in subsequent urease assay procedures (see procedures). Preliminary studies showed that *B. pasteurii* urease was potently inhibited by K-phosphate but not by K-citrate or K-HEPES (data not shown). The inhibition by phosphate appears to be most potent at low pH, resulting in an upward shift of the pH optimum (Figure 2).

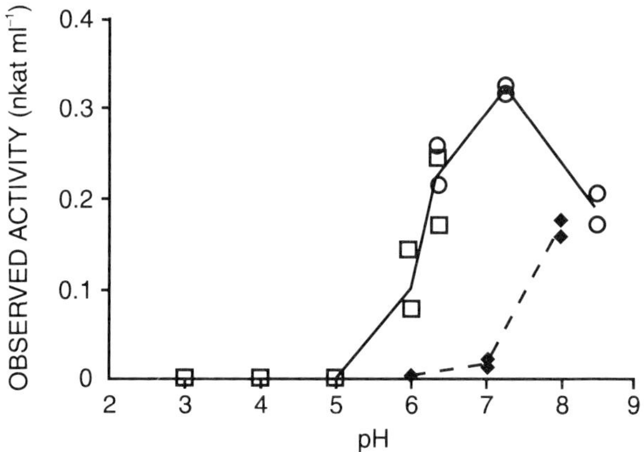

Figure 2. Observed activity of the urease from *Bacillus pasteurii* in the pH range 3.0 to 8.5 in K-HEPES, K-Citrate and K-Phosphate at I = 0.90 M, 30°C (See Procedures) ◆: K-Phosphate; □: K-Citrate; ○: K-HEPES —— Activity in K-Citrate and K-HEPES; - - - activity in K-Phosphate.

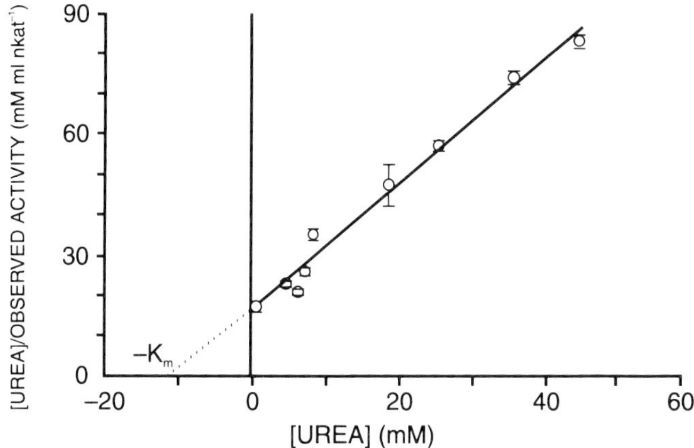

Figure 3. Hanes plot (Wilson, 1986) fitted by linear regression of the relationship between urea concentration and the activity of *Bacillus pasteurii* urease in 150 mM K-HEPES, pH = 6.5, I = 0.90 M, 30°C (See Procedures). Error bars represent ± 95% confidence limits.

This implies that the acidic, protonated forms are the active inhibitory phosphate species, as observed for other ureases (Kistiakowsky and Thompson, 1956; Mobley and Hausinger, 1989). The K_m of this urease was found to be 10.5 ± 3.2 mM at pH = 6.5, I = 0.90 M, 30°C in K-HEPES (Figure 3). The previously reported K_m of 100 mM in the pH range 5.7 to 6.7 appears to be unreliable as that determination was performed in phosphate buffers (Larson and Kallio, 1953).

Both urease and β–galactosidase were weakly inhibited by De-Odorase in a fashion linearly related to the concentration of the preparation in the enzyme reaction medium (Figure 4). Linear regression analysis of the data presented in Figure 4 showed that the amounts of observed activity inhibited per volume of De-Odorase preparation were 3.2 ± 0.4 and 5.4 ± 1.6 nkat ml preparation^{-1} for urease and β–galactosidase, respectively. Inhibition was significant only at concentrations several orders in excess of typical feed inclusion levels and appears to be non-specific. Similar results were obtained, under less rigorous conditions of ionic strength, with the urease of *C. ensiformis* (data not shown).

Similarly, no significant inhibition ($P > 0.10$) of urease by 4 ml l^{-1} of De-Odorase preparation (16–40 fold typical feed inclusion levels) was observed over the pH range 3.0 to 8.5 in the presence of the two non-inhibiting buffers used, K-citrate and K-HEPES (Figure 5).

Discussion and conclusions

The Berthelot reaction of ammonia with phenol and hypochlorite to give phenolindophenol dye (Chaney and Marbach, 1962) is often used to

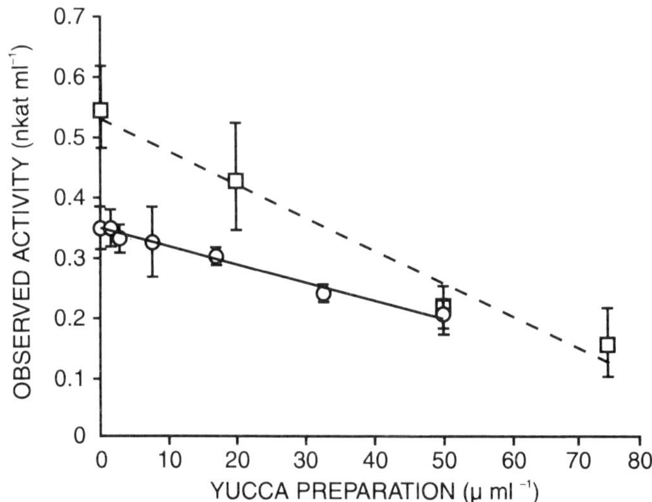

Figure 4. Effect of varying concentrations of De-Odorase preparation on the observed activities of urease from *Bacillus pasteurii* and β–galactosidase from *Aspergillus oryzae* at pH = 6.5 and 4.5 respectively (See Procedures). Error bars represent ± 95% confidence limits. ○: Urease; □: β–galactosidase.

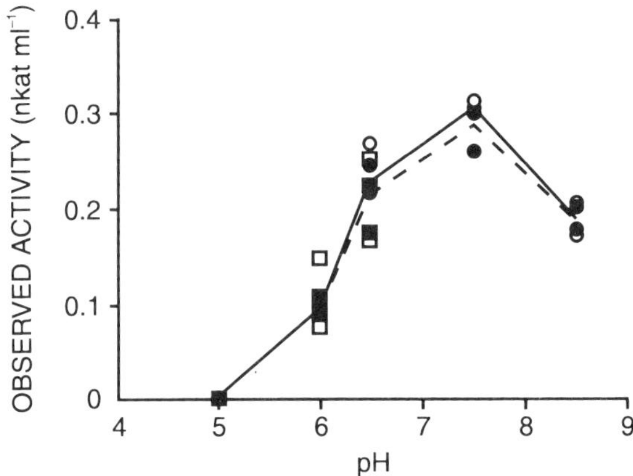

Figure 5. Effect of 4 ml l^{-1} De-Odorase preparation on the activity of urease from *Bacillus pasteurii* in the pH range 5.0 to 8.5 in 150 mM K-HEPES and K-Citrate at I = 0.90 M, 30°C (See Procedures).

□: K-Citrate; ■: K-Citrate plus *Y. schidigera*; ○: K-HEPES; ●: K-HEPES plus *Y. schidigera*.

—— Control (K-Citrate and K-HEPES buffers without *Y. schidigera* preparation.)
- - - - analytical (K-Citrate and K-HEPES buffers with 4 ml l^{-1} *Y. schidigera* preparation.)

determine ammonia concentration and hence urease activity. However De-Odorase has been shown to reduce levels of measurable ammonia, as determined by the Berthelot reaction method (see Headon *et al.*, 1991 and Results). Thus, as the concentration of such preparations in urease assay incubates is increased, the level of measurable ammonia decreases. This effect is often mistakenly interpreted as inhibition of urease.

The observed potent inhibition by phosphate (Figure 2), particularly at low pH, is in agreement with studies of other ureases (Kistiakowsky *et al.*, 1952; Kistiakowsky and Thompson, 1956; Mobley and Hausinger, 1989; Martin and Hausinger, 1992) and clearly demonstrates the need for careful choice of non-inhibitory buffers and rigorous pH control in studies of urease effectors. Bacterial (Mobley and Hausinger, 1989; Martin and Hausinger, 1992) and plant (Kistiakowsky *et al.*, 1952; Kistiakowsky and Thompson, 1956) ureases are inhibited by several acidic metabolites, including phosphate. Phosphate inhibition of plant (Kistiakowsky *et al.*, 1952; Kistiakowsky and Thompson, 1956) and bacterial (Mobley and Hausinger, 1989; Martin and Hausinger, 1992) ureases is competitive in nature. Therefore, although the published K_m values for *B. pasteurii* urease (Larson and Kallio, 1953) are much higher than for other bacterial ureases (Mobley and Hausinger, 1989), these K_m determinations were carried out in phosphate buffers (Larson and Kallio, 1953) and accordingly should be interpreted with a degree of caution. The observed phosphate inhibition and the K_m value determined in K-HEPES buffer for *B. pasteurii* urease confirms that the previous kinetic characterization of this urease (Larson and Kallio, 1953), carried out in phosphate buffers, may be unreliable and explains why this previously reported K_m is much higher than typical values for plant (Kistiakowsky and Rosenberg, 1952; Fishbein *et al.*, 1965; Sundaram, 1973) and other microbial ureases (Mobley and Hausinger, 1989). The urease of *B. pasteurii* may therefore be regarded as a typical urease and is a suitable model enzyme for studying bacterial urease effectors. Subsaturating urea concentrations were used to emphasize the effects of competetive and mixed inhibitors. Based on the observed K_m, the concentration of urea in the incubates and the Michaelis-Menten equation (Wilson, 1986), the activity observed in these experiments is calculated to be 32% of total activity under saturating conditions.

In spite of its inhibitory properties, phosphate and several other urease inhibiting acids (Kistiakowsky *et al.*, 1952) are commonly used as buffers in procedures alleged to demonstrate genuine urease inhibition by extracts and preparations of *Y. schidigera*.

Ionic strength also influences the kinetics of urea hydrolysis by *C. ensiformis* urease (Kistiakowsky *et al.*, 1952; Kistiakowsky and Thompson, 1956). Whether increasing ionic strength has a positive or negative effect on activity is dependent upon pH. However the relationship between activity and ionic strength consistently behaves according to the Debye-Hückel limiting law (Kistiakowsky and Thompson, 1956). Hence, solutions of high ionic strength should be used to negate the electrolyte effects of added effectors in any study of urease inhibitors or stimulators, including De-Odorase preparations.

Another common error in the design of urease effector studies is the use of weakly buffered reaction media and/or *Y. schidigera* preparations of unadjusted pH. Such preparations, particularly from commercial sources, are often acidic and may have strong buffering capacity. As the pH optima of bacterial ureases range from 6.8 to 8.0 in non-inhibiting buffers (Mobley and Hausinger, 1989), *Y. schidigera* preparations can cause misleading drops in urease activity simply by acidification of the reaction medium. Furthermore, the effect of acidification may be amplified by use of phosphate buffers which are reported to be inhibitory only when in their acidic, protonated forms (Kistiakowsky and Thompson, 1956; Mobley and Hausinger, 1989).

No biologically significant inhibitory action by *Y. schidigera* preparation occurs at any pH falling within the range encountered at the *in vivo* sites of action (see Figure 5). Based on our study at pH = 6.5, ignoring the diluent effect of digestive secretions, an inclusion of the preparation at 11 kg tonne^{-1} in feed would be required in order to inhibit the small urease concentration used in this study by a mere 10%. Furthermore, comparison of the observed inhibition of urease per volume of *Y. schidigera* preparation with reported *in vivo* urea degradation rates further highlights the insignificance of the inhibitory properties of such extracts. The whole body urea degradation rates in man (Wrong and Vince, 1984), sheep (Whitelaw and Milne, 1991; Whitelaw *et al.*, 1991), rabbits (Forsythe and Parker, 1985) and pigs (Mosenthin *et al.*, 1992) are 19, 37–54, 97 and 50 nkat kg^{-1}, respectively. Assuming such degradation rates to represent the maximum urea degradation capacity of the animal, and that the observed urease inhibition capacity represents only 32% of the total urease-inhibiting potential of the *Y. schidigera* preparation (see procedures), then to achieve 10% inhibition such animals would have to contain, at any given time, an amount of *Y. schidigera* preparation equivalent to between 0.02 and 0.10% of their total body weight, representing between 80 and 400% of their own body weight of feed containing the maximum recommended dose.

In fact such an assumption vastly underestimates the total activity of microbial urease in the gastrointestinal tract. *In vivo* urea degradation rates in pigs (Mosenthin *et al.*, 1992) and man (Walser and Bodenlos, 1959) have been shown to increase in direct proportion to plasma urea concentrations. Urease is therefore in significant excess at the proximal colon and terminal ileum where the bulk of hydrolysis occurs. Urea concentration is also therefore significantly below saturating conditions. Even in ruminants the proportion and concentration of ammonia derived from the hydrolysis of endogenous urea is greater in the large intestine than in the rumen (Dixon and Nolan, 1983). In saturating concentrations of urea, bovine intestinal epithelia and rumen liquor possess *in vitro* urease activities of up to 0.19 nkat cm^{-2} and 83 nkat ml^{-1} respectively (Abdullah and Hutagalung, 1988). Similarly, *in vitro* urease activities in saturating urea concentrations of 7.4 to 22.3 nkat ml^{-1} have been reported for ovine rumen liquor (Whitelaw and Milne, 1991, Whitelaw *et al.*, 1991). Such high activities clearly demonstrate that gastrointestinal urease is greatly in excess of any effectors present in *Y. schidigera* preparations, at credible feed inclusions, and is not the rate limiting step of urea hydrolysis in the mammalian gastrointestinal tract. On this basis it is clear that *Y. schidigera* preparations do not inhibit urease *in vitro* sufficiently to account for their *in vivo* effects as animal feed additives and therapeutic agents.

References

Abdullah, N. and R.I. Hutagalung. 1988. Rumen fermentation, urease activity and performance of cattle given palm kernel cake-based diets. Anim. Feed Sci. Tech. 20:79–86.

Bingham, R., D.H. Harris and T. Laga. 1978. Yucca plant saponin in the treatment of hypertension and hypercholesterolemia. J. Appl. Nutr. 30:127–136.

Chaney, A.L. and E.P. Marbach. 1962. Modified reagents for determination of urea and ammonia. Clin. Chem. 8:130–132.

Christians, S. and H. Kaltwasser. 1986. Nickel content of urease from *Bacillus pasteurii*. Arch. Microbiol. 145:51–55.

Cromwell, G.L., T.S. Stahly and A.T. Monegue. 1984. Efficacy of sarsaponin for weanling and growing-finishing swine housed at two animal densities. J. Anim. Sci. 61(Supp.):111.

Dixon, R.M. and J.V. Nolan. 1983. Studies of the large intestine of sheep; 3. Nitrogen kinetics in sheep given chopped lucerne (*Medicago sativa*) hay. Br. J. Nutr. 50:757–768.

Ellenberger, M.A., W.V. Rumpler, D.E. Johnson and S.R. Goodall.

1984. Evaluation of the extent of ruminal urease inhibition by sarsaponin and sarsaponin fractions. J. Anim. Sci. 61(Suppl.):491–492.

Faye, L., J.S. Greenwood and M.J. Chrispeels. 1986. Urease in Jack Bean (*Canavalia ensiformis (L.) DC*) seeds is a cytosolic protein. Planta 168:579–585.

Fishbein, W.N., S.W. Thorne and J.D. Davidson. 1965. Urease catalysis: I. Stoichiometry, specificity and kinetics of a second substrate: hydroxyurea. J. Biol. Chem. 240:2402–2406.

Forsythe, S.J. and D.S. Parker. 1985. Urea turnover and transfer to the digestive tract in the rabbit. Br. J. Nutr. 53:183–190.

Foster, J.R. 1983. Sarsaponin for growing-finishing swine alone and in combination with an antibiotic at different pig densities. J. Anim. Sci. 57(Suppl.):245

Friedrich, B. and B. Magasanik. 1977. Urease of *Klebsiella aerogenes*: control of its synthesis by glutamate synthetase. J. Bacteriol. 131: 446–452.

Gibson, M.L., R.L. Preston, R.H. Pritchard and S.R. Goodall. 1984. Effect of sarsaponin and monensin on rumen ammonia levels and *in vitro* dry-matter digestibilities. J. Anim. Sci. 61(Suppl.):492.

Goodall, S.R. and J.K. Matsushima. 1979. Sarsaponin and monensin effects upon ruminal VFA concentrations and weight gains of feedlot cattle. J. Anim. Sci. 49(Suppl.):371.

Goodall, S.R., J.D. Eichenblaum and J.K. Matsushima. 1979. Sarsaponin and monensin effects upon *in vitro* VFA concentration, gas production and feedlot performance. J. Anim. Sci. 49(Supp.):370–371.

Headon, D.R., K.A. Buggle, A.B. Nelson and G.F. Killeen. 1991. Glycofractions of the Yucca plant and their role in Ammonia control. In: Biotechnology in the Feed Industry-7. T.P. Lyons (Ed). Alltech Technical Publications, Nicholasvile, Kentucky, 95–108.

Jeffries, C.D. 1964a. Urease activity in intact and disrupted bacteria. Arch. Pathol. 77:544–547.

Jeffries, C.D. 1964b. Intracellular microbial urease. Nature (London) 202:930.

Johnston, N.L., C.L. Quarles, D.J. Fagerberg and D.D. Cavens. 1981. Evaluation of Yucca saponin on broiler performance and ammonia suppression. Poultry Sci. 60:2289–2292.

Jones, B.D. and H.L.T. Mobley. 1988. *Proteus mirabilis* urease: genetic organisation, regulation and expression of structural genes. J. Bacteriol. 170: 3342–3349.

Killeen, G.F, K.A.B. Buggle, M.J. Hynes, G.A. Walsh, R.F. Power and D. Headon. 1994. Influence of *Yucca schidigera* preparations on the activity of urease from *Bacillus pasteurii*. J. Sci. Food Agric. 65:433–440.

Kistiakowsky, G.B. and A.J. Rosenberg. 1952. The kinetics of urea hydrolysis by urease. J. Am.Chem. Soc. 74:5020–5025.

Kistiakowsky, G.B., P.C. Mangelsdorf, A.J. Rosenberg and W.H.R. Shaw. 1952. The effects of electrolytes on urease activity. J. Am. Chem. Soc. 74:5015–5021.

Kistiakowsky, G.B. and W.E. Thompson. 1956. Kinetics of the urease catalysed hydrolysis of urea at pH 4.3. J. Am. Chem. Soc. 78:4821–4829.

Larson, A.D., R.E. Kallio. 1953. Purification and properties of bacterial urease. J. Bacteriol. 68:67–76.

Mader, T.L., M.C. Brumm. 1987. Effect of feeding sarsaponin in cattle and swine diets. J. Anim. Sci. 65:9–15.

Martin, P.R. and R.P. Hausinger. 1992. Site directed mutagenesis of the active site cysteine in *Klebsiella aerogenes* urease. J. Biol. Chem. 267:20024–20027.

McCoy, D.D., A. Cetin and R.P. Hausinger. 1992. Characterisation of urease from *Sporosarcina ureae*. Arch. Microbiol. 157:411–416.

Mobley, H.L.T. and R.P. Hausinger. 1989. Microbial ureases: significance, regulation and molecular characterisation. Microbiol. Rev. 53:85–108.

Mobley, H.L.T., B.D. Jones and A.E. Jerse. 1986. Cloning of urease gene sequences from *Providencia stuartii*. Infect. Immun. 54:161–169.

Mosenthin, R., W.C. Sauer and C.F.M. de Lange. 1992. Tracer studies of urea kinetics in growing pigs: I. The effect of intravenous infusion of urea on urea cycling and the site of urea secretion into the gastrointestinal tract. J. Anim. Sci. 70:3458–3466.

Myles, A.D., W.C. Russell, I. Davidson and D. Thirkell. 1991. Ultrastructure of ureaplasma urealyticum serotype-8 and the use of immunogold to confirm the localisation of urease and other antigens. FEMS Microbiol. Lett. 80:19–22.

Oser, B.L. 1966. An evaluation of *Yucca mohavensis* as a source of food grade saponin. Fd. Cosmet. Toxicol. 4:57–61.

Peekstock, L.A. 1979. An investigation into the application of *Y. schidigera* extracts to biological waste treatment. M.Sc. Thesis, Miami University, Oxford, Ohio.

Preston, R.L., S.J. Bartle and S.R. Goodall. 1984a. Influence of sarsaponin on growth, feed and nitrogen utilisation in growing male rats fed diets with and without urea. J. Anim. Sci. 61(Suppl.):301.

Preston, R.L., R.H. Pritchard and S.R. Goodall. 1984b. Determination of the optimum dose of Sevarin (sarsaponin) for feedlot steers. J. Anim. Sci. 61(Suppl.):464.

Rowland, L.O., J.E. Plyler and J.W. Bradley. 1976. *Yucca schidigera* extract effect on egg production and house ammonia levels. Poultry Sci. 55:2086.

Sundaram, P.V. 1973. The kinetic properties of microencapsulated urease. Biochim. Biophys. Acta 321:319–328.

Tanaka, Y., A. Kagamiishi, A. Kiuchi and T. Horiuchi. 1975. Purification and properties of β–galactosidase from *Aspergillus oryzae*. J. Biochem. 77:241–247.

Visek, W.J. 1978. The mode of growth promotion by antibiotics. J. Anim. Sci. 46:1447–1469.

Visek, W.J. 1984. Ammonia: its effects on biological systems, metabolic hormones and reproduction. J. Dairy Sci. 67:481–498.

Walser, M., L.J. Bodenlos. 1959. Urea metabolism in man. J. Clin. Invest. 38:1617–1626.

Whitelaw, F.G. and J.S. Milne. 1991. Urea degradation in sheep nourished by intragastric infusion: Effect of level and nature of inputs. Exp. Physiol. 76:77–90.

Whitelaw, F.G., J.S. Milne and X.B. Chen. 1991. The effect of a rumen microbial fermentation on urea and nitrogen metabolism of sheep nourished by intragastric infusion. Exp. Physiol. 76:91–101.

Wilson, K. 1986. Enzyme techniques. In: A Biologist's Guide to the Principles and Techniques of Practical Biochemistry. K. Wilson and K.H. Goulding (Eds). Arnold, London, 87–100.

Wrong, O.M. and A. Vince. 1984. Urea and ammonia metabolism in the human large intestine. Proc. Nutr. Soc. 43:77–86.

Zerner, B. 1991. Recent advances in the chemistry of an old enzyme, urease. Bioorganic Chem. 19:116–131.

REDUCING AMMONIA AND FLIES IN LAYER OPERATIONS WITH YUCCA

DON CROBER

Nova Scotia Agricultural College, Truro, Nova Scotia, Canada

Introduction: ammonia and health

Environmental quality in livestock facilities has become a very important factor, both from the human and the animal points of view. Not only does poor environment negatively affect people looking after the animals, it also adversely affects both health and productivity of livestock and poultry. Table 1 illustrates that even relatively low levels of ammonia adversely affect human health.

Table 1. The effects of ammonia exposure on humans.

Ammonia, ppm	
5	Lowest level detectable
7–10	Recommended maximum level
6–20	Eye irritation and respiratory problems
40	Headache, nausea and reduced appetite
Ammonia, ppm/h	
100	Irritation to mucosal surfaces
400	Irritation to nose and throat

In northern climates the amount of ventilation in livestock facilities is very much dependent on outside temperature. When outside temperature is low, ventilation involves moving cold outside air into warm facilities. Thus, increasing ventilation rates will decrease inside temperatures. In facilities such as layer barns where animal body heat is the sole source of heat, energy intake must increase. Alternatively, bringing in cold air increases the amount of supplemental heat (and thus cost) in facilities where heat is provided, e.g. broiler barns. As the ventilation rate increases, the amount of ammonia removed also increases. Conversely, if heat is conserved by decreasing ventilation rate, ammonia levels increase.

Yucca schidigera extract: impact on ammonia and temperature control

In 1989, the opportunity arose to follow the inside and outside environment at a commercial layer operation housing 25 000 birds whose environment was controlled by the Farmax Ventilation System. This computerized ventilation control system monitors at 5 min intervals the inside temperature in multiple locations, outside temperature, inside and outside relative humidity and ammonia level in the barn. Since birds are housed at the same time each year in this facility, year-to-year comparisons can be made to determine the influence of various treatments. Table 2 shows the outside temperature and the inside temperature during the months of January over 4 years. The year 1989–90 provides baseline measurements as no treatment was given. In subsequent years, De-odorase (*Yucca schidigera* extract, Alltech Inc.) was added to the feed. It is evident from these data that when De-odorase was added ammonia level in the house decreased. De-odorase was used for a shorter period of time in 1990–91 than in subsequent years, thus the ammonia levels that year were relatively higher.

Table 2. Relative ammonia vs temperature inside and outside the layer house.

Year	Actual mean temperature, °C		Relative ammonia (actual mean, ppm)
	Outside	Inside	
1989–90 (Control)	−8.79	21.2	63.56
1990–91 (Treatment)	−8.60	22.2	56.33
1991–92 (Treatment)	−8.12	21.9	33.15
1992–93 (Treatment)	−7.89	22.8	29.68

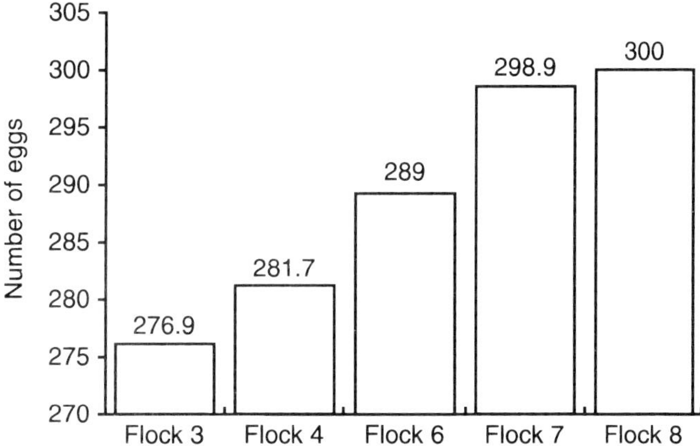

Figure 1. Number of eggs per hen housed.

Egg production parameters were also affected. It can be seen that in each of the years De-odorase was used there was an increase in egg production per hen housed (Figure 1). Similarly, feed consumption per dozen eggs declined in response to a lower ventilation demand (Figure 2). Decreased feed consumption can be attributed both to the higher house temperature maintained when De-odorase was present and also to the improved environment.

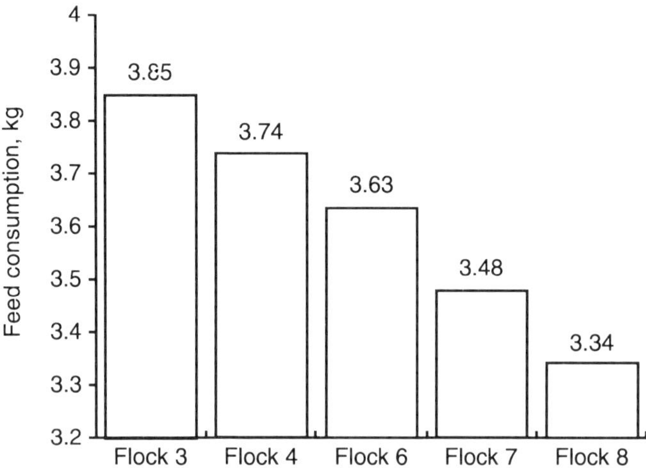

Figure 2. Feed consumption per dozen eggs produced.

In economic terms, with eggs valued at $1.00 per dozen, income per hen housed increased in each of the years De-odorase was used (Figure 3). Feed cost per hen housed decreased in each of the years as would be expected with the improved temperatures (Figure 4). As a result, there was a consistent increase in margin per hen housed (Figures 5 and 6).

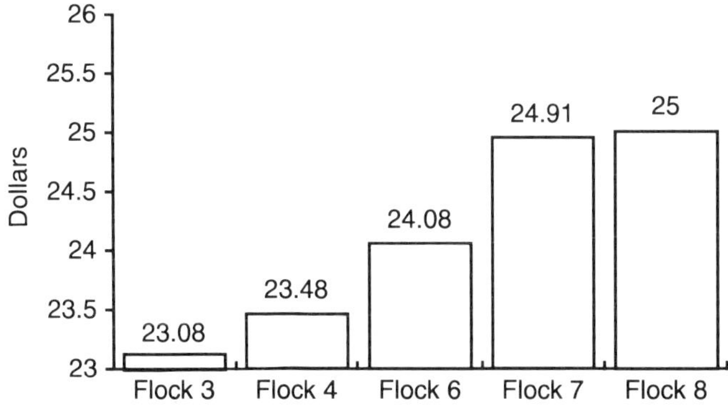

Figure 3. Sales income per hen housed ($1.00/dozen eggs).

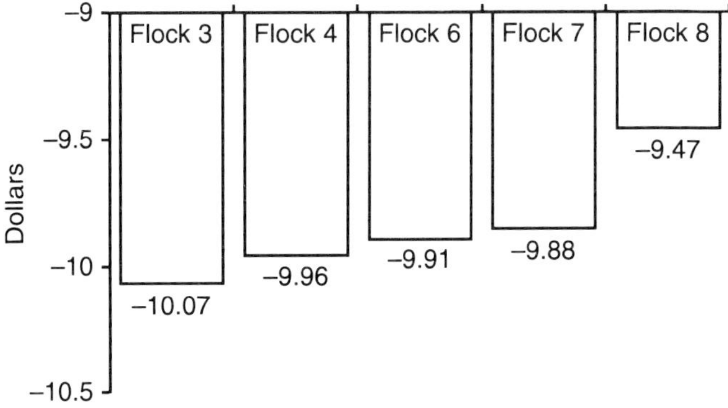

Figure 4. Cost of feed per hen housed (shown as negative).

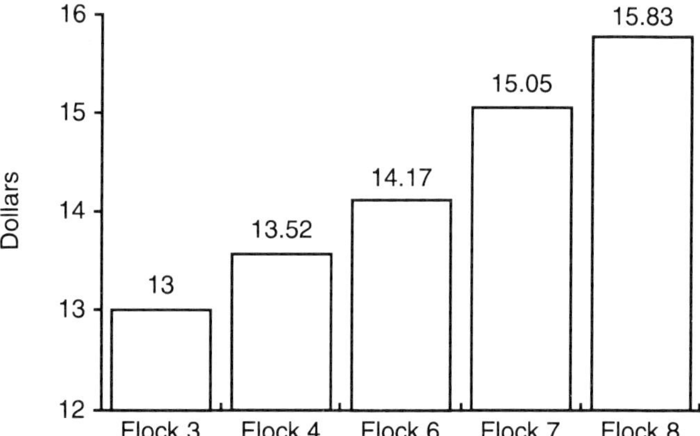

Figure 5. Gross margin per hen housed.

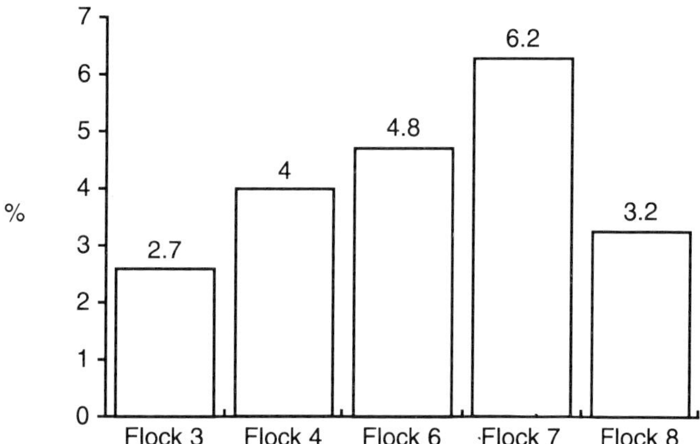

Figure 6. Percent increase in gross margin.

Yucca schidigera extract: effect on fly problems

While difficult to document, the producer/owner noted that before he began using De-odorase he had a problem with flies. He was spending each year approximately $3000.00 on fly control with less than complete success. Following the introduction of De-odorase, insecticides were no longer needed and he currently has little or no problem with flies. The local Department of Agriculture entomologist has identified seven different species of parasitic insects living in the deep pit droppings. These insects are responsible for the decrease in flies. A problem with water leaks and wet droppings caused the fly problem to return. However, as the leak was repaired and the droppings dried, the parasitic insects returned and the flies disappeared. It is an ongoing challenge to try to document the factors contributing to the presence of the parasitic insects. It is felt that De-odorase has played a role, as has ventilation to keep the droppings dry.

Conclusion

In summary, De-odorase at a relatively low level per tonne has contributed to the environmental well-being of this commercial poultry operation and during the period that it has been used the producer has seen a large improvement in the farm's economic profile. It will be interesting to analyse the latest data from this farm since the producer has increased the level of De-odorase to 100 g per tonne and increased the temperature an additional 2°C. The reasoning behind these changes is that increased temperature will further decrease feed consumption while De-odorase and the Farmax ventilation control system work to maintain low levels of ammonia in the barn.

USING YUCCA TO IMPROVE PIG PERFORMANCE WHILE REDUCING AMMONIA

D.J.A. COLE[1] and K. TUCK[2]

[1]*University of Nottingham, Sutton Bonington Campus, Loughborough, Leicestershire, UK*
[2]*Alltech (Ireland), Tallaght, Dublin, Ireland*

Ammonia and odour have long been recognized as unfortunate consequences of animal production. In recent years considerable effort has been made to reduce their effects on both animal performance and on the environment. In this context extracts of *Yucca schidigera* have been used throughout the world. The material is used at low level inclusion (e.g. 60–120 g/tonne) in the diet in order to reduce the ammonia and odour emissions from excreta. A principle mode of action is the binding of ammonia, although other effects may also be of importance in relation to fermentation and the digestive processes. The inhibition of ammonia is simply demonstrated in the laboratory where the effects are instantaneous. In fact, the quality control test (B_{50}) is based on the binding property of the Yucca extract.

The farm environment represents a very different situation from laboratory conditions. Firstly, no two farms are exactly alike and account needs to be taken of this. On any one farm the unique combination

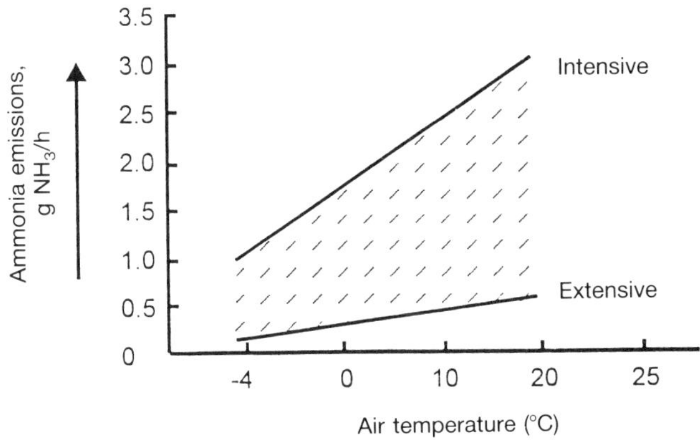

Figure 1. The influence of air temperature and type of system on ammonia emissions. Based on Mannebeck and Oldenburg (1991).

of animals, buildings, ventilation system, slurry collection/storage etc. determines the level of odour and noxious gases emitted.

In addition, the climate in which the farm is situated will be important. For example, the higher the temperature, the more ammonia is volatilized (Figure 1). In parts of the world having naturally high climatic temperatures large amounts of ammonia and odour result. However, because of the climate, building design may allow rapid clearance of the gases to the atmosphere. While this may benefit the animal and stockperson, it can be argued that it is not beneficial to the environment at large. In some warm climates with fluctuations in temperature, it is common to close the sides of the building at night and it is important in establishing basal ammonia levels in a building to monitor them at various times of the day and night.

In cooler climates the problems are often reported to be greater in winter. This is again the result of reduced ventilation in order to maintain temperature within the house. Figure 2 shows the large difference in ammonia levels in broiler houses in summer and winter; a problem that is exacerbated by the increased body mass of the birds.

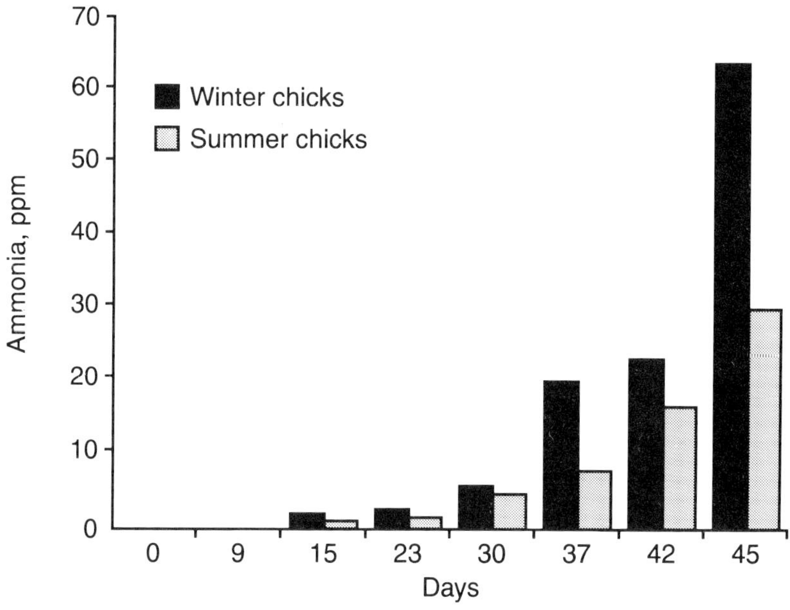

Figure 2. Ammonia levels in broiler houses in summer and winter.

The individual unit

A trial was conducted to study the pattern of atmospheric ammonia in a modern pig house and how this might be influenced by the addition of a Yucca extract (De-Odorase) to the diet. The work was conducted in houses of identical design in Ireland. They had similar stocking densities and slurry levels, and contained pigs of similar genotype. In the treated house De-Odorase was included in the diet at 120 g/tonne. Ammonia levels were monitored weekly at six different points in each house. The trial covered the period of growth from 30 to 90 kg liveweight.

ENVIRONMENT AND PERFORMANCE

The effects of De-Odorase addition are given in Figure 3 and Table 1 and are similar to the response pattern found on many farms. It appears that unlike laboratory conditions, in the farm situation it takes some time for ammonia to be reduced. There was no significant difference between the treatments in the first 4 weeks but the levels of ammonia in the house containing the De-Odorase treated pigs were significantly ($P<0.001$) lower during the second 3 weeks than in the house having untreated pigs. The trial finished after 7 weeks as the pigs had reached market weight. The treated pigs grew 9.4% more quickly than the untreated controls.

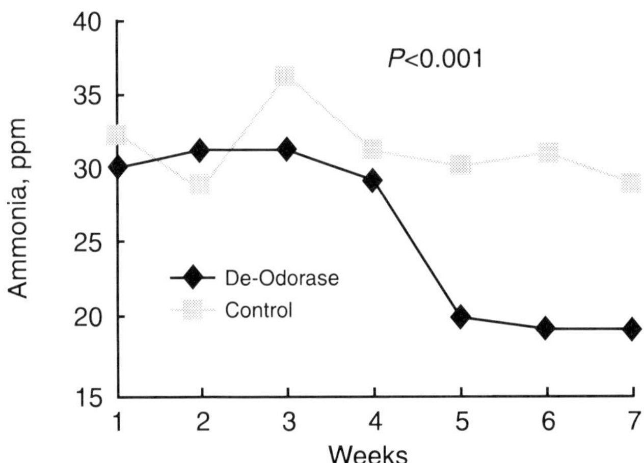

Figure 3. Effect of De-Odorase on ammonia levels on a large pig unit.

Table 1. Environmental ammonia levels and pig performance with and without dietary De-Odorase, on a large pig unit.

	Control	De-Odorase
NH_3 at start (ppm)	31.2	30.1
NH_3 at finish (ppm)	30.0	19.6
Average NH_3 in first 4 weeks (ppm)	30.5	32.1
Average NH_3 in second 3 weeks (ppm)	30.7	19.9
Daily liveweight gain (g/day)	817.3	847.3

Ammonia and performance

In the United Kingdom the levels of substances in the workplace (e.g. an animal house) are regulated by law in order to protect the respiratory health of the worker (Health and Safety Executive, 1989). Ammonia is severely controlled by this legislation with levels of 25–35 ppm allowing the presence of the worker for only 10 minutes per day. When the level of ammonia exceeds 35 ppm then the worker is not allowed in that building without a respirator. Ammonia is an even greater danger in the presence of dust.

A number of surveys have shown the high incidence of respiratory conditions among pig workers. Similarly poor air quality is associated with poor performance and ill health in pigs. Consequently, it is tempting to speculate that the improved performance of pigs is simply a reflection of them spending a large period of time in an environment with better air quality. Certainly this should account for a large part of the effect, but other effects may be occurring at gut level.

While exposure to long periods of poor air quality is clearly detrimental to the animal, recent work at the Royal Veterinary and Agricultural University of Denmark has shown that very short periods of exposure to high ammonia levels can seriously affect performance. They showed that exposure of pigs to 50 ppm ammonia, for 20 minutes/day, on four occasions only, seriously reduced performance (growth rate was reduced by more than 9%) between 37 and 90 kg liveweight (Table 2). Parallel situations are easily seen in practice. For example, some farms do not apparently have an ammonia problem when tests are conducted during the day but report that occasionally there is a big build up overnight. The seriousness of these less frequent occurrences is now brought into focus.

Table 2. Average daily gain and feed conversion of pigs exposed to ammonia on four occasions only during the growing/finishing period.

NH_3 (ppm)	Daily gain (g)	FCR (g LW/kg feed)
5	946	476
50	869	385

Andreason *et al.* (1994)

The Danish group conducted a further experiment in which pigs subjected to ammonia treatments were challenged with *Pasturella multocida* which is associated with the mycoplasma-induced respiratory disease complex. Respiratory health was reduced at the lowest level of ammonia used (50 ppm) (Table 3). This extends findings that ammonia damages the respiratory lining and reduces bacterial clearance from the lungs.

Table 3. Response of pigs exposed to ammonia and challenged with *Pasturella multocida*.

	Pneumonia			
NH_3 (ppm)	Frequency	Extent*	Cough index†	Daily gain (g)
<5	13/17	3.6	6.3	641
50	9/10	7.5	7.9	590
100	9/10	6.5	10.7	609

Andreason *et al.* (1994)
* % surface with consolidation;
† no. coughing per day/no. pigs in group

Conclusions

Ammonia pollution is a characteristic of the individual animal unit and is influenced by many factors. Even short-term exposure to high levels of ammonia can reduce pig performance substantially. However, the reduction of ammonia by the use of a dietary *Yucca schidigera* extract (De-Odorase) is particularly effective in long-term control and this has been associated with improvements in growth performance. Whether the long-term benefits are solely the result of reduced environmental ammonia or whether the Yucca extract has other benefits is open to question.

References

Andreason, M., P. Baekbo and K. Nielsen (1994). Effect of aerial ammonia on the MIRD complex Proc. Int. Pig Vet. Soc. Congr., 429.

Health and Safety Executive 1989. Occupational exposure limits 1989. Guidance Note 40/89. HMSO: London.

Mannebeck, H. and J. Oldenburg (1991). Comparison of different systems of ammonia emissions. In: Odour and Ammonia Emissions from Livestock Farming. V.C. Nielsen, J.H. Voorburg and P.L'Hermite (Eds) Elsevier Applied Science, London and New York.

SOME CONSIDERATIONS TO REDUCE ASCITES SYNDROME IN BROILERS

JOSÉ ARCE MENOCAL

Instituto Nacional de Investigaciones Forestales y Agropecuarias, Morelia, Michoacan, México

Introduction

Ascites is a pathologic manifestation characterized by liquid accumulation in the abdominal cavity and can be part of a generalized syndrome. This is the case for the ascites syndrome in broilers. The highest mortality usually occurs between the sixth and seventh weeks of age. The ascites syndrome is called by several names which reflect either its gross symptoms or etiology including avian edema, idiopathic ascites, toxic fat syndrome, altitude edema, polyserousitis, edema disease, water belly, feed toxemia, and round heart disease, to name just a few.

The first ascites cases in broilers were described by American poultrymen in 1890. The cases reported before the 1970s were isolated and identified with etiologies such as salt poisoning (Edward, 1918), mercurial compounds in feed (Gallager, 1919), *Crotalia spectabilis* (Thomas, 1934), toxic factors in feed and low energy diets (Sanger *et al.*, 1958), dioxins (Wooton and Alexander, 1959), polychlorinated biphenyls (Flick and Odell, 1965). However, after the marked increase in incidence of the ascites syndrome in many countries and in Mexico since the 1970s, research intensified in order to solve this problem. Cueva *et al.* (1974) reported that ascites was caused by low oxygenation in the bird. Estudillo (1976) mentioned that the problem was strongly related to insecticides and pesticides present in the feed. Hernández (1979) related it to altitude above sea level. Renjifo (1979) correlated ascites with low temperatures; López and Barbosa (1981) with feed; Agudelo (1981) associated ascites with aspergillosis, vitamin E and selenium deficiency; López *et al.* (1982) saw interactions between environment and feed; Velazco (1982) related ascites to sex and strain; Rosiles (1982) thought ascites to be associated with aflatoxins; Machorro and Passch (1983) related ascites to cardiac insufficiency; Ortega (1984) noted a relationship with hypoproteinemia and respiratory diseases; Germán and López (1985) thought wrong management procedures were responsible; Arce *et al.* (1986a,b) found a relationship between ascites and physical form of feed; Arce *et al.* (1987) found ammonia levels to affect ascites; and Arce *et al.* (1988) related ascites incidence to high and low growth strains.

Environmental effects on ascites mortality

Birds, like humans, are sensitive to environment. Because of this, it is essential to maintain environment stability throughout the year in poultry houses. Adequate environmental management is necessary in order to meet optimum requirements for best growth, reduce stress, promote optimum economic production, provide adequate heat and fresh air, and remove humidity along with ammonia, dust, carbon monoxide and carbon dioxide from the house. Temperature, humidity, light, altitude, wind speed, solar energy, air quality, building materials, orientation and thermal insulation are interrelated and all have a great impact on production. In recent years, technical knowledge of these variables has grown considerably due to increased research.

Homeothermal organisms like birds maintain relatively constant body temperature by balancing metabolic heat excess with environmental heat. This thermal equilibrium is established through behavioral and thermoregulatory mechanisms. If there is fast or slow heat loss, the bird must make significant physiological changes that, if they persist, will adversely affect egg and meat production. In general, the ideal temperature in a poultry house is considered to be between 18°C and 20°C for birds 4 weeks of age or older. Temperatures from 30°C to 32°C are considered stressful; and 42°C is classified as sufficient to produce acute stress. It is important to recognize these differences in stress as related to temperature and age. Chicks from 0 to 5 days of age should be kept at around 33°C because thermoregulatory mechanisms are developed after the fifth day of age (Freeman, 1984). For a bird over 3 weeks of age temperatures beyond 26°C could be excessive (North and Bell, 1990).

In addition to oxygen, the bird notices the presence of irritant gases like ammonia in the environment. In open sided houses ammonia can cause considerable metabolic and performance trouble. Although it has been demonstrated that ammonia has a bacteriocidal effect in litter, it can produce adverse effects on productivity. Ammonia can produce keratoconjunctivitis (Carnaghan, 1958), reduce body weight gain (Bullis *et al.*, 1950), reduce feed efficiency (Canvey and Quarles, 1978; Reece and Lott, 1980), cause lung, liver and kidney degeneration (Cristopher, 1935), cause breast blisters and delayed sexual maturity (Charles and Payne, 1960) and increase coccidial damages (Quarles and Faberger, 1979).

Ammonia is produced in the litter by bacterial decomposition of nitrogen substances. Severity of lesions in the upper respiratory tract depend on the atmospheric concentration of ammonia and the length of exposure. Negative effects have been noted with ammonia concentrations of 10 and 40 ppm (Nagaraja *et al.*, 1983, 1984). It is possible that high ammonia levels interfere with the utilization of iron to produce hemoglobin.

Much work has been done to prove the effectiveness of increasing the temperature during the first few weeks of life in order to control incidence of hypoxic ascites. This has been found to be an economic, practical and effective approach to ascites control (Hernández, 1985). A

Table 1. Maximum and minimum temperature registered the previous week in relation to ascites mortality in broilers.

High temperatures and ascites		Low temperatures and ascites	
Temperature range, °C	Ascites mortality, %	Temperature range, °C	Ascites mortality, %
23.0–26.75	1.40±1[a]	9.0–12.5[a]	4.20±2[a]
26.7–30.51	2.45±2[a]	12.6–16.1[a]	2.45±2[ab]
30.5–34.26	3.50±3[a]	16.2–19.7[a]	1.40±1[ab]
34.2–38.0	2.45±2[a]	19.8–23.3[a]	1.05±1[b]

[ab]Means in a column with different superscripts differ significantly (P<0.05)

recent study investigated effects of temperature, humidity and ammonia levels in broiler houses on the incidence of ascites. Ascites mortality in this study began at 3 weeks of age. Maximum temperatures investigated ranged from 23°C to 38°C. Mortality patterns showed no significant difference in ascites (P>0.05) among the ranges established (Table 1). This was not the case for the minimum temperature ranges which oscillated from 9°C to 23.3°C. As the temperature decreased, ascites mortality increased with mortality higher (P<0.05) below 12.5°C.

The ranges between minimum and maximum temperatures significantly affected ascites mortality (P<0.05) as the temperature variation was considerable (Table 2). Additionally, ascites mortality was significantly decreased (P<0.05) when medium temperature was above 23°C. There was a numeric tendency (P>0.05) for decreasing ascites mortality when relative humidity increased (Table 3). This was probably due to

Table 2. Temperature changes between minimum and maximum and median temperature registered the previous week in relation to ascites mortality.

Temperature changes between minimum and maximum		Medium temperature ranges	
Temperature range, °C	Ascites mortality, %	Temperature range, °C	Ascites mortality, %
9.0 –11.25[a]	1.05±1[a]	16.5–20.0[a]	3.5 ±1[a]
11.2 –16.25[a]	2.10±1[b]	20.1–23.0[a]	3.15±1.7[a]
16.51–21.75[a]	4.20±1.7[c]	23.1–27.0[a]	1.05±1[b]

[abc]Values in the same column with different superscripts differ significantly (P<0.05)

Table 3. Relative humidity registered the previous week in relation to ascites mortality.

Humidity range, %	Ascites mortality, %
49.0[a]–59.5	4.20±5[a]
59.6[a]–70.0	5.60±2[a]
70.1[a]–80.5	3.50±4[a]
80.6[a]–91.0	2.45±2[a]

Table 4. Effect of ammonia level registered the previous week on ascites mortality in broilers.

Ammonia levels (range, ppm)	Ascites mortality, %
0.0–3.40[a]	1.40±1[a]
3.5–6.90[a]	3.50±3[ab]
7.0–10.4[a]	3.50±2[ab]
10.5–13.4[a]	4.20±4[ab]
14.0–17.4[a]	7.00±3[b]

[ab]Values in the same column with different superscripts differ significantly ($P<0.05$)

the presence of dust. With low humidity levels dust could predispose birds to respiratory problems and as a consequence of hypoxia this leads to ascites. Ammonia levels were comparatively low; however ammonia above 13 ppm significantly increased ascites (Table 4).

Ascites syndrome in broilers is mainly due to hypoxia and a metabolic compensation between musculoskeletal and cardiopulmonary systems. Any factor that produces hypoxia is likely to cause ascites. Environmental factors play a major role, including an increase in oxygen requirements due to low temperatures, lung tissue damage due to infection, chemical and toxic factors that diminish oxygen uptake, carbon monoxide and carbon dioxide levels that stimulate an increase in plasma levels of carboxyhemoglobin and decrease the ability to capture oxygen, or a low partial pressure of oxygen because of high altitude or deficient ventilation.

The results of this study agree with the findings of Hernández (1985) who reported that maintaining a higher temperature in the poultry house is a practical, effective and economical method to reduce ascites. However, we should not overlook wide temperature variations, dust presence, and keeping poultry houses at minimum temperatures over 12°C and ammonia levels below 13 ppm.

Experiments with *Yucca schidigera* extract

It has been demonstrated that chronically high environmental ammonia severely affects the respiratory tract of poultry. Difficulty in gas interchange in turn produces hypoxia which leads to an increase in ascites syndrome mortality.

The *Yucca shidigera* plant is a desert plant native to the southwestern regions of the United States and northern parts of Mexico. An extract of this plant in a commercial product (De-Odorase) has been shown to contain glycoproteins that bind ammonia. For this reason De-Odorase has been used in animal feeds and slurry pits to reduce ammonia arising from decomposing manure. A study was designed to evaluate effects of different inclusion levels of Yucca extract in commercial feed on ascites syndrome mortality in broilers.

Table 5. Effect of De-Odorase inclusion rate on performance and mortality of broilers at 56 days.

Treatment	Body weight, g	Feed consumption, g	Feed conversion, g/g	General mortality, %	Ascites mortality, %
Control	2811[a]	5606[a]	2.01[a]	24.00[a]	20.85[a]
90 grams/ton	2818[a]	5644[a]	2.02[a]	17.42[b]	12.57[b]
120 grams/ton	2862[a]	5638[a]	1.99[a]	20.85[ab]	16.85[ab]

[ab]Values within the same row with different scripts differ significantly ($P<0.05$)

Table 6. Effect of De-Odorase inclusion level in broiler feed on ammonia levels at 33 and 53 days of age in poultry litter.

	Days of age	
	33	53
Control	13.6	13.7
90 grams/ton	12.0	12.5
120 grams/ton	10.7	11.2

Addition of Yucca extract to commercial feed did not affect body weight, feed consumption or feed conversion; however there was a significant decrease ($P<0.05$) in general and ascites mortality (Table 5). There was a tendency toward ammonia reduction in litter with the use of Yucca extract (Table 6). It is possible that this small reduction was enough to produce significant reduction ($P< 0.05$) in ascites with De-Odorase.

Feed restriction programs and ascites

Feed restriction programs in broilers have been investigated for their effects on a wide range of criteria including abdominal fat (Griffiths *et al.*, 1977; Arafa *et al.*, 1983; Mollison *et al.*, 1984), compensatory weight gain (Washburn and Bondari, 1978; Calvert *et al.*, 1987; Summers *et al.*, 1990), feed efficiency (Plavnik and Hurwitz, 1985; Plavnik *et al.*, 1986; Robinson *et al.*, 1992), and metabolic disease control (Buckland *et al.*, 1976; Ononiwn *et al.*, 1979; Wilson *et al.*, 1984; Classen and Riddel, 1989; Arce *et al.*, 1992b; Robinson *et al.*, 1992). The results have not been consistent, so application of feed restriction programs has been controversial. Broiler feed restriction programs in Mexico have been used in general to control ascites incidence since it is recognized as one of the major metabolic problems affecting the national poultry industry in recent years. It is known that genetics plays a major role in development of the problem; however it should be noted that nutritional, pathological and environmental factors affect incidence (Arce *et al.*, 1986ab; 1987; 1988). Simple feed restriction, dilution of commercial feed with grain, reduction of feed energy levels and use of egg-type bird growing feeds have been some of the approaches used by farmers. Methods of application vary among farms, and some methods are unproductive. In the last five years use of feed restriction methods to reduce ascites incidence has become more widespread (Palos *et al.*, 1988; Suárez and Rubio, 1988; Téllez *et al.*, 1989; Arce *et al.*, 1990; Berger, 1990; Villagómez, 1990), however virtually all agree that these programs are only palliatives. Much research into how sex, strain, severity, and feed restriction time interacts with several ecological conditions is needed.

There are some encouraging experimental reports that demonstrate that a reduction in metabolic demand of the bird, through a reduction in body weight gain in early ages, it is possible to significantly reduce ascites incidence during the production cycle (Arce *et al*, 1989a; Arce, 1990; Castellaños and Berger, 1992).

Some studies have shown a positive correlation between ascites incidence and body weight (Nick and Villacres, 1990). Body weight at 21 days of age can be an important signal to predict ascites incidence toward the end of the broiler cycle. The greatest metabolic change in a broiler is within the 3 first weeks of life (Albers *et al.*, 1990). Therefore a reduction in body weight in the first 21 days has been an effective method of reducing ascites.

There are several ways to reduce the metabolic rate of the bird, two of which are the qualitative and quantitative methods (Yu and Robinson, 1992). The qualitative method consists of using low protein and energy levels through formulation or diet dilution. Quantitative restriction involves reducing time of access to feed, or photoperiodic alternative forms, use of chemicals to inhibit feed consumption such as glycolic acid (Pinchasov and Jensen, 1989) or high doses of tryptophan in feed (Lacy *et al.*, 1982). Whatever the method, a number of variables must be taken into account for example, chick quality, ammonia levels, respiratory problems, bird density, equipment, house design, altitude above sea level, temperature, physical form of feed. These factors, among others, must be considered in order to define a good restriction program under the particular production conditions. Below are summarized a series of experiments on several restriction programs designed to reduce ascites syndrome. The experiments were done by a group of Mexican researchers from the National Institute of Forest and Animal Research (INIFAP-SARH), the National University of Mexico (UNAM) and the private sector. The work was done on INIFAP-SARH experimental farm in Morelia (Michoacan state) at an altitude of 1940 m above sea level.

One of the first studies done at INIFAP-SARH Michoacan confirmed that feed restriction reduced ascites as stated by Heras and Lopez (1984) (Arce, 1990). However feed restriction in these studies was so severe that results were not economically viable. At this point the investigation of economic feed restriction models began.

In general, feed restriction programs begin in the fourth week of life

Table 7. Treatment design

Treatment	Days of age at total fast
1	Ad libitum access
2	7–9–11–13
3	15–17–19–21
4	22–24–26–28

or from the point where the producer recognizes the beginning of ascites mortality. Some producers use severe programs that drastically reduce body weight and delay the production cycle by a week or more. Basic research has resulted in preventative programs that work to reduce the incidence of ascites (Arce *et al.*, 1989b; Odom *et al.*, 1989; Paasch, 1988; 1991; Table 7).

The preventative program, which reduced feed in the early production stages, resulted in acceptable production parameters compared with the control (Table 8). There was no significant difference in body weight, feed intake or feed conversion at 51 days of age. Ascites mortality represented less than 50% in all treatments compared with the control. Economic analysis revealed that the treated group had the highest productivity index with a cost decrease of 9.8, 7 and 6.7%, respectively, for restriction of 7–14, 15–21 and 22–28 days. These data indicated that feed restriction in the early stages was successful regarding production parameters as well as ascites control.

Another study was designed that instead of a total fasting restriction restricted feed only on alternate days (Table 9). The effectiveness of early

Table 8. Effect of feed restriction on performance and mortality of broilers at 51 days of age.

	Body weight, g	Feed intake, g	FCR	Mortality, %	
				General	Ascites
Control	2146[a]	4273[a]	2.02[a]	43[b]	37[b]
7–14[a]	2143[a]	4400[a]	2.09[a]	18[a]	15[a]
15–21[a]	2034[a]	4190[a]	2.09[a]	21[a]	17[a]
22–28[a]	2058[a]	4469[a]	2.20[a]	13[a]	8[a]

[a,b]($P<0.01$)

Table 9. Treatment design.

Treatment	Age, days From	To	Feed access, h/day
Control	Free access		
2	8	42	8
3	8	56	8
4	8	21	5
5	8	21	5
	22	42	8

Table 10. Effects of feed restriction on performance and mortality of broilers at 51 days.

	Body weight, g	Feed intake, g	FCR	Mortality, %	
				General	Ascites
Control	2246[a]	5287[a]	2.39[d]	9.3[a]	5.21[a]
2	2091[b]	4593[bc]	2.23[e]	2.3[b]	0.5[b]
3	1995[c]	4587[bc]	2.34[de]	1.3[b]	–[b]
4	2145[b]	4811[b]	2.28[de]	5.1[ab]	0.79[b]
5	2077[bc]	4522[c]	2.21[e]	3.1[b]	0.83[b]

[a,b,c]($P<0.01$)
[d,e]($P<0.05$)

feed restriction was clear; however as feed restriction time lengthens, body weight is considerably affected (Table 10). Feed conversion was improved in all treatments when feed restriction was applied due to low feed consumption observed. The most common cause of death was ascites which is again reduced by the use of these programs. Control, along with early and continued feed restriction through 56 days of age showed the highest economical benefits, attaining reductions in production costs of 3.5, 5.1 and 5.4%.

Growth rate reduction in broilers through feed restriction has proven an effective means of reducing ascites incidence. Weight reduction is more important during the first weeks of age when it seems that feed restriction has a major impact on broiler physiology. When body weight is reduced, oxygen demand is also reduced thereby avoiding hypoxia that could lead to pulmonary hypertension.

For all this, another experiment was designed using feed restriction starting at one day of age (Table 11). Using the prevention approach in another manner, feed restriction at one day of age worked to reduce ascites incidence (Table 12). The longer the restriction, the lower the body weight. The reduction in feed consumption improved feed conversion in feed restricted treatments in relation to the control ($P<0.01$). This is probably due to a more efficient use of the nutrients after some hours of rest. The control group had the higher economic index by 1.6 and 1.3% in the restriction models from 1 to 21 days for 5 hours and early continuous restriction through 42 days of age, respectively.

Table 11. Treatment design.

	Days of age		Feed access,
	From	To	h/day
Control	Free access		
2	1	14	5
3	8	21	5
	22	42	8

Table 12. Effects of feed restriction on performance and mortality of broilers at 53 days*.

	Body weight, g	Feed intake, g	FCR	Mortality, %	
				General	Ascites
Control	2241[a]	4777[a]	2.16[a]	8.25[a]	3.88[a]
2	2135[a]	4429[b]	2.11[ab]	5.00[ab]	1.75[b]
3	2024[b]	4164[c]	2.09[b]	3.25[b]	0.38[b]

a,b,c (P<0.01)
*Arce *et al.*, 1992a

Feed restriction in manual form as presented in the above studies is not the only way to reduce metabolic demand in the first weeks of life. Feeding low density feeds has been successful in field experiences (Albers *et al.*, 1990). It is important to compare different methods and be able to fit them with different production systems. With this in mind, the following studies were conducted. The objective of the first trial was to provide relatively low levels of protein and amino acids with low energy diets for the first 21 days of life compared with manual restriction (5 h/day access) from 1 to 21 days of age using normal feed. The control group was offered normal feed *ad libitum*. From 21 days of age all treatments were fed with the same feed in two phases (22 to 42 and 43 to 53 days of age). Calculated analyses of the diets are shown in Table 13.

All treatments had reduced body weight (P<0.01) when feed or energy was restricted during the first 21 days of life compared with the control group (Table 14). Birds with manual restriction had the lowest feed consumption and therefore lowest energy intake (Table 15, P<0.01); this also made feed utilization more efficient (P<0.01). The improved feed conversion in animals with limited access to feed has been noted by other authors (McMurtry *et al.*, 1988) demonstrating that fasting stimulates enzyme activity related to lipid synthesis, increasing body weight with less feed when feeding restarts *ad libitum*. The control group had the highest general and ascites mortality (P<0.05); however feed conversion was low and more economically attractive.

The second study involved feeding relatively high levels of protein and amino acids with low energy diets for the first 21 days of age compared with normal protein and amino acid levels, low energy and manual feed restriction (5 h daily) for the first 21 days of age. The control group had commercial feed *ad libitum*. Days 21 through finish all treatments had the same feed in two stages, 22 to 35 and 36 to 54 days of age. Calculated analyses are shown in Tables 16.

Manual restriction, as well as normal protein and low energy level diets, once again reported the most attractive economic index in comparison to the control group (Table 17). This is because of low general mortality and better feed efficiency. The control group had highest energy consumption (P<0.01) at the end of the trial; together with

Table 13. Calculated analysis of broiler diets.

	Days 1–21				Days 22–42	Days 43–53
	Treatment 1	Treatment 2	Treatment 3	Treatment 4		
Protein, %	20	21	23	23	20.5	18.5
Metabolizable energy, kcal/kg	2700	2700	2700	3100	3100	3100
Lysine, %	1.10	1.17	1.25	1.25	1.05	0.95
Methionine, %	0.40	0.44	0.47	0.47	0.45	0.38
Met+Cys, %	0.80	0.87	0.93	0.93	0.83	0.73
Calcium, %	1.00	1.00	1.00	1.00	1.00	0.90
Phosphorus, %	0.50	0.50	0.50	0.50	0.50	0.45

Table 14. Effects of low energy diets and feed restriction on performance and mortality of broilers.

Protein/kcal CP/ME	Body weight, g	Feed intake, g	FCR	Mortality, %	
				General	Ascites
20/2700	2178[b]	4962[a]	2.32[a]	3.33[d]	0[d]
21/2700	2178[b]	4929[ab]	2.30[a]	3.33[d]	1.33[d]
23/2700	2233[b]	4777[b]	2.18[b]	7.00[ef]	1.33[d]
Control	2355[a]	4970[a]	2.14[bc]	9.33[e]	4.33[e]
Restricted	2191[b]	4559[c]	2.12[c]	5.00[df]	1.00[d]

a,b,c($P<0.01$); d,e,f ($P<0.05$). *Arce *et al.*, 1993

Table 15. Energy consumption by stage in broilers with different nutrient concentrations.

Protein/kcal CP/ME	Days of age 21	42	53	Total
20/2700	2189.7[b]	7781.0[a]	5087.1[a]	15057.8[ab]
21/2700	2241.0[b]	7588.8[a]	5118.1[a]	14947.9[ab]
23/2700	2176.2[b]	7173.4[b]	5136.7[a]	14486.3[bc]
Control	2399.4[a]	7672.5[a]	5335.1[a]	15407.0[a]
Rest.	1487.0[c]	7449.3[ab]	5195.6[a]	14132.9[c]

[a,b,c]($P<0.01$)

Table 16. Calculated analyses of broiler diets days 1–21 and days 22–35 and days 36–54.

	Treatment 1 Days 1–21	Treatment 2	Control	Days 22–35	Days 36–54
Protein, %	27	23	23	19.5	19.0
Metabolizable energy, kcal/kg	2700	2700	3000	3150	3240
Lysine, %	1.56	1.30	1.30	1.12	1.00
Methionine, %	0.67	0.57	0.59	0.54	0.49
Met+Cys, %	1.08	0.94	0.94	0.84	0.78
Calcium, %	1.00	1.10	1.00	1.00	1.00
Phosphorus, %	0.45	0.50	0.45	0.42	0.40

the high protein and low energy group (Table 18). These data help us to design in the future energy requirements that a broiler can consume without affecting performance while reducing metabolic upsets like ascites syndrome.

Nutrient restriction before 21 days of age is only a palliative method of reducing ascites syndrome incidence; however, not all restriction methods are economical for the producer. Energy dilution in diets

Table 17. Effect of low energy diets and feed restriction on performance and mortality of broilers through 54 days.

CP/ME	Body weight, g	Feed consumption, g	FCR	Mortality, %	
				General	Ascites
Control	2505[c]	5214[a]	2.10[a]	12.17[c]	8.42[c]
Restricted	2364[d]	4734[b]	2.03[a]	5.91[d]	3.78[d]
27/2700	2476[c]	5143[a]	2.10[a]	12.30[c]	8.62[c]
23/2700	2482[c]	5005[ab]	2.04[a]	9.80[cd]	5.86[cd]

ab ($P<0.01$)
cd ($P<0.05$)

Table 18. Energy consumption in different stages in broilers with different nutrient concentrations.

Protein:energy	21	35	54	Total
Control	2334.75[a]	4467.45[a]	9832.6[d]	16635[a]
Restricted	1463.25[c]	4269.15[a]	9416.3[de]	15149[b]
27/2700	2080.35[b]	4342.62[a]	9673.8[de]	16097[ab]
23/2700	1986.52[b]	4427.16[a]	9323.9[e]	15738[b]

[a,b,c] ($P<0.01$)
[c,d] ($P<0.05$)

reduces body weight and increases feed consumption which affects feed efficiency. Perhaps this is the most important productive difference between the qualitative and quantitative methods, increasing production costs even with reduction of ascites syndrome mortality. Research must focus on finding suitable protein:energy ratios in different production stages. Manual restriction is still the best option and can be a valuable resource to determine the metabolic demand for a broiler in different ages and environmental conditions. Development of new genetic strains resistant to ascites syndrome remains the most attractive alternative.

Recommendations

We must keep in mind that restriction programs do not represent a magic solution for the ascites syndrome. Any problems in house temperature, ventilation (dust concentration and ammonia), cardiopulmonary diseases, toxicity, chick quality or poor management can adversely affect feed restriction programs.

References

Agudelo, G.L. 1981. Causas posibles de edema aviar. Ind. Avicola pp. 47–50.

Albers, G.A., Z. Barran, Zurita and Ortíz. 1990. Correct feed restriction prevents Ascites. Poultry 6(2):22–23.

Anderson, D.P., R.P. Wolfe, F.L. Cherms and W.E. Roper. 1968. Influence of dust and ammonia on the development of air sac lesions in turkeys. Am. J. Vet. Res. 29:1049–1058.

Arafa, A.S., M.A. Boone, D.M. Janky, M.R. Wilson, R.D. Miles and M.S. Harms. 1983. Energy restriction as a means of reduction of fat pads in broilers. Poultry Sci. 62:314–420.

Arce, M. J. 1990. El uso de restricción de alimento en edades tempranas en el pollo de engorda para reducir la incidencia del síndrome ascitico. 2. Mesa Redonda, ANECA. pp. 1–12.

Arce M. J., P.C. Vásquez and C.C. López. 1986a. Concentración de amoníaco, temperatura y humedad ambiental sobre la mortalidad del síndrome ascitico en zonas de mediana altitud; Memorias XI Convención anual de ANECA, Puerto, Vallarta, México. pp. 6–12.

Arce, M.J, C.G. Soto, and G.E. Avila. 1986b. Efecto de la presentacion fisica del alimento con relacion a la incidencia del sindrome ascitico en el pollo de engorda. Tecnica Pecuaria en México. Num 51.

Arce, M.J., C.C. López and P.C. Vásquez. 1987. Análisis de la incidencia del síndrome ascitico en el Valle de México. Técnica Pecuaria en México 25(3):338–346.

Arce, M.J., C.A. Magaña, C.C. López, P.C. Vásquez and G.E. Avila. 1988. Constantes fisiológicas y parámetros productivos de tres líneas comerciales de pollo de engorda y su relación con el síndrome ascitico. Memorias XIII Convención Nacional de la Asociación Nacional de Especialistas en Ciencias Avícolas. Acapulco Gro., pp. 11–135.

Arce, M. J., C.C. López, P.C. Vásquez and G.E. Avila. 1989a. Efecto de la reducción de ganancia de peso en edades tempranas del pollo de engorda sobre la incidencia del síndrome ascitico; Memorias IV Congreso Nacional AMENA, Acapulco, México. pp. 78–84.

Arce, J.M., P.C. Vásquez, G.E. Avila, and C.C. López. 1989b. Respuesta hematológica del pollo de engorda criado en zonas de media altitud. Memorias XIV Convención Nacional de la Asociación Nacional de Especialistas en Ciencias Avícolas. Puerto Vallarta, Jal. pp. 1–8.

Arce, M.J., G.F. Castellaños, M.M. Berger and C.C. López. 1990. Programas de alimentación para el control del síndrome ascitico. Memorias XV Convención Nacional de la Asociación Nacional de Especialistas en Ciencias Avícolas. Cancún, Q.R., pp. 169–177.

Arce, M.J., C.C. López, and G.E. Avila 1992a. Restricción de alimento al día de edad en pollos de engorda para el control del síndrome ascitico. Memorias XVII Convención Nacional de la Asociación Nacional de Especialistas en Ciencias Avícolas, pp. 27–32.

Arce, M.J., M. Berger, and C.C. López. 1992b. Control of ascites syndrome by feed restriction techniques. J. Appl. Poultry Res. 1:1–5

Arce, M.J., G.G. Peñalva, C.C. López and G.E. Avila. 1993. Densidad de energía y proteína en dietas del pollo de engorda sobre los parámetros productivos y la mortalidad del síndrome ascitico. Memorias XVIII Convención Nacional de la Asociación Nacional de Especialistas en Ciencias Avícolas.

Berger, M. 1990. Implementación de programas de restricción alimentica para el control del síndrome ascitico; 2 Mesa Redonda, ANECA.

Buckland, R.B., D.E. Bernon and A. Goldrosen. 1976. Effect of four lighting regimens on broiler performance, leg abnormalities and plasma corticod levels. Poultry Sci. 55:1072–1076.

Bullis, K.L., G.H. Snayenbos and H. Van Roekel. 1950. A keratoconjuntivitis in chickens. Poultry Sci. 29:386–389.

Calvert, C.C., J.P. McMurty, R.W. Rosebrough and R.G. Campbell. 1987. Effect of energy level on the compensatory growth response of broilers following early feed restriction. Poultry Sci. 66(Suppl):75.

Canveny, D.D. and C.L. Quarles. 1978. The effect of atmospheric ammonia stress on broiler performance and carcass quality. Poulty Sci. 57:1124–1125.

Carnaghan, R.B. 1958. Keratoconjuntivitis in broiler chickens. Vet. Res. 70:35–37.

Castellaños, G.F. and M.M. Berger. 1992. Modulación temprana del peso corporal para el control del síndrome ascitico en pollo de engorda. Memorias XVII Convención Nacional de la Asociación Nacional de Especialistas en Ciencias Avícolas, pp. 47–54.

Charles, D.R. and C.G. Payne. 1966. The influence of graded levels of atmospheric ammonia on chickens. Effects of respiration on the performance of broiler and replacement growing stock. Br. Poult. Sci. 7: 177–187.

Christopher, J. 1975. Effects of excess ammonia gas on the chickens. Ind. J. Anim. Res. 9(2):3–86.

Classen, H.L. and C. Riddell. 1989. Photoperiodic effects of performance and leg abnormalities in broiler chickens. Poultry Sci. 68: 873–879.

Cueva, S., H. Sillau, A. Valenzuela and H. Ploog. 1974. High altitude induced pulmonary hypertension and right heart failure in broiler chickens. Res. Vet. Sci. 16:370–374.

Edward, U.T. 1918. Salt poisoning in pig and poultry. J. Camp. Pathol. Therap. pp 31–40.

Estudillo, L.J. 1976. Edema aviar, ascítis ideopática, enteritis no especifica, síndrome de las grasas tóxicas, lipoidosis tóxica, edema de las alturas. Memorias del Primer Congreso Nacional de ANECA. pp 96–104.

Flick, D.F and R.G. Odell. 1965. Studies of the chick edema disease 3. Similarity of symptoms produced by feeding chlorinated biphenol. Poultry. Sci. 44:1460–1467.

Freeman, B.M. 1984. Trasporte de las aves. Revista de la Sección Española de la asociación Mundial de Avicultura Científica 40 (1): 3–11.

Gallagher, B.A. 1919. Experiments in avian toxicology. J.A.V.M.A. 54.

Germán, C.C. and C.C. López. 1985. Evaluación de una formula alimenticia y manejo especial para disminuir la incidencia del síndrome ascítico. Tesis de Licenciatura. Fac. de Med. Vet. y Zoot. UNAM. Méx. D.F.

Griffiths, L., S. Leeson and J.D. Summers. 1977. Fat deposition in broilers: effect of dietary energy to protein balance, and early life calotic restriction on productive performance and abdominal fat pad size. Poultry Sci. 56:638–646.

Heras, P.A. and C.C. López. 1984. Efectos de programas alimenticios para el control del síndrome ascitico sobre los parámetros productivos del pollo de engorda. Memorias IX Convención Nacional de la Asociación Nacional de Especialistas en Ciencias Avícolas. Guanajuato, Gto., pp. 152–157.

Hernandez, V.A. 1979. Comprobacion de la ascitis hipoxica (un tipo de edema aviar). Revista ACOVEZ 3:44–47.

Hernández, A.V. 1985. Ascítis aviar de origen hipóxico: estudios en Colombia. Rev. Avi. Profesional 54–55.

Lacy, M.P., H.P. Van Krey, D.M. Denbow, P.B. Siegel and J.A. Cherry. 1982. Amino acid regulation of food intake in domestic fowl. Nutr. Behav. 1:65–74

López, C.C., L.C. Casas and M.L. Paasch. 1982. Efecto de la altura sobre la presentación del síndrome ascítico. Memorias de la Reunión de Investigación Pecuaria en México 1982. Fac. de Med. Vet. y Zoot. Universidad Nacional Autónoma de México. pp. 214–217.

López, C.C. and E.J. Barbosa. 1981. Ascitis in a high altitude pheasant. Proceedings 30th Western Poultry Disease Conference and 15th Poultry Health Symposium. Davis, California, pp. 80–82.

Machorro, E. and M.L. Paasch. 1983. Evaluacion del efecto de la hipertension pulmonar en la presentacion del sindrome ascitico. Memorias de la VIII Convencion Anual de ANECA. Ixtapa Méx.

McMurtry, J.P., R.W. Rosebrough, I. Plavnik and A.L. Cartwright. 1988. Influence of early plane of nutrition on enzyme systems and subsequent tissue deposition. Biomechanisms Regulating Growth and Development. Beltsville Symposia in Agricultural Research. pp 329–341.

Mollison, B., W. Guenter and B.R. Boycott. 1984. Abdominal fat deposition and sudden death syndrome in broilers: the effect of restricted intake, early life caloric (fat) restriction, and calorie:protein ratio. Poultry Sci. 63:1190–1200.

Nagaraja, K.V., D.A. Emery, K.A. Jordan, J.A. Newman and B.S. Pomeroy. 1983. Scanning electron microscopic studies of adverse effects of ammonia on tracheal tissues of turkeys. Amer. J. Vet. Res. 44:1530–1536.

Nagaraja, K.V., D.A. Emery, K.A. Jordan, V. Sivanandan, J.A. Newman and B.S. Pomeroy. 1984. Scanning electron microscopic studies of adverse effects of ammonia on tracheal tissues of turkeys. Amer. J. Vet. Res. 45:392–395.

Nick, D. and A. Villacres.1990. Effect of changes in energy and nutrient density on the incidence of ascites in broilers. PSA and SPSS Abstracts 69 (Suppl. 1): 162.

Odom, T.W., B.M. Hargis, Y. Ono, C.C. López and M.J. Arce. 1989. Time course changes in electrocardiographic and hematological variables during the development of ascites in broiler chickens. Poultry Sci. 68:107

Ononiwu, J.C., R.G. Thomson, H.C. Carlson and R.J. Julian. 1979. Studies on the effect of lighting on "sudden death syndrome" in broiler chickens. Can. Vet. J. 20:74–77.

Ortega, S.J. 1984. Importancia económica de la ascítis y su interrelación con aflatóxinas y otros factores. Memorias VII Ciclo Internacional de Conferencias Sobre Avicultura. pp. 157–187.

Paasch, M.L. 1988. Síndrome ascitico: aspectos fisiopatológicos. Correo Avícola, Año 1, Vol. 1; Febrero, pp.16–28.

Paasch, L.M. 1991. Desarrollo de algunas investigaciones sobre el síndrome ascítico en México. Ciencia Veterinaria. Volumen 5.

Palos, R.N., P.A. Suárez, P.C. Vásquez, M.J. Arce and G.E. Avila. 1988. Efecto de la restricción alimenticia sobre la presencia del síndrome ascítico en pollo de engorda comercial. Memorias de la Reunión de Investigación Pecuaria en México, 1988, México, D. F. p. 90.

Pinchasov, Y. and L.S. Jensen. 1989. Comparsion of physical and chemical means of feed restriction in broiler chickens. Poultry Sci. 68:61–69.

Plavnik, I., J.P. McMurtry and R.W. Rosebrough. 1986. Effects of early feed restriction in broilers.I. Growth performance and carcass composition. Growth 50:68–76.

Plavnik, I., and Hurwitz, 1985. The performance of broiler chicks during and following a severe feed restriction at an early age. Poultry Sci. 64:348–355.

Quarles, C.L. 1979. Evaluation of ammonia stress and coccidiosis on broiler performance. Poultry Sci. 58:465–468.

Reece, F.N. and D.B. Lott. 1980. The effect of ammonia and carbon dioxide during brooding on the performance of broiler chickens. Poultry Sci. 59:1654.

Renjifo L. l. J. 1979. Poliserositis en pollos parrilleros del Valle Central de Cochabamba. Memorias del VI Congreso Latinoamericano. Cochabamba Bolivia. pp. 125–147.

Robinson, F.E, H.L. Classen, J.A. Hanson and D.K. Onderka. 1992. Growth performance, feed efficiency and incidence of skeletal and metabolic disease in full-fed and feed restricted broiler and roaster chickens. J. Appl. Poultry Res. 1:33–41.

Sanger V.L., L. Scott, L. Hamdy, C. Gale and W.D. Pounden. 1958. Alimentary toxemia in chickens. J.A.V.M.A.:172–176.

Suárez, O.M.E. and R.M. Rubio. 1988. Efecto de la restricción de alimento en la incidencia de ascítis. Memorias de la Reunion de Investigacion Pecuaria en México, México, D. F. p. 90.

Summers, J.D., D. Spratt and J.L. Atkinson. 1990. Restricted feeding and compensatory growth for broilers. Poultry Sci. 69:1855–1861.

Téllez, G., M. Galvan, J. Fuentes and L. Paasch. 1989. Experiencias de campo en el control del síndrome ascítico en pollo de engorda explotado a 2,600 m.s.n.m.; Memorias de la XIV Convención Nacional de la Asociación de Especialistas en Ciencias Avicolas de México, pp. 221–224.

Velazco N.L. 1982. Correlación fenotípica entre mortalidad por mal de altura y peso a la quinta semana en parrilleros. Vet. Inv. Pec. (IVITA) 2(1):43–48.

Washburn, K.W. and K. Bondari. 1978. Effects of timing and duration of restricted feeding on compensatory growth in broilers. Poultry Sci. 57:1013–1021.

Wilson, J.L., W.D. Weaver, Jr., W.L. Beane and J.A. Cherry. 1984. Effect of light and feeding space on leg abnormalities in broilers. Poultry Sci. 63:565–567.

Wooton, J.C. and J.C. Alexander. 1959. Some chemical characteristics of the chick edema disease factor. J. Assoc. Ag. Chem. 42.

Yu, M.W. and F.E. Robinson. 1992. The application of short-term feed restriction to broiler chicken production: a review. J. Appl. Poultry Res. 1:147–153

EFFECTS OF *YUCCA SCHIDIGERA* EXTRACT ON MANURE ODOR ON A DAIRY FARM

D.E. WEAVER

Cornell Cooperative Extension, E. Aurora, NY, USA

Introduction

Nutrient management and odor control are major concerns of animal producers and their neighbors. As dairy, livestock, and poultry farms expand in an effort to be competitive and profitable, manure nutrients and odors become more concentrated. Odors also intensify when manure is stored in an effort to optimize the timeliness of applying the manure nutrient resource. In this experiment a 16-week odor control project was conducted at a 300 cow dairy farm from August to December, 1992 using De-Odorase, a natural product extracted from *Yucca schidigera* plant.

Yucca schidigera extract has been used for many years in pig and poultry operations, but only recently in dairy applications. Components in the extract bind ammonia and hydrogen sulfide present in the manure, which results in a reduction in odor evolved. Although there are other odorous compounds in manure, ammonia and hydrogen sulfide are among the most objectionable. Results from the project showed a 40–50% reduction in ammonia released into the air.

Methods

The project was conducted at the J. Rob Dairy Farm in LeRoy, NY, in cooperation with Ag Network, a feed and dairy supply firm. A neighboring 300 cow farm which shares the manure irrigation equipment served as the control.

At the beginning of our trial, 25 gallons of liquid De-Odorase were used to charge the storage lagoon. Cows were then fed three grams of De-Odorase per head per day to reduce the rate of ammonia released both in the barn and when manure is spread.

During the study, ammonia was measured in the atmosphere in the free stall barn, near the lagoon, and at the site where treated manure was applied to the soil. Dräger tubes and En/Met Kitagawa precision gas detector tubes were used to measure ammonia. Our biggest problem was measuring the low levels of ammonia found on dairy farms. The Dräger tubes measured ammonia from 0 to 1500 ppm, however the highest ammonia level recorded in the barns was less than 10 ppm. Humans can detect ammonia at about 1 ppm; and at levels of 25–30 ppm, ammonia will make the eyes water. However, people become desensitized with prolonged exposure and may not be able to detect ammonia at levels lower than 5–10 ppm.

En/Met toxic gas detector tubes graduated from 0 to 260 ppm were also used, however these were not sufficiently sensitive to measure minor differences in ammonia. During the last month of study, En/Met tubes that measure ammonia from 0.2 to 20 ppm were used. These tubes improved the sensitivity of measurements.

Results

The study results showed that ammonia levels in the free stall barn dropped from an average of 9.5 ppm pre-treatment to an average of 2.2 ppm by the end of the trial. It should be noted, however, that temperatures were cooler toward the end of the trial, a factor which tends to lower ammonia (Figures 1 and 2).

Dräger tubes were also used to measure ammonia levels at the lagoon. De-Odorase appeared to increase bacterial activity in the lagoon as indicated by the increase in gas bubbles on the surface. This reduced crust thickness and time necessary for agitation. Agitation may not be necessary at every cleanout unless manure is being irrigated. Ammonia was never above 2.0 ppm while the lagoon was crusted over. Ammonia on the downwind side of the lagoon was 5 ppm during agitation (Figures 3 and 4). Ammonia readings were much lower using the En/Met Kitagawa tubes than with the Dräger tubes. However, the results showed that

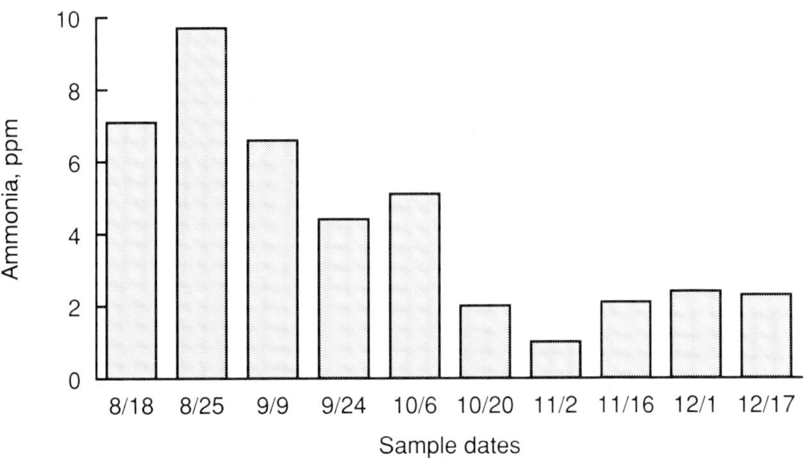

Figure 1. Ammonia levels in the free stall barn measured using Drager tubes.

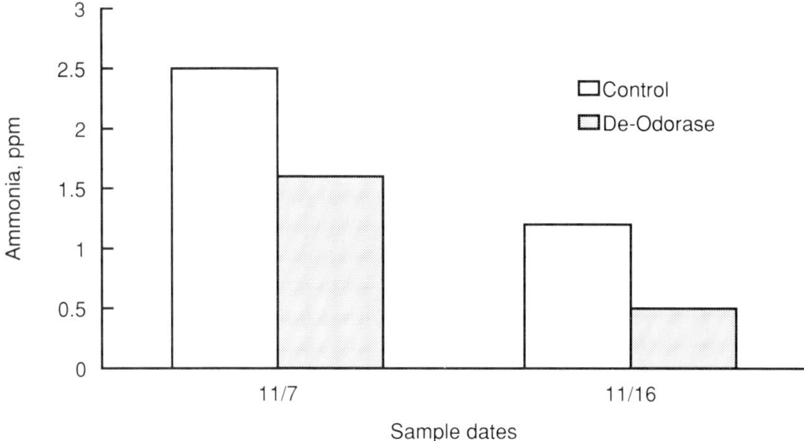

Figure 2. Ammonia levels in control and De-Odorase-treated free stall barns measured using En/Met Kitagawa tubes.

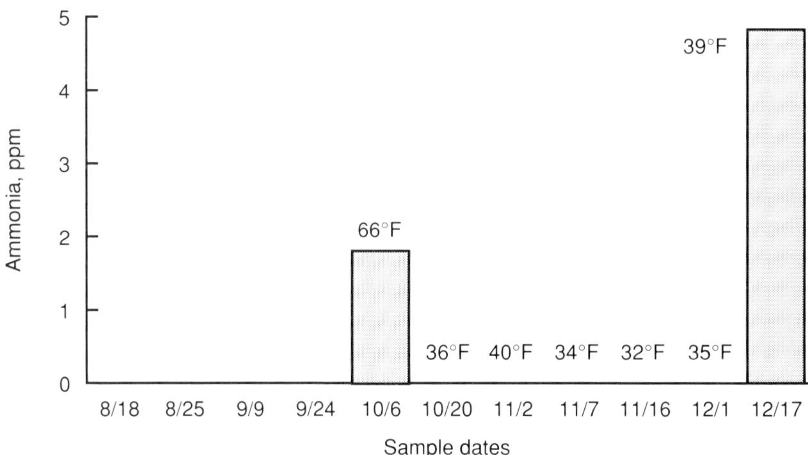

Figure 3. Ammonia levels measured at the lagoon over the test period and at agitation.

ammonia levels were reduced about 50% as a result of De-Odorase application.

Sample results from the manure irrigation sites were not very conclusive because of variability in weather conditions (Figure 5). Ammonia at the untreated site the day of manure application was 0.4 ppm under very windy conditions. Samples collected at the De-Odorase treated irrigation site contained 0.1 ppm and 0.35 ppm ammonia. However, a 1.4 ppm ammonia reading was found in a low sheltered spot in the field following application.

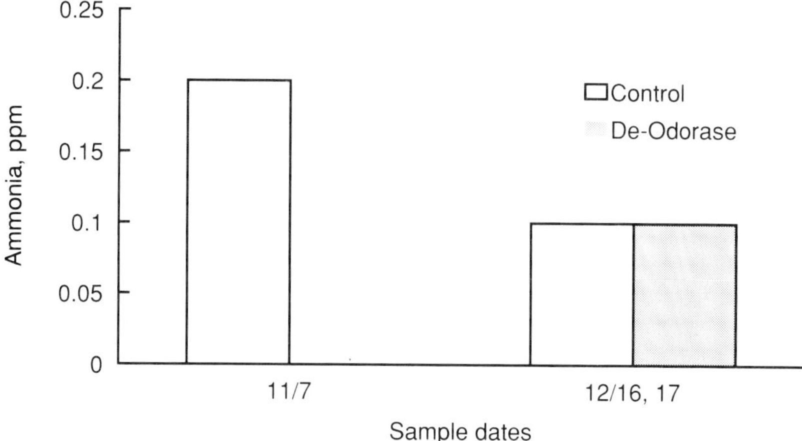

Figure 4. Ammonia levels during agitation at control and treated sites measured using En/Met Kitagawa tubes.

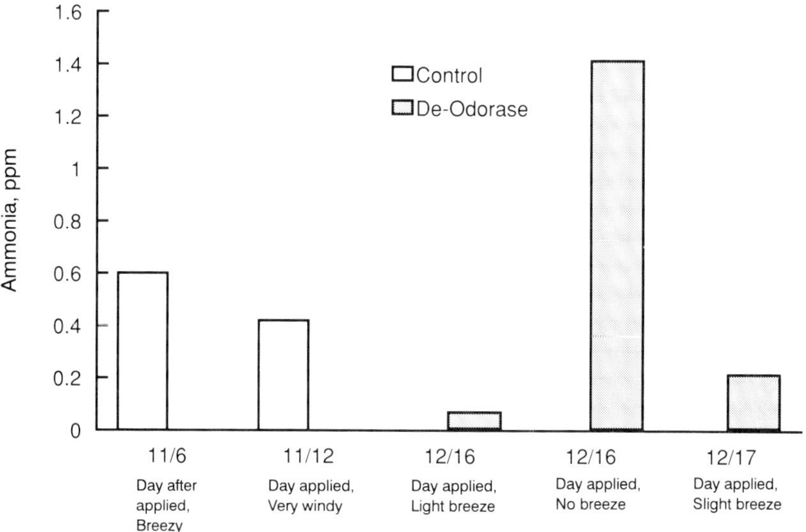

Figure 5. Ammonia levels at the irrigation site for control and treated farms measured using En/Met Kitagawa tubes.

Conclusions

In conclusion, De-Odorase appeared to reduce ammonia by 40–50%. Odor control products hold the promise of reducing ammonia and other odorous products in manure and could reduce odor complaints. A possible fringe benefit of feeding De-Odorase is the reduction of

rumen ammonia which leads to reduced blood urea nitrogen (BUN). The cost of the three gram treatment is about $1.50–2.00 per cow per month. Two grams of De-Odorase daily may be effective in improving lagoon workability while reducing odor. Further evaluation may be necessary to refine odor sampling techniques, measure effectiveness at various treatment levels and evaluate economics.

A NOVEL BACTERIAL INOCULANT TO REDUCE AEROBIC DETERIORATION OF SILAGES

SHELLEY A. GOODMAN[1], PHILIP J. WARNER[2] and COLIN ORR[1]

[1]*Agricultural Consultant, Preston, UK*
[2]*Biotechnology Centre, Cranfield University, Cranfield, Bedfordshire, UK*

Introduction

Since the late 1970s commercial inoculant products for production of silage (grass, maize, and whole crop cereal silages) have concentrated almost exclusively upon fermentation pathways with the goal of increasing fermentation activity through the addition of lactic acid-producing microorganisms. The subsequent increase in lactic acid in the ensiled crop has been considered desirable and an extensive volume of research data has accumulated demonstrating benefits in the form of increased dry matter retention and intake potential in comparison to untreated silages or with acid-treated silages.

The years since 1978 have brought about a considerable change in the types of crops and methods used for winter forage conservation in the United Kingdom. Although bacterial products have come to dominate the silage additive market, the predominant silage crops are now the higher sugar grasses and maize along with intermittent experimentation with whole crop cereals particularly wheat. More efficient harvesting and storage methods have tended to lower the incidence of problems associated with inefficient fermentation while increasing the difficulties encountered from secondary fermentation and aerobic spoilage. To date, although claims have been made for efficacy of products based on everything from propionibacteria to garlic extract, there appears to have been little real success in improving matters regarding aerobic spoilage. Yeasts in particular are identified as primarily responsible for breakdown of soluble carbohydrates into alcohols, as well as for producing conditions in the ensilage that allow for secondary development of molds with toxic damage potential.

Particularly in Ag-Bag™ (large, densely packed plastic bags containing up to 400 tonnes of forage) silages, it was noted that when crops of varying buffering capacity and varying water soluble carbohydrate (WSC) content fermented 'ideally' (i.e., in totally anaerobic conditions, with the addition of a lactic acid bacteria inoculum), extreme levels of deterioration often occurred when the ensilage was exposed to air. The suspicion was raised that in some instances the addition of lactic acid producers, particularly *Lactobacillus plantarum*, actually stimulated yeast growth by providing more substrate (lactic acid, preserved WSC) and a better growth environment.

Research on additives to decrease aerobic spoilage

LACTIC ACID BACTERIAL ADDITIVES, CHEMICAL AGENTS

Since 1988 a serious attempt has been made to identify methods of inhibiting yeast and preventing aerobic spoilage organisms from lowering the nutritive value of the ensiled crop while improving palatability and intake. Research was undertaken at the Biotechnology Centre, Cranfield Institute of Technology* to examine the potential for existing inoculant product bacteria to produce anti-fungal agents. Over a period of time, *Lactobacillus* and *Pediococcus* strains were tested along with chemical agents including *trans*-cinnamic acid and sodium *meta*-periodate (respectively at 0.01% and 0.05% w/v). Some success was achieved in inhibiting yeasts *in vitro* on MRS medium. Subsequent field trials in 5 tonne mini-silos were less encouraging, even though one particular combination of bacteria with the sodium *meta*-periodate made a fairly significant improvement in the quality of the silage. Table 1 shows the size of inhibition zones on MRS agar supplemented with 0.01% (w/v) chemical agents.

Table 1. Comparative effects of chemical agents in inhibiting silage yeasts.

Indicator organism	Control	Sodium propionate	*Trans*-cinnamic acid	D-L phenylalanine	Sodium *meta*-periodate
R. glutinis	1.5	1.5	9	2	2
T. candida	1	2	6.5	1	NG
R. mucilaginosa NCYC 1659	2	4	7	2	7
R. mucilaginosa NCYC 1660	1.5	2	8	1	4
H. anomola	+	+	3	+	–
C. krusei	–	–	1.5	+	2
Saccharomyces cerevisiae	0.5	–	4	–	NG
S. exiguus	–	–	1.5	–	8

Numbers show the size of the inhibition zone in mm, average of 6 tests.
– Indicates no inhibition; + indicates immeasurable zone of inhibition; NG indicates no growth of indicator organism.

BACILLUS SPP.

Further studies were undertaken in an attempt to develop a 'natural' additive with the ability to produce anti-fungal compounds which could

*Now Cranfield University

Table 2. Comparative effects of *Bacillus* spp. in inhibiting silage yeasts.

	B. circulans	*B. cereus*	*B. licheniformis* strain	*B. subtilis* strain
R. glutinis	–	–	–	5
T. candida	–	–	–	12
R. mucilaginosa	–	–	–	3
S. cerevisiae	4	–	11	3
S. exiguus	4	–	12	5

Numbers indicate the size of zone of inhibition in mm.
– No inhibition observed.

be characterized and purified. In particular, *Bacillus* spp. which are known to produce anti-fungal substances were examined. Some strains were selected for further examination and the inhibitory effect on yeasts and molds tested in various environments.

Specific strains of *B. subtilis* and *B. licheniformis* were found to produce a substance ('zymocin') which inhibited the growth of some yeasts without negative effects upon growth of lactic acid-producing bacteria. Inhibitory activity was particularly marked against *S. exiguus, S. cerevisiae, T. candida, and R. glutinis* – all yeasts which had been identified in, and taken from, deteriorating silages.

Bacteriophages in *Bacillus* spp. are well documented, particularly those in *B. subtilis*. Several whole and defective bacteriophages have been shown to possess biological activity analogous to that of bacteriocins. Although there have been no reports of such agents being active against yeasts, it was satisfactorily demonstrated in this study that the inhibitory agent was able to diffuse through agar and that inhibition was not due to either bacteriophage or contact of the producing cells with the indicator organism.

The effectiveness of zymocin against some filamentous fungi implicated in silage spoilage was determined. *Penicillium aurantio, Aspergillus flavus, Aspergillus niger,* and *Trichoderma harzanium* were selected. Zymocin inhibited growth of these fungi (Figures 1–4).

Zymocin production has been shown to be medium dependent – the initial experiments were carried out on malt extract agar, as zymocin activity was not demonstrated when *B. subtilis* were grown in 'Bacillus medium' even with the addition of maltose or glucose at 1% (w/v). Zymocin is produced as a response to the environment (a characteristic of secondary metabolites) very much dependent upon the pH, carbon and nitrogen source.

FURTHER FIELD WORK

As a result of this study, several variations of the selected *Bacillus* strains, along with an enzyme complex capable of increasing the necessary substrate, were used on commercial farms to produce mainly grass silages as early as 1990. In the seasons of 1990–91, 91–92, 92–93, to the present date, over 200 000 tonnes of silage were treated with the

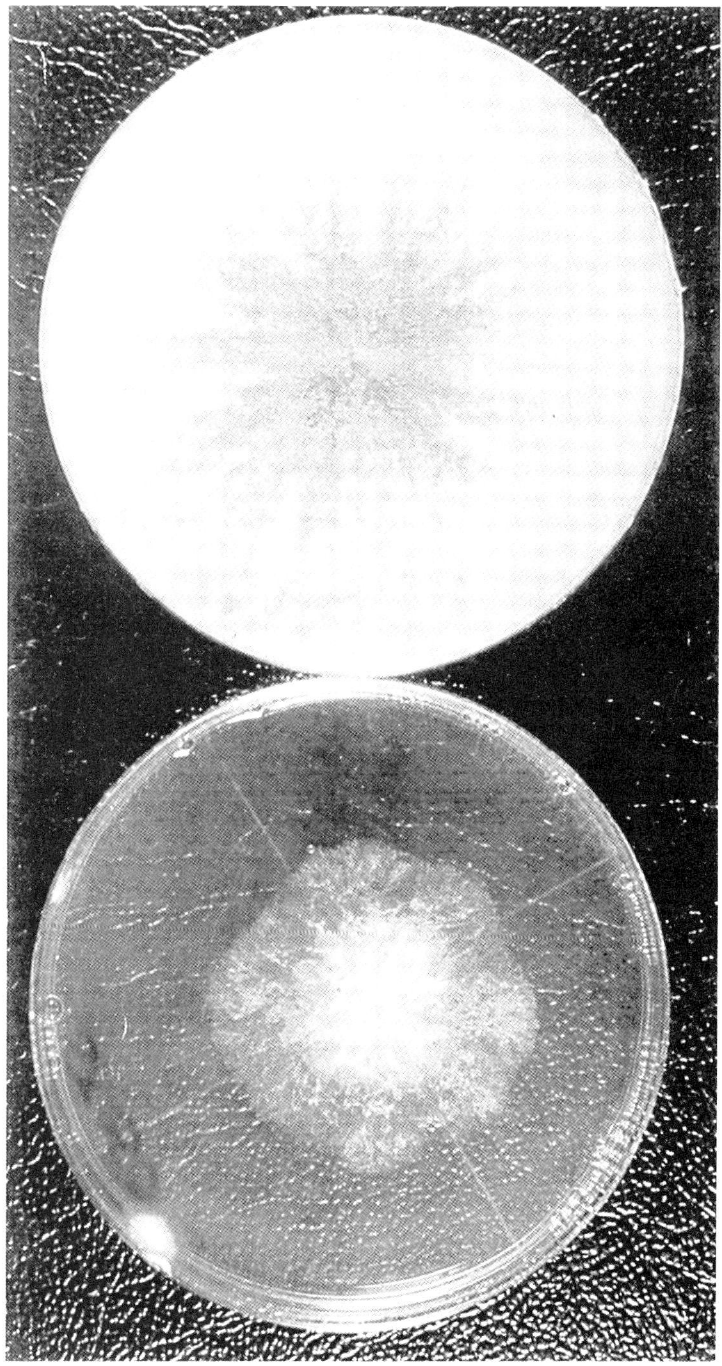

Figure 1. Growth of *Trichoderma harzanium* on malt extract medium in the absence (top) and presence (bottom) of zymocin (130 AU/ml).

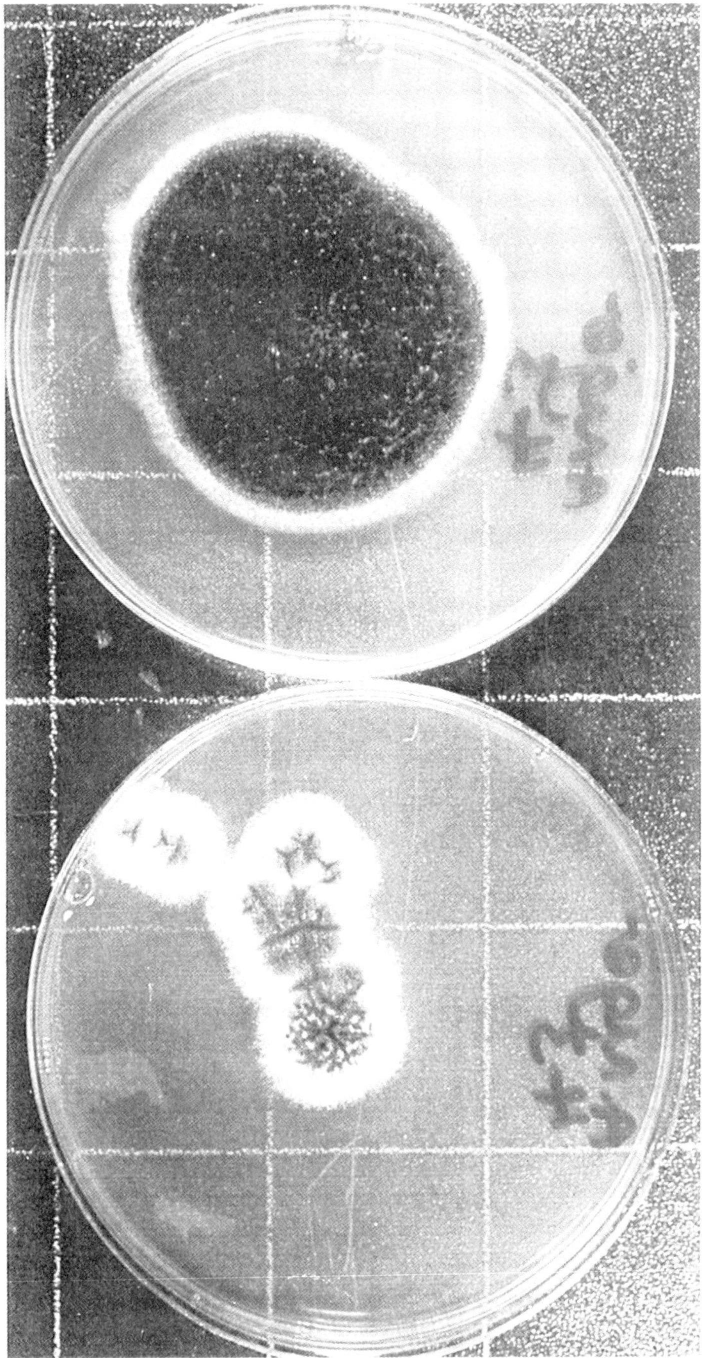

Figure 2. Growth of *Aspergillus niger* on malt extract medium in the absence (top) and presence (bottom) of zymocin (130 AU/ml).

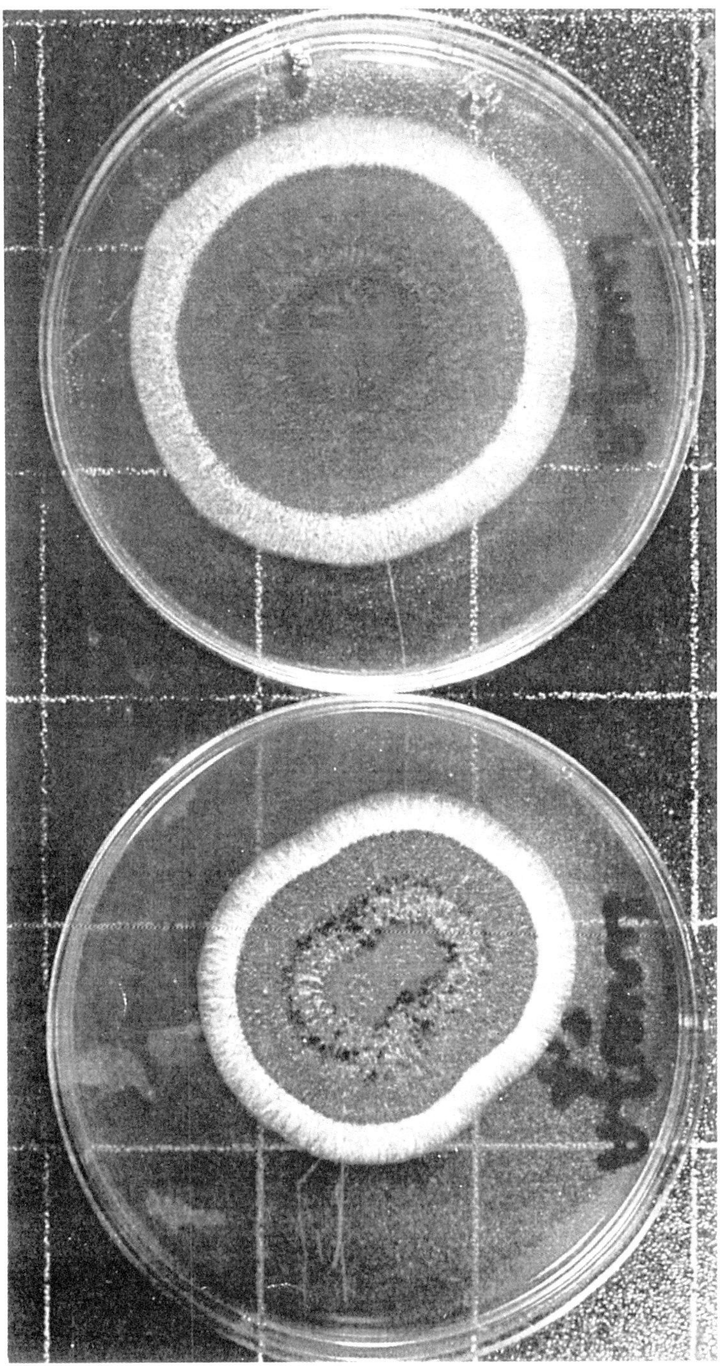

Figure 3. Growth of *Aspergillus flavus* on malt extract medium in the absence (top) and presence (bottom) of zymocin (130 AU/ml).

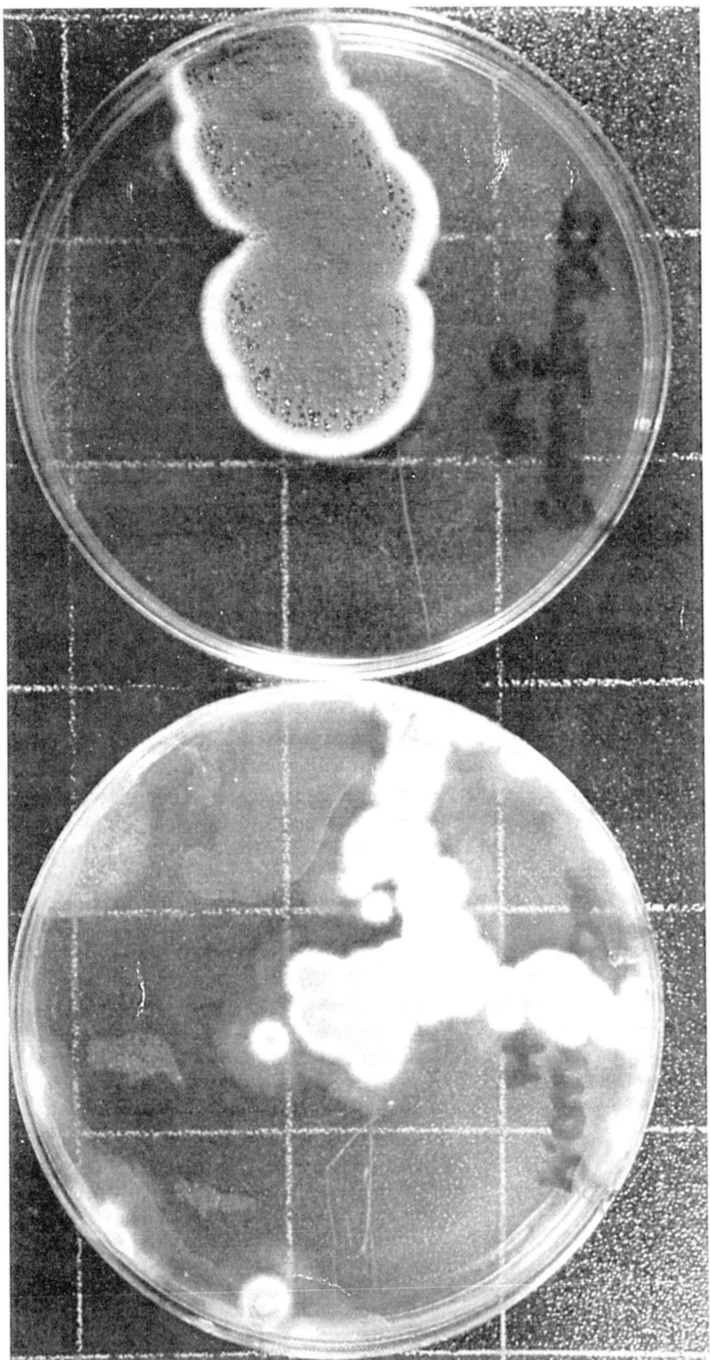

Figure 4. Growth of *Pencillium awanko* on malt extract medium in the absence (top) and presence (bottom) of zymocin (130 AU/ml).

'experimental' product to gauge commercial viability and response in extreme circumstances with differing silages (varying from 14% to 70% dry matter).

Two unexpected benefits were also shown with the product that clearly demonstrated improved stability:

1. The predominantly hemicellulase enzyme activity produced a very soft, sweet, silage which, despite increasing 'sink' in the ensiled material, found immediate favour with farmers.
2. One of the *Bacillus* strains showed residual activity both in the clamp and on feed-out, which may well provide additional digestive enzymes and enhance the nutritive value of the silage. This is currently under research.

Summary and conclusions

The results of both the study at Cranfield, further investigative work by Alltech Inc., and the considerable commercial work are quite clear – zymocin is capable of inhibiting yeasts and molds in silages of all types.

The anti-fungal substances produced by one of the *Bacillus* strains was studied in detail and was found to be produced on a specific range of sugars in a chemically defined medium. This work also suggested that activity was not repressed by high concentrations of glucose. Zymocin production is affected by external pH, and zymocin is only detected in acidic conditions. It is likely that there is a requirement for a proton motive force for the secretion of the zymocin, and this is probably linked to the transport of the sugars on which it is produced. To date, the zymocin peptide has not been purified due to its size (some 400–500 Da), but work is continuing.

One or two commercial products have appeared utilizing *Bacillus* strains, but it is likely that claims for efficacy against yeast/molds have more to do with expression of zymocin-like substances than inhibition through competition. Our view is that the expression of zymocin is directly related to environment, and that manipulation of the environment is a prime factor in the volume (and efficacy) of inhibitory agent produced.

Acknowledgement

Much of the above work was submitted by Shelley Goodman in 1993 as part of a Ph.D. thesis at Cranfield Institute of Technology.

'CROP SET': STIMULATING CROP PRODUCTION WITH AN ORGANIC FOLIAR SPRAY

J. GEOFFREY FRANK

ImproCrop, South Africa

Introduction

With the current popularity of biotechnology involving some 1200 companies, it is a natural conclusion that biotechnology will infiltrate the agronomic and horticultural world at an ever increasing rate. The influence of the escalating world population, pollution and chemical contamination of the planet are of major concern in producing sufficient quantities of quality food crops. As a result of this, regulations governing the use of currently accepted chemicals, in the future, will doubtlessly become increasingly restrictive. The one method of combatting these restrictions will be by stimulating plant growth and plant health with the aid of biotechnology and its inherently environmentally safe products.

Microbial and viral pathogens and unavailable essential nutrients are problems in the agronomic world as in the animal production sector. Currently Alltech is involved in finding non-chemical means of both combatting these problems and stimulating plant growth. So far observations with Crop Set have been of a practical nature, however results to date indicate this is an extremely promising area for research.

Crop Set application and mode of action

Crop Set is a sweet smelling brown/green liquid with a low pH comprised of biologically active plant nutrients, minerals and natural root development enhancers in combination with a liquid by-product derived from Alltech's unique fermentation process. Crop Set is thought to increase plant growth by providing bioactive forms of nutrients in their relative proportions. The result of this increased activity is that the plant has an increase of glucose which, in turn, is a source of energy. The increased energy would allow for plant growth and cell differentiation.

With increased plant growth there is an increased uptake and utilization of available nutrients. These factors are borne out by recent observations in a number of potato crops sprayed with Crop Set:

1. A more vigorous plant.
2. Darker green foliage.
3. Improved root development.
4. Increased setting of crop.
5. Improved disease resistance.

Crop Set can be applied at different plant morphological stages to achieve different goals. If Crop Set is sprayed on to a crop during the crop/fruit set stage (cell differentiation), increased flowering and an increase in amount of crop or fruit set by the plant or tree (in the instance of fruit trees) have been observed. Additionally, in fruit crops such as citrus, Crop Set application has resulted in less fruit drop in times of stress. Applied during the crop growth stage (cell enlargement after the fruit has set), Crop Set aids the plant in obtaining nutrients to meet demand during this stress period. It has also been observed that Crop Set plays a major role in assisting plants during stressful events such as hail, wind or in drought situations. In such instances, observations have been that a second application of 600 ml of Crop Set per hectare has assisted the plant to overcome stress, as seen in Figure 1.

The potato plants on the left and centre of Figure 1, received 600 ml of Crop Set at 75% tuber initiation, while the plant on the right received no Crop Set. The plant on the left received an additional application of 600 ml of Crop Set 4 weeks after the first application. It should be noted that the above plants were representative samples from a severely drought stressed crop.

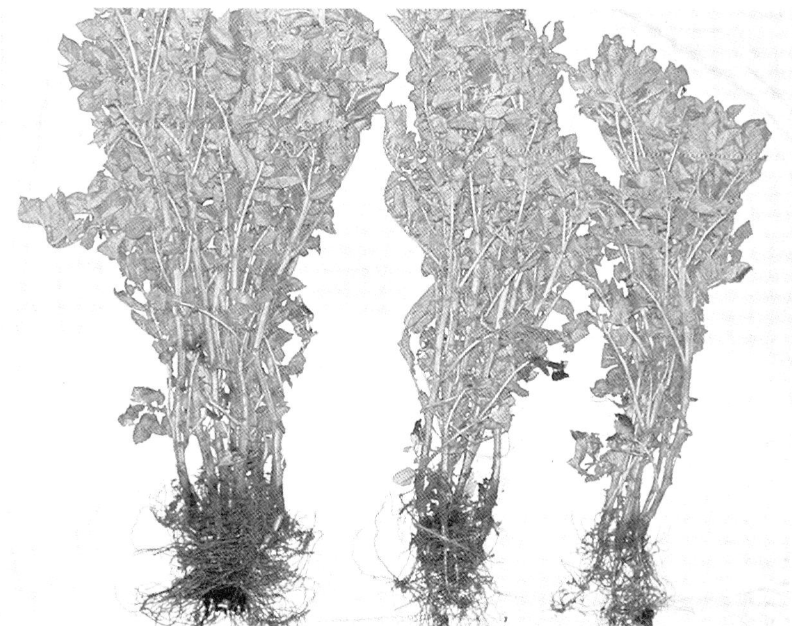

Figure 1. Potato plants treated (left and center) or not treated (right) with Crop Set.

The plants with two applications of 600 ml per hectare each of Crop Set showed a significant increase in root development, while top growth was more healthy and robust. Plants that had a single application of 600 ml per hectare showed good balance of root development in relation to top growth, with healthy looking plants, while the control sector showed bad drought stress with poor root development and significantly less top growth.

Performance tests

POTATO CROPS TESTED

A number of field trials have been performed on potato crops, all of which have produced similar results. Trial No 1 (Figure 2) was performed on a high-yielding variety of potato called BP 1 that has a medium length growing season of 80–110 days from emergence to haulm die-back. The crops used in this trial were grown under supplementary irrigation.

The Van der Plank variety used in Trial No 2 (Figure 3) is a low yielding, short season potato used in the potato chip industry. The growing season is between 80 and 90 days from emergence to haulm die-back. The trial was carried out under irrigation.

A high yielding, medium to short growing season (80–90 days) variety of potato called Buffelspoort was grown in Trial No 3 (Figure 4). This trial was carried out under dryland conditions.

TRIAL LAYOUT AND MEASUREMENTS

All the trials were conducted in summer rainfall areas, receiving 750–850 mm of rain per annum. The trials were carried out under commercial conditions; and in all three trials the land was divided into treated and untreated sections. Ten sample sites of 5 square metres row length with an inter-row spacing of 92 cm were drawn from each of the trials, with five from the treated and five from the control. Each sample site was chosen at random. After harvest, the potatoes from each trial were sorted into groups and categorized based on weight (Table 1).

Table 1. Weight-based potato grades used to evaluate Crop Set effects

Trial 1		Trials 2 and 3	
Category	Weight range	Category	Weight range
Chats	< 30 grams	Chats	< 30 grams
Seed	28–170 grams	Seed	30–170 grams
Large seed	170–240 grams	Large	> 170 grams
Large	> 240 grams		

TRIAL RESULTS AND OBSERVATIONS

Trial 1
In the chats samples, a reduction from 14.1 kg to 7.1 kg was observed (Figure 2). In the seed category, there was an increase in weight from 45.7 kg in the untreated area to 55.8 kg in the treated plot. In both the large seed and the large categories, the treated areas showed increases in weight over the untreated areas of 16.1 kg and 6 kg, respectively.

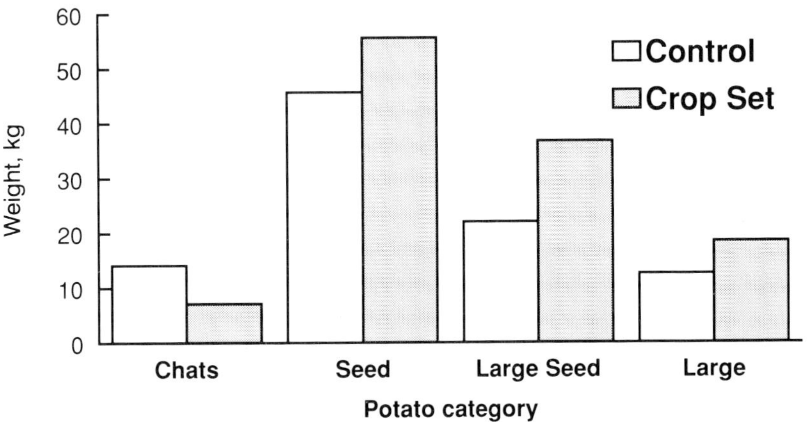

Figure 2. Effect of Crop Set treatment on distribution of potatoes graded by size in Trial 1 (Chats, <30 g; Seed, 30–170 g; Large seed, 170–240 g; Large, >240 g).

The total weight of samples from the untreated area was 94.3 kg. The total weight of samples from the treated area was 120 kg. The total increase in weight due to application of Crop Set was 25.7 kg or 5.51 tonnes per hectare. Additionally, tubers from the treated plot were of a far more even sizing than the tubers from the control.

Trial 2
In this trial it can be seen that in all three size groupings there was an increase in weight where Crop Set was applied (Figure 3). There were 2.2 kg, 9.4 kg and 7 kg increases in weight for chat, seed and large grade potatoes, respectively.

Total weight from treated areas was 73 kg. Total weight from untreated areas was 54.8 kg. Converted into tonnes this figure is equivalent to an increase in yield of 23.7 tonnes per hectare in the untreated area and 31.6 tonnes per hectare in the treated area. A total increase in yield of 7.9 tonnes per hectare was obtained.

This variety of potato has a poor yield of seed owing to the number of tubers normally set and the very large and uneven nature of this variety. Crop Set had the effect of increasing the numbers of tubers set and strikingly improved the uniformity of sizing. There were significantly

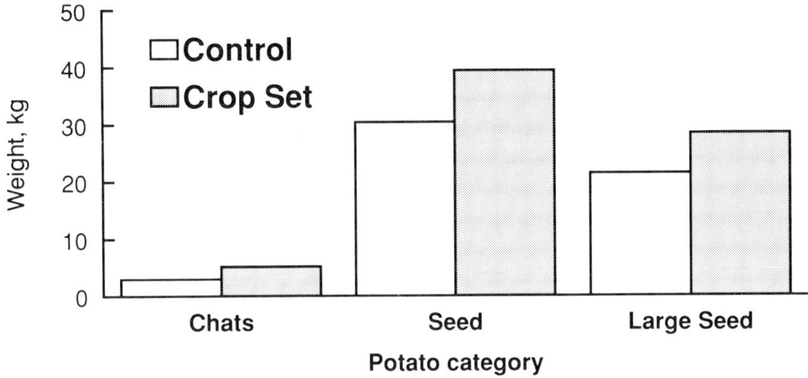

Figure 3. Effect of Crop Set application on distribution of potato grades in Trial 2 (Chats, <30 g; Seed, 30–170 g; Large seed, 170–240 g).

more flowers in the section where Crop Set was applied than typically observed and plants appeared more healthy.

Trial 4

In this trial there was an overall increase of 6.3 tonnes per hectare as a result of Crop Set application (Figure 4). There were fewer chats in the treated section when compared with the untreated section. Seed potatoes increased by 7 kg and large potatoes increased by 8.6 kg. Crop Set had the effect of setting more tubers in addition to producing a more uniform sized crop. An increased flowering and more healthy plant with less blight was also observed.

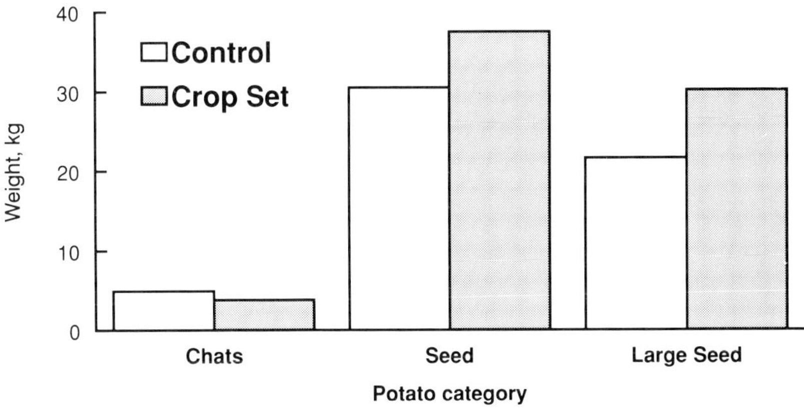

Figure 4. Effect of Crop Set on yield of Buffelspoort (seed crop) potatoes in Trial 3 (Chats, <30 g; Seed, 30–170 g; Large seed, 170–240 g).

Summary and conclusions

Similar trials to those discussed above have been conducted in a numerous countries, all of which have produced similar results and observations. These countries include Holland, Ireland and Canada.

Crop Set has also been applied to wheat at various stages of growth. It was found to produce less lodging, increased yields and generally healthier plants. Where Crop Set was applied to tomatoes, strawberries and peas it was found to increase flowering, fruit set and fruit size. In the case of peas, observations were that pods were filled more evenly where Crop Set was applied, and the crop appeared more healthy and stayed greener for a longer period than in the untreated areas. These facts all contributed to yield differences of 26% in favour of the treated areas.

Where Crop Set has been applied to floral plants, observations are of increased and deeper blooms.

Crop Set has been applied by tractor drawn boom sprayers, knapsack, air and through centre pivots, with success. It has produced consistently good results with only one trial to date which produced no conclusive result. The trial concerned was carried out in Canada on potatoes; and all crops, inclusive of the trial plot, in that area produced double the normal yield. This was owing to abnormally good climatic conditions experienced.

It is concluded that there would be great benefit in utilizing Crop Set as an organic foliar spray on a large variety of cereals, fruit and underground crops to stimulate yields and to produce healthier plants which will be more able to withstand specific periods of stress.

ROLE OF BIOLOGICAL ADDITIVES IN CROP CONSERVATION

RAYMOND JONES

Institute of Grassland and Environmental Research, Aberystwyth, UK

Introduction

Silage production in Western Europe has been estimated to account for almost 70% (105 million tonnes DM) of the forage preserved as winter feed for ruminants. This includes silages from grass and maize in the main, plus some other relatively minor contributions from crops such as legumes and whole crop cereals. There has also been a trend, particularly in Western areas of the UK, to move towards direct cut or minimum wilt herbage of high digestibility, resulting in problems of greater effluent production, particularly in the first few weeks of ensilage (Jones and Jones, 1992).

Silage quality on commercial farms remains variable as a consequence of weather, crop species, chemical composition of the herbage, and practical aspects of ensiling (McDonald *et al.*, 1991). The most significant factors, however, include (a) the variability of the natural population of lactic acid bacteria, and (b) the amount and availability of water soluble carbohydrate (WSC) in the crop necessary to drive the fermentation process.

The future direction in crop conservation will be dictated by changes towards lower intensification with reduced stocking rates and nitrogen applications as a consequence of CAP/GATT policy to lower export subsidies by 36% and the volume of agricultural exports by 21% (Aggregate Measure of Support). Environmental considerations may also play a much greater part in government policies. In the UK, government strategy to reduce environmental pollution comprises legislation. The Water Act (1989), Environmental Protection Act (1990) and Control of Pollution (silage, slurry and agricultural fuel) Regulations (1991) all place additional constraints on farming activities to avoid pollution to the wider environment.

As a result of these economic and environmental pressures grassland farming will have to address the problems of over stocking and nitrogen applications, move away from intensive use of grassland monocultures and make more use of species-rich pastures and alternative crops.

This chapter will be concerned with the use of biological additives applied to forage crops, their strengths and their weaknesses to cope with future demands for low input and environmentally friendly forage conservation.

The epiphytic microflora

A major factor determining the initiation and outcome of the silage fermentation both in the presence or absence of an inoculant is the magnitude and composition of the epiphytic population of lactic acid bacteria on herbage at the time of cutting. Reports of viable numbers of bacteria on standing crops range from < 10 CFU per gram to in excess of 10^3 CFU per gram (Figure 1, Andrieu and Gouet, 1991).

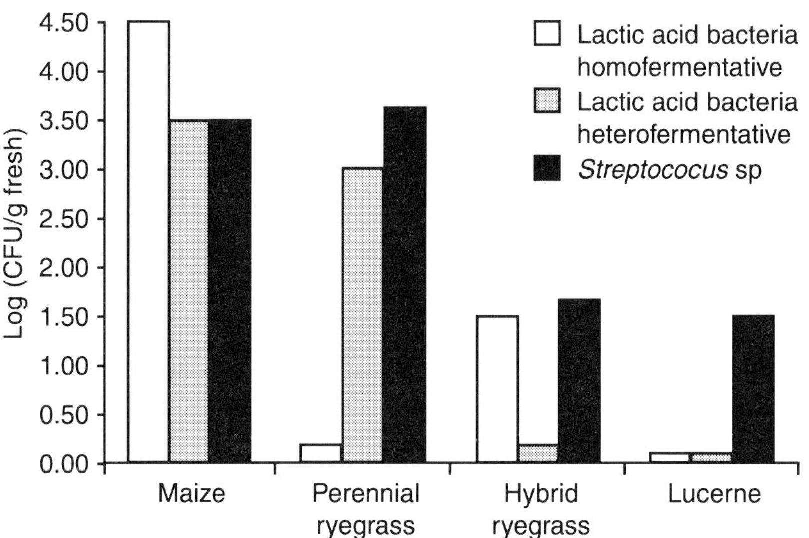

Figure 1. Range of epiphytic bacteria on forage crops.

The enormous range and variation in numbers of lactic acid bacteria detected on the standing crop has been attributed to many factors including temperature, ultraviolet radiation, relative humidity and factors associated with the crop itself (Pahlow, 1991). A hypothesis of 'chopper inoculation' of herbage was suggested (Fenton, 1987; Muck, 1987, 1990) relating to increase in numbers of lactic acid bacteria often seen after the forage is harvested for ensilage. Pahlow (1991) attempted to explain these observations in silage. He suggested that in response to unfavourable conditions on the outer surface of intact plants (e.g. UV radiation, desiccation, nutrient depletion or toxic oxygen species) lactic acid bacteria have developed a survival strategy similar to the 'somnicell' state described by Roszak and Colwell (1987). These authors proposed that a proportion of bacteria in natural environments are in a viable but non-culturable state and exhibit living attributes, other than the ability to grow in culture media. Pahlow (1991) suggested that the harvesting process releases nutrients and enzymes that can resuscitate the dormant lactobacilli. The level and composition of commercial inoculants applied to crops for ensiling is therefore an important aspect and the reader should refer to Merry *et al.* (1993).

Biological additives

The variability in quantity of epiphytic bacteria on forage crops, especially lactic acid bacteria, has led research and development and commercial companies to develop a range of biological additives that will augment the natural population of bacteria. By ensuring adequate levels of desirable bacteria, the carbohydrate source within the crops can be rapidly fermented to lactic acid to preserve the ensiled crop.

Availability of substrates in the form of fermentable sugars (WSC) is a major requirement for lactic acid bacteria strains. The principal sugars in temperate grasses are glucose, fructose, sucrose and fructans (Smith, 1973). The total concentration of these sugars can range from 5 to greater than 300 g/kg herbage DM (Henderson, 1974) depending on cultivar, maturity, lighting, temperature, time of day and N fertilizer applied (McDonald *et al.*, 1991), and newer cultivars can contain in excess of this (Humphries, 1989).

Wilkinson *et al.* (1981) and O'Kiely (1990) have proposed the requirement that at least 3.0% WSC in the fresh crop is required to give adequate substrate for good fermentation in the absence of applied inoculants. However, data from our trials would indicate that a lower level of WSC (2.0% fresh matter) is adequate to give good fermentation with applied inoculants as probably the added strains of bacteria are more efficient at converting WSC to lactic acid than the epiphytic strains.

Fructan is a major carbohydrate in grass species and has been shown to be degraded in the silo (Lunden *et al.*, 1990). Its rate and extent of degradation can vary considerably (Nesbakken and Broch-Due, 1991) and depend on the release of plant carbohydrases during the early stages of ensilage as indicated by Charmley and Veira (1991) for hemicellulase activity. Recently, Muller *et al.* (1992) observed that although 90% of the lactic acid bacteria that they had isolated from grass could ferment glucose, only 5% were able to ferment fructan. Future development of lactic acid bacteria selections may be advantageous to make full use of the fructan substrate without the need for enzyme hydrolysis.

Enzyme preparations are also classified as biological additives; and their main goal is to degrade cell wall fractions of crops to liberate additional fermentable carbohydrates and to improve digestibility of the silage by predigestion of the plant cell wall. Since most of the commercial enzymes used as silage additives are crude fermentation products from fungi, their precise definition is extremely complex and at best their main activities may be measured only on model substrates. Correlation to activities on crop cell wall degradation has been lacking. Usually the effect of enzymes on the degradation of cell wall constituents of silage has been measured in the laboratory on the basis of neutral detergent fibre (NDF), acid detergent fibre (ADF), cellulose and hemicellulose changes between freeze-dried standing crop analysis and silage.

Goering *et al.* (1970) showed that natural degradation of cell wall occurs in untreated silage with as much as 30% of hemicellulose and 5% of cellulose produced from natural plant enzymes. The optimum pH for enzymes is 4–5 and the activity of added enzymes on intact plant material at pH 6 must be questionable. The main activity of added enzymes will probably occur after the first week of ensiling. Different species cell walls such as legumes and maize are more difficult to degrade than grass (Henderson *et al.*, 1982). Maize also has an additional complication in that amylolytic enzyme activity will degrade the high starch content to fermentable sugars for yeasts (Beuvink and Spoelstra 1990).

It appears that enzymes applied at commercial rates do not liberate sufficient additional sugars during the onset of silage fermentation (Honig and Pahlow, 1990) to produce additional lactic acid. More recently, Selmer-Olsen *et al.* (1993) observed a 10–15% disappearance of both cellulose and hemicellulose from herbages treated with a fungal enzyme preparation, but did not observe any utilization of additional sugar to produce lactic and/or acetic acids.

It must also be noted that producing additional carbohydrate substrate in the silage fermentation process will not improve preservation unless adequate levels of epiphytic bacteria are available.

SILAGE FERMENTATION

Application of inoculants on herbage was first introduced in the 1920s. Kuchler (1926) described a system developed by a Bavarian Institute which included the growing of inoculant on the farm, but preservation of live bacteria was difficult and hindered further by selection of unsuitable strains. Today technological advances in freeze drying and the better understanding of microbiological techniques have led to the development of improved strains of commercial inoculants. Research data on fermentation characteristics from inoculated silages and fibrolytic enzyme products (39 experiments conducted world-wide) by Spoelstra (1991) are shown in Table 1. A further set of data extracted from work presented at Eurobac (1990) and published data post 1991 are shown in Table 2.

Table 1. Mean chemical composition of inoculated and enzyme treated silages.

	% of untreated LAB inoculated	% of untreated fibrolytic enzymes
pH	–0.16*	–0.2*
Ammonia-N	84.0	83.0
Lactic acid	128.0	350.0
Acetic acid	78.0	87.0

*pH treated – pH untreated; LAB = lactic acid bacteria

Table 2. Silage fermentation characteristics from inoculated silages.

Reference	Crop DM	Product	Crop	pH	Ammonia	Lactic	Acetic
				% of treated			
1	<18%	Mixture	Grass	100	97	114	89
1	18–25%	Mixture	Grass	100	98	134	69
1	>25%	Mixture	Grass	95	84	116	50
2	14%	Ecosyl	Grass	100	95	96	98
3	17%	Lactisil	Grass	100	89	106	88
3	26%	Lactisil	Grass	100	80	105	50
4	18%	Sil-All	Grass	100	100	100	82
5		Sil-All	Grass	89	60	190	55
5		Sil-All(p)	Grass	95	85	102	80
5		Sil-All	Grass	92	94	118	71
1	>30%	Mixture	Maize	94	65	122	62

1 = Eurobac (1990); 2 = Keady *et al.* (1994); 3 = Martinsson (1992);
4 = Patterson (1993); 5 = Jones (IGER, unpublished)

The research data in Tables 1 and 2 confirm that inoculation of forage crops with bacterial inoculants or use of fibrolytic enzyme products can improve fermentation characteristics by reducing ammonia nitrogen content and acetic acid while enhancing lactic acid content providing that adequate water soluble sugars are available in the fresh crop. Most ryegrass crops have inherent high WSC content and similarly maize crops have adequate substrate for lactic acid production. However, changes in grassland management using alternative crops for silage such as grass/clover, lupins, kale, sunflower, etc. may require further investigation of the range and content of epiphytic bacteria prevalent on these crops and also specific activities of cellulolytic enzymes.

Further research on the basic microbiology and plant chemistry of these crops will be required before new bacterial inoculants can be developed and evaluated.

SILAGE QUALITY AND ANIMAL PERFORMANCE

The efficacy of any biological additive will ultimately be assessed by improvements in the nutritive quality of the silage and effect on animal production. A considerable body of evidence now exists (Mayne and Steen, 1990; Gordon, 1992; Jones, 1992; Jones and Woolford, 1992) to suggest that the new generation of inoculants is capable of markedly influencing silage quality and animal performance, even when low dry matter herbage is ensiled. Data extracted from recent European trials are presented in Table 3.

The data in Table 3 were extracted from approximately 50 different experiments and would reflect a considerable variation in types and quantity of inoculants applied as well as variation in crops, season and maturity. The trends, however, show a positive improvement in silage quality as well as animal performance.

Table 3. Data extracted from European published research on effect of inoculants on silage quality and animal performance relative to untreated.

	Silage quality, % of untreated					Performance, % of untreated				
Reference	MADF	ME	OMD	CP	Enterprise	DMI	LWG	Milk yield	Milk fat	Milk protein
1	103	na	105	101	dairy	114	101	105	104	104
2	95	102	105	102	beef	100	na	na	na	na
3	99	92	na	99	beef	103	113	na	na	na
3	93	101	na	91	beef	97	92	na	na	na
4	101	103	na	99	beef	107	111	na	na	na
5	na	100	na	100	dairy	104	na	100	99	103
5	na	104	104	100	dairy	107	na	103	102	98
6	102	na	na	100	dairy	107	na	102	102	98
7	na	na	102	na	dairy	103	107	103	100	na
7	enz		99	na	dairy	107	100	103	98	na
Mean	98	100	103	99	105	104	103	101	101	

1 = Mayne (1990); 2 = Keady *et al.* (1994); 3 = Kennedy *et al.* (1989); 4 = Steen *et al.* (1989); 5 = Martinsson (1992); 6 = Patterson (1993); 7 = Spoelstra (1991). MADF = modified acid detergent fibre; ME = metabolizable energy; OMD = organic matter digestibility; CP = crude protein; DMI = dry matter intake; LWG = liveweight gain.

Silage nutritive value is mainly dependent on the type and species of crops as well as the stage of maturity. The reader is referred to a previous publication by the author on this topic Jones (1994). An improvement in digestibility from inoculated silages (mean 103%) compared with untreated in the above data has been further substantiated by recent work conducted in Ireland (Keady *et al.*, 1994) where a 5% higher apparent digestibility of organic matter was observed from inoculated silages. These improved differences would suggest a more efficient utilization of nutrients in the rumen.

Trials conducted over a 3-year period at this Institute with beef steers fed untreated and inoculant treated hybrid ryegrass silages have also shown improvements in animal performance (Table 4). Different harvesting conditions and maturity/growth stages of the ensiled herbage were inevitable due to seasonal variations. Significant improvements in fermentation characteristics were observed in bunker silages produced using two of the inoculants (Agros and Sil-All) where poor preservation of the untreated silage was found. Liveweight gains of steers fed these inoculated silages were significantly higher (> 20%) than those fed untreated silages in the two experiments (Table 4) where improvements in fermentation characteristics were observed, while dry matter intakes were similar for both untreated and treated silage. However a third inoculant treatment (HM inoculant) did show additional improvement in animal performance when compared with formic acid treated silage, however the fermentation characteristics of the control formic treatment were good.

In another trial (Table 4) with baled silage no differences were observed in the fermentation characteristics of the inoculant treated

Table 4. Effect of inoculant additives on silage fermentation characteristics and performance of beef cattle.

Additive	Process of harvesting	pH		Ammonia/Total N		ME, MJ/kg/ DM		DMI kg/head/day (% of control)	LWG kg/head/day (% of control)
		Control	Treated	Control	Treated	Control	Treated		
1	Bunker	4.6	3.9	6.8	6.4	10.7	10.8	103	138
2*	Bunker	3.9	3.8	6.4	4.3	11.0	10.8	91	91
3	Bunker	4.5	3.9	10.4	4.0	10.3	10.7	102	120
3	Bale	4.1	3.9	6.4	5.9	10.8	10.9	102	133
Mean		4.2	3.8	7.5	5.1	10.8	10.8	99	121

* = positive control formic acid treated; 1, Sil–All, Alltech; 2, HM Inoculant, Nutrimix, Lytham, UK; 3, Agros, Interprise, Port Talbot, UK; (Jones, R., unpublished data)

compared with untreated silages. Production response in beef cattle, however, showed a significant improvement (32%) compared with the control. This latter finding supports the observations of Mayne and Steen (1990) and Gordon (1992).

Improvements in animal production from inoculated silages in the absence of changes in fermentation characteristics are difficult to explain and it seems likely that the current parameters used to predict silage fermentation and quality may need some re-evaluation. There may be a need for more detailed chemical analysis to take account of structural differences in cell walls, the degradation of cell walls during ensiling and changes in protein fractions. More sophisticated analysis of the cell wall and protein fractions may be required rather than the conventional detergent fibre and Kjeldahl N methods. In addition, a large number of fermentation products are not determined in conventional silage analysis. Further information on these aspects could help to interpret differences in in-silo losses and nutritive quality of silage.

PROTEIN AND NITROGEN UTILIZATION

Many undesirable processes which directly affect silage quality are pH dependent. For example, extent of proteolysis is closely related to rate of pH decline (Heron *et al.*, 1988) and it appears to be largely mediated by plant proteases whose activity decline with decreasing pH, although some activity occurs even below pH 4 (Heron *et al.*, 1989). Coupled with this, ammonia-N or crude protein ($N \times 6.25$) values are used extensively as indicators of silage protein status. However, Williams *et al.* (1992) have demonstrated clearly using the more sophisticated fast protein liquid chromatography (FPLC) technique to fractionate and quantitate soluble herbage proteins (mainly Fraction 1 leaf protein), that ammonia-N is sometimes poorly correlated with extent of protein degradation and may be more closely related to bacterial deamination of amino acids, which presumably occurs after hydrolysis of protein. They ensiled perennial ryegrass after treatment with a number of additives which influence the fermentation in different ways, and although intact soluble protein in the mature (90 d) silages ranged from 40 to 60% of the concentration found in the original herbage, no significant differences in the corresponding ammonia-N concentrations were observed between silages. In this experiment the highest residual protein content was observed in the inoculant treated silage. In further work at this Institute (Cussen *et al.*, unpublished data) two inoculants which had been shown to promote different initial rates of pH decline in silage were examined to determine their potential for reducing proteolysis in perennial ryegrass/white clover silages. Pure stands of ryegrass and clover were mown, chopped, and mixed to give ratios of ryegrass:clover of 100:0, 70:30 or 40:60 on a fresh matter basis, in order to manipulate WSC content and buffering capacity and provide a range of ensiling conditions. The herbages were ensiled after treatment with freshly cultured *Lactobacillus plantarum* (10^6 CFU g^{-1} FM) or a mixture of *L. plantarum* (10^6 CFU g^{-1} FM) and *L. lactis* (5×10^5 CFU g^{-1} FM), in comparison with untreated or formic acid treated (3 litres per tonne). All silages were well preserved after 60 days with similar pH values of less than 4. Nevertheless, soluble protein content measured by FPLC was considerably higher in inoculant treated silages than in either the untreated or formic acid treated controls, which confirmed the earlier findings of Williams *et al.* (1992).

For all types of herbage the highest residual protein content was observed with the mixed strain inoculant treatment where a faster initial rate of pH decline may have suppressed plant enzyme mediated proteolytic activity. If protein status of silage is of nutritional importance as suggested by the results of Charmley and Veira (1990), more detailed characterization of protein will be needed in future silage quality assessments. Recent research data presented by Keady *et al.* (1994) on inoculated silages suggest a positive correlation between animal performance and digestibility of the silage. Furthermore, nitrogen retention in the rumen was higher for inoculated silages indicating a more efficient utilization of nutrients.

Further research must be directed towards better methods of characterizing the energy content of silages, in particular the cell wall fraction, and of protein characterization.

SILAGE LOSSES AND EFFLUENT PRODUCTION

The amount of effluent produced during the ensiling of different crops is influenced by a number of factors. Crop dry matter content exerts a major influence, but the nature of the crop (species, age, fertilizer regime) and additive use also influence effluent production. The degree of consolidation, silo height and the type of mechanical pretreatment before ensiling may also be important (Woolford, 1978). The actual amount of effluent produced during ensiling varies from 500 litres per tonne for crops such as beet tops to little or no effluent for crops wilted to a dry matter content of 300 g/kg or more (Table 5).

There is a characteristic pattern to the rate of effluent production with maximum flow occurring in the first few days after sealing the clamp. Bastiman (1976) showed peak production to be related to crop dry matter content. Wet crops (160 g/kg DM) showed a peak production of 29 litres/day with 90% of the total effluent produced in the first 20

Table 5. Amounts of effluent produced from different crops.

Crop	Dry matter content (g/kg)	Effluent produced (l/t)
Sugarbeet leaves	120–180	500–200
Forage maize	250–300	100–0
Grass or grass clover		
(fresh)	170–200	290–180
(wilted)	>280	0
Catch crops	100–150	440–330

(After Kuntzel, 1991)

days of ensiling while drier crops (250 g/kg DM) produced only 45% of the total effluent in the same period. Similar patterns of production have been shown by Stewart and McCullough (1974) and McDonald *et al.* (1960). Jones *et al.* (1990) found peak effluent production in ensiled ryegrass (160 g/kg DM) to occur on the second day after sealing. Half the total effluent was produced in the first week and 80% in the first 6 weeks. A slow seepage of effluent continued, however, throughout the 19 week measurement period.

Silage additives and effluent production

Some additives alter the structural integrity of plant material and may affect its water holding capacity (Woolford, 1978). Changes in structural and cell wall characteristics may also enable plant juices to escape more readily. Several studies have shown that addition of formic acid, usually at rates of 2.5 to 6 litres/t, to crops at ensiling increases effluent production particularly in the early stages. Winter *et al.* (1987) attributed the high initial release of effluent from formic acid treated crops to a more rapid disruption of mesophyll cell membranes causing release of the cell contents. Additionally, the increased permeability resulting from the reaction of formic acid on the waxy leaf surface could be significant. Pedersen *et al.* (1973) found the addition of formic acid, ammonium formate formic acid mixtures or formalin to increase effluent flow from grass silages particularly in the first 24 hours. They pointed out that formic acid is known to induce a rapid release of cell sap from fresh plant material and postulated that it induces changes in lipophilic components. Organic acids are normally present in the cells in the dissociated form. Acid addition lowers the pH below their pKa thereby causing undissociation and inducing cell leakage. Similar effects of formic acid increasing effluent production have been noted in laboratory silos by Henderson and McDonald (1971) when ensiling different grass species with and without formic acid at rates equivalent to 2–5 litres/t of crop.

Formic acid caused a more rapid initial release and a higher total production of effluent than from untreated crops in a number of clamp silo experiments reported by Bastiman (1976). Treatment with formalin, in contrast to the results of Pedersen *et al.* (1973), reduced effluent suggesting that preservatives suppressing fermentation may reduce effluent. In one experiment quoted with low dry matter grass (160 g/kg DM), formic acid treatment increased effluent production by 17% while formalin treatment reduced it by 30% compared with untreated grass. Harkess (1986) found the use of an acid additive on wet grass (160 g/kg DM) in bunker silos to markedly increase effluent flow. At the end of the first day's filling, flow from formic acid treated grass was 42 litres/hour compared with only 5 litres for untreated grass. The rate of flow from the formic acid treatment did not decline to that of the untreated grass until 4 days after sealing the clamp. Jones *et al.* (1990) found the application of 5 l of formic acid per tonne to autumn harvested ryegrass (161 g/kg DM) to increase the peak flow rate of effluent by over 60% compared with untreated grass. Overall, formic acid increased total effluent production by 17% (Table 6).

Table 6. Effect of formic acid treatment on effluent production (l/t) from late cut ryegrass.

Treatment	Days post-ensiling					
	1	2	5	9	21	136
Untreated	1.7	13.5	19.1	23.4	32.0	51.4
Formic acid (5 l/t)	6.4	21.8	30.4	36.3	42.9	60.2

(Jones *et al.*, 1990)

There appears to be little information on the effects of mineral acids on effluent production. Woolford (1978) however, quotes evidence for sulphuric acid having similar effects to formic acid in total effluent production. There were indications of a less rapid but more prolonged release of effluent from the inorganic acid.

The use of cell wall digesting enzymes, i.e., crude cellulase and hemicellulase preparations, has been shown by McAllan *et al.* (1991) to increase effluent production from ryegrass in 100 kg silos, particularly at low crop dry matter contents. Similarly, O'Kiely (1990) found effluent production to be increased by 50% by the application of cellulolytic enzymes before ensiling. McAllan *et al.* (1991) could find no consistent effect of inoculant application on effluent production, but Fisher *et al.* (1981) noted inoculant treated grass to produce more effluent than untreated grass. Jones (1988), using clamp silos, found inoculant treatment of ryegrass to only give half the effluent of the formic acid control in the initial stages of ensiling. Overall the inoculant application resulted in a 30% reduction in effluent over the 16 week experimental period. Table 7 shows effluent losses and silage dry matter losses from inoculated silages and enzyme treated silage as percentage of untreated.

It appears that effluent production from inoculated silages is less than from untreated with exception of one inoculated silage experiment – an autumn cut grass crop with a dry matter content of 14.5%. In contrast, however, fibrolytic enzyme treated silages appear to increase the rate and accumulative total of silage effluent release.

Total in-silo losses, a major contributor to silage losses on commercial

Table 7. Silage effluent and dry matter loss from inoculated silages as a percentage of untreated.

% DM	Reference	Effluent production	Dry matter loss, %
15.7	1	98.00	103.00
15.7	2 (spring)	99.00	129.00
14.5	2 (autumn)	200.00	103.00
14.0	3 (enzyme)	125.00	na
18.0	4 (Sil-All)	87.00	95.00

1 = Mayne(1990); 2 = Kennedy *et al.* (1989); 3 = McAllan *et al.* (1991); 4 = Jones R (unpublished)

farms, may be attributed mainly to respiration and effluent. Zimmer (1967) analysed dry matter losses from 504 experiments involving herbage and found that DM losses ranged between 0.8 and 71% with a mean value of 19.4%. Similar losses were reported by Bastiman and Altman (1985) from 205 silages in the UK. Current data with inoculated silages appear to show that in-silo dry matter losses are extremely variable depending on inoculated product and maturity of crops. In most of the European studies these losses are similar to untreated silage. This is a major area to investigate further if input costs are to be lowered on farms.

SILAGE DRY MATTER LOSSES

Maintaining anaerobic conditions is the main prerequisite for successful ensilage of herbage. Oxidative losses fall into three main phases (a) field and during silo filling, (b) infiltration of air during ensilage and (c) aerobic deterioration during feed out. Field respiration losses during wilting periods may be extremely variable depending on the weather during harvesting; but generally these DM losses are comparatively small, ranging from 1 to 3%. Many workers have investigated the effect of a prolonged aerobic phase in the silo on the outcome of fermentation by delaying the sealing times. Yoder *et al.* (1960) demonstrated that a delay of 12 h in sealing lucerne silage effectively replaced a lactic acid fermentation with a butyric one. Weise (1968) made similar observations on grass silage. The extent of aerobic deterioration of silage after its removal from the silo has not been submitted to much detailed investigation, although a small-scale technique for the estimation of aerobic stability was proposed by Woolford *et al.* in 1977. A range of 0.8–20% DM losses as a result of a 7-day aerobic exposure period was reported by Henderson *et al.* (1979). Silages particularly at risk from aerobic deterioration are those made from water soluble carbohydrate rich fodders such as maize and those herbages retaining high levels of residual sugars (e.g. formic acid treated) because of restricted fermentation.

LOSSES IN BALED SILAGE

Baled herbage ensiled in large plastic bags or film wrapped to provide a rapid and adequate seal will normally have low DM losses. In a study by Morrison *et al.* (1981) DM losses from grass ensiled in plastic bags ranged between 3.4% and 8.9%. Trials conducted at IGER (Jones, unpublished data) have also shown in-bale DM losses to be less than 10%. Kennedy *et al.* (1989) found that wrapped or bagged baled silage incurred less than 10% DM loss compared with 27.2% from conventional clamp silages. The use of silage additives such as antibiotics, formaldehyde and propionic acid have shown variable degrees of success in reducing losses and more recently the use of biological additives such as inoculants or dried feed concentrate incorporation have also been shown to reduce losses (Table 8).

Table 8. In-silo dry matter loss.

	% in-silo loss	
	Control	Treated
Sil-All	9.3	8.9
Single strain LAB (clamp silage)	11.0	8.0
Singe strain LAB (baled silage)	7.0	9.0
Sugar beet incorporated 25 kg/tonne	13.2	7.7
Commercial concentrate, addition 50 kg/tonne (Fodder match)	23.1	15.2

LAB = lactic acid bacteria
Source: Jones R (unpublished data)

AEROBIC STABILITY

Our Institute has invested considerable efforts to develop techniques such as aerobic stability vessels (ASV) to study changes in temperature of silages during aerobic exposure. This technology will allow replicated mini silo tests to be conducted to compare treated and untreated silages. A major study was set up during 1994 to assess aerobic stability of maize crops treated with a range of biological additives. We are unable to publish the full data in this chapter because of commercial confidentiality, but will indicate the usefulness of the technique by presenting data from acid treated, Maize-All biological inoculated and untreated maize silages.

The present studies with maize silage indicate a very favourable effect of Maize-All inoculant treatment in improving aerobic stability (Table 9) with accumulative temperature changes during a 48 hour recording

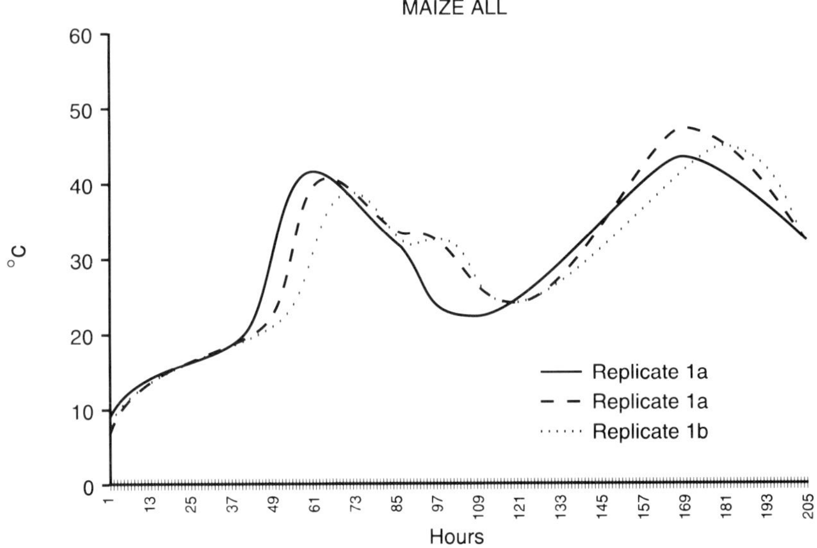

Figure 2. Temperature profile of maize silage subjected to aerobic exposure.

Table 9. Cumulative temperature changes (48 hours) during aerobic exposure of maize silages (Annova: single-factor).

Groups	Replicates	Average
Maize All (inoculant)	3	833
Control (untreated)	3	1013
Maxmais (acid)	3	791

period showing an increase of 20% in stability (Figure 2). There are very limited data available on aerobic stability of grass crops treated with inoculant, however, Kung *et al.*, 1987 and Rust *et al.*, 1989 have shown a reduction in the stability of silages treated with bacterial inoculants. It is likely that this reduction in stability is mediated by changes in microbiology and organic acid ratios. Research is continuing at our Institute to determine factors that influence stability of silages during aerobic exposure.

References

Andrieu, J.P. and J. Gouet. 1991. Proceedings of Forage Conservation Towards 2000. G. Pahlow and H. Honig (Eds) Landbauforschung Volkenrode 123:287–288.

Bastiman, B. 1976. Factors affecting silage effluent production. Exp. Husb. 31:40–46.

Bastiman, B. and J.F.B. Altman. 1985. Losses at various stages in silage making. Research and Development in Agric. 21: 19–25.

Beuvink, J.M.W. and S.F. Spoelstra. 1990. Cell wall degrading enzymes in grass–whole crop maize silage. In: Proceedings of the Ninth Silage Conference 1990. Fac. Agric University Newcastle upon Tyne, 94–95.

Charmley, E. and D.M. Veira. 1990. The effect of heat treatment and gamma irradiation on the composition of unwilted and wilted lucerne silages. Grass and Forage Sci. 46:381–390.

Eurobac 1990. Proceedings of the Eurobac Conference. S. Lindgren and K. Lunden Pettersson (Eds). Swedish University of Agricultural Sciences.

Fenton, M.P. 1987. An investigation into the sources of lactic acid bacteria in grass silage. J. Appl. Bacteriol. 62:181–188.

Fisher, L.J., P. Zurcher, J.A. Shelford and J. Skinner. 1981. Quantity and nutrient content of effluent losses from ensiled high moisture grass. Can. J. Plant Sci. 61:307–312.

Goering, H.K., L.W. Smith, S. Laksmanan, C.H. Gordon. 1970. Fate of carbon-14-labeled cell walls in silage fermentation. Agron. J. 62:532–535.

Gordon, F. 1992. Improving the feeding value of silage through biological control. Proceedings of the Alltech European Lecture Tour, Birmingham. February 1992. Alltech UK, Wrexham, Clwyd, UK. pp 2–17.

Harkess, R.D. 1986. Silage effluent – prevention, retention and collection. The West of Scotland Agric. Coll. Tech. Note No 267: 1–10.

Henderson, A.R., P. McDonald and D. Anderson. 1982. The effect of a cellulase preparation derived from *Trichoderma viride* on the chemical changes during ensilage of grass, lucerne and clover. J. Sci. Food Agric. 33:16–20.

Henderson, A.R. and P. McDonald. 1971. Effect of formic acid on the fermentation of grass of low dry matter content. J. Sci. Food Agric. 22:157–163.

Henderson, A.R., M.J. Ewart and G.M. Robertson. 1979. Aerobic stability in grass silage. J. Sci. Food Agric. 30:223–228.

Henderson, A.R. 1974. Chemical changes during the ensilage of grass with particular reference to carbohydrates. PhD Thesis. University of Edinburgh, UK.

Heron, S.J.E., R.A. Edwards and P. McDonald. 1988. The effects of inoculation, addition of glucose and mincing on fermentation and proteolysis in ryegrass ensiled in laboratory silos. Anim. Feed Sci. Technol. 19:85–96.

Heron, S.J.E., R.A. Edwards and P. Phillips. 1989. Effect of pH on the activity of ryegrass (*Lolium multiflorum*) proteases. J. Sci. Food Agric. 46:267–277.

Honig, H. and G. Pahlow. 1990. The effect of an enzyme preparation on the fermentation of grass silage. In: Proceedings of the Ninth Silage Conference. Fac. Agric University Newcastle upon Tyne pp. 18–19.

Humphries, M.O. 1989. Water soluble carbohydrates in perennial ryegrass breeding. II. Cultivar and hybrid progeny performance in cut plots. Grass and Forage Sci. 44:237–244.

Jones, D.I.H. 1988. Effect of additives on effluent production and silage quality. I.G.E.R. Welsh Plant Breeding Station Annual Report. pp. 37.

Jones, R. 1992. Effect of a biological additive on silage quality, effluent production and animal performance. Irish J. Agric. Food Res. 31:89.

Jones, R. 1994. The importance of quality fermentation in grass silage. Alltech Eight European Lecture Tour, pp. 33–59.

Jones, D.I.H. and R. Jones. 1992. In: Review of silage effluent. ADAS Trawsgoed, and Reading, and IGER Aberystwyth. pp. 76–138.

Jones, R. and M.K. Woolford. 1992. Effect of biological additives on silage quality, effluent production and animal performance. Proceedings of the British Grassland Society Third Annual Conference. Northern Ireland, Session 5, p. 101.

Jones, D.I.H., R. Jones and G. Mosely. 1990. Effect of incorporating rolled barley in autumn cut ryegrass silage on effluent production, silage fermentation and cattle performance. J. Agric. Sci., Camb.115: 399–408.

Keady, T.W.J., R.W.J. Steen, D.J. Kilpatrick and C.S. Mayne. 1994. Effect of biological additives on silage quality and animal performance. Grass and Forage Sci. 49:284–294.

Kennedy, S.J., H.I. Gracey, E.F. Unsworth, R.W.J. Steen and R. Anderson. 1989. Evaluation studies in the development of a commercial bacterial inoculant as an additive for grass silage. Grass and Forage Sci., 44:371–380

Kuchler, L.F. 1926. Die Zeitgemasse Grunfutterkonservierung. Datter, Friesing-Munchen.

Kung, L., L.D. Satter, B.A. Jones, K.W. Genin, A.L. Sudoma, G.L. Enders, J.R. Kim and H.S. Kim. 1987. Microbial inoculation of low moisture alfalfa silage. J. Dairy Sci. 70:2069–2077.

Kuntzel, U. 1991. Silage effluent – an environmental problem. In: Forage Conservation Towards 2000. Landbauforschung Volkenrode 123:364–367.

Lunden Pettersson, K. and S. Lindgren. 1990. The influence of the carbohydrate fraction and additives on silage quality. Grass and Forage Sci. 45:223–233.

Martinsson, K. 1992. A study of the efficiency of a bacterial inoculant and formic acid as additives for grass silage in terms of milk production. Grass and Forage Sci. 47:189–198.

Mayne, C.S. 1990. An evaluation of an inoculant (*Lactobacillus plantarum*) as an additive for grass silage for dairy cattle. Anim. Prod. 51:1–13.

Mayne, C.S. and R.W.J. Steen. 1990. Recent research on silage additives for milk and beef production. In: Annual Report No. 63. Agricultural Research Institute, Northern Ireland (1989/90). pp. 31–42.

McAllan, A.B., J.L. Jacobs and R.J. Merry. 1991. In: Proceedings of Forage Conservation Towards 2000. G. Pahlow and H. Honig (Eds). Landbauforschung Volkenrode 123:368–370.

McDonald, P., A.C. Stirling, A.R. Henderson, W.A. Dewar, G.H. Stark, W.G. Davie, H.T. Macpherson, A.M. Reid and J. Slater. 1960. Studies on ensilage. The Edinburgh School of Agriculture. Tech. Bull. No 24, 1–83.

McDonald, P., A.R. Henderson and S.J.E. Heron. 1991. The biochemistry of silage, 2nd ed. Chalcombe Publications, Marlow, UK.

Merry, R.J., R.F. Cussen-MacKenna and R. Jones. 1993. Biological silage additives. Ciencia E, Investigacion Agraria 20:372–401.

Morrison, R.R., A.R. Henderson and C.E. Hinks. 1981. Proceedings of the 6th Silage Conference, Edinburgh, 85–86.

Muck, R.E. 1987. Factors affecting numbers of lactic acid bacteria on lucerne prior to ensiling. In: Proceedings on the Ninth Silage Conference. IGAP, Hurley, UK. pp 3–4.

Muck, R.E. 1990. Initial bacterial numbers on lucerne prior to ensiling. Grass and Forage Sci. 44:19–26.

Muller, T., M. Muller and W. Seyfarth. 1992. Fermentation of grass fructans by epiphytic lactic acid bacteria. Proceedings of the Second International Symposium on Fructan. IGER, Aberystwyth, UK.

Nesbakken, T. and M. Broch-Due. 1991. Effects of a commercial inoculant of lactic acid bacteria on the composition of silages made from grasses low of dry matter content. J. Sci. Food Agric. 54:177–190.

O'Kiely, P. 1990. Factors affecting silage effluent production. Farm and Food Res. 21 (2):4–6.

Pahlow, G. 1991. In: Proceedings of Forage Conservation Towards 2000. G. Pahlow and H. Honig (Eds) Landbauforschung Volkenrode 123:26–36.

Patterson, D.C. 1993. A report of a study to evaluate an inoculant (Sil-All) as an additive for grass silage offered to dairy cattle. Alltech UK 16–17 Adenbury Way, Wrexham Ind. Est., Wrexham, Clwyd, Wales, LL13 9UZ.

Pederson, T.A., R.A. Olsen and D.M. Guttormsen. 1973. Numbers and types of microorganisms in silage and effluent from grass ensiled with different additives. Acta Agric. Scand. 23:109–120.

Roszak, D.B. and R.R. Colwell. 1987. Metabolic activity of bacterial cells enumerated by direct viable count. Appl. Environ. Microbiol. 53:2889–2893.

Rust, S.R., H.S. Kim and G.L. Enders. 1989. Effects of microbial inoculation on fermentation characteristics and nutritional value of corn silage. J. Prod. Agric. 2:235–241.

Selmer-Olsen, I., A.R. Henderson, S. Robertson and A. McGinn. 1993. Cell wall degrading enzymes for grass silage. 1. The fermentation of enzyme treated silage in laboratory silos. Grass and Forage Sci. 48:45–54.

Spoelstra, S.F. 1990. Comparison of the content of clostridial spores in wilted grass silage ensiled in either laboratory, pilot-scale or farm silos. Neth. J. Agric. Sci. 38:423–434.

Spoelstra, S.F. 1991. Biological additives. In: Proceedings of Forage Conservation Towards 2000. G. Pahlow and H. Honig (Eds). Landbauforschung Volkenrode 123:48–70.

Smith, D. 1973. The non-structural carbohydrates. In: Chemistry and Biochemistry of Herbage. G.W. Butler and R.W. Bailey (Eds). Academic Press, London, UK. pp. 105–212.

Steen, R.W.J., E.F. Unsworth, H.I. Gracey, S.J. Kennedy, R. Anderson and D.J. Kilpatrick. 1989. Evaluation studies in the development of a commercial bacterial inoculant as an additive for grass silage. 3. Responses in growing cattle and interaction with protein supplementation. Grass and Forage Science 44:381–390.

Stewart, T.A. and I.I. McCullough. 1974. Silage effluent-quantities produced, composition and disposal. Agriculture in Northern Ireland 48:368–372.

Weise, F. 1968. The influence of chopping on the fermentation process in direct cut silage. Das Wirtschaftseigene Futter 14:294–303.

Wilkinson, J.M., P.F. Chapman, R.J. Wilkins and R.F. Wilson. 1981. Interrelationships between pattern of fermentation during ensilage and initial crop composition. Proceedings of the Sixteenth International Grassland Congress. pp 631–634.

Williams, A.P., R.J. Merry, J.K.S. Tweed and D.K. Leemans. 1992. The effect of different additives on proteolysis during ensilage of ryegrass. Anim. Prod. 54:487.

Winter A.L., P.A. Whittaker and R.K. Wilson. 1987. Microscopic and chemical change during the first 22 days in Italian ryegrass and cocksfoot silages made in laboratory silos. Grass and Forage Sci. 42:191–196.

Woolford, M.K. 1978. The problem of silage effluent. Herbage Abstr. 10,48: 397–403.

Woolford, M.K., H. Honig and J.S. Fenton. 1977. Studies on the aerobic deterioration of silage using small scale technique. Wirtschaftseigene Futter 23(1):10–22.

Yoder, J.M., D.L. Hill and V.S. Lunquist. 1960. Aerobic stability in silages. J. Anim. Sci. 19:1315 (abstract).

Zimmer, E. 1967. Dry matter losses in commercial farm silages. Das Wirtschaftseigne Futter 13: 271–286.

EFFECT OF CROP SET ON PRODUCTION PARAMETERS OF WHEAT, PEAS AND CANOLA

BRIAN HUTTON

Alltech Canada, Winnipeg, Manitoba, Canada

Introduction

Crop Set, a foliar-applied supplement from Alltech, was initially evaluated in seed potato production. Investigations continued in commercial potato production. Results obtained from studies in South Africa demonstrated yield increases along with improvements in uniformity and tuber size. Additionally, there was an improved distribution of tuber sizes in plots treated with Crop Set (Table 1). These and other observations in potato crops indicated that Crop Set is best applied at the time of most rapid cell differentiation and cell enlargement. It is not believed Crop Set works as a plant growth regulator (i.e., a plant hormone). However, the plant would appear to be more able to utilize the nutrients available to it during these critical stages in its development.

Table 1. Effect Crop Set applied at tuber initiation on yield and size distribution of potatoes*.

	Yield, tonnes per hectare		
Size category	Crop Set	Control	Difference
Under 30g (chats)	1.63	2.13	–22.5%
30–170 g (seed size)	16.3	13.26	+18.7%
≥170 g (table size)	13.13	9.39	+28.5%
Yield	31.09	24.78	+20.3%

*Variety BP 13

Materials and methods

Investigations were carried out by Dr Ottmar Philipp of Agriprocess Inc. to evaluate the effects of Crop Set on production parameters in a number of significant field crops grown in western Canada. The crops evaluated included peas and lentils, canola, and wheat. Observations were made on the effects of Crop Set on crop growth habits. In all trials Crop Set was applied at 600 ml/ha with a bicycle sprayer equipped with SS 8002 Teejet nozzles at 275 kPa in 200 l/ha of water. Plot size was 2 m × 5 m. Crops were seeded with a press drill with 15 cm between rows. Each trial comprised a randomized complete block design with four replications. This discussion will concentrate on effects on wheat, peas and canola.

Wheat

Crop Set was applied at various growth stages to Sceptre durum wheat. The stages of application included Zadok's stages: 3.0, 3.1, 3.7, 4.5, and 5.8 (Table 2). No effect on height of plants was observed. A darker green color in the leaves was observed some 14 days post application in the plants from the area treated with Crop Set. This darker green would be similar to an added nitrogen effect. It is thought this indicated that nitrogen was more available to plants in plots treated with Crop Set. No differences were noted in number of spikes or spikelets. There were no significant differences in protein content among treatments (Table 3).

Table 2. Plant morphological stages at treatment.

Zadok's 3.0	Tillering leaf sheaths strongly erect
Zadok's 3.1	Stem extension stage, first node visible
Zadok's 3.7	Stem extension stage, last leaf just visible
Zadok's 4.5	Stem extension stage, in the boot
Zadok's 5.8	Heading or flowering stage (wheat)

Table 3. Effect of Crop Set application on yield and % crude protein of Sceptre durum wheat when applied at four different morphological stages.

Application stage (Zadok's scale)	3.1	3	4.5	3.7	Control
Protein, %	13.1	13	13	12.6	13
Yield, kg/ha	4441	4382	4320	4305	4238
Yield, % of control	105	103	102	102	100

Although differences in yield were not statistically significant, these trends would translate into a increase in returns of $26.11 USD/ha. In terms of the cash inputs in today's crop production, this type of return would be considered significant. With this type of improvement, continued evaluation is warranted.

Peas

Crop Set was applied to Magda peas at two growth stages, the five to six trifoliate leaf stage and at first flower. Magda peas are a conventional variety of peas, i.e., it is not semi-leafless. As in the wheat study, the pea crop had a darker green color when treated with Crop Set. The observation was again that the color difference was similar to a nitrogen effect. No observations were made in relation to any relative differences in degree of nodulation between the plants from treated and untreated areas.

An apparent shortening of the internode area on the stem of the peas was noted in those areas treated with Crop Set. This change did not appear to lead to significantly different numbers of pods between plants from treated and untreated plots. However, plants from the treated areas tended to have more uniform numbers and sizes of peas per pod.

Yields from the areas treated either at first flower or at the five to six trifoliate stage were significantly greater than the control $P<0.05$, Table 4). Apparent differences in nodulation, internode length, and uniformity of numbers of seeds per pod combined to clearly demonstrate a significant benefit from Crop Set treatment.

Table 4. Effect of Crop Set on yield of Magda peas.

	Crop Set		Control
Stage	First flower	5–6 trifoliate leaf	
Yield, g/plot	1202	1197	953
Yield, kg/ha	1502	1496	1191
Yield, bushels/acre	22	22	17

Canola

Canola has become the most important oilseed crop in Canada in the past 20 years. Canola is also becoming important in crop production in a number of countries due to its extremely high quality oil. This crop is very susceptible to various environmental and nutrient stress factors. In recent years, hybrid canola production has started to become important. One of the methods used in hybrid seed production is the CMS (cytoplasmic male sterility) system. This system is characterized by extremely poor seed yield. In canola the four major yield components are (in decreasing order of importance): branching of the plant, pods per branch (of much less importance), seeds per pod, and size of seeds.

Trials were carried out to evaluate the application of Crop Set to hybrid canola. Crop Set was applied at the late rosette stage of development before bolting of the female plants in seed production. At the late flowering stage, plants from the areas treated with Crop Set had about 60% more branching and 30% more pods per branch. In spite of the promising indications, plots treated with Crop Set produced significantly lower yield (Table 5).

Such a dramatic decline in yield of canola seed is not unheard of. In 1981 Manitoba Agriculture reported a similar situation in open-

Table 5. Effect of Crop Set on yield of hybrid canola seed.

Treatment	Control	Crop Set
Yield, kg/ha	2282	2025
bushels/acre	41	36
Yield, % of control	100	89

pollinated canola. Yield components apparently indicated a very large yield potential. However, the producer experienced a very average yield. The indications were that yield was limited by fertility factors. In future trials it will be important to provide fertility potential adequate for the yield potential to be expressed. Tissue analysis should be used to evaluate possible deficiencies.

Crop Set in crop production in the future

Strip plot trials in a number of field crops continue to give encouraging results. These crops include carrots, onions, grapes, tomatoes, and strawberries. In general, all of the results are similar to those experiences in potatoes, i.e, more uniform and larger fruit. More attention will be paid to the intricacies of the field crops summarized earlier. However, it is easy to see that such improvements to yield and quality due to Crop Set hold great potential for crop production in the future.

INDEX

DISTRIBUTORS AROUND THE WORLD

ARGENTINA
LABORATORIOS SCOPE S.A.
Av. San Juan 2866
1932 Buenos Aires
TEL:54–1–941–6774
FAX:54–1–941–5016

AUSTRALIA
RHONE-POULENC ANIMAL NUTRITION
19–23 Paramount Road
West Footscray
Victoria 3012
TEL:61–3–316–9750
FAX:61–31–314–9386

AUSTRIA
ALLTECH AUSTRIA GMB
A-2094 Pingendorf 20
TEL:43–2912–6217–0
FAX:43–2912–6217–30

BRAZIL
ALLTECH DO BRAZIL
Caixa Postal 10808
Cep 81170–610
CIC –Curitiba –Paraná
TEL:55–41–246–6515
FAX:55–41–246–5188

CANADA
ALLTECH CANADA
1 Air Park Place
Guelph, Ontario
N1H 6H8
TEL:519–763–3331
FAX:519–763–5682

CHILE
ALLTECH CHILE
Atenas 7542
Las Condes
Santiago
TEL/FAX:56–2–201–2986

CHINA
ALLTECH CHINA
Institute of Animal Science
Chinese Academy of Agricultrual Science
CASS
Malianwa Haidian
100094 Beijing
TEL:86–1–258–2225–2003
FAX:86–1–258–4349

COLOMBIA
INVERSIÓWES AMAYA
Calle 85 No. 20–25
Oficina 401A
Bogotá
TEL:57–1–218–2829
FAX:57–1–218–5317

COSTA RICA
NUTEC, S.A.
Apartado 392
P.O. Box 392
Tibas
TEL:506–236–3110
FAX:506–233–3110

CYPRUS
CHRONEL BIOTECHNOLOGY Ltd.
P.O. Box 2792
Larnaca
TEL:357–4–638082
FAX:357–4–638082

CZECH REPUBLIC
ALLTECH CZECH REPUBLIC
Mezirka 13
60200 Brno
TEL:42–5–41–21–57–40
FAX:42–5–41–21–57–41

DENMARK
NUTRISCAN A/S
Rørsangervej 8
Postboks 141
DK-8300 Odder
TEL:45–86–542–488
FAX:45–86–560–359

DOMINICA REPUBLIC
SANUT, S.A.
Km 10 1/2 Aut. Duarte
Santo Domingo, Dominica Republic
TEL:809–560–5840
FAX:809–564–4070

ECUADOR
PROPEC, S.A.
Av. 12 de Noviembre
24–85 y av. el Rey
Ambato
TEL:593–382–3180
FAX:593–382–9267

EGYPT
EGYTECH
13, El-Solouly St
P.O. Box 442
Dokki
Cairo 12311
TEL:202–361–0605
FAX:202–361–5909

FINLAND
BERNER Ltd.
Eteläranta 4B
SF 00130
FINLAND
TEL:358–0–134–511
FAX:358–0–134–51380

FRANCE
ALLTECH FRANCE
2–4, Avenue du 6 juin 1944
95190 Goussainville
FRANCE
TEL:33–1–398–86351
FAX:33–1–398–80778

GERMANY
ALLTECH DEUTSCHLAND GmbH
Esmarchstrasse 6
23795 Bad Segeberg
GERMANY
TEL:49–4551–88700
FAX:49–4551–887099

GREAT BRITIAN
ALLTECH U.K.
Unit 16–17
Abenbury Way
Wrexham Ind. Estate
Wrexham, Clwyd
LL13–9UZ
TEL:44–1–97–8–660–198
FAX:44–1–97–8–661–136

GREECE
LAPAPHARM, INC.
73 Menandrou Str.
10437 Athens
TEL:30–1–522–7208
FAX:30–1–522–7152

HONDURAS
S.B.F. INTERNATIONAL
Colonia Palermo
No. 1862
Tegucigalpa
TEL/FAX:504–32–39–64

HONG KONG
PING SHAN ENTERPRISE Co. Ltd.
21–24 Connaught Road West,
2/F Seaview Commerical Building,
TEL:852–859–9999
FAX:852–858–1452

HUNGARY
ALLTECH HUNGARY
Kresz Geza utca 16
H-1132 Budapest
TEL:36–1–269–5384
FAX:36–1–201–5215

INDIA
VETCARE
No. 90, 3rd Cross
2nd Main, Ganganagar
Gangenahalli
Bangalore 560–032
TEL:91–80–3332–174
FAX:91–80–3334–041

INDONESIA
P.T. ROMINDO PRIMAVETCOM
Dr. Saharjo No. 266
Jakarta 12870
TEL:62–21–830–0300
FAX:62–21–828–0678

IRELAND
ALLTECH IRELAND
Unit 28, Cookstown Industrial Estate
Tallaght, Dublin 24
TEL:353–14–510276
FAX:353–14–510131

ITALY
ALLTECH ITALY
ASCOR CHIMICI
Via Piana, 265
47032 Capocolle (Forlì)
TEL:39–543–448070
FAX:39–543–448644

JAMAICA
MASTER BLEND FEEDS Ltd.
P.O. Box 24
Old Harbour
St. Catherine
TEL:809–983–2305
FAX:809–983–9241

JAPAN
MITSUI & CO. Ltd.
2–1 Ohtemachi 1–Chome
Chiyoda-Ku
Tokyo
TEL:81–33–285–5026
FAX:81–33–285–9958

KOREA
YOONEE CHEMICAL CO.,Ltd.
C.P.O. Box 6161
Seoul
TEL:82–2–585–1801 (S.D. 76)
FAX:82–2–584–2523 (S.D. 88)

KOREA
I.E. SUNG INTERNATIONIAL
IL Bok Bldg. 2F
1602–4, SeoCho-Dong
SeoCho-Ku, Seoul
TEL:82–2–521–0501–4
FAX:82–2–521–1300

MALAYSIA
FARM CARE SDN. BHD.
No. 48–3, Jalan Radin Tengah
Seci Petaling
5700 Kuala Lumpur
TEL:60–3–957–3669
FAX:60–3–957–3648

MEXICO
APLIGEN SA DE CV
Palestina 67–A
Col. Claveria 02800
TEL:525–3963840
FAX:525–3963565

NEPAL
NEPA PHARMAVET PVT Ltd.
GA-1–481, Wotu Tole
Kathmandu-3
TEL:977–1–217–952
FAX:977–1–224–627

NETHERLANDS
ALLTECH NETHERLANDS
Holandsch Diep 63
2904 EP Capelle aan den Ijssel
TEL:31–10–450–1038
FAX:31–10–442–3798

NEW ZEALAND
CUNDY TECHNICAL SERVICES
P.O. Box 69
170 Glendene
Auckland 8
TEL:64–9–837–3243
FAX:64–9–837–3214

PERU
ALLTECH PERU
Av. Comandante Espinar 260
3er Piso, Oficina 301
Miraflores
Lima 18
TEL/FAX:51–14–47–69–82

PHILIPPINES
FERMENTATION INDUSTRIES CORP.
P.O. Box 440
Greenhills, Metro Manila
TEL:63–2–241–0870 or 241–0846
FAX:63–2–241–0840

POLAND
ALLTECH POLAND
ul. Szczesliwicka 29/31
02–353– Warszawa
TEL:48–90–216–293
FAX:48–90219939

PORTUGAL
ALLTECH PORTUGAL
ALLTECH ADITIVOS
ALIMENTACO ANIMAL LDA
Rua Álvaro de Brée No. 6
Leceia, P-2745 Queluz
TEL:351–1–421–8029
FAX:351–1–421–8100

SOUTH AFRICA
ALLTECH SOUTH AFRICA
P.O. Box 241
Somerset West
7129
TEL:27–24–517052
FAX:27–24–517000

SPAIN
PROBASA
c/o Argenters, 9 Nave 3
Pol Ind Santiga, Sta
Perpetua de la Moguda
Barcelona
TEL:34–3718–2215
FAX:34–3718–1307

SWEDEN
VETPHARAM AB
Thord Bengtsson
Annedalsvgen 9
S-227 64 LUND
TEL:46–46–12–81–00
FAX:46–46–14–65–55

SWITZERLAND
INTERFERM AG
Postfach 112
Hardturmstrasse 175
CH-8037 Zürich
SWITZERLAND
TEL:41–1–272–8024 (S.D. 36)
FAX:41–1–273–1844 (S.D. 42)

TAIWAN
JARSEN CO., Ltd.
12th floor, No. 1337
Chung Cheng Road
Tao-Yuan City
Tao Yuan Hsien
TEL:886–3–356–6678
FAX:886–3–356–5527

THAILAND
DIETHELM TRADING CO., Ltd.
2533 Sukhumvit Road
Bangchack, Prakhnong
Bangkok, 10250
TEL:66–2–332–7140
FAX:66–2–332–7164

TRINIDAD/ WEST INDIES
ALLTECH TRINIDAD
44 St. Michaels Terrace
Blue Range Diegos, Martin
TEL:809–632–4519
FAX:809–628–0971

VENEZUELA
SIDECA-SIDELAC
P.O. Box 1813
Maracaibo
TEL:58–62–47079
FAX:58–61–918889

VIETNAM
AGRITECH SAIGON
13 Xom Mol Hamlet
Phuong Long Village
Thu Duc District
Ho Chi Minh City
TEL:84–8–960–127
FAX:84–8–961–523